Recent Advances in Reservoir Simulation and Carbon Utilization and Storage

Recent Advances in Reservoir Simulation and Carbon Utilization and Storage

Guest Editors
Wenchao Liu
Hai Sun
Daobing Wang

Basel • Beijing • Wuhan • Barcelona • Belgrade • Novi Sad • Cluj • Manchester

Guest Editors

Wenchao Liu
School of Resources and
Safety Engineering
University of Science and
Technology Beijing
Beijing
China

Hai Sun
College of petroleum
Engineering
China University of
Petroleum (East China)
Qingdao
China

Daobing Wang
School of Mechanical
Engineering
Beijing Institute of
Petrochemical Technology
Beijing
China

Editorial Office
MDPI AG
Grosspeteranlage 5
4052 Basel, Switzerland

This is a reprint of the Special Issue, published open access by the journal *Energies* (ISSN 1996-1073), freely accessible at: www.mdpi.com/journal/energies/special_issues/Reservoir_Simulation_Carbon_Utilization_Storage.

For citation purposes, cite each article independently as indicated on the article page online and using the guide below:

Lastname, A.A.; Lastname, B.B. Article Title. *Journal Name* **Year**, *Volume Number*, Page Range.

ISBN 978-3-7258-3568-3 (Hbk)
ISBN 978-3-7258-3567-6 (PDF)
https://doi.org/10.3390/books978-3-7258-3567-6

© 2025 by the authors. Articles in this book are Open Access and distributed under the Creative Commons Attribution (CC BY) license. The book as a whole is distributed by MDPI under the terms and conditions of the Creative Commons Attribution-NonCommercial-NoDerivs (CC BY-NC-ND) license (https://creativecommons.org/licenses/by-nc-nd/4.0/).

Contents

About the Editors . . . . **vii**

Shan Yuan, Lianjin Zhang, Tao Li, Tao Qi and Dong Hui
Study of Gas–Liquid Two-Phase Flow Characteristics at the Pore Scale Based on the VOF Model
Reprinted from: *Energies* **2025**, *18*, 316, https://doi.org/10.3390/en18020316 1

Youssef E. Kandiel, Gamal Attia, Farouk Metwalli, Rafik Khalaf and Omar Mahmoud
Innovative Role of Magnesium Oxide Nanoparticles and Surfactant in Optimizing Interfacial Tension for Enhanced Oil Recovery
Reprinted from: *Energies* **2025**, *18*, 249, https://doi.org/10.3390/en18020249 18

Qingdong Zeng, Taixu Li, Long Bo, Xuelong Li and Jun Yao
Comprehensive Investigation of Factors Affecting Acid Fracture Propagation with Natural Fracture
Reprinted from: *Energies* **2024**, *17*, 5386, https://doi.org/10.3390/en17215386 32

Zuochun Fan, Mei Tian, Man Li, Yidi Mi, Yue Jiang and Tao Song et al.
Assessment of CO_2 Sequestration Capacity in a Low-Permeability Oil Reservoir Using Machine Learning Methods
Reprinted from: *Energies* **2024**, *17*, 3979, https://doi.org/10.3390/en17163979 49

Ming Yue, Quanqi Dai, Haiying Liao, Yunfeng Liu, Lin Fan and Tianru Song
Prediction of ORF for Optimized CO_2 Flooding in Fractured Tight Oil Reservoirs via Machine Learning
Reprinted from: *Energies* **2024**, *17*, 1303, https://doi.org/10.3390/en17061303 62

Shiyuan Li, Jingya Zhao, Haipeng Guo, Haigang Wang, Muzi Li and Mengjie Li et al.
A Viscoplasticity Model for Shale Creep Behavior and Its Application on Fracture Closure and Conductivity
Reprinted from: *Energies* **2024**, *17*, 1122, https://doi.org/10.3390/en17051122 82

Fenglan Wang, Binhui Li, Sheng Cao, Jiang Zhang, Quan Xu and Qian Sang
Experimental Study of the Fluid Contents and Organic/Inorganic Hydrocarbon Saturations, Porosities, and Permeabilities of Clay-Rich Shale
Reprinted from: *Energies* **2024**, *17*, 524, https://doi.org/10.3390/en17020524 105

Sheng Cao, Qian Sang, Guozhong Zhao, Yubo Lan, Dapeng Dong and Qingzhen Wang
CO_2–Water–Rock Interaction and Its Influence on the Physical Properties of Continental Shale Oil Reservoirs
Reprinted from: *Energies* **2024**, *17*, 477, https://doi.org/10.3390/en17020477 124

Dmitriy Shirinkin, Alexander Kochnev, Sergey Krivoshchekov, Ivan Putilov, Andrey Botalov and Nikita Kozyrev et al.
High Permeability Streak Identification and Modelling Approach for Carbonate Reef Reservoir
Reprinted from: *Energies* **2024**, *17*, 236, https://doi.org/10.3390/en17010236 144

Ligen Tang, Guosheng Ding, Shijie Song, Huimin Wang, Wuqiang Xie and Jiulong Wang
A Case Study on the CO_2 Sequestration in Shenhua Block Reservoir: The Impacts of Injection Rates and Modes
Reprinted from: *Energies* **2023**, *17*, 122, https://doi.org/10.3390/en17010122 163

Wenchao Liu, Chen Liu, Yaoyao Duan, Xuemei Yan, Yuping Sun and Hedong Sun
Fracture Spacing Optimization Method for Multi-Stage Fractured Horizontal Wells in Shale Oil Reservoir Based on Dynamic Production Data Analysis
Reprinted from: *Energies* **2023**, *16*, 7922, https://doi.org/10.3390/en16247922 182

Ping Wang, Wenchao Liu, Wensong Huang, Chengcheng Qiao, Yuepeng Jia and Chen Liu
Dynamic Productivity Prediction Method of Shale Condensate Gas Reservoir Based on Convolution Equation
Reprinted from: *Energies* **2023**, *16*, 1479, https://doi.org/10.3390/en16031479 204

Qitao Zhang, Wenchao Liu, Jiaxin Wei, Arash Dahi Taleghani, Hai Sun and Daobing Wang
Numerical Simulation Study on Temporary Well Shut-In Methods in the Development of Shale Oil Reservoirs
Reprinted from: *Energies* **2022**, *15*, 9161, https://doi.org/10.3390/en15239161 224

Hongyang Chu, Tianbi Ma, Zhen Chen, Wenchao Liu and Yubao Gao
Well Testing Methodology for Multiple Vertical Wells with Well Interference and Radially Composite Structure during Underground Gas Storage [†]
Reprinted from: *Energies* **2022**, *15*, 8403, https://doi.org/10.3390/en15228403 248

Daobing Wang, Zhan Qu, Zongxiao Ren, Qinglin Shan, Bo Yu and Yanjun Zhang et al.
Numerical Simulation on Borehole Instability Based on Disturbance State Concept
Reprinted from: *Energies* **2022**, *15*, 6295, https://doi.org/10.3390/en15176295 268

Yunqi Shen, Zhiwen Hu, Xin Chang and Yintong Guo
Experimental Study on the Hydraulic Fracture Propagation in Inter-Salt Shale Oil Reservoirs
Reprinted from: *Energies* **2022**, *15*, 5909, https://doi.org/10.3390/en15165909 286

Tingting Cai, Lei Shi, Yulong Jiang and Zengchao Feng
A Core Damage Constitutive Model for the Time-Dependent Creep and Relaxation Behavior of Coal
Reprinted from: *Energies* **2022**, *15*, 4174, https://doi.org/10.3390/en15114174 308

About the Editors

Wenchao Liu

Dr. Wenchao Liu is an associate professor at the University of Science and Technology Beijing (USTB) in China. In 2007, Dr. Wenchao Liu received a Bachelor's degree in Information and Computing Science from the China University of Petroleum (East China). In 2010, he received a Master's degree in Mathematics from the China University of Petroleum (East China). In 2013, Dr. Liu received a Doctorate in Petroleum Engineering from the China University of Petroleum (East China). From 2012 to 2013, he studied at the University of Calgary in Canada with the support of the China National Study Abroad Fund. From 2014 to 2017, he worked as a postdoctoral researcher at the Institute of Mechanics, Chinese Academy of Sciences. Since 2017, he has been engaged in teaching and scientific research at the School of Resources and Safety Engineering at the University of Science and Technology Beijing.

Dr. Wenchao Liu primarily conducts research in unconventional oil and gas seepage mechanics, numerical simulation methods, well-test interpretations, and production performance data analysis. He has led multiple scientific projects including the National Natural Science Foundation of China, China Postdoctoral Science Foundation, National Foreign Expert Program of the Ministry of Science and Technology, and PetroChina Innovation Foundation. Dr. Liu has published over 40 academic papers in his research fields, authored three academic monographs as the first author, and secured six Chinese invention patents as the primary inventor. He serves as an editorial board member for journals such as *Petroleum Science* and *Reservoir Evaluation and Development*. His research achievements have earned prestigious awards such as the First Prize of Science and Technology Progress Award from the Ministry of Education, the Innovation Team Award from the China Petroleum and Chemical Industry Federation, and so on.

Hai Sun

Hai Sun is a professor from the China University of Petroleum (East China). His research focuses on the transport phenomena in porous media such as hydrocarbon reservoirs. The multi-scale simulation of fluid flow in porous media is his main research area presently. Molecular simulation, digital rock, and the Lattice Boltzmann method (LBM) are the main research methods in the micro- and nano-scale simulation of fluid flow in porous media at the pore scale, while the dual continuum media model and discrete fracture model (DFM) are the main models in the macro-scale simulation of fluid flow in fracture reservoirs at the field scale. The numerical simulation of CO_2 sequestration and enhanced natural gas recovery (CSEGR) is also his research area. Prof. Hai Sun published over 200 papers, including more than 100 indexed by *SCI*, with a total of over 6,000 citations and an H-index of 44. He holds 20 granted invention patents and software copyrights. He serves as an editorial board member for *SCI* journals and is a reviewer for over 30 international journals. He received one First Prize in Natural Science from the Ministry of Education and three Second Prizes for Scientific and Technological Progress at the provincial and ministerial levels.

Daobing Wang

Dr. Daobing Wang is an Associate Professor, PhD Supervisor, and Youth Editorial Board Member of the Journal of *Advances in Geo-Energy Research* at the Beijing Institute of Petrochemical Technology, where he has been teaching at the School of Mechanical Engineering since October 2019. He earned his Ph.D. in Oil and Gas Well Engineering from the China University of Petroleum, Beijing (2017), with a CSC-funded joint Ph.D. fellowship at Universitat Politècnica de Catalunya (UPC, Spain, 2015–2016) specializing in Computational Mechanics. His academic trajectory includes postdoctoral research in Solid Mechanics through the Beihang University–Beijing Engineering Research Center of Safety Engineering collaboration (2017–2019) and a domestic visiting scholar position at Beihang University (2023–2024).

A leading researcher in sustainable energy systems, Dr. Wang's expertise spans geothermal energy utilization, hydrogen technologies, unconventional hydrocarbon development, and deep/ultra-deep reservoir extraction. He has led three NSFC projects, authored over 50 *SCI*/EI-indexed publications, and earned three provincial/ministerial-level science awards (two first prizes as lead researcher) alongside two collaborative recognitions. His honors include the Green Mining Society's Youth Science & Technology Award, the Outstanding Youth Contribution Award in Petroleum & Chemical Automation, and a national nomination for excellence in mining and petroleum education, reflecting his interdisciplinary impact on energy innovation.

Article

Study of Gas–Liquid Two-Phase Flow Characteristics at the Pore Scale Based on the VOF Model

Shan Yuan *, Lianjin Zhang, Tao Li, Tao Qi and Dong Hui

Research Institute of Exploration and Development, PetroChina Southwest Oil & Gas Field Company, Chengdu 610041, China
* Correspondence: yuanshan2018@petrochina.com.cn

Abstract: To study the effects of liquid properties and interface parameters on gas–liquid two-phase flow in porous media. The volume flow model of gas–liquid two-phase flow in porous media was established, and the interface of the two-phase flow was reconstructed by tracing the phase fraction. The microscopic imbibition flow model was established, and the accuracy of the model was verified by comparing the simulation results with the classical capillary imbibition model. The flow characteristics in the fracturing process and backflow process were analyzed. The influence of flow parameters and interface parameters on gas flow was studied using the single-factor variable method. The results show that more than 90% of the flowing channels are invaded by fracturing fluid, and only about 50% of the fluid is displaced in the flowback process. Changes in flow velocity and wetting angle significantly affect Newtonian flow behavior, while variations in surface tension have a pronounced effect on non-Newtonian fluid flow. The relative position of gas breakthrough in porous media is an inherent property of porous media, which does not change with fluid properties and flow parameters.

Keywords: porous media; volume flow model; two-phase flow; non-Newtonian fluid; OpenFOAM

Academic Editor: Hossein Hamidi

Received: 5 December 2024
Revised: 6 January 2025
Accepted: 7 January 2025
Published: 13 January 2025

Citation: Yuan, S.; Zhang, L.; Li, T.; Qi, T.; Hui, D. Study of Gas–Liquid Two-Phase Flow Characteristics at the Pore Scale Based on the VOF Model. Energies 2025, 18, 316. https://doi.org/10.3390/en18020316

Copyright: © 2025 by the authors. Licensee MDPI, Basel, Switzerland. This article is an open access article distributed under the terms and conditions of the Creative Commons Attribution (CC BY) license (https://creativecommons.org/licenses/by/4.0/).

1. Introduction

With the impact of renewable resources such as wind and solar energy on conventional fossil energy, the demand space for oil and gas as primary energy sources is further compressed [1]. As the cost of renewable resource development and utilization technologies continues to decline, exploring high-quality development and high return has become the key to sustainable development of oil and gas resources. China is endowed with substantial natural gas resources, and as a clean energy source, natural gas plays a pivotal role in facilitating the country's goals of carbon peaking and carbon neutrality. It also supports the transition towards a low-carbon, high-quality national energy system [2]. The process of oil and gas exploration is a complex systems engineering challenge, with the flow of hydrocarbons through porous media representing one of its most intricate aspects [3]. Due to the presence of primary water in the reservoir and the intrusion of fracturing fluid, the gas flow in the reservoir behaves as a gas–liquid two-phase flow in a porous media. The storage and flow spaces at the micro and nanoscale significantly enhance interphase interactions, leading to lower gas well flowback rates and accelerated production decline during the development of unconventional gas reservoirs [3,4]. Therefore, understanding the characteristics of gas–liquid two-phase flow in porous media at the microscale is crucial for enhancing the development of unconventional gas reservoirs and promoting their comprehensive utilization.

For the simulation of multiphase flow in porous media, extensive research has been conducted by researchers using both laboratory experiments and numerical simulations. Laboratory experiments are categorized into sand-filled model experiments [5,6] and micro-etching experiments [7,8] based on the method of experimental model construction. The sand-filled experiment uses actual reservoir rock particles to form experimental samples by pressing them in certain ratios. Using this method, physical models with different parameters are constructed for flow analysis. By adjusting the mineral composition and particle size distribution, the effect of particle composition, wetting characteristics, and pore size distribution on the flow process can be reflected to some extent. However, this method mainly studies the flow pattern by monitoring the pressure differences between the inlet and outlet of the rock sample and the fluid flow rate. It does not reflect the real flow process inside the rock sample. Micro-etching experiments combined with CT scanning technology allow for the microscopic scanning of rock samples to obtain the internal flow channel characteristics of rock samples. The fabrication of microscopic flow channels on characteristic plates is achieved using etching techniques. Controlling the injected pore volume flow rate with a microflow pump and simulating the flow processes at the microscale can detect the fluid flow patterns within the microscale channels in real-time. Unfortunately, this method cannot manually modify the interface parameters during the flow process.

For unconventional gas reservoirs, the presence of ultra-low-scale flow channels poses significant challenges for laboratory experimental studies. However, advancements in computational fluid dynamics (CFD) technology now offer a crucial methodology for simulating multiphase flow within porous media at the micro- and nanoscales. At present, the numerical simulation of multiphase flow in porous media in the field of petroleum engineering is divided into two research directions [3,9,10]: one is to study the coupling of continuous and discrete phases between different phases, and to analyze the effects of the boundary layer, surface tension, and wetting angle on the flow process. Secondly, this study investigates the dynamics of particle collisions, mixing processes, and post-collision motion in liquid–solid and gas–solid systems relevant to proppant transport. In this work, we concentrate on gas–liquid two-phase flow within microscale flow spaces. Given the rapid changes at the phase interface between gas and liquid phases, it is essential to employ numerical models capable of accurately capturing these interface dynamics. Some commonly used simulation methods include the Lattice Boltzmann Method (LBM) [11–13], the Phase Field Method (PFM) [14,15], the Level Set Method (LSM) [16–18], the Interface Tracking Method (FT) [19], and the Volume of Fluid (VOF) method [20,21]. The LBM method is more effective for simulating low-velocity small-scale multiphase flow without tracking the phase interface, but the stability is poor under high-density gas–water ratio conditions. The PFM method has shortcomings in simulating the dissolution of small droplets or bubbles. The LSM has a good ability to describe complex flow channel topology, but the volume conservation on the grid is poor, and there is excessive smoothness in the treatment of phase interfaces at the sharp points of solid particles. The LSM can be seen as a simplification of the PFM, but the physical meaning of the phase field is lost [22]. The FT can prevent the overly-smoothed reconstruction of interfaces at sharp features, but its sensitivity to the topological complexity of the flow channel can lead to instability in computational outcomes. The VOF method reconstructs the phase interface for simulation purposes, allowing it to manage the complex topologies of porous media while maintaining superior volume conservation [23,24].

The simulation dimensions of porous media are typically small, with pore-throat sizes generally in the micron to nanometer range. Unlike at the macroscopic scale, this small size amplifies the influence of factors such as the wettability, flow velocity, and viscosity ratio on flow morphology. Porous media models can be categorized into two-dimensional (2D) and three-dimensional (3D) models. Two-dimensional models, due to their lesser number of computational grids, are convenient for studying flow patterns. In contrast, 3D models have significantly larger computational dimensions and a greater number of discrete grid elements compared to 2D models; their spatial attributes make them more suitable for analyzing the spatial aggregation distribution characteristics of fluids [25,26]. Various numerical simulation methods for multiphase flow have been applied to both types of models. The choice of model does not affect the analysis of microscale flow characteristics, although the selection of the numerical simulation method and the setting of computational parameters are critical [10].

In our study, based on the VOF model in computational fluid dynamics combined with the modeling method of porous media at the microscale, a numerical simulation research method of gas–liquid two-phase flow in porous media is formed. On this basis, the flow differences between gas-driven water and water-driven gas processes were studied. The influence of liquid properties on the flow process was discussed, and the effects of flow rate, wetting angle, and surface tension on the gas–liquid flow shape were analyzed.

2. Two-Phase Flow Simulation Method

2.1. VOF Model

Based on the conventional flow equation, VOF introduces the concept of phase fraction to form a phase fraction equation. During the solution process, pressure, velocity, and phase fraction are solved, and the phase interface shape is reconstructed using the numerical value of the phase fraction. The numerical value of the phase fraction is related to the proportion of fluid in the grid. The defined phase fraction is represented by the symbol α. The value range of α is between [0, 1]; when $\alpha = 0$, it indicates that the cell is all gas phase, when $\alpha = 1$, it indicates that the cell is all liquid phase. When the value of α ranges from 0 to 1 it indicates the presence of gas and water phases in the cell, as shown in Figure 1.

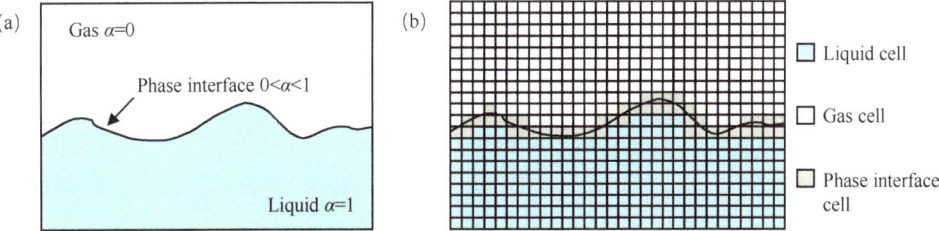

Figure 1. (a) Phase fraction interface in the VOF model; (b) cell type of the computational region.

The equation of continuity for the flow of a mutually immiscible two-phase fluid in a porous media and the equation of momentum considering the effect of gravity is expressed as follows:

$$\nabla \cdot (\mathbf{U}) = 0 \tag{1}$$

$$\frac{\partial \rho \mathbf{U}}{\partial t} + \nabla \cdot (\rho \mathbf{U}\mathbf{U}) - \nabla \cdot \tau = -\nabla p + \rho \mathbf{g} + \mathbf{F} \tag{2}$$

where $\mathbf{U}$ represents the velocity tensor, m/s; p represents the pressure in the grid, Pa; ρ represents the average density of phases in the grid, kg/m^3; τ represents the shear stress at the two-phase interface, Pa^{-1}; $\mathbf{F}$ represents the surface tension of the two-phase interface, N/m; $\mathbf{g}$ represents the acceleration of gravity, taken as 9.8 m/s^2.

The average density is defined as follows:

$$\rho = \alpha \rho_l + (1-\alpha)\rho_g \tag{3}$$

Regarding the shear stress at the interface between two phases, when the wetting phase fluid is a Newtonian fluid

$$\tau = \mu\left(\nabla \mathbf{U} + \nabla \mathbf{U}^T\right) \tag{4}$$

where μ represents the average viscosity in a grid, Pa·s.

The calculation formula for the average viscosity in the cell is as follows:

$$\mu = \alpha_l \mu_l + \alpha_g \mu_g \tag{5}$$

where μ_l represents the viscosity of the liquid phase in the grid, Pa·s; μ_g represents the viscosity of the gas phase in the grid, Pa·s.

When the wetting phase fluid is a non-Newtonian fluid, the shear stress and fluid viscosity are related to the shear rate. The power rate model is used to calculate the non-Newtonian fluid viscosity [27,28]:

$$\tau = \mu \gamma \tag{6}$$

$$\mu = k(\gamma)^{n-1} \tag{7}$$

$$\gamma = \sqrt{\frac{1}{2}\left(\nabla \mathbf{U} + \nabla \mathbf{U}^T\right)_{ij}\left(\nabla \mathbf{U} + \nabla \mathbf{U}^T\right)_{ji}} \tag{8}$$

where k represents the power coefficient, dimensionless; n represents parameters related to fluid properties, dimensionless; γ represents the shear rate, s^{-1}; the subscript i represents the i-th column in the x-axis direction; the subscript j represents the j-th row in the y-axis direction.

Any cell for the computational region satisfies $\Sigma \alpha = \alpha_l + \alpha_g = 1$.

The surface tension in Equation (2) is defined by the continuous surface force model [29] as follows:

$$\mathbf{F} = \sigma\left[\frac{\rho \kappa_N \nabla \alpha}{\frac{1}{2}(\rho_g + \rho_w)}\right] \tag{9}$$

where κ_N represents the curvature at the two-phase interface, m^{-1}; σ represents the interfacial tension coefficient, N/m; ρ_w represents the liquid phase density in the grid, kg/m^3; ρ_g represents the gas phase density in the grid, kg/m^3.

The value of κ_N is related to the divergence of the unit normal vector $\mathbf{n}$ at the phase interface [25]:

$$\kappa_N = -\nabla \cdot \mathbf{n} = -\nabla \cdot \left(\frac{\nabla \alpha}{|\nabla \alpha|}\right) \tag{10}$$

The value of the phase fraction α is related to the fluid properties, independent of the flow process. The phase fraction field equation for incompressible two-phase flow can be expressed as follows:

$$\frac{\partial \alpha}{\partial t} + \mathbf{U} \cdot \nabla \alpha = 0 \tag{11}$$

Equation (11) is the phase equation of the VOF model.

For incompressible systems, where pressure acts as a relative value and the differential pressure serves as the primary driving force, the cell pressure is defined to simplify the momentum equation, Equation (2), as follows:

$$p_{rgh} = p - \rho \mathbf{g} \cdot \mathbf{h} \tag{12}$$

where **h** represents the center position vector of a grid.

Gradient calculation for Equation (12):

$$\nabla p_{rgh} = \nabla p - \mathbf{g} \cdot \mathbf{h} \nabla \rho - \rho \mathbf{g} \tag{13}$$

Bringing Equation (13) to Equation (2) for simplification:

$$\frac{\partial \rho \mathbf{U}}{\partial t} + \nabla \cdot (\rho \mathbf{U}\mathbf{U}) - \nabla \cdot (\mu \nabla \mathbf{U}) - \nabla \mathbf{U} \cdot \nabla \mu = -\nabla p_{rgh} - \mathbf{g} \cdot \mathbf{h} \nabla \rho + \sigma \kappa_N \nabla \alpha \tag{14}$$

Equations (1), (11), and (14) together form the mathematical model for the VOF model of the two-phase flow system.

2.2. Numerical Method

The VOF model is implemented using the open-source CFD toolkit OpenFOAM-11, and the VOF model is solved using the interFoam solver. OpenFOAM uses the finite volume method to discretize the VOF model, the second-order upstream interpolation function to discretize the spatial terms, and the first-order Eulerian format to discretize the time terms. The phase equations are solved utilizing the Multi-Dimensional Universal Limiter with Explicit Solution (MULES) algorithm. For the coupled pressure-velocity field, the solution procedure employs the Pressure-Implicit with Splitting of Operators (PISO) method.

Berea sandstone and Fontainebleau sandstone serve as two frequently adopted models in the field of porous media and are extensively utilized for investigating fluid flow behaviors within such media. The primary differentiations between these models reside in their respective porosities and permeabilities. Tailored to specific research goals, the Berea sandstone model is predominantly employed to examine the flow dynamics of hydrocarbons within porous media. Conversely, the Fontainebleau sandstone model is favored for studies concerning carbon dioxide sequestration and nuclear waste disposal. For the purposes of this study, data from the standard Berea sandstone model have been selected; CINEMA 4D-R17 professional modeling software was used to build a 10 mm × 3 mm model of porous media, as shown in Figure 2a. The gray areas depict solid particles and the white areas depict flow channels. A structured hexahedral mesh is generated for the pore medium model using the snappyHexMesh toolbox in OpenFOAM, as shown in Figure 2b. The left side of the model is designated as the inlet, the right side is the outlet, and the solid particles are the walls. The inlet and outlet and wall boundary conditions are set as shown in Table 1. The pore walls were subjected to complete wetting conditions, and the Coulomb number C_0 was set to 0.3 with a fixed time step. The values of the fluid parameters and computational parameters in the model are listed in Table 2, and the gas phase properties use the same parameters as for methane gas.

Table 1. Boundary condition settings in OpenFOAM.

Boundary Type	Velocity Boundary Condition	Pressure Boundary Condition
inlet	fixedValue	zeroGradient
outlet	zeroGradient	fixedValue
wall	noSlip	zeroGradient

Table 2. Fluid parameters and calculation parameters.

Fluid Parameters	Value	Fluid Parameters	Value
Gas density ρ_g (kg/m³)	0.67	Wetting angle θ (°)	0
Water density ρ_l (kg/m³)	1000	Interfacial tension σ (N/m)	0.032
Slickwater density ρ_l (kg/m³)	1020	Calculation duration t (s)	1
Gas viscosity μ_g (Pa·s)	1.11×10^{-5}	Time step Δt (s)	1×10^{-5}
Water viscosity μ_w (Pa·s)	1×10^{-3}	Iterative residual ε (-)	1×10^{-5}
Flow index n (-)	0.8	Outlet pressure p_{out} (Pa)	100
Consistency coefficient k (Pa·s^n)	1×10^{-5}	Inlet velocity v (m/s)	0.01

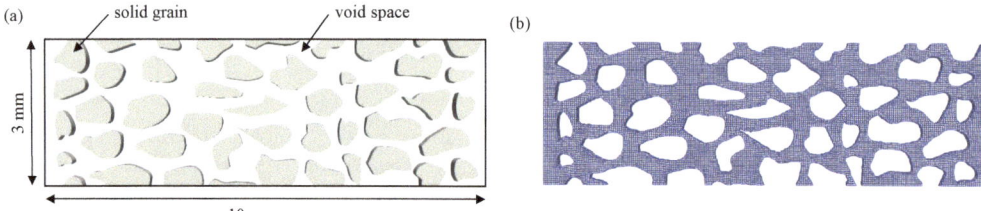

Figure 2. (a) Porous media model; (b) generation of hexahedral mesh (16,100 cells).

2.3. Model Validation

To validate the accuracy of the VOF model calculations, a microscale capillary model, as depicted in Figure 3, was established to simulate the position of the gas–liquid interface during the capillary action. The accuracy and applicability of the numerical simulation method were verified by comparing the simulation results with the theoretical solutions of classical capillary. Neglecting gravity and inertia forces, the two-phase interface position is determined by the capillary force and viscous force, and its mathematical expression is as follows [25]:

$$\sigma \cos(\theta) = \frac{6}{r}[\mu_w x + \mu_n(L-x)]\frac{dx}{dt} \tag{15}$$

where σ represents the interfacial tension, N/m; θ represents the wetting angle, °; r represents the capillary radius, m; μ_w represents the viscosity of the wetting phase, Pa·s; μ_n represents the non-wetting phase viscosity, Pa·s; L represents the length of the capillary, m; x represents the position of the interface between the two phases, m; t represents the capillary action time, s.

As depicted in Figure 3, the capillary model features a fluid inlet on the left and a fluid outlet on the right. At the initial moment, the capillary is entirely filled with the non-wetting phase fluid. Subsequently, driven by capillary forces, the wetting phase fluid intrudes from the left-hand side, progressively displacing the non-wetting phase fluid. A critical aspect of this process is the dynamic movement of the contact interface separating the two phases. The capillary length on the right side L is 1000 μm and the capillary diameter r is 30 μm. The boundary conditions are set to no-slip boundary conditions and wetting boundary conditions, and the sides are set to periodic boundary conditions. The wetting phase fluid density is 1000 kg/m³, viscosity 1000 mPa·s. The non-wetting phase fluid density is 0.7 kg/m³, viscosity 10 mPa·s. The surface tension is 0.03 N/m and the wall is set to be fully wetted. The Canny algorithm in MATLAB R2018b software was used to batch process the obtained flow pictures in the capillary at different flow times to obtain the contour lines of the non-wetting phase and to measure the change in the position of the interface between the two phases using the relative size of the pixels.

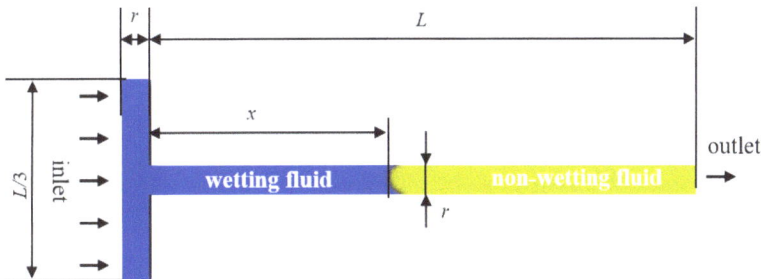

Figure 3. Capillary percolation model (blue is the wetting phase, yellow is the non-wetting phase).

Figure 4 presents a comparison between the simulation results for the two-phase interface displacement obtained from the capillary percolation model and the theoretical values derived using Equation (15). The data indicate a substantial agreement between the simulation outcomes and the theoretical predictions, validating the applicability of the VOF model for simulating two-phase flow processes at the pore scale.

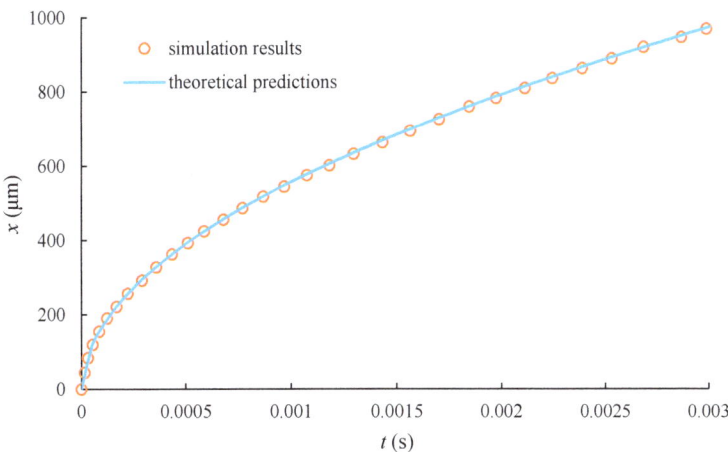

Figure 4. Comparison of the simulation results of the two-phase interface displacement of the capillary percolation model with the theoretical values.

3. Fracturing and Rejection Process

Fracturing horizontal wells for the hydraulic fracturing and rejection process is a typical gas–liquid two-phase flow process, as shown in Figure 5. The fracturing fluid is pumped into the horizontal well through the surface high-pressure sink and filtered out to the reservoir through the hydraulic fracture; the fluid invades the reservoir matrix pore space under the actions of pumping pressure and capillary force to drive out the gas stored in the pore space. After hydraulic fracturing, the well is opened and the fracturing fluid is returned to the reservoir; the gas in the reservoir repels the fluid due to the production pressure difference, forming a gas flow channel and replacing some of the intruded fluid in the matrix pores. Using the porous media model established in Figure 2, the microscopic flow characteristics of the two stages of fracturing and re-discharge are simulated with the calculated parameters in Tables 1 and 2 to study the effects of two types of fluids, Newtonian fluid water and non-Newtonian fluid slickwater, on gas flow.

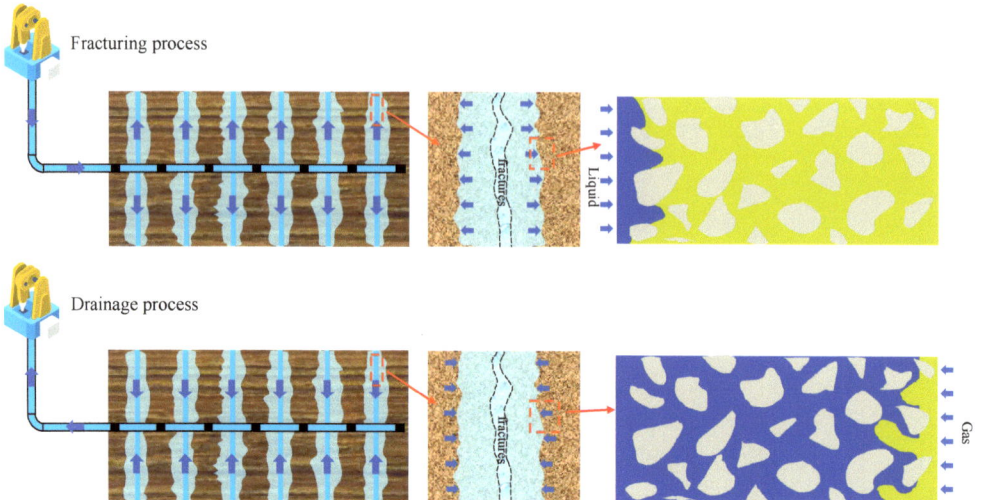

Figure 5. Schematic diagram of the fracturing and rejection process of a multi-stage fractured horizontal well.

3.1. Fracturing Process Simulation

Figure 6 illustrates the evolution of liquid-phase saturation as water intrudes into the porous medium, displacing the gas. With the injection of water, due to the additional capillary force, the water phase breaks through the small pore size preferentially, and the liquid-phase saturation in the porous medium gradually increases; the increasing relationship between saturation and time is nearly linear. When the water phase breaks through, the liquid-phase saturation reaches a maximum value of 90.42% and remains stable. The gas that could not be displaced becomes primarily concentrated in the dead-end regions of the porous medium. During the simulation, large bubbles were also observed splitting into two small bubbles.

Figure 7 illustrates the variation in liquid-phase saturation during slickwater intrusion into the porous medium to repel the gas. With the injection of slickwater, the liquid-phase saturation increases linearly and reaches a maximum saturation of 91.47% by the time the liquid phase breaks through and remains stable. Comparing the simulation results depicted in Figures 6 and 7, it is observed that within the time interval of 0–0.002 s, the saturation of slickwater in porous media is slightly higher than that of water, indicating that slickwater more readily enters dead-end positions during the initial stages of flow. Within the subsequent time interval of 0.002–0.005 s, the water occupies a greater proportion of the larger pores, while the slickwater displaces more gas from the dead-end regions of the porous medium. The displacement velocity of the water is marginally faster than that of the slickwater, resulting in a higher water saturation within the porous medium. The water achieves a stable flow earlier than the slickwater. Given that slickwater has a propensity to enter dead-end locations more easily, it exhibits a higher degree of gas displacement compared to water. Ultimately, upon reaching stable flow conditions, the saturation of slickwater in the porous medium is slightly higher than that of water.

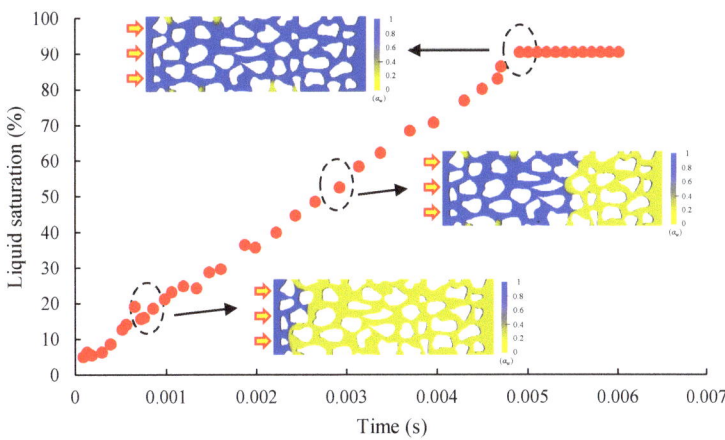

Figure 6. Simulation of liquid phase saturation changes during water intrusion into porous media.

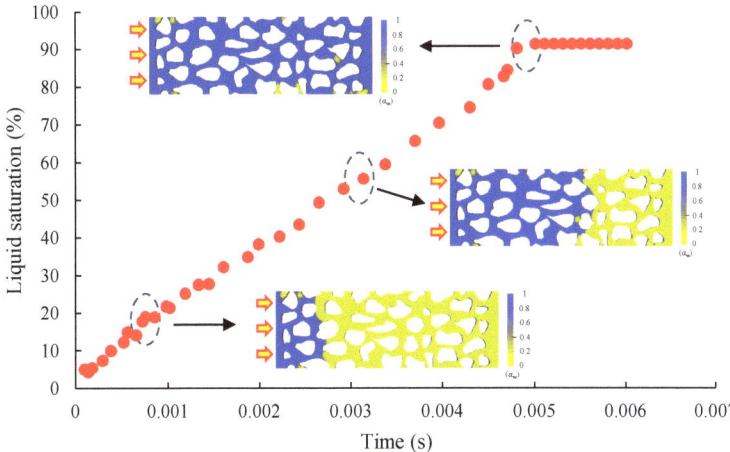

Figure 7. Simulation of liquid-phase saturation changes during slickwater intrusion into porous media.

In simulations of fracturing fluid intrusion into porous media, the liquid phase acts as the wetting phase, repelling the gas phase along the pore walls and displacing more than 90% of the gas while occupying a larger percolation area. Non-Newtonian fluids are more likely to penetrate the porous medium compared to Newtonian fluids. For non-Newtonian fluids, due to the high initial flow velocity, shear thinning occurs. However, due to complete wetting of the walls, there is little difference in the final liquid-phase saturation between the two types of fluids, as shown in Figure 8. The variation in liquid-phase saturation over time is approximately linear, and the process of liquid-phase intrusion into the porous medium is characterized by "stable displacement" at the microscopic scale.

The research phenomenon mentioned above occurs mainly because water is a pressurized fluid, which causes a significant difference in the microscopic flow of water and gas. The liquid as a carrier has a pressure-bearing role; when the liquid percolation is blocked, the agitated pressure will propagate throughout the water column, seeking a breakthrough to form a new percolation channel under the action of the agitated pressure, or break through the narrow throat bundle under the action of the pressure difference, which will appear as an unstable displacement leading edge in the microscopic flow [26]. The liquid phase as the wetting phase is more likely to infiltrate the pore wall to form a water film; the presence of the water film weakens the adsorption of gas molecules onto the wall and increases the content of free gas in the pore [30]. The intrusion of the liquid phase into the formation accelerates the desorption of gas to a certain extent; however, with an increase in liquid-phase saturation in the pore, the thickness of the water film gradually increases, which can lead to the appearance of the gas phase in the liquid-phase trapped pore [31].

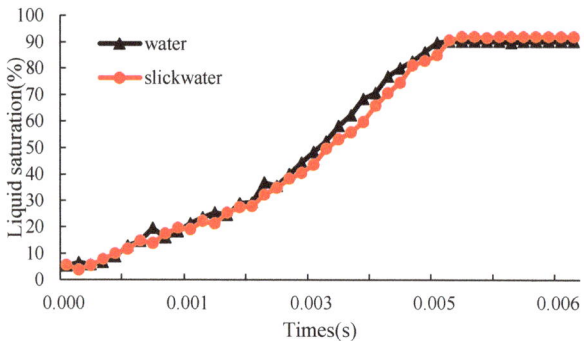

Figure 8. Comparison of liquid saturation in porous media at different time.

3.2. Simulation of the Return Process

Figure 9 shows the variation in the gas saturation when gas breaks through the water phase blockage during the rejection stage. With the injection of gas, the gas saturation in the porous medium increases linearly; the gas phase breaks through the water phase blockade to form a dominant seepage channel when the gas saturation reaches 50.26%. Unlike the fracturing process where the liquid phase intrudes into the porous medium, the gas phase preferentially breaks through in the large-pore diameter pore channel due to the additional capillary force acting as a resistance.

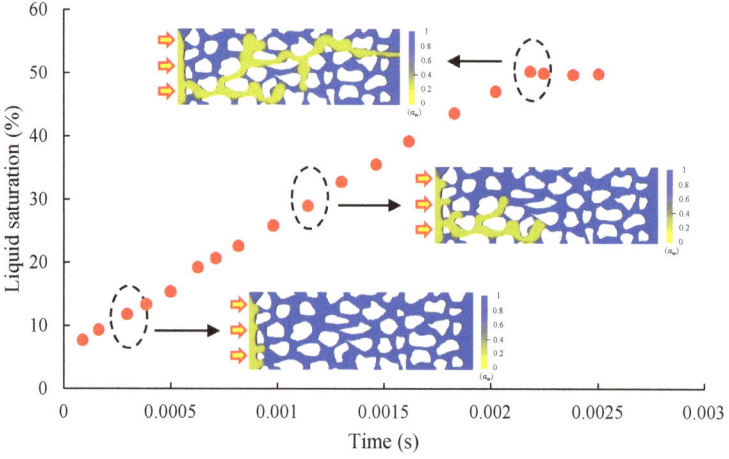

Figure 9. Simulation of gas saturation changes during gas displacement water.

Figure 10 shows the change in gas saturation during the gas breakthrough of the slickwater closure. With the injection of gas, the gas saturation in the porous medium shows a linear increase; the gas saturation reaches 48.54% when the gas breaks through the slickwater closure to form a dominant seepage channel. The gas–liquid interface of the rejection process shows an obvious fingering phenomenon, and this instability in the flow significantly reduces the rejection efficiency of the fracturing fluid [32]. Without considering the reservoir matrix subjected to compression pore closure, about 50% of the fluid is retained in the porous medium and cannot be discharged; this part of the fluid is retained in the pore space and cannot be discharged, which is the main reason for the low rejection rate of the fracturing fluid. In both fluids with liquid phase properties, the gas breakthrough point is at a high level due to gravity and the breakthrough location is the same in both cases, indicating that the gas will automatically choose the direction of least seepage resistance to flow; this optimal path selection process is not affected by the nature of the fluid [14].

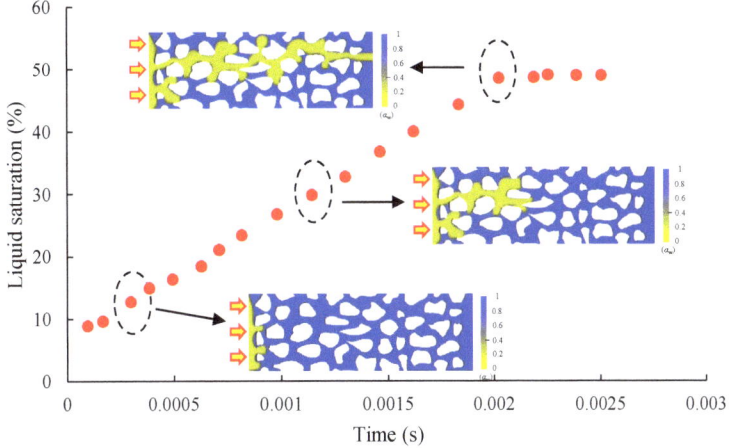

Figure 10. Simulation of gas saturation changes during gas displacement of slickwater.

Comparing the fracturing process and the rejection process, the time required for gas breakthrough to form a stable seepage flow is half that required for the liquid phase intrusion to form a stable seepage flow, and it therefore takes longer for liquid to enter the porous medium. If we want to improve the efficiency of liquid percolation to replace the gas, we need to consider a reasonable soak time [33].

4. Gas–Liquid Two-Phase Flow Characteristics

Fluid flow in porous media is a complex interaction process; fluid flow velocity affects the flow pattern, wetting angle affects the interface relationship between the fluid and solids, and surface tension affects the interaction between different fluids. In this section the values of flow velocity, wetting angle, and surface tension are varied to study the flow characteristics of gas in Newtonian and non-Newtonian fluids; the porous medium model and simulation parameters are referred to in Figure 2 and Table 2.

4.1. Flow Rate

Gas wells experience frequent changes in their production regime during production, resulting in changes in fluid flow rates which are ultimately reflected as changes in the daily gas production in the gas well. In simulating the effect of gas flow rate changes, three different flow rates are set, as shown in Figure 11. As the percolation velocity increases, the time for the gas to break through the porous medium decreases. Within the porous medium, the gas breaks through in the direction of the minimum seepage resistance, and after breaking through the seepage channel the gas will expand the previous seepage channel and turn to strengthen the already formed seepage channel when a new one cannot be formed.

The variation in gas saturation as the gas breaks through the porous medium at different flow rates is presented in Figure 12. As the flow rate increases, gas saturation within the porous medium gradually rises, with this trend being particularly pronounced in the water phase. However, the final gas saturation in the slickwater shows little change [34]. Given that most gas wells in the field are fractured using non-Newtonian fracturing fluids, the increase in the fracturing fluid rejection rate achieved by raising the production pressure differential after re-establishing gas flow following a shut-in is not significantly noticeable [4,35,36].

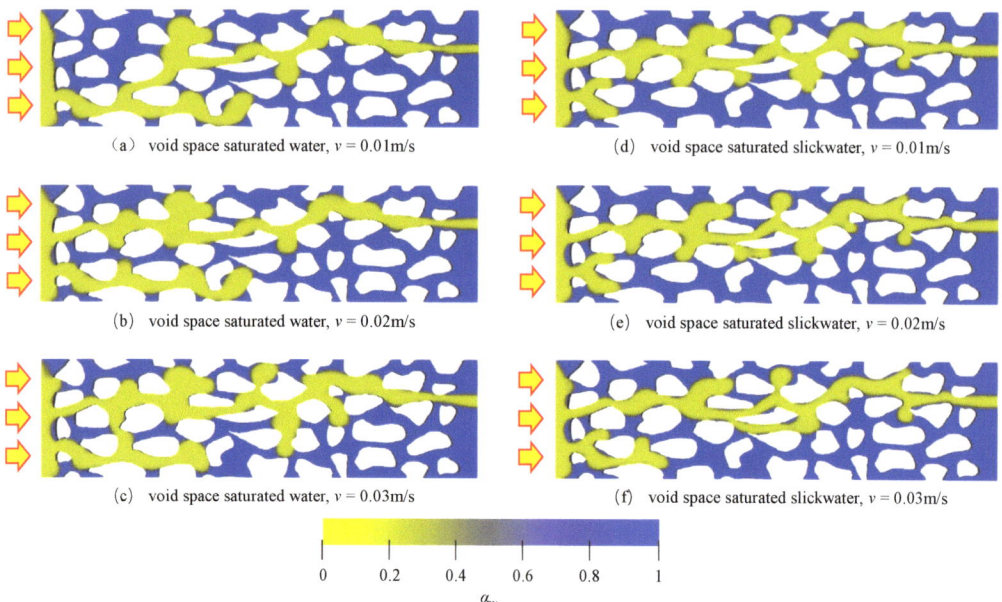

Figure 11. Effect of flow rate on gas flow pattern in liquid phase fluids of different nature.

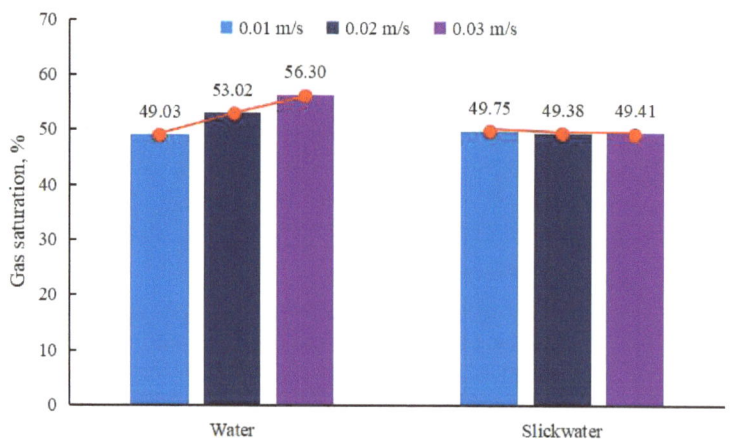

Figure 12. Variation in gas saturation at different flow rates during gas breakthrough in porous media

4.2. Wetting Angle

The wetting angle changes the spreading state of the wetting phase on the solid surface in porous media. From the simulation results in Figure 13, it can be seen that a change in the wetting angle has a large effect on the flow pattern of the gas. As the wetting angle increases, the porous medium changes from strongly hydrophilic to neutral wetting, reducing the instability of the leading edge of the gas flow; the pore wall has a weaker ability to adsorb the wetting fluid, enhancing the gas repellent effect and increasing the relative permeability of the gas phase. Similar flow pattern characteristics are shown in both Newtonian and non-Newtonian fluids.

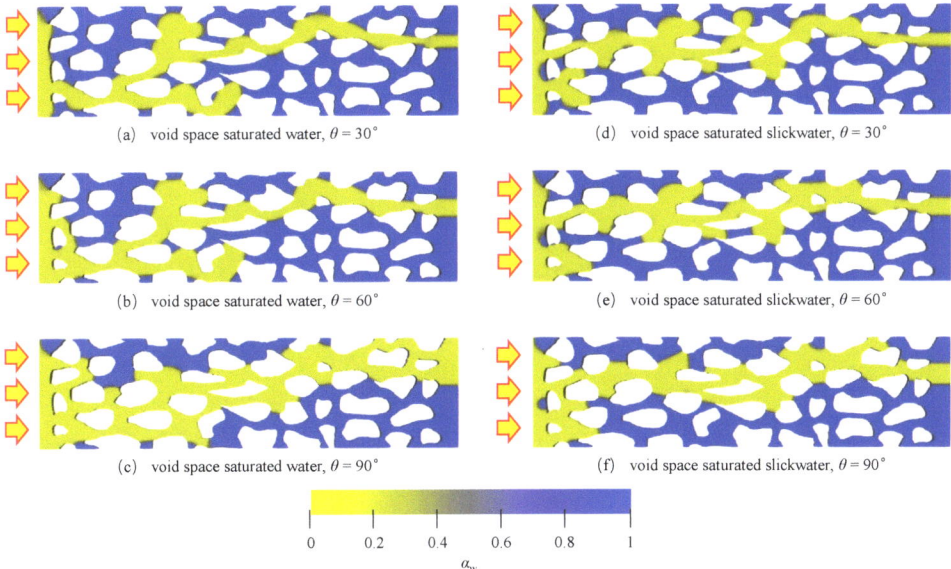

Figure 13. Effect of wetting angle on gas flow pattern in liquid phase fluids of different nature.

The change in gas saturation within the porous medium at the moment of gas breakthrough for various wetting angle values is presented in Figure 14. It can be seen that, with an increase in the wetting angle, the change in gas saturation in the water phase is larger; the gas-containing saturation increases by 12.01% during the process of changing the water phase from the wetting phase to the non-wetting phase. However, with an increase in the wetting angle, the change in gas saturation in slickwater is not obvious. As the pore radius of an unconventional reservoir is small, it is easy to form a "water lock effect" after contact with the liquid phase. When using a wetting reversal agent to release a water lock, we should consider the nature of the liquid phase that produces the water lock and choose an appropriate wetting reversal agent [37].

4.3. Surface Tension

There is a role for surface tension between the interface of two immiscible fluids, the liquid phase and the gas phase. Figure 15 illustrates the flow state of a gas inside a porous medium as the surface tension increases. As the surface tension increases, the liquid becomes more capable of contracting between the phase interfaces; the force required to change the liquid phase morphology is greater, and the flow of the gas at the phase interface is more biased towards being compressed, and therefore the flow pattern is more regular; this change is more pronounced at the gas edges [38]. In slippery water, as the surface tension decreases, the Haynes step phenomenon is evident [39,40], and the gas phase is more likely to form small bubbles that snap off. A comparison of the change in the wetting angle reveals that changing the gas–liquid interaction or the solid–liquid interaction has a greater effect on the choice of gas percolation path, with the gas flowing in the aqueous phase shifting from low to high breakthrough, but preferring high for breakthrough in slippery water, a feature that does not appear in the flow velocity change study.

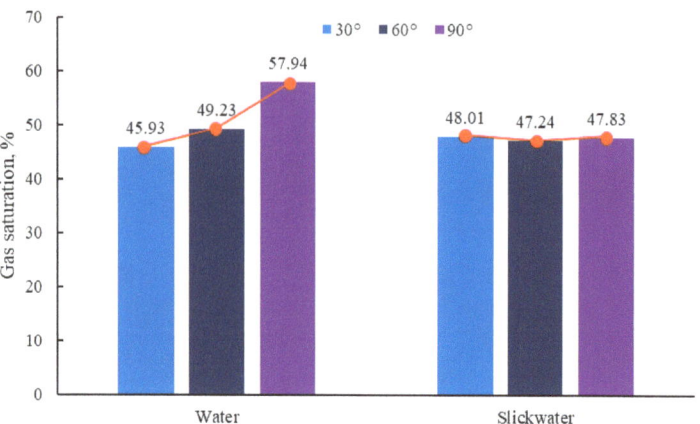

Figure 14. Variation in gas saturation during gas breakthrough in porous media at different wetting angles.

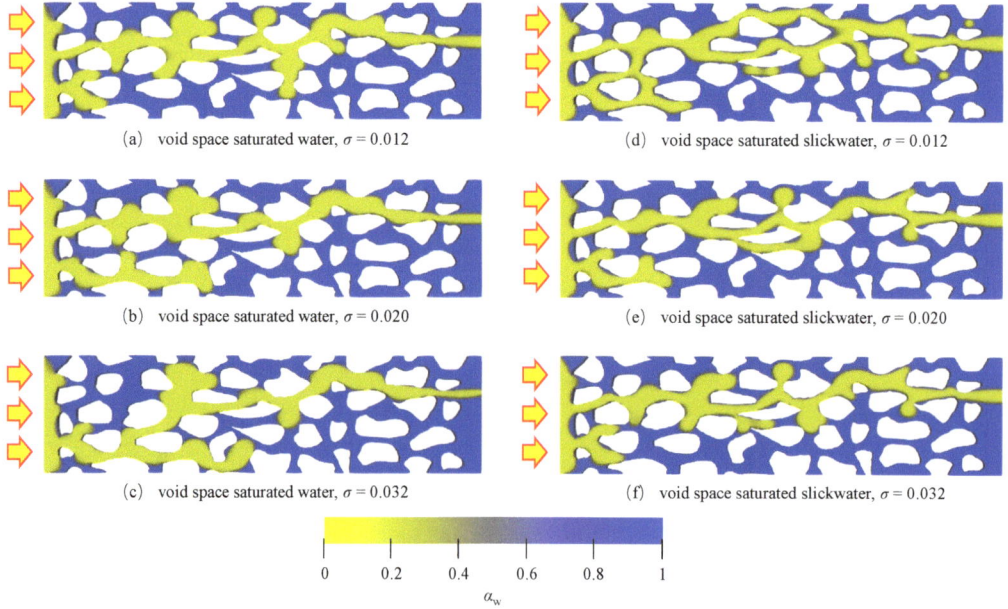

Figure 15. Effect of surface tension on gas flow patterns in liquid phase fluids of different nature.

Figure 16 shows the variation in gas saturation at the moment of gas breakthrough for different surface tensions. As the surface tension decreases, the gas saturation increases in both liquid-phase fluids, but the increase in the gas saturation in slickwater is more pronounced. Since the lower the surface tension the lower the force required for the gas to repel the liquid phase, the amount of fluid repelled under the same conditions is larger. Selecting fracturing fluids with lower surface tensions is more conducive to fracturing fluid rejection.

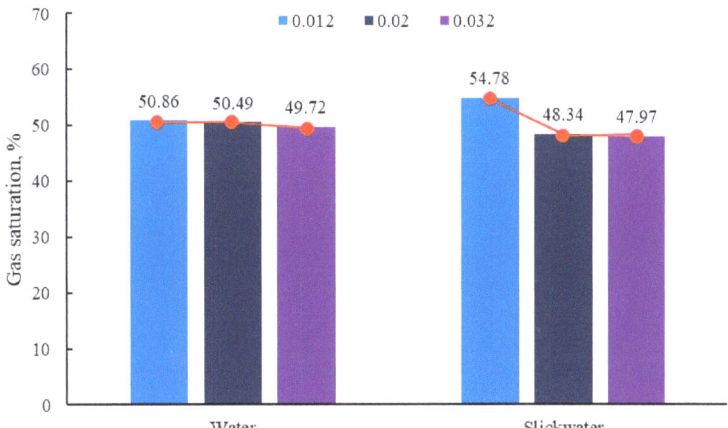

Figure 16. Variation in gas saturation at different surface tensions during gas breakthrough in porous media.

5. Conclusions

This paper develops a gas–liquid two-phase analysis methodology for microscale fractures, employing fracture modeling and numerical simulations to investigate the behavior of gas-phase flow under various conditions. The key conclusions in this study are summarized as follows:

1. The VOF model and porous media modeling techniques can realize the dynamic simulation of gas–liquid two-phase flow at the microscale, and the accuracy of the simulation results is verified using the classical percolation model.
2. Liquid-driven gas processes preferentially break through small-pore channels in porous media, occupying more than 90% of the percolation channels in the wetting phase. The gas-driven liquid process preferentially breaks through large-pore channels in porous media and occupy about 50% of the seepage channels after gas breakthrough. The effect of wettability is the main reason for the lower fracturing fluid rejection rate.
3. The impact of the flow velocity and wetting angle on gas saturation is pronounced in gas-driven Newtonian fluid flow processes, while surface tension significantly affects gas saturation in gas-driven non-Newtonian fluid flow processes. Gas flow within the porous medium follows the path of least resistance, which is determined by the internal structure of the porous medium and remains unchanged regardless of variations in flow parameters and fluid properties.

This study significantly enhances our understanding of the microscale characteristics of gas–liquid flow and fracturing fluid flowback. By providing novel insights into these phenomena, this research establishes a foundational reference for future investigations into gas–liquid flow dynamics within complex porous media.

Author Contributions: S.Y.: Writing-original draft preparation, software; L.Z.: supervision, methodology; T.L.: conceptualization, writing-reviewing; T.Q.: investigation, data curation; D.H.: visualization, validation. All authors have read and agreed to the published version of the manuscript.

Funding: This work was supported by the China National Petroleum Corporation Major Science and Technology Project (Grant No. 2023ZZ16-03).

Data Availability Statement: The original contributions presented in the study are included in the article, further inquiries can be directed to the corresponding author.

Conflicts of Interest: The authors declare that this study received funding from China National Petroleum Corporation. The funder was not involved in the study design, collection, analysis, interpretation of data, the writing of this article or the decision to submit it for publication. All Authors were employed by the company PetroChina Southwest Oil & Gas Field Company. The authors declare no conflict of interest.

References

1. Zou, C.N.; Xiong, B.; Xue, H.Q.; Zheng, D.W.; Ge, Z.X.; Wang, Y.; Jiang, L.Y.; Pan, S.Q.; Wu, S.T. The role of new energy in carbon neutral. *Pet. Explor. Dev.* **2021**, *48*, 411–420. [CrossRef]
2. Zou, C.N.; Ma, F.; Pan, S.Q.; Lin, M.J.; Zhang, G.S.; Xiong, B.; Wang, Y.; Liang, Y.B.; Yang, Z. Earth energy evolution, human development and carbon neutral strategy. *Pet. Explor. Dev.* **2022**, *49*, 411–428. [CrossRef]
3. Liu, Y.W.; Gao, D.P.; Li, Q.; Wan, Y.Z.; Duan, W.J.; Zeng, X.G.; Li, M.Y.; Su, Y.W.; Fan, Y.B.; Li, S.H.; et al. Mechanical frontiers in shale-gas development. *Adv. Mech.* **2019**, *49*, 201901.
4. Ghanbari, E.; Dehghanpour, H. The fate of fracturing water: A field and simulation study. *Fuel* **2016**, *163*, 282–294. [CrossRef]
5. Xiao, Y.H.; Zheng, J.; He, Y.M.; Zhou, B. Experimental study on seepage law of fractional wet porous media. *J. Chengdu Univ. Technol. (Sci. Technol. Ed.)* **2021**, *48*, 9–18.
6. Long, Y.Q.; Zhu, W.Y.; Huang, X.H.; Wu, J.J.; Song, F.Q. Experimental Study on Oil Displacement of Aqueous Dispersion System of Nano/Micron-sized Polymer Particles in Heterogeneous Reservoir. *J. Southwest Pet. Univ.* **2015**, *37*, 129–137.
7. Zhao, Y.L.; Lu, G.; Zhang, L.H.; Yang, K.; Li, X.H.; Luo, J.X. Physical simulation of waterflooding development in large-scale fractured-vuggy reservoir considering filling characteristics. *J. Pet. Sci. Eng.* **2020**, *191*, 10732. [CrossRef]
8. Guo, F.; Aryana, S.A. An Experimental Investigation of Flow Regimes in Imbibition and Drainage Using a Microfluidic Platform. *Energies* **2019**, *12*, 1390. [CrossRef]
9. Yao, J.; Sun, H.; Li, A.F.; Yang, Y.F.; Huang, C.Q.; Wang, Y.Y.; Zhang, L.; Kou, J.L.; Xie, H.J.; Zhao, J.L.; et al. Modern system of multiphase flow in porous media and its development trend. *Chin. Sci. Bull.* **2018**, *63*, 425–451. (In Chinese) [CrossRef]
10. Woerner, M. Numerical modeling of multiphase flows in microfluidics and micro process engineering: A review of methods and applications. *Microfluid. Nanofluid.* **2012**, *12*, 841–886. [CrossRef]
11. Mora, P.; Morra, G.; Yuen, D.A.; Juanes, R. Influence of Wetting on Viscous Fingering Via 2D Lattice Boltzmann Simulations. *Transp. Porous Media* **2021**, *138*, 511–538. [CrossRef]
12. Zhao, Y.L.; Liu, X.Y.; Zhang, L.H.; Shan, B.C. A basic model of unconventional gas microscale flow based on the lattice Boltzmann method. *Pet. Explor. Dev.* **2021**, *48*, 156–165. [CrossRef]
13. Zhang, L.H.; Xiong, Y.; Zhao, Y.L.; Tang, H.M.; Guo, J.J.; Jia, C.S.; Lei, Q.; Wang, B.H. A reservoir drying method for enhancing recovery of tight gas. *Pet. Explor. Dev.* **2022**, *49*, 125–135. [CrossRef]
14. Liu, J.H.; Ju, Y.; Zhang, Y.Q.; Gong, W.B. Preferential Paths of Air-water Two-phase Flow in Porous Structures with Special Consideration of Channel Thickness Effects. *Sci. Rep.* **2019**, *9*, 16204–16217. [CrossRef]
15. Cogswell, D.A.; Szulczewski, M.L. Simulation of incompressible two-phase flow in porous media with large timesteps. *J. Comput. Phys.* **2017**, *345*, 856–865. [CrossRef]
16. Osher, S.; Sethian, J.A. Fronts propagating with curvature-dependent speed: Algorithms based on Hamilton-Jacobi formulations. *J. Comput. Phys.* **1988**, *79*, 12–49. [CrossRef]
17. Amiri, H.A.A.; Hamouda, A.A. Pore-scale modeling of non-isothermal two phase flow in 2D porous media: Influences of viscosity, capillarity, wettability and heterogeneity. *Int. J. Multiph. Flow* **2014**, *61*, 14–27. [CrossRef]
18. Zhao, Y.L.; Zhou, H.J.; Li, H.X.; Wen, T.; Zhang, R.H. Gas-water Two-phase Flow Simulation of Low-permeability Sandstone Digital Rock: Level-set Method. *Chin. J. Comput. Phys.* **2021**, *38*, 585–594.
19. Chern, I.L.; Glimm, J.; Mcbryan, O.; Plohr, B.; Yaniv, S. Front tracking for gas dynamics. *J. Comput. Phys.* **1986**, *62*, 83–110. [CrossRef]
20. Hirt, C.W.; Nichols, B.D. Volume of fluid (VOF) method for the dynamics of free boundaries. *J. Comput. Phys.* **1981**, *39*, 201–225. [CrossRef]
21. Minakov, A.V.; Guzei, D.V.; Pryazhnikov, M.I.; Filimonov, S.A.; Voronenkova, Y.O. 3D pore-scale modeling of nanofluids-enhanced oil recovery. *Pet. Explor. Dev.* **2021**, *48*, 825–834. [CrossRef]
22. Ubbink, O. *Numerical Prediction of Two Fluid Systems with Sharp Interfaces*; University of London: London, UK, 1997.
23. Weller, H.G.; Tabor, G.; Jasak, H.; Fureby, C. A tensorial approach to computational continuum mechanics using object-oriented techniques. *Comput. Phys.* **1998**, *12*, 620–631. [CrossRef]

24. OpenFOAM. The Open Source CFD Toolbox, 2012, User Guide. Available online: https://www.openfoam.com/documentation/guides/v2012/doc/guide-fvoptions-sources.html (accessed on 25 September 2024).
25. Liu, H.H.; Valocchi, A.J.; Werth, C.; Kang, Q.J.; Oostrom, M. Pore-scale simulation of liquid CO_2 displacement of water using a two-phase lattice Boltzmann model. *Adv. Water Resour.* **2014**, *73*, 144–158. [CrossRef]
26. Liu, H.H.; Sun, S.L.; Wu, R.; Wei, B.; Hou, J. Pore-Scale Modeling of Spontaneous Imbibition in Porous Media Using the Lattice Boltzmann Method. *Water Resour. Res.* **2021**, *57*, e2020WR029219. [CrossRef]
27. Sontti, S.G.; Atta, A. CFD analysis of microfluidic droplet formation in non-Newtonian liquid. *Chem. Eng. J.* **2017**, *330*, 245–261. [CrossRef]
28. Bao, K.; Lavrov, A.; Nilsen, H.M. Numerical modeling of non-Newtonian fluid flow in fractures and porous media. *Comput. Geosci.* **2017**, *21*, cp-494. [CrossRef]
29. Brackbill, J.U.; Kothe, D.B.; Zemach, C. A continuum method for modeling surface tension. *J. Comput. Phys.* **1992**, *100*, 335–354. [CrossRef]
30. Hu, Y.Q.; Zhao, C.N.; Zhao, J.Z.; Wang, Q.; Zhao, J.; Guo, D.; Fu, C.H. Mechanisms of fracturing fluid spontaneous imbibition behavior in shale reservoir: A review. *J. Nat. Gas Sci. Eng.* **2020**, *82*, 103498. [CrossRef]
31. Yang, Y.F.; Wang, K.; Lv, Q.F.; Askari, R.; Wang, C.C. Flow simulation considering adsorption boundary layer based on digital rock and finite element method. *Pet. Sci.* **2021**, *18*, 183–194. [CrossRef]
32. Suo, S.; Gan, Y.X. Tuning capillary flow in porous media with hierarchical structures. *Phys. Fluids* **2021**, *33*, 034107–034116. [CrossRef]
33. Akbarifard, M.G.; Azdarpour, A.; Aboosadi, Z.A.; Honarvar, B.; Nabipour, M. Numerical simulation of water production process and spontaneous imbibition in a fractured gas reservoir—A case study on homa gas field. *J. Nat. Gas Sci. Eng.* **2020**, *83*, 103603. [CrossRef]
34. Cheng, Z.L.; Ning, Z.F.; Wang, Q.; Zeng, Y.; Qi, R.R.; Huang, L.; Zhang, W.T. The effect of pore structure on non-Darcy flow in porous media using the lattice Boltzmann method. *J. Pet. Sci. Eng.* **2018**, *172*, 391–400. [CrossRef]
35. Shen, Y.H.; Ge, H.K.; Meng, M.M.; Jiang, Z.X.; Yang, X.Y. Effect of water imbibition on shale permeability and its influence on gas production. *Energy Fuels* **2017**, *31*, 4973–4980. [CrossRef]
36. Zhang, T.; Li, X.F.; Yang, L.F.; Li, J.; Wang, Y.H.; Feng, D.; Yang, J.; Li, P.H. Effects of shut-in timing on flowback rate and productivity of shale gas wells. *Nat. Gas Ind.* **2017**, *37*, 48–60.
37. Zeng, F.H.; Zhang, Q.; Guo, J.C.; Zeng, B.; Zhang, Y.; He, S.G. Mechanisms of shale hydration and water block removal. *Pet. Explor. Dev.* **2021**, *48*, 646–653. [CrossRef]
38. Bui, B.T.; Tutuncu, A.N. Interfacial tension induced-transport in shale: A pore-scale study. *J. Pet. Sci. Eng.* **2018**, *171*, 1409–1419. [CrossRef]
39. Haines, W.B. Studies on the physical properties of soil. *J. Agric. Sci.* **1930**, *20*, 97–116. [CrossRef]
40. Zacharoudiou, I.; Boek, E.S. Capillary filling and Haines jump dynamics using free energy lattice Boltzmann simulations. *Adv. Water Resour.* **2016**, *92*, 43–56. [CrossRef]

Disclaimer/Publisher's Note: The statements, opinions and data contained in all publications are solely those of the individual author(s) and contributor(s) and not of MDPI and/or the editor(s). MDPI and/or the editor(s) disclaim responsibility for any injury to people or property resulting from any ideas, methods, instructions or products referred to in the content.

Article

Innovative Role of Magnesium Oxide Nanoparticles and Surfactant in Optimizing Interfacial Tension for Enhanced Oil Recovery

Youssef E. Kandiel [1], Gamal Attia [2], Farouk Metwalli [2], Rafik Khalaf [2] and Omar Mahmoud [1,*]

[1] Department of Petroleum Engineering, Faculty of Engineering and Technology, Future University in Egypt (FUE), Cairo 11835, Egypt; youssef.kandiel@fue.edu.eg

[2] Department of Geology, Faculty of Science, Helwan University, Cairo 11795, Egypt; gamalattia1949@gmail.com (G.A.); farouk.metwalli@science.helwan.edu.eg (F.M.); rafik.khalaf@science.helwan.edu.eg (R.K.)

* Correspondence: omar.saad@fue.edu.eg

Academic Editor: Hossein Hamidi

Received: 5 December 2024
Revised: 26 December 2024
Accepted: 7 January 2025
Published: 8 January 2025

Citation: Kandiel, Y.E.; Attia, G.; Metwalli, F.; Khalaf, R.; Mahmoud, O. Innovative Role of Magnesium Oxide Nanoparticles and Surfactant in Optimizing Interfacial Tension for Enhanced Oil Recovery. *Energies* 2025, 18, 249. https://doi.org/10.3390/en18020249

Copyright: © 2025 by the authors. Licensee MDPI, Basel, Switzerland. This article is an open access article distributed under the terms and conditions of the Creative Commons Attribution (CC BY) license (https://creativecommons.org/licenses/by/4.0/).

Abstract: Enhancing oil recovery efficiency is vital in the energy industry. This study investigates magnesium oxide (MgO) nanoparticles combined with sodium dodecyl sulfate (SDS) surfactants to reduce interfacial tension (IFT) and improve oil recovery. Pendant drop method measurements revealed a 70% IFT reduction, significantly improving nanoparticle dispersion stability due to SDS. Alterations in Zeta Potential and viscosity, indicating enhanced colloidal stability under reservoir conditions, were key findings. These results suggest that the MgO-SDS system offers a promising and sustainable alternative to conventional methods, although challenges such as scaling up and managing nanoparticle–surfactant dynamics remain. The preparation of MgO nanofluids involved magnetic stirring and ultrasonic homogenization to ensure thorough mixing. Characterization techniques included density, viscosity, pH, Zeta Potential, electric conductivity, and electrophoretic mobility assessments for the nanofluid and surfactant–nanofluid systems. Paraffin oil was used as the oil phase, with MgO nanoparticle concentrations ranging from 0.01 to 0.5 wt% and a constant SDS concentration of 0.5 wt%. IFT reduction was significant, from 47.9 to 26.9 mN/m with 0.1 wt% MgO nanofluid. Even 0.01 wt% MgO nanoparticles reduced the IFT to 41.8 mN/m. Combining MgO nanoparticles with SDS achieved up to 70% IFT reduction, enhancing oil mobility. Changes in Zeta Potential (from −2.54 to 3.45 mV) and pH (from 8.4 to 10.8) indicated improved MgO nanoparticle dispersion and stability, further boosting oil displacement efficiency under experimental conditions. The MgO-SDS system shows promise as a cleaner, cost-effective Enhanced Oil Recovery (EOR) method. However, challenges such as nanoparticle stability under diverse conditions surfactant adsorption management, and scaling up require further research, emphasizing interdisciplinary approaches and rigorous field studies.

Keywords: MgO nanoparticles; interfacial tension reduction; enhanced oil recovery; sustainable nanotechnology

1. Introduction

Enhanced Oil Recovery (EOR) techniques are increasingly critical as conventional oil extraction methods, including primary and secondary recovery, near their operational limits in terms of efficiency [1,2]. These methods leave a substantial proportion of oil unrecovered within reservoirs, posing significant challenges for meeting rising global energy demands [3]. Consequently, the petroleum industry has prioritized the development of advanced methods to access and extract this remaining trapped oil. EOR serves as a pivotal phase in this pursuit by employing various strategies to modify the physical and chemical properties of reservoir fluids and rocks, thereby mobilizing residual oil [4].

Among the most prevalent EOR methods are chemical injection, thermal recovery, and gas-based techniques, each with distinct mechanisms and applications [5]. However, these methods face challenges concerning environmental sustainability and economic feasibility. For instance, the use of surfactants and polymers has been associated with environmental degradation and elevated operational costs [6,7]. These limitations highlight the pressing need for innovative and efficient EOR solutions that are both economically viable and environmentally friendly [8].

A critical factor in many EOR processes involves reducing the interfacial tension (IFT) between the displacing fluid and the trapped oil. By lowering the IFT, the capillary forces that trap oil within reservoir pores can be mitigated, thereby enhancing oil mobility. In particular, the following aspects are considered:

Capillary Forces: The interfacial tension (IFT) between the displacing fluid and the oil strongly influences the capillary forces that trap the oil in the pores of the reservoir rock. Mobilization of Trapped Oil: Reducing the IFT lowers the capillary forces, which in turn enhances the mobility of the trapped oil. This reduction in IFT promotes the coalescence of oil droplets and allows the displacing fluid to sweep more oil toward the production well. Improved Sweep Efficiency: By improving the fluid's ability to displace oil, the overall sweep efficiency is increased, leading to higher oil recovery factors.

Nanotechnology has emerged as a promising solution to overcome these limitations in EOR applications [9]. Nanoparticles, characterized by their high surface-area-to-volume ratio, exhibit unique physical and chemical properties that enable superior interactions with reservoir fluids and rock surfaces [10]. Additionally, their nanoscale dimensions allow them to traverse the porous structures of reservoirs, potentially accessing regions that are inaccessible to conventional agents [11]. Over the past decade, silica-based nanoparticles have received significant attention for EOR applications, demonstrating promising results in altering the wettability of reservoir rocks and reducing interfacial tension (IFT) [12,13].

Despite these advancements, diverging hypotheses exist regarding the effectiveness of different nanoparticle types. While silica nanoparticles have shown success, some studies suggest that metal oxide nanoparticles, such as magnesium oxide (MgO) and aluminum oxide (Al_2O_3), may offer superior results [14–16]. Other metal oxide nanoparticles, such as zirconium dioxide (ZrO_2), cerium oxide (CeO_2), titanium dioxide (TiO_2), zinc oxide (ZnO), and iron oxide (Fe_2O_3), exhibit unique properties that make them excellent candidates for EOR [17,18].

Recent studies have highlighted the potential of these metal oxide nanoparticles to enhance oil recovery via multiple mechanisms and formulations [19,20]. For example, MgO nanoparticles have demonstrated particular efficacy in reducing fine migration by modifying surface properties, a critical factor in maintaining reservoir permeability and oil flow [21,22]. Similarly, Ogolo et al. [23] explored the role of metal oxide nanoparticles in mitigating clayey fines migration, reporting that aluminum oxide nanoparticles could immobilize migrating fines by influencing reservoir pH levels. Nonetheless, challenges persist, including nanoparticle agglomeration, which can lead to pore blockage and compromise permeability, as observed by [9]. This finding, however, contrasts with evidence suggesting challenges in the stability and dispersion of such nanoparticles under reservoir conditions [18].

Building upon this context, this study aims to evaluate the potential of MgO nanoparticles, both independently and in conjunction with an SDS surfactant, to reduce IFT and improve oil recovery. SDS is hypothesized to enhance nanoparticle dispersion stability, enabling uniform distribution at the oil–water interface. By combining MgO nanoparticles with SDS, this study explores a synergistic approach to optimizing nanoparticle performance, increasing stability, and maximizing oil recovery under reservoir conditions. This investigation seeks to address critical knowledge gaps in the field, offering insights that contribute to the broader goal of achieving sustainable and efficient EOR technologies.

2. Materials and Methods

2.1. Nanoparticles and Base Fluids

This study employed high-purity MgO nanoparticles, supplied by MKnano (MK Impex Corp., Mississauga, ON, Canada), characterized by its specific surface area, bulk density, and purity (Table 1). To verify the characteristics of the MgO nanoparticles used in this study in terms of the purity and nanoscale size, additional characterization was performed. Dynamic Light Scattering (DLS) was employed to determine the particle size distribution, while X-ray Diffraction (XRD) analysis verified the crystalline structure and composition. The results demonstrated that the MgO nanoparticles are within the expected size range and exhibit high purity, as illustrated in Figure 1. These findings validate the suitability of the nanoparticles for the intended applications. Synthetic brine was prepared by dissolving 3.0 wt.% sodium chloride (NaCl) in deionized water, simulating reservoir salinity (approximately 30,000 ppm). The oil phase used in all experiments was paraffin oil, selected for its stable and reproducible properties. SDS, sourced from Sigma Aldrich (St. Louis, MO, USA), was used as a stabilizing agent in nanofluid preparations.

Table 1. Properties of used materials.

Properties	MgO Nanopowder	SDS
Specific surface area (m^2/g)	90	N.A
Bulk density (g/L)	100–150	490–560
Average molecular weight (g/mol)	40.3	288.38
Chemical formula	MgO	$CH_3(CH_2)_{11}OSO_3Na$
Purity	>99.9%	>99.9%

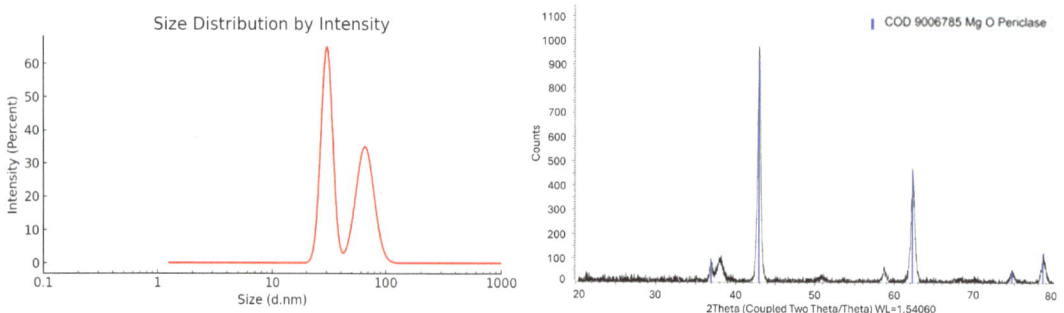

Figure 1. MgO nanoparticle characterization using DLS (**left**) and XRD (**right**).

2.2. Instrumentation and Characterization Methods

2.2.1. Density and pH Measurements

Density was measured using ISOLAB 50 mL pycnometers (ISOLAB Laborgeräte GmbH, Eschau, Bavaria, Germany), ensuring high precision in mass-to-volume calculations. The pH and surface conductivity of the prepared systems, including brine, nanofluids, and surfactant-enhanced nanofluids, were measured with a Hach IQ240 digital pH meter (Hach Company, Loveland, CO, USA) under ambient conditions.

2.2.2. Ultrasonic Mixing

Consistent nanoparticle mixing and dispersion were achieved using the Sonics Vibracell VCX 750 Ultrasonic Homogenizer (Sonics & Materials, Inc., Newtown, CT, USA), operated at 50–80% amplitude with a power output of 750 watts. Additional mixing during preparation was facilitated by an IKA RCT Basic magnetic hot plate stirrer (IKA-Werke GmbH & Co. KG, Staufen, Germany).

2.2.3. Zeta Potential Analysis

The Malvern Zetasizer Nano ZS (Malvern Panalytical Ltd., Worcestershire, UK) was employed to evaluate nanofluid stability. This instrument measures Zeta Potential, electric conductivity, and electrophoretic mobility, providing insights into nanoparticle interactions and dispersion stability. High absolute Zeta Potential values signify greater nanoparticle stability due to strong electrostatic repulsion.

2.3. IFT Measurements

IFT was measured using the Core Lab (Tulsa, OK, USA) "Temco" Pendant Drop IFT-10-P system, capable of replicating reservoir conditions at pressures up to 10,000 psi and temperatures up to 350 °F. Pendant drop formation was captured using an 8 MP HD CCD camera, and IFT values were calculated using Axisymmetric Drop Shape Analysis (ADSA) software (https://www.ramehart.com/diadv.htm accessed on 6 January 2025). This system ensures precise measurements (uncertainty: ±0.2–0.5 mN/m), validated in prior studies [24]. This approach enables accurate characterization of fluid interactions (IFT) under reservoir conditions. Figures 2 and 3 show the schematic of the IFT system and measurements using the Pendant drop method, respectively.

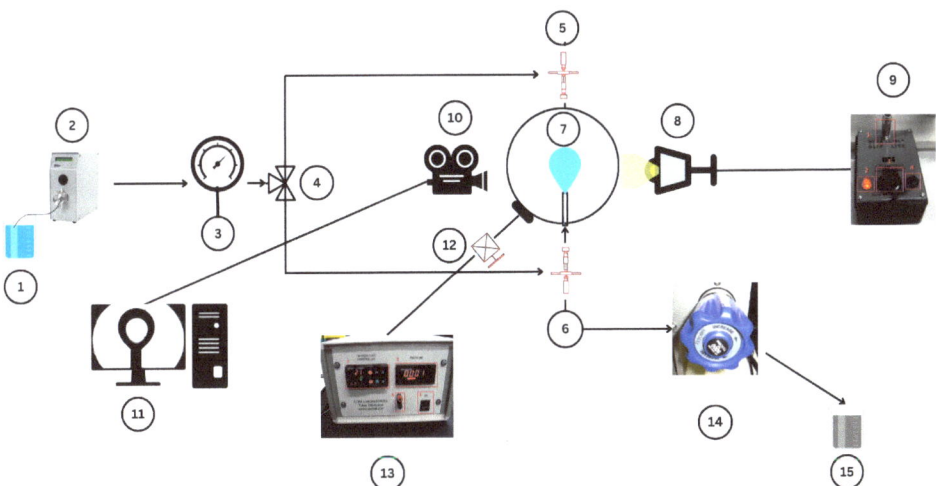

Figure 2. IFT Cell System scheme: (1) injecting/or filling fluid; (2) Prep HPLC Pump 0.1–24.0 mL/min; (3) pressure gauge, with 6000 psi and 300 °F; (4) three-way valve; (5) top cell attachment; (6) bottom cell attachment (including needle); (7) IFT cell; (8) light source; (9) light control unit; (10) camera; (11) computer containing software; (12) valve; (13) transducer, with pressure of 6000 psi and temp. of 300 °F; (14) back pressure regulator with 6000 psi and 300 °F; (15) collecting beaker.

Figure 3. IFT measurement using Pendant drop method.

2.4. Research Method

A comprehensive overview of the experimental procedure followed in this study is summarized in Figure 4. The experimental procedure flowchart systematically outlines the preparation, characterization, and IFT measurements of the fluids investigated, including the base fluid (synthetic brine), nanofluids, and surfactant nanofluids. The figure highlights the critical steps in the process, such as the assessment of fluid density, viscosity, pH, Zeta Potential, and electrophoretic mobility, which are key parameters for evaluating the physical and chemical behavior of fluids. Additionally, it illustrates the focus areas of the IFT measurements, emphasizing the effects of nanofluid concentrations and the surfactant (SDS) on IFT reduction.

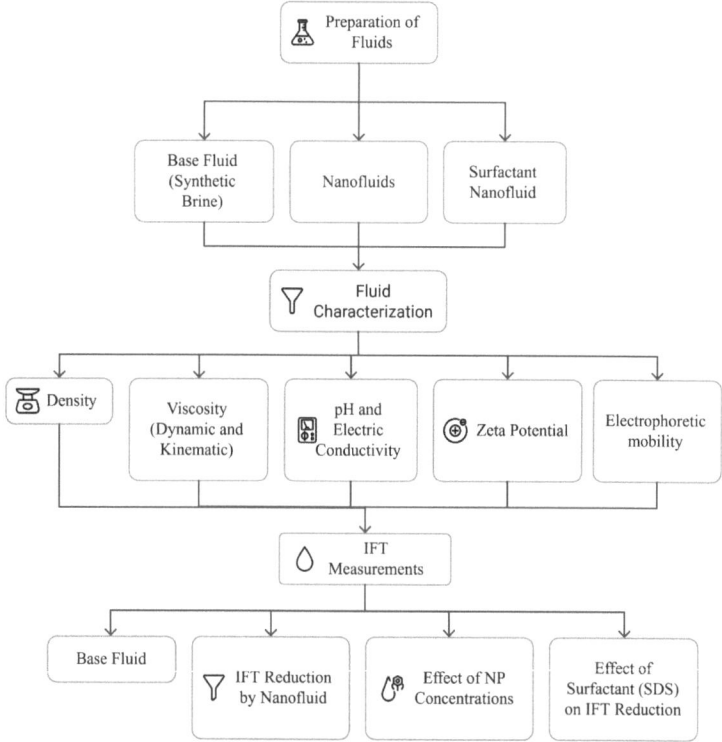

Figure 4. Flowchart of experimental procedure.

2.4.1. Nanofluid Preparation and Characterization

Nanofluid Synthesis: MgO nanofluids were synthesized at varying concentrations (0.01, 0.03, 0.05, 0.1, and 0.5 wt.%) in synthetic brine solution following a two-step protocol:

1. High-speed magnetic stirring for 15 min.
2. Ultrasonic homogenization for 30 min at 50–80% amplitude to ensure uniform nanoparticle dispersion.

SDS was incorporated at a constant concentration of 0.5 wt.% across all nanofluid compositions. An additional 15 min of sonication was performed after SDS addition to further stabilize the nanofluids. Figure 4 shows the workflow of the experimental procedure.

2.4.2. Nanofluid Stability Analysis

Theoretical Background

Stability is critical in EOR applications, as particle aggregation can alter fluid properties and reservoir performance. Electrostatic repulsion between nanoparticles, quantified by Zeta Potential, governs this stability [25].

Stabilization Techniques

To mitigate particle aggregation, the following techniques were employed:

Surfactant Addition: SDS acts as a surfactant, creating electrostatic and steric barriers to prevent aggregation [26].
pH Control: Adjusting pH optimizes nanoparticle surface charges, enhancing repulsion forces and reducing aggregation risks [27].
Ultrasonic Vibration: High-frequency sound waves break up particle clusters and improve dispersion uniformity [28].
Surface Modification: Chemical treatment of nanoparticle surfaces can introduce functional groups that enhance stability by creating additional repulsive mechanisms between particles [29].

Stability Assessment

Zeta Potential serves as a critical indicator of nanofluid stability in this experimental study Figure 5.

A high absolute Zeta Potential value (positive or negative) signifies strong repulsive forces between particles, resulting in a well-dispersed and stable nanofluid that is critical for EOR applications. Conversely, low Zeta Potential values indicate weak inter-particle repulsion, which increases the risk of aggregation and potential system destabilization.

Multiple techniques have been developed to assess nanofluid stability, each offering unique insights into particle behavior:

1. Sedimentation Balance Method: This tracks particle settling rates and suspension stability over time [30].
2. UV-Vis Spectrophotometry: This monitors particle concentration and dispersion through light absorption [31].
3. Zeta Potential Analysis: This measures the electrical charge at particle interfaces [32].
4. Light Scattering Method: This evaluates particle size distribution and aggregation [33].
5. Direct Observation: This provides visual confirmation of nanofluid stability [34].

In this study, the Malvern Zetasizer Nano ZS was employed to comprehensively characterize the fluid systems by measuring the following:

Zeta Potential (mV);
Electric conductivity (mS/cm);
Electrophoretic mobility (μmcm/Vs).

By systematically measuring these parameters, researchers can optimize nanofluid formulations, evaluate nano-surfactant effectiveness, and ensure stable suspension properties critical for EOR applications.

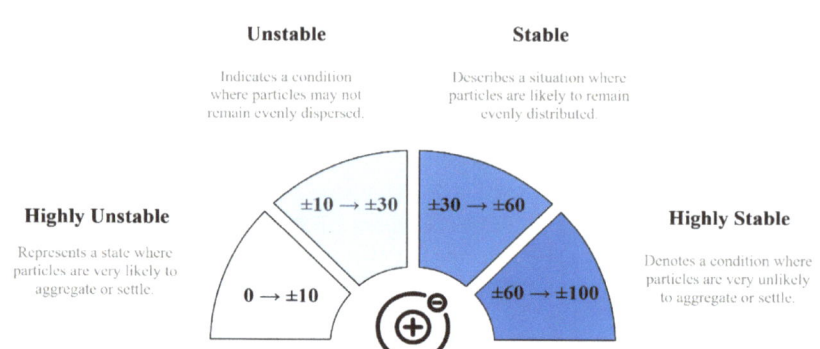

Figure 5. Zeta Potential as a function of fluid stability (data from [30]).

2.4.3. Fluid Characterization

Comprehensive Characterization: Fluid properties were measured using standardized laboratory techniques:

1. Density was determined using ISOLAB 50 mL calibrated pycnometers.
2. Viscosity was measured with the Brookfield KF30 Falling Ball Viscometer (Brookfield Engineering Laboratories, Inc., Middleborough, MA, USA).
3. pH and Surface Conductivity were measured using a digital pH meter (Hach IQ240, Hach Company, Loveland, CO, USA).

The results for each fluid system, including brine, nanofluids, and SDS-enhanced nanofluids, are summarized in Table 2, offering a detailed overview of density, viscosity, pH, and conductivity for all compositions.

Table 2. Fluid properties of all fluids at ambient temperature.

Fluid	Density, (g/cc)	Dynamic Viscosity (cP)	Kinematic Viscosity (mm^2/s)	pH	Surface Conductivity (mV)
Paraffin Oil	0.85	2.03	2.01	N.A	N.A
Brine (3 wt% NaCl)	1.014	0.75	0.74	8.4	−83.05
MgO 0.01 wt%	1.015	0.70	0.69	10	−176.85
MgO 0.03 wt%	1.015	0.83	0.82	10.3	−194
MgO 0.05 wt%	1.015	0.83	0.82	10.2	−186.7
MgO 0.1 wt%	1.016	0.85	0.84	10.4	−199.4
MgO 0.5 wt%	1.019	0.92	0.90	10.8	−223.05
MgO 0.01 wt% + SDS	1.018	0.76	0.75	9.7	−163.7
MgO 0.03 wt% + SDS	1.018	0.82	0.80	9.9	−172.9
MgO 0.05 wt% + SDS	1.017	0.82	0.80	10.2	−189.6
MgO 0.1 wt% + SDS	1.017	0.83	0.81	10.3	−192.6
MgO 0.5 wt% + SDS	1.017	0.88	0.86	10.4	−199

3. Results

3.1. Stability

The stability assessment of MgO nanofluids highlighted notable disparities between the brine-based and SDS-based systems across varying nanoparticle concentrations. In brine-based fluids, Zeta Potential values ranged from −12.35 mV at 0.01 wt% MgO to 3.45 mV at 0.5 wt% MgO, as depicted in Figure 5. These values fall within the unstable range (−30 mV to +30 mV), signifying inadequate colloidal stability. Conversely, the SDS-based fluids exhibited a marked improvement in stability, with Zeta Potential values spanning from −28.75 mV to −39.7 mV over the same concentration range (Figure 5). The increasingly negative Zeta Potential values in the SDS-enhanced system suggest stronger electrostatic stabilization.

Additionally, the pH of both systems increased with rising MgO concentration, ranging from 8.4 to 10.8 in brine and from 9.7 to 10.4 in SDS-based fluids, consistent with the basicity of MgO (Figure 6). A slight decrease in electric conductivity was observed with increasing MgO concentrations in both systems, declining from 60.95 mS/cm to 50.9 mS/cm (Figure 6). This reduction likely results from ion adsorption onto the nanoparticle surfaces. Overall, the inclusion of SDS significantly improved the stability of MgO nanofluids, likely due to a combination of steric and electrostatic stabilization mechanisms. Dynamic and kinematic viscosities followed a similar trend, increasing from 0.75 cP to 0.92 cP in brine and from 0.76 cP to 0.88 cP in SDS systems (Figure 7).

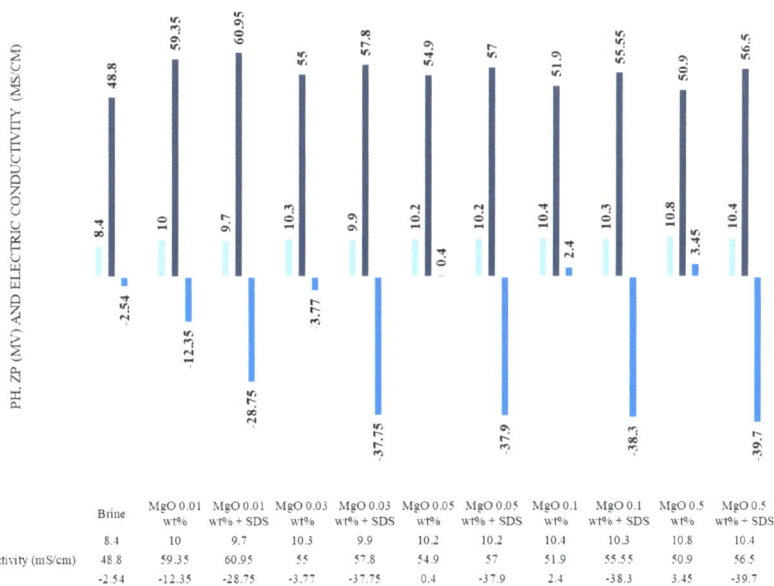

Figure 6. pH, Zeta Potential, and electric conductivity as a function of concentration for both systems.

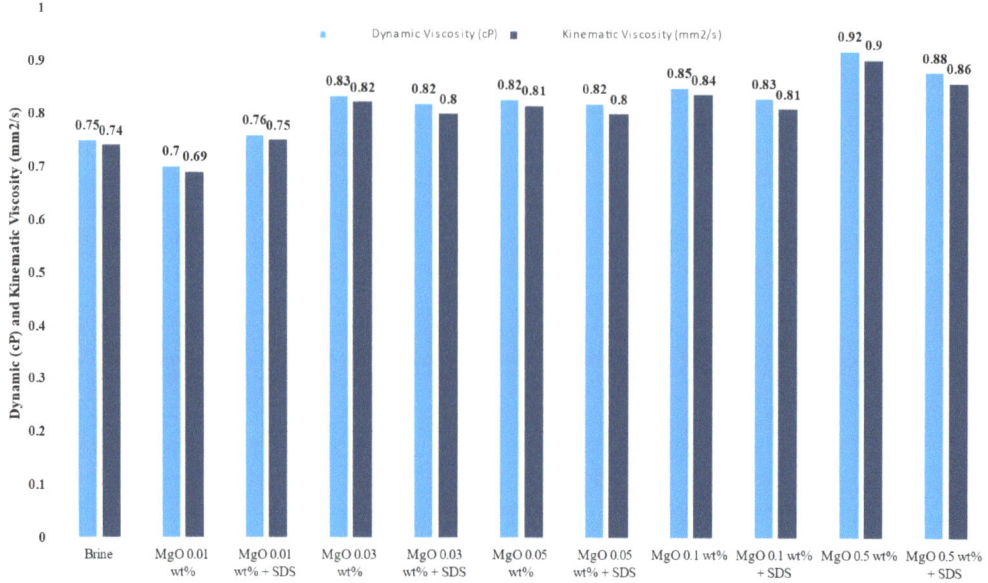

Figure 7. Dynamic and kinematic viscosity for nanofluid and surfactant–nanofluid systems.

3.2. IFT Reduction

The IFT behavior of MgO nanofluids revealed distinct differences between brine-based and SDS-based systems, with significant implications for EOR. In the brine-based system, IFT values ranged inconsistently from 26.9 mN/m to 47.9 mN/m, showing no definitive trend (Figure 8). This variability suggests that MgO nanoparticles alone, when dispersed in brine, are insufficient to achieve substantial IFT reduction, potentially due to inadequate stabilization or ineffective interaction at the oil–water interface.

Conversely, the SDS-based system exhibited a consistent and pronounced reduction in IFT with increasing MgO concentration. Specifically, the IFT decreased from 15.66 mN/m at 0.01 wt% MgO to an exceptionally low 5.6 mN/m at 0.5 wt% MgO, representing a 64.2% reduction (Figure 8). This dramatic improvement strongly indicates a synergistic interaction between the SDS surfactant and MgO nanoparticles.

The synergism is hypothesized to arise from the adsorption of SDS molecules onto the surface of MgO nanoparticles, which enhances their dispersibility and stability at the oil–water interface. This adsorption modifies the surface energy of the nanoparticles, enabling their alignment and effective participation in forming a robust interfacial barrier. The SDS molecules further stabilize the nanoparticles in the aqueous phase, preventing aggregation and ensuring their active involvement in reducing capillary forces. These combined effects result in a significant decrease in IFT, which is particularly advantageous for EOR processes.

Such low IFT values are instrumental in reducing capillary forces, enhancing oil mobilization, and improving sweep efficiency during oil recovery operations. The revised understanding of this interaction mechanism underscores its practical implications and highlights the potential of MgO nanoparticles in SDS-based systems as effective agents for improving EOR performance.

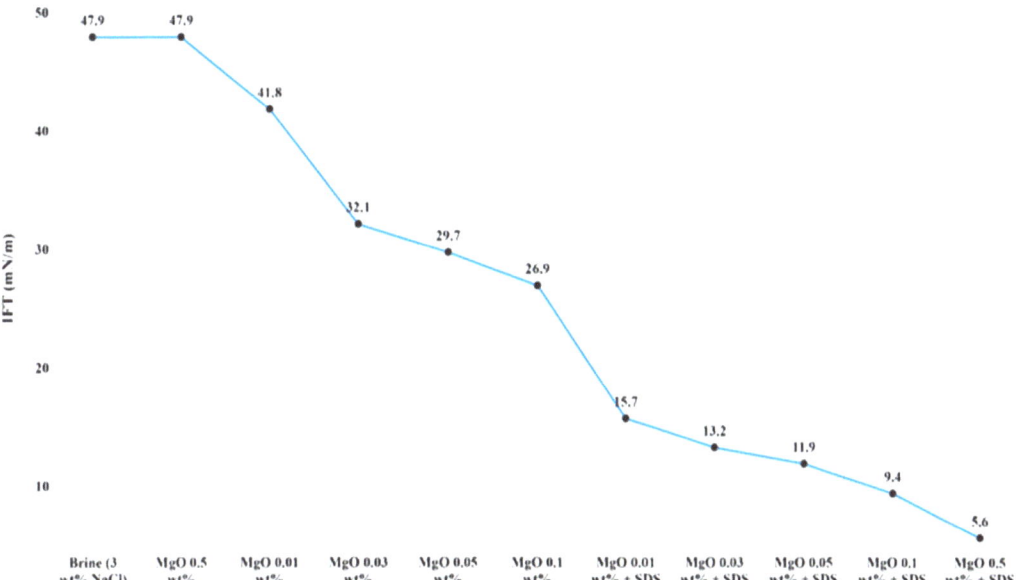

Figure 8. IFT as a function of nanoparticle concentration.

4. Discussion

4.1. Significance of Results

The findings emphasize the critical role of surfactants in enhancing the stability and IFT-reducing performance of MgO nanofluids. The enhanced Zeta Potential values in the SDS-based system reflect strong electrostatic repulsion, preventing nanoparticle agglomeration and ensuring colloidal stability. This observation supports the compatibility of MgO-SDS nanofluids for reservoir conditions. Coupled with the observed pH and viscosity trends, this stability underscores the potential of these nanofluids in EOR applications. The observed 64.2% IFT reduction achieved with the SDS-MgO system corroborates the findings of [35], which demonstrated similar concentration-dependent reductions in IFT.

Compared to traditional EOR techniques, this study highlights a promising pathway for achieving significant oil displacement with nanoparticle–surfactant systems. The MgO-SDS combination not only matches but potentially enhances the IFT-lowering capabilities of previously studied surfactant-based EOR methods as shown in Table 3. This result aligns with the findings of [5,35] which demonstrate the importance of ultra-low IFT values in boosting recovery rates. While ultra-low IFTs (10^{-3} mN/m) were not achieved here, the reduction to 5.6 mN/m still represents a significant advancement for EOR technology.

Table 3. Comparative analysis with other chemical EOR technologies.

Criteria	MgO-SDS Combination	Alternative Nanoparticles (e.g., ZnO, SiO$_2$)	Alternative Surfactants (e.g., CTAB, Triton X-100)
IFT Reduction Efficiency	Superior due to synergy	Moderate, often requires surface modification	Moderate, depending on salinity and temperature
Cost	Low	Moderate to high due to synthesis and functionalization	Moderate

Table 3. *Cont.*

Criteria	MgO-SDS Combination	Alternative Nanoparticles (e.g., ZnO, SiO$_2$)	Alternative Surfactants (e.g., CTAB, Triton X-100)
Environmental Impact	Low, biodegradable	Potentially harmful depending on the material	Variable, some are non-biodegradable
Thermal/Salinity Tolerance	High	High (varies by nanoparticle)	Moderate
Scalability	High	Moderate (complex preparation)	High

4.2. Implications

The ability to achieve IFT values as low as 5.6 mN/m with MgO-SDS nanofluids highlights their potential for efficient oil displacement. This result suggests that such systems can effectively mobilize trapped oil, leading to improved recovery rates. Importantly, the synergistic interaction between MgO nanoparticles and the SDS surfactant reflects the broader potential of integrating nanotechnology with chemical EOR techniques [36,37].

These findings also indicate that nanofluid-based EOR methods could offer an alternative to conventional methods with higher economic and environmental feasibility. While challenges remain, the observed compatibility of MgO-SDS nanofluids with reservoir conditions points to their potential scalability for industrial applications.

4.3. Limitations

While the results demonstrate the potential of MgO-SDS nanofluids, several challenges remain:

1. Testing Conditions: This study was conducted under ambient conditions, whereas reservoir environments exhibit higher temperatures and pressures that may alter fluid behavior.
2. Long-Term Stability: The durability of nanofluid stability under prolonged storage or operational conditions remains unclear.
3. Conductivity Trends: The observed decline in electrical conductivity with increasing MgO concentration warrants further investigation into its implications for ionic interactions in reservoirs.

Additionally, potential environmental and economic concerns, such as the scalability of SDS-based nanofluids and the impact of surfactant degradation, should be considered in future studies.

4.4. Future Directions

Future research should focus on the following areas to build upon the current findings:

1. Reservoir Conditions: Evaluate the performance of MgO-SDS nanofluids under elevated pressures and temperatures to replicate realistic reservoir environments.
2. Surfactant Alternatives: Investigate the effects of alternative surfactants or co-surfactant systems to optimize IFT reduction and stability.
3. Field-Scale Validation: Conduct field-scale tests to validate laboratory findings and assess the economic viability of MgO-SDS nanofluids for EOR applications.
4. Sustainability Studies: Explore the environmental impact and potential biodegradability of MgO-SDS nanofluids to ensure sustainable deployment in oil recovery operations [38,39].

Finally, interdisciplinary collaborations could explore integrating nanofluid EOR systems with other advanced recovery technologies, such as CO_2 injection or thermal recovery, to achieve synergistic benefits.

5. Conclusions

The evaluation of MgO nanofluids demonstrates that incorporating the SDS surfactant significantly enhances their performance for EOR applications. The SDS-MgO nanofluid system exhibited a pronounced synergistic effect, achieving a substantial reduction in IFT from 15.66 mN/m to 5.6 mN/m as the MgO concentration increased from 0.01 wt% to 0.5 wt%. This 64.2% reduction in IFT is critical for EOR, as it directly improves oil mobilization and sweep efficiency by reducing capillary forces.

Stability analysis further validated the effectiveness of the SDS-based system. Zeta Potential values became increasingly negative (-28.75 mV to -39.7 mV) with rising MgO concentrations, reflecting enhanced electrostatic stabilization and superior colloidal stability. Additionally, the modest increases in viscosity (from 0.76 cP to 0.88 cP) and pH (from 9.7 to 10.4) indicate a stable and well-dispersed nanofluid system without the excessive thickening that could hinder flow in reservoir conditions. The slight reduction in electric conductivity (from 60.95 mS/cm to 56.5 mS/cm) likely results from ion adsorption onto nanoparticle surfaces, contributing to the system's improved interfacial and stability characteristics.

In contrast, the brine-based MgO nanofluid displayed inconsistent IFT values and significantly poorer stability, underscoring its limited applicability for EOR purposes. The combined advantages of substantial IFT reduction and enhanced stability in the SDS-MgO system highlight its potential as a promising formulation for EOR applications. These findings emphasize the critical role of surfactant–nanoparticle interactions in optimizing nanofluids for advanced oil recovery processes. Future work should focus on refining this formulation, exploring its performance under reservoir conditions, and conducting field-scale tests to assess its practical viability in improving oil recovery rates.

Author Contributions: Conceptualization, Y.E.K. and O.M.; methodology, Y.E.K. and O.M.; software, Y.E.K.; validation, Y.E.K., F.M. and O.M.; formal analysis, Y.E.K.; investigation, Y.E.K.; resources, O.M.; writing—original draft preparation, Y.E.K.; writing—review and editing, O.M. and G.A.; visualization, Y.E.K.; supervision, G.A., F.M., R.K. and O.M.; project administration, Y.E.K.; funding acquisition, O.M. All authors have read and agreed to the published version of the manuscript.

Funding: This research received no external funding.

Data Availability Statement: Data are contained within the article.

Acknowledgments: The authors would like to express their sincere gratitude to Future University in Egypt (FUE) for materials support and for providing access to the laboratory facilities that were essential for the successful completion of this study. Additionally, heartfelt thanks are extended to the Geology Department at Helwan University for the invaluable educational foundation and academic support that has significantly contributed to the development of this work.

Conflicts of Interest: The authors declare that there are no conflicts of interest regarding the publication of this manuscript.

References

1. Lake, L.W. *Enhanced Oil Recovery*; Prentice Hall: Englewood Cliffs, NJ, USA, 1989; ISBN 978-0-13-281601-4.
2. Green, D.W.; Willhite, G.P. *Enhanced Oil Recovery*; SPE Textbook Series; Henry, L., Ed.; Doherty Memorial Fund of AIME, Society of Petroleum Engineers: Richardson, TX, USA, 1998; ISBN 978-1-55563-077-5.
3. Alvarado, V.; Manrique, E. Enhanced Oil Recovery: An Update Review. *Energies* **2010**, *3*, 1529–1575. [CrossRef]
4. Thomas, S. Enhanced Oil Recovery—An Overview. *Oil Gas Sci. Technol. Rev. IFP* **2008**, *63*, 9–19. [CrossRef]
5. Sheng, J.J. *Modern Chemical Enhanced Oil Recovery*; Elsevier: Amsterdam, The Netherlands, 2011; ISBN 978-1-85617-745-0.

6. Hendraningrat, L.; Engeset, B.; Suwarno, S.; Torsæter, O. Improved Oil Recovery by Nanofluids Flooding: An Experimental Study. In Proceedings of the SPE Kuwait International Petroleum Conference and Exhibition, Kuwait City, Kuwait, 10–12 December 2012; p. SPE-163335-MS.
7. Sorbie, K.S. *Polymer-Improved Oil Recovery*; Springer: Dordrecht, The Netherlands, 1991; ISBN 978-94-011-3044-8.
8. Nwidee, L.N.; Lebedev, M.; Barifcani, A.; Sarmadivaleh, M.; Iglauer, S. Wettability Alteration of Oil-Wet Limestone Using Surfactant-Nanoparticle Formulation. *J. Colloid Interface Sci.* 2017, 504, 334–345. [CrossRef] [PubMed]
9. Ogolo, N.A.; Olafuyi, O.A.; Onyekonwu, M.O. Enhanced Oil Recovery Using Nanoparticles. In Proceedings of the SPE Saudi Arabia Section Technical Symposium and Exhibition, Al-Khobar, Saudi Arabia, 8 April 2012; p. SPE-160847-MS.
10. Yu, W.; Xie, H. A Review on Nanofluids: Preparation, Stability Mechanisms, and Applications. *J. Nanomater.* 2012, 2012, 435873. [CrossRef]
11. Li, S.; Hendraningrat, L.; Tors, O. Improved Oil Recovery by Hydrophilic Silica Nanoparticles Suspension: 2- Phase Flow Experimental Studies. In Proceedings of the IPTC 2013: International Petroleum Technology Conference, Beijing, China, 26–28 March 2013; European Association of Geoscientists & Engineers: Bunnik, The Netherlands, 2013.
12. Giraldo, J.; Benjumea, P.; Lopera, S.; Cortés, F.B.; Ruiz, M.A. Wettability Alteration of Sandstone Cores by Alumina-Based Nanofluids. *Energy Fuels* 2013, 27, 3659–3665. [CrossRef]
13. Roustaei, A.; Moghadasi, J.; Iran, A.; Bagherzadeh, H.; Shahrabadi, A. An Experimental Investigation of Polysilicon Nanoparticles' Recovery Efficiencies through Changes in Interfacial Tension and Wettability Alteration. In Proceedings of the SPE International Oilfield Nanotechnology Conference and Exhibition, Noordwijk, The Netherlands, 12 June 2012; p. SPE-156976-MS.
14. Hendraningrat, L.; Li, S.; Torsæter, O. Effect of Some Parameters Influencing Enhanced Oil Recovery Process Using Silica Nanoparticles: An Experimental Investigation. In Proceedings of the SPE Reservoir Characterization and Simulation Conference and Exhibition, Abu Dhabi, United Arab Emirates, 16 September 2013; p. SPE-165955-MS.
15. Emadi, S.; Shadizadeh, S.R.; Manshad, A.K.; Rahimi, A.M.; Mohammadi, A.H. Effect of Nano Silica Particles on Interfacial Tension (IFT) and Mobility Control of Natural Surfactant (Cedr Extraction) Solution in Enhanced Oil Recovery Process by Nano-Surfactant Flooding. *J. Mol. Liq.* 2017, 248, 163–167. [CrossRef]
16. Kandiel, Y.E.; Metwalli, F.I.; Khalaf, R.E.; Attia, G.M.; Mahmoud, O. Synergistic Effect of MgO Nanoparticles and SDS Surfactant on Interfacial Tension Reduction for Enhanced Oil Recovery. In Proceedings of the Mediterranean Offshore Conference, Alexandria, Egypt, 20 October 2024; p. D011S004R008.
17. Hendraningrat, L.; Torsaeter, O. Unlocking the Potential of Metal Oxides Nanoparticles to Enhance the Oil Recovery. In Proceedings of the Offshore Technology Conference-Asia, Kuala Lumpur, Malaysia, 25 March 2014; p. OTC-24696-MS.
18. Suleimanov, B.A.; Ismailov, F.S.; Veliyev, E.F. Nanofluid for Enhanced Oil Recovery. *J. Pet. Sci. Eng.* 2011, 78, 431–437. [CrossRef]
19. Salem, K.G.; Tantawy, M.A.; Gawish, A.A.; Salem, A.M.; Gomaa, S.; El-hoshoudy, A.N. Key Aspects of Polymeric Nanofluids as a New Enhanced Oil Recovery Approach: A Comprehensive Review. *Fuel* 2024, 368, 131515. [CrossRef]
20. Salem, K.G.; Salem, A.M.; Tantawy, M.A.; Gawish, A.A.; Gomaa, S.; El-hoshoudy, A.N. A Comprehensive Investigation of Nanocomposite Polymer Flooding at Reservoir Conditions: New Insights into Enhanced Oil Recovery. *J. Polym. Environ.* 2024, 32, 5915–5935. [CrossRef]
21. Maghzi, A.; Mohammadi, S.; Ghazanfari, M.H.; Kharrat, R.; Masihi, M. Monitoring Wettability Alteration by Silica Nanoparticles during Water Flooding to Heavy Oils in Five-Spot Systems: A Pore-Level Investigation. *Exp. Therm. Fluid Sci.* 2012, 40, 168–176. [CrossRef]
22. Ahmadi, M.; Habibi, A.; Pourafshary, P.; Ayatollahi, S. Zeta-Potential Investigation and Experimental Study of Nanoparticles Deposited on Rock Surface To Reduce Fines Migration. *SPE J.* 2013, 18, 534–544. [CrossRef]
23. Ogolo, N.A.; Onyekonwu, M.O.; Akaranta, O. Trapping Mechanism of Nanofluids on Migrating Fines in Sand. In Proceedings of the SPE Nigeria Annual International Conference and Exhibition, Lagos, Nigeria, 5 August 2013; p. SPE-167502-MS.
24. Arashiro, E.Y.; Demarquette, N.R. Use of the Pendant Drop Method to Measure Interfacial Tension between Molten Polymers. *Mat. Res.* 1999, 2, 23–32. [CrossRef]
25. Rao, Y. Nanofluids: Stability, Phase Diagram, Rheology and Applications. *Particuology* 2010, 8, 549–555. [CrossRef]
26. Wang, J.; Yang, X.; Klemeš, J.J.; Tian, K.; Ma, T.; Sunden, B. A Review on Nanofluid Stability: Preparation and Application. *Renew. Sustain. Energy Rev.* 2023, 188, 113854. [CrossRef]
27. Ghadimi, A.; Saidur, R.; Metselaar, H.S.C. A Review of Nanofluid Stability Properties and Characterization in Stationary Conditions. *Int. J. Heat Mass Transf.* 2011, 54, 4051–4068. [CrossRef]
28. Chang, H.; Jwo, C.S.; Fan, P.S.; Pai, S.H. Process Optimization and Material Properties for Nanofluid Manufacturing. *Int. J. Adv. Manuf. Technol.* 2007, 34, 300–306. [CrossRef]
29. Li, Y.; Zhou, J.; Tung, S.; Schneider, E.; Xi, S. A Review on Development of Nanofluid Preparation and Characterization. *Powder Technol.* 2009, 196, 89–101. [CrossRef]
30. Zhu, H.; Zhang, C.; Tang, Y.; Wang, J.; Ren, B.; Yin, Y. Preparation and Thermal Conductivity of Suspensions of Graphite Nanoparticles. *Carbon* 2007, 45, 226–228. [CrossRef]

31. Lee, K.; Hwang, Y.; Cheong, S.; Kwon, L.; Kim, S.; Lee, J. Performance Evaluation of Nano-Lubricants of Fullerene Nanoparticles in Refrigeration Mineral Oil. *Curr. Appl. Phys.* **2009**, *9*, e128–e131. [CrossRef]
32. Daungthongsuk, W.; Wongwises, S. A Critical Review of Convective Heat Transfer of Nanofluids. *Renew. Sustain. Energy Rev.* **2007**, *11*, 797–817. [CrossRef]
33. Hong, K.S.; Hong, T.-K.; Yang, H.-S. Thermal Conductivity of Fe Nanofluids Depending on the Cluster Size of Nanoparticles. *Appl. Phys. Lett.* **2006**, *88*, 031901. [CrossRef]
34. Wei, X.; Zhu, H.; Kong, T.; Wang, L. Synthesis and Thermal Conductivity of Cu_2O Nanofluids. *Int. J. Heat Mass Transf.* **2009**, *52*, 4371–4374. [CrossRef]
35. Nowrouzi, I.; Khaksar Manshad, A.; Mohammadi, A.H. Effects of TiO_2, MgO and γ-Al_2O_3 Nano-Particles on Wettability Alteration and Oil Production under Carbonated Nano-Fluid Imbibition in Carbonate Oil Reservoirs. *Fuel* **2020**, *259*, 116110. [CrossRef]
36. Xu, F.; Zhong, X.; Li, Z.; Cao, W.; Yang, Y.; Liu, M. Synergistic Mechanisms Between Nanoparticles and Surfactants: Insight Into NP–Surfactant Interactions. *Front. Energy Res.* **2022**, *10*, 913360. [CrossRef]
37. Ahmadi, Y.; Hemmati, M.; Vaferi, B.; Gandomkar, A. Applications of Nanoparticles during Chemical Enhanced Oil Recovery: A Review of Mechanisms and Technical Challenges. *J. Mol. Liq.* **2024**, *415*, 126287. [CrossRef]
38. Elsaid, K.; Olabi, A.G.; Wilberforce, T.; Abdelkareem, M.A.; Sayed, E.T. Environmental Impacts of Nanofluids: A Review. *Sci. Total Environ.* **2021**, *763*, 144202. [CrossRef]
39. Lechuga, M.; Fernández-Serrano, M.; Ríos, F.; Fernández-Arteaga, A.; Jiménez-Robles, R. Environmental Impact Assessment of Nanofluids Containing Mixtures of Surfactants and Silica Nanoparticles. *Environ. Sci. Pollut. Res.* **2022**, *29*, 84125–84136. [CrossRef]

Disclaimer/Publisher's Note: The statements, opinions and data contained in all publications are solely those of the individual author(s) and contributor(s) and not of MDPI and/or the editor(s). MDPI and/or the editor(s) disclaim responsibility for any injury to people or property resulting from any ideas, methods, instructions or products referred to in the content.

Article

Comprehensive Investigation of Factors Affecting Acid Fracture Propagation with Natural Fracture

Qingdong Zeng [1], Taixu Li [1], Long Bo [1], Xuelong Li [1,*] and Jun Yao [2]

[1] College of Energy and Mining Engineering, Shandong University of Science and Technology, Qingdao 266590, China; upc.zengqd@163.com (Q.Z.); litaixv2025@163.com (T.L.); bl_sdkjdx@163.com (L.B.)
[2] Research Center of Multiphase Flow in Porous Media, China University of Petroleum (East China), Qingdao 266580, China; upc_rcogfr@163.com
* Correspondence: lixlcumt@126.com

Abstract: Acid fracturing is a crucial stimulation technique to enhance hydrocarbon recovery in carbonate reservoirs. However, the interaction between acid fractures and natural fractures remains complex due to the combined effects of mechanical, chemical, and fluid flow processes. This study extends a previously developed hydro-mechano-reactive flow coupled model to analyze these interactions, focusing on the influence of acid dissolution. The model incorporates reservoir heterogeneity and simulates various scenarios, including different stress differences, approaching angles, injection rates, and acid concentrations. Numerical simulations reveal distinct propagation modes for acid and hydraulic fractures, highlighting the significant influence of acid dissolution on fracture behavior. Results show that hydraulic fractures are more likely to cross natural fractures, whereas acid fractures tend to be arrested due to wormhole formation. Increasing stress differences and approaching angles promote fracture crossing, while lower angles favor diversion into natural fractures. Higher injection rates facilitate fracture crossing by increasing pressure accumulation, but excessive acid concentrations hinder fracture initiation due to enhanced wormhole formation. The study demonstrates the importance of tailoring fracturing treatments to specific reservoir conditions, optimizing parameters to enhance fracture propagation and reservoir stimulation. These findings contribute to a deeper understanding of fracture mechanics in heterogeneous reservoirs and offer practical implications for improving the efficiency of hydraulic fracturing operations in unconventional reservoirs.

Keywords: acid fracturing; natural fracture; hydro-mechano-reactive flow model; phase field method; fracture propagation modes

Citation: Zeng, Q.; Li, T.; Bo, L.; Li, X.; Yao, J. Comprehensive Investigation of Factors Affecting Acid Fracture Propagation with Natural Fracture. *Energies* **2024**, *17*, 5386. https://doi.org/10.3390/en17215386

Academic Editor: Efstathios E. Michaelides

Received: 12 October 2024
Revised: 23 October 2024
Accepted: 28 October 2024
Published: 29 October 2024

Copyright: © 2024 by the authors. Licensee MDPI, Basel, Switzerland. This article is an open access article distributed under the terms and conditions of the Creative Commons Attribution (CC BY) license (https://creativecommons.org/licenses/by/4.0/).

1. Introduction

By injecting acidic solutions at high pressures, acid fracturing creates and propagates fractures while simultaneously dissolving the carbonate rock, resulting in increased permeability and improved connectivity within the reservoir. The presence of natural fractures adds complexity to this process, as the interaction between induced acid fractures and pre-existing natural fractures can significantly influence fracture propagation paths and, consequently, the efficiency of the stimulation, as shown in Figure 1. A comprehensive understanding of the factors affecting acid fracture propagation in naturally fractured reservoirs is essential for optimizing treatment designs and maximizing production.

Numerical studies have investigated the interactions between induced fractures and natural fractures in the context of hydraulic fracturing [1–4]. Yi et al. [5] proposed a coupled fluid flow and fracture phase field evolution model to simulate hydraulic fracture propagation in porous media with natural fractures. Their results demonstrated that smaller approaching angles, lower natural fracture strength, and smaller in situ stress differences make natural fractures more likely to initiate and propagate. Zhou et al. [6] advanced this understanding by developing a fully coupled hydro-mechanical model based on the extended finite element method (XFEM) to handle different fracture interaction behaviors. Introducing a new weakly discontinuous junction enrichment function and adopting a tensile stress criterion for new fracture initiation, they investigated the effects of natural fracture properties, treatment parameters, matrix permeability, anisotropic stress, and intersection angle on fracture interaction. Xiong and Ma [7] developed a hydraulic fracture random propagation method using mesh node splitting and zero-thickness cohesive elements. Their study demonstrated that natural fractures could open and slip even without direct contact with hydraulic fractures, highlighting the influence of stress shadow effects and the importance of formation properties and injection rates on fracture complexity. Sun et al. [8] proposed a quantitative model to predict hydraulic fracture propagation across cemented natural fractures based on numerical simulation and developed a probability function to predict crossing behavior using logistic regression.

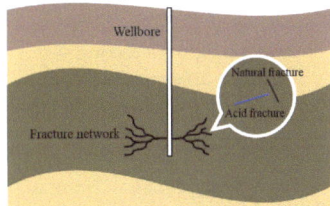

Figure 1. Schematic diagram of acid fracturing in naturally fractured reservoirs.

Experimental investigations have provided valuable insights into fracture propagation behavior in naturally fractured reservoirs [9–11]. Zhou et al. [12] conducted experimental studies using a tri-axial fracturing system, revealing complex interactions under various stress conditions and rock properties. Zhang et al. [13] performed experiments on hydraulic fracture propagation in tight sandstone formations with closed cemented natural fractures. By creating artificial tight sandstone specimens with controlled properties and embedded natural fractures, they showed that factors such as approach angle, in situ stress state, and natural fracture properties significantly influence the interaction between hydraulic and natural fractures. Qiu et al. [14] conducted laboratory tests to investigate the interaction between hydraulic fractures and natural fractures in deep unconventional reservoirs under high stress conditions and identified various interaction modes between hydraulic and natural fractures.

The topology and characteristics of natural fractures play a significant role in fracture network development. Wang et al. [15] investigated the role of natural fractures with different topology structures on hydraulic fracture propagation in continental shale reservoirs. Utilizing data from the Yanchang Formation, they classified natural fractures into three types based on node structures. Their study found that natural fractures with Type II nodes had the most substantial effect on inducing hydraulic fracture propagation and creating complex fracture networks. Injection parameters and fluid properties also critically impact fracture propagation. Li et al. [16] conducted a sensitivity analysis using a discrete fracture network (DFN) model based on outcrop data from the Ordos Basin, China. Their findings indicated that increasing natural fracture aperture decreased fracture complexity, while higher friction coefficients improved hydraulic fracturing efficiency. Higher injection rates led to more complex fracture networks, whereas higher fluid viscosity reduced fracturing efficiency. Tong et al. [17] proposed a new criterion for predicting the interaction between hydraulic fractures and natural fractures at non-orthogonal angles. Considering fluid flow, stress shadow effects, and poroelastic responses, they revealed that hydraulic fractures are more likely to cross natural fractures under conditions of short fracture half-length, high friction coefficient, high fluid viscosity, and high fracture toughness.

While significant progress has been made in understanding hydraulic fracture propagation, there is a relative paucity of studies specifically addressing acid fracture propagation in the presence of natural fractures. Acid fracturing introduces additional complexities due to chemical reactions between the acid and the rock, which can alter fracture surfaces, change rock mechanical properties, and influence fracture propagation paths [18–20]. Dai et al. [21] conducted true triaxial acid fracturing experiments to study steering acid fracturing in carbonate reservoir and found that the complexity of fractures is influenced by natural fracture and the fluid viscosity. Chen et al. [22] addressed some of these complexities by simulating acid transport and dissolution in fracture networks using a 3D unified pipe-network method, allowing for the modeling of complex fracture geometries and their evolution due to acid dissolution. Zhu et al. [23] presented a discretized virtual internal bond approach for acid fracturing in complex fractured-vuggy carbonate reservoirs. By considering full hydro-mechanical-chemical coupling effects in large fractures and partial coupling in small fractures, their method could simulate large fracture propagation, interactions with natural fractures and cavities, and the activation of inactive natural fractures. Recently, we developed a hydro-mechano-reactive flow coupled model to simulate acid fracturing in heterogeneous reservoirs [24]. Despite these advancements, a comprehensive investigation integrating the various factors affecting acid fracture propagation with natural fractures remains lacking. Therefore, this study aims to fill this gap by conducting a comprehensive investigation into the factors affecting acid fracture propagation in the presence of natural fractures by extending the previously developed hydro-mechano-reactive flow coupled model.

This paper is structured as follows: Section 2 outlines the primary governing equation of the model, while Section 3 presents the numerical solution. Section 4 validates the model and provides an in-depth analysis of acid fracture propagation in conjunction with natural fractures. Finally, Section 5 offers some conclusions.

2. Governing Equations

Consider the scenario of an acid fracture approaching a natural fracture at an angle β, as illustrated in Figure 2. Acid fracture propagation encompasses acid transport, fluid flow, rock deformation, and fracture propagation. We present a hydro-mechano-reactive flow coupled model previously developed in [24]. This model is now extended to analyze the interaction between acid fractures and natural fractures. The study is based on the following assumptions: the fracturing process occurs under quasi-static and isothermal conditions; the reservoir is heterogeneous and saturated, accounting for variations in porosity, mineral inclusions, and elastic modulus; the fluid is compressible, and acid dissolution occurs instantaneously, following a first-order kinetic mechanism; principal stresses are considered. Limitations of the model include the following: it is a two-dimensional representation; the size of the research domain is constrained due to the small mesh scale required for accurate acid dissolution modeling. The main governing equations are provided for model completeness as follows.

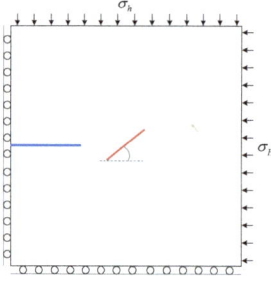

Figure 2. Schematic of interaction between acid fracture and natural fracture.

2.1. Reservoir Heterogeneity Characterization

Reservoir rock, at the microscopic level, comprises a solid matrix, pores, and mineral inclusions. Due to the random distribution of pores and minerals, the rock exhibits inherent heterogeneity, affecting its macroscopic mechanical properties. Two scales are considered: the micro-scale, where pores disperse in the solid matrix forming a porous matrix, and the meso-scale, where mineral inclusions are distributed. A two-step linear homogenization approach, based on the Mori–Tanaka scheme [25,26], estimates the macroscopic mechanical parameters. This approach accounts for the heterogeneity of Young's modulus and Poisson's ratio of the rock. The acid dissolution modeling necessarily considers the heterogeneity of porosity, which in turn influences the heterogeneity of the rock's mechanical parameters. Furthermore, acid dissolution degrades the rock's mechanical modulus, significantly affecting fracture propagation. Therefore, it is crucial to account for the initial heterogeneity of the rock's mechanical parameters.

The first homogenization step accounts for the micro-scale pore effect, estimating the effective elastic stiffness of the porous medium as follows [27]:

$$C^{mp} = C^m : (I - f_p A^p) \qquad (1)$$

where C^m is the elastic stiffness tensor of the solid matrix. I denotes the four-order unit tensor. F_p is the pores to matrix volume ratio, and A_p is the strain concentration tensor relating uniform strain to local strain:

$$A^p = (I - P^p : C^m)^{-1} : \left[(1 - f_p)I + f_p(I - P^p : C^m)^{-1}\right]^{-1} \qquad (2)$$

where P_p represents the porous matrix's Hill tensor, determined by integrating the Green function [25,27].

The second homogenization step considers mineral inclusions, simplifying them into an equivalent phase to estimate rock's macroscopic effective elastic stiffness:

$$C^{hom} = C^{mp} + f_i(C^i - C^{mp}) : A^i \qquad (3)$$

where A_i links macroscopic strain to local strain in inclusions:

$$A^i = \left[I + P^i : (C^i - C^{mp})\right]^{-1} : \left[(1 - f_i)I + f_i\left(I + P^i : (C^i - C^{mp})\right)^{-1}\right]^{-1} \qquad (4)$$

where P_i is the macroscopic medium's Hill tensor, calculated similarly to P_p with C^{mp}.

2.2. Fluid Flow and Acid Transport

As fractures initiate and propagate, the properties of the damaged reservoir area change including compressibility, permeability, and Biot's coefficient. These are updated using Lee et al.'s linear interpolation approach [5]. Considering porosity alterations due to mineral dissolution, the fluid flow continuity equation is as follows:

$$\frac{\partial \phi}{\partial t} + S_t \frac{\partial p}{\partial t} + \nabla \cdot \left(-\frac{k_t}{\mu} \nabla p\right) = Q_0 - \alpha_t \frac{\partial \varepsilon_{vol}}{\partial t} \qquad (5)$$

where S_t, k_t, and α_t are the medium's equivalent compressibility, permeability, and Biot's coefficient, calculated via interpolation [28]. Φ symbolizes porosity, μ is the fluid viscosity, and Q_0 is the source term.

Acid transport involves convection, dispersion, and transfer. The transfer term, describing acid flow to the fluid–solid interface, is a first-order kinetic reaction [29]:

$$R(C_f) = k_c(C_f - C_s) = k_s C_s = \frac{k_c k_s}{k_c + k_s} C_f \qquad (6)$$

where k_c is the transfer coefficient, and k_s is the surface reaction rate. C_f and C_s are acid concentrations in the fluid phase and at the fluid–solid interface.

Acid transport's continuity equation is as follows:

$$\frac{\partial(\phi C_f)}{\partial t} + \nabla \cdot (vC_f) = \nabla \cdot (\phi D_e \cdot \nabla C_f) - a_v R(C_f) \tag{7}$$

where v is the fluid velocity. D_e denotes the effective dispersion tensor. Av denotes the specific surface area.

The transfer term changes porosity:

$$\frac{\partial \phi}{\partial t} = \frac{R(C_f) a_v \alpha_d}{\rho_s} \tag{8}$$

where α_d is the dissolving coefficient, and ρ_s is the rock's density.

Based on the two-scale continuum model [29], the pore structural parameters and related transfer coefficient and dispersion tensor are analyzed at the pore scale. Semi-empirical relations estimate the pore radius, permeability, and surface area due to porosity change:

$$\frac{r_p}{r_0} = \sqrt{\frac{k_r \phi_0}{k_0 \phi}} \tag{9}$$

$$\frac{k_r}{k_0} = \frac{\phi}{\phi_0} \left(\frac{\phi(1-\phi_0)}{\phi_0(1-\phi)}\right)^{2\gamma} \tag{10}$$

$$\frac{a_v}{a_0} = \frac{\phi r_0}{\phi_0 r_p} \tag{11}$$

where r_0, k_0, and α_0 denote initial average pore radius, permeability and surface area when porosity equals ϕ_0. Γ is the pore broadening parameter.

The transfer coefficient kc is derived via the Sherwood number:

$$Sh = \frac{2k_c r_p}{D_m} = Sh_\infty + 0.7 Re_p^{1/2} Sc^{1/3} \tag{12}$$

where D_m is molecular diffusivity, and Sh_∞ is the asymptotic Sherwood number. Rep is the Reynold's number, and Sc is the Schmidt number (Sc = v_k/D_m), where v_k is the kinetic velocity.

The dispersion tensor is characterized by longitudinal and transverse dispersion coefficients [29]:

$$D_{eX} = (\alpha_{os} + \lambda_X Pe_p) D_m \tag{13}$$

$$D_{eT} = (\alpha_{os} + \lambda_T Pe_p) D_m \tag{14}$$

where X and T denote acid injection and perpendicular directions. α_{os}, λ_X, and λ_T are pore-structure constants. Pep is the Peclet number, defined as Pep = $|v| d_h/(\phi D_m)$, where d_h is the pore's diameter.

2.3. Rock Deformation and Fracture Propagation

Considering elastic strain energy, crack surface energy, and fluid pressure dissipation energy, the porous medium's total energy functional is as follows:

$$\Psi(u, \Gamma, p) = \int_\Omega \psi(\varepsilon) d\Omega + \int_\Gamma G_c d\Gamma - \int_\Omega \alpha_t p \cdot (\nabla \cdot u) d\Omega - \int_\Omega b \cdot u d\Omega - \int_{\partial \Omega_t} \bar{t} \cdot u d\partial \Omega_t \tag{15}$$

where u and p are displacement tensor and fluid pressure. ψ represents elastic strain energy, and ε is the strain tensor. G_c is the critical energy release rate. b is the body force, and $\bar{t}$ is the traction on the boundary $\partial \Omega t$.

Acid dissolution degrades the rock's elastic modulus [30,31], with Young's modulus exhibiting exponential decline with increasing dissolution-induced porosity. It is used to a chemical damage variable [32,33]:

$$d_{chem} = e^{-r|\Delta\phi|} \quad (16)$$

where r is a degradation coefficient, derived by fitting experimental results [24,34].

Using the phase field method, the total energy functional is as follows [24]:

$$\Psi(u,d,p) = \int_\Omega g(d,d_{chem})\psi_+(\varepsilon) + \psi_-(\varepsilon)d\Omega + \int_\Omega \frac{G_c}{2}\left[l_0 \nabla d \cdot \nabla d + \frac{d^2}{l_0}\right]d\Omega \\ - \int_\Omega \alpha p \cdot (\nabla \cdot u)d\Omega - \int_\Omega b \cdot u d\Omega - \int_{\partial\Omega_t} \bar{t} \cdot u d\partial\Omega_t \quad (17)$$

where g is the degradation function. $\psi+$ and $\psi-$ are tensile and compressive elastic strain energies. d denotes the phase field, and l_0 is a characteristic length parameter.

Integrating crack phase field and chemical damage field influences, the degradation function is as follows:

$$g(d,d_{chem}) = (1-k_0)[d_{chem}(1-d)]^2 + k_0 \quad (18)$$

where k_0 ensures the stiffness matrix remains well conditioned as the phase field approaches 1.

Using the variational approach, the governing equations for rock deformation are as follows:

$$\nabla \cdot (\sigma - \alpha p I) + b = 0 \quad (19)$$

Similarly, the phase field governing equation, with local history variable $\mathcal{H}$ is as follows:

$$\left[\frac{2l_0(1-k_0)d_{chem}^2 \psi_+}{G_c} + 1\right]d - l_0^2(\nabla d \cdot \nabla d) = \frac{2l_0(1-k_0)d_{chem}^2 \mathcal{H}}{G_c} \quad (20)$$

Initial and natural fractures are modeled by prescribing strain field history, as proposed by Borden et al. [35]:

$$\mathcal{H}_0(x) = \begin{cases} \frac{BG_c}{4l_0}\left(1 - \frac{s(x,l)}{l_0}\right), & s(x,l) \leq l_0 \\ 0, & s(x,l) > l_0 \end{cases} \quad (21)$$

where B is a constant. $s(x,l)$ represents the distance of point x to line l.

3. Numerical Methods

The governing equations are highly nonlinear, necessitating an iterative algorithm for solutions. A hybrid method utilizes the finite element method for stress and phase field discretization and the finite volume method for fluid pressure and acid concentration discretization. The fixed stress split method accelerates convergence due to stress and pressure coupling.

Within the finite element framework, weak formulations of stress and phase field equations are obtained by applying the virtual work principle:

$$\int_\Omega (\sigma - \alpha p I) : \varepsilon(\delta u) d\Omega = \int_\Omega b \cdot \delta u d\Omega + \int_{\partial\Omega_t} \bar{t} \cdot \delta u d\partial\Omega_t \quad (22)$$

$$\int_\Omega \left[2(1-k_0)d_{chem}^2 \mathcal{H} + \frac{G_c}{l_0}\right]d\delta d d\Omega + \int_\Omega G_c l_0 \nabla d \cdot \nabla(\delta d) d\Omega = 2\int_\Omega (1-k_0)d_{chem}^2 \mathcal{H}\delta d d\Omega \quad (23)$$

where δu denotes the virtual displacement tensor, and δd represents the virtual phase field variable.

Within the finite volume framework, the fluid pressure integration equation over elements yields the following:

$$\int_{\Omega_e} S_t \frac{\partial p}{\partial t} dV - \int_{\Omega_e} \nabla \cdot \left(\frac{k_t}{\mu} \nabla p\right) dV = \int_{\Omega_e} Q_0 - \frac{R(C_s) a_v \alpha_d}{\rho_s} dV - \int_{\Omega_e} \alpha_t \frac{\partial \varepsilon_{vol}}{\partial t} dV \quad (24)$$

The flow flux between elements relies on the flow conductivity at the interaction boundary, calculated by the harmonic mean.

Similarly, the acid concentration integration equation is as follows:

$$\int_{\Omega_e} \frac{\partial(\phi C_f)}{\partial t} dV + \int_{\Omega_e} \nabla \cdot (vC_f) dV = \int_{\Omega_e} \nabla \cdot (\phi D_e \cdot \nabla C_f) dV - \int_{\Omega_e} \frac{k_c k_s a_v}{k_c + k_s} C_f dV \quad (25)$$

Using the upwind scheme for advection flux and harmonic mean for diffusion flux, time derivation is generally estimated by the backward scheme.

Discretized field equations are solved using an efficient iterative algorithm. At each new time step, fluid pressure, displacement, and phase field are iterated using staggered manner until convergence, and then acid concentration is solved based on fluid pressure updates in a decoupled manner. Porosity and property parameters are updated for the next time step. Detailed methods can be found in the previous study [24].

4. Simulation Results

The hydro-mechano-reactive flow coupled model for simulating acid fracture propagation has been validated in a previous study [24], including the calculations of acid dissolution and crack propagation. To further verify its applicability to fracture interaction problems, we examined the scenario of a hydraulic fracture approaching a natural fracture, as this configuration has been extensively studied both experimentally and theoretically. Subsequently, we employed the model to investigate the interaction modes between acid fractures and natural fractures, followed by a comparative analysis of the propagation patterns of acid fractures and hydraulic fractures. Finally, we conducted a comprehensive analysis of the combined effects of acid dissolution and other factors (including stress differential, approach angle, injection rate, and acid concentration) on fracture propagation.

4.1. Verification of Interaction Between Hydraulic Fracture and Natural Fracture

The interaction between hydraulic fractures and pre-existing natural fractures has been a subject of significant research in the field of hydraulic fracturing. Blanton's seminal work [36], which involved experimental studies, led to the development of an interaction criterion that examines the effects of approaching angle and stress difference on fracture behavior. To validate our numerical model against Blanton's established criterion, we simplified our hydro-mechano-reactive flow coupled model to simulate hydraulic fracture propagation by omitting acid transport processes. This allows for a direct comparison between our numerical solutions and Blanton's experimental criterion, thereby providing a robust verification of our model's capability to accurately simulate fracture interactions in the absence of chemical reactions.

Our simulation setup comprises a square sample containing an initial hydraulic fracture and an inclined natural fracture, similar to the configuration illustrated in Figure 2. The domain is discretized into a uniform grid of square elements, with 100 elements along each axis, resulting in a 100 × 100 mesh. The key distinction in this verification study is the absence of acid in the fracturing fluid. By systematically varying the approaching angle and stress difference in our simulations, we aim to reproduce the interaction behaviors observed in Blanton's experiments. Table 1 presents the model parameters used in these simulations.

Numerical solutions reveal two primary modes when a hydraulic fracture approaches a natural fracture: crossing and diversion. We obtained propagation modes under various scenarios with different approaching angles and stress differences, comparing them with Blanton's criterion as illustrated in Figure 3. In this figure, the region to the right of Blanton's criterion represents crossing, while the left represents no crossing. Our solution demonstrates consistency with Blanton's criterion, thereby validating the model's application for hydraulic fracture and natural fracture interaction.

Table 1. Input parameters for hydraulic fracture propagation with natural fracture.

Parameter	Symbol	Value	Unit
Length of domain	L	0.5	m
Height of domain	H	0.5	m
Length of initial hydraulic fracture	l_{hf}	0.15	m
Length of natural fracture	l_{nf}	0.1	m
Young's modulus	E	25	GPa
Poisson's ratio	v	0.25	-
Characteristic length parameter	l_0	0.01	-
Critical energy release rate	G_c	50	Pa·m
Maximum principal stress	σ_H	8	MPa
Minimum principal stress	σ_h	5	MPa
Matrix permeability	k_r	1.0×10^{-15}	m^2
Fracture permeability	k_f	1.0×10^{-8}	m^2
Injection rate	v_0	7.5×10^{-4}	m^2/s
Fluid viscosity	μ	1.0×10^{-3}	Pa·s

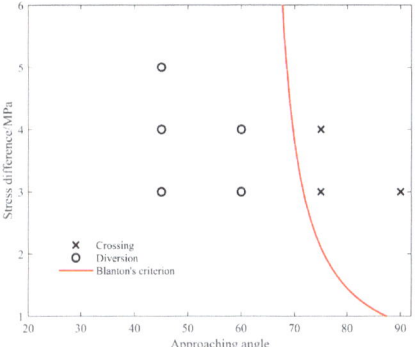

Figure 3. Comparison of fracture propagation mode between our solution and Blanton's criterion.

The phase field method employed in our model automatically determines the hydraulic fracture propagation direction without additional criteria. The fracture always propagates along the path of least energy dissipation. Figure 4 illustrates fracture propagation paths under different approaching angles with a stress difference of 3 MPa.

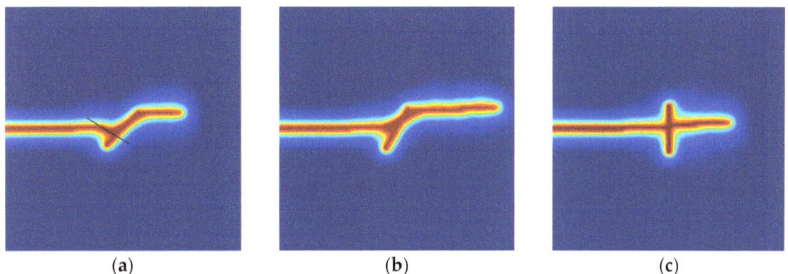

Figure 4. Fracture propagation paths under different approaching angles: (**a**) $\beta = 45°$; (**b**) $\beta = 60°$; (**c**) $\beta = 90°$.

When the approaching angle is 45°, the hydraulic fracture exhibits minimal diversion before contacting the natural fracture (denoted by the black line). It then diverts into the natural fracture, ultimately reinitiating from the tip of the natural fracture and continuing to extend. At a 60° approaching angle, the hydraulic fracture propagates towards the natural fracture and diverts into it. When the approaching angle is 90°, the hydraulic fracture crosses the natural fracture without diversion. These results can be explained by the principle of least energy dissipation. Two main energies are dissipated during fracture propagation: fracture surface energy and elastic strain energy. When a hydraulic fracture approaches a natural fracture at a small angle, it diverts into the natural fracture due to significantly lower fracture surface energy, despite potentially increased dissipated elastic strain energy compared to crossing the natural fracture. However, as the approaching angle increases, the dissipated elastic strain energy along the natural fracture direction increases, making fracture crossing the path of least total energy dissipation.

Figure 5 presents the injection pressures for different approaching angles. The injection pressure for the 45° approaching angle increases first due to hydraulic fracture diversion before contacting the natural fracture. Injection pressures for 60° and 90° approaching angles increase simultaneously. When the hydraulic fracture diverts into the natural fracture, the dissipated energy is less than that of crossing the natural fracture. Consequently, the peak injection pressure is highest when the hydraulic fracture crosses the natural fracture at a 90° approaching angle.

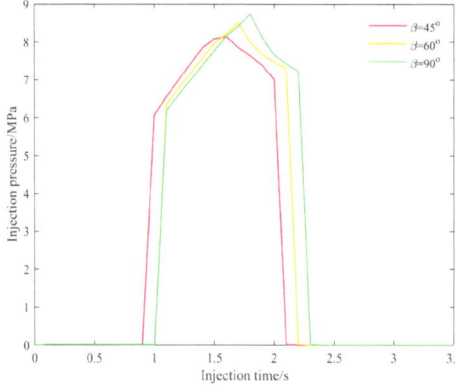

Figure 5. Evolution of injection pressures for different approaching angles.

4.2. Interaction Mode Between Acid Fracture and Natural Fracture

To investigate the interaction mode between acid fractures and natural fractures, we employ a model comprising a square sample with an initial acid fracture and an inclined fracture, as illustrated in Figure 2. It takes approximately 30 min of computation time on an Intel Core i7-6700 CPU.

Our model incorporates reservoir heterogeneity by describing porosity and mineral inclusions using the Weibull distribution. We estimate the macroscopic mechanical parameters using a two-step homogenization method. Table 2 provides the model parameters used in the simulation. Figure 6 presents the distribution of porosity, the volume fraction of mineral inclusions, and the resulting elastic modulus. The heterogeneity in porosity influences acid dissolution and wormhole formation, which, in combination with the heterogeneity in elastic modulus, significantly affects fracture propagation.

Table 2. Input parameters for acid fracture propagation with natural fracture.

Parameter	Symbol	Value	Unit
Length of domain	L	0.5	m
Height of domain	H	0.5	m
Length of initial hydraulic fracture	l_{hf}	0.15	m
Length of natural fracture	l_{nf}	0.1	m
Approaching angle	β	90	°
Characteristic length parameter	l_0	0.01	
Critical energy release rate	G_c	50	Pa·m
Elastic modulus of solid matrix	E_m	12	GPa
Poisson's ratio of solid matrix	v_m	0.25	-
Elastic modulus of mineral inclusion	E_i	98	GPa
Poisson's ratio of mineral inclusion	v_i	0.15	-
Scale parameter of f_p		0.12	-
Shape parameter of f_p		10	-
Scale parameter of f_i		0.4	-
Shape parameter of f_i		10	-
Maximum principal stress	σ_H	8	MPa
Minimum principal stress	σ_h	5	MPa
Acid concentration	C_{f0}	0.15	-
Acid injection rate	v_0	1.05×10^{-3}	m^2/s
Acid surface reaction rate	k_s	2.0×10^{-3}	m/s
Molecular diffusion coefficient	D_m	3.6×10^{-9}	m^2/s
Asymptotic Sherwood number	Sh_∞	3.66	-
Constants	$\alpha_{os}, \lambda_X, \lambda_T$	0.5, 0.5, 0.1	-
Initial average permeability	kr	1.0×10^{-15}	m^2
Initial specific surface area	α_0	5.0×10^3	m^{-1}
Initial pore diameter	d_0	1.0×10^{-5}	M
Pore broadening parameter	β	1	-
Chemical degradation coefficient	r	5	-
Fluid viscosity	μ	1.0×10^{-3}	Pa·s
Rock density	ρ_s	2.71×10^3	kg/m^3

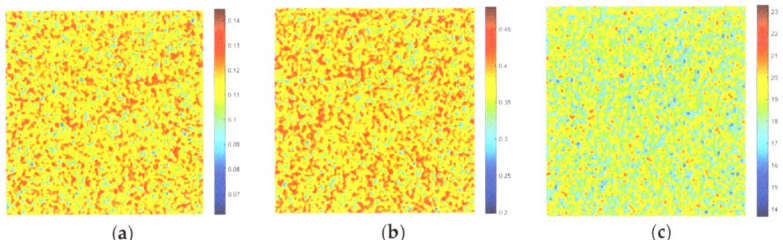

Figure 6. Distribution of reservoir parameters: (**a**) porosity; (**b**) volume fraction of mineral inclusion; (**c**) elastic modulus (GPa).

The acid fracture propagation path, determined using the input data from Table 2, is illustrated in Figure 7a. In this scenario, the acid fracture crosses the natural fracture. To explore additional interaction modes between acid fractures and natural fractures, we adjusted certain parameters. When the stress difference is reduced to 0, the acid fracture diverts into the natural fracture, as shown in Figure 7b. When the injection rate is decreased to 0.75×10^{-3} m^2/s, the acid fracture is arrested by the natural fracture, as depicted in Figure 7c.

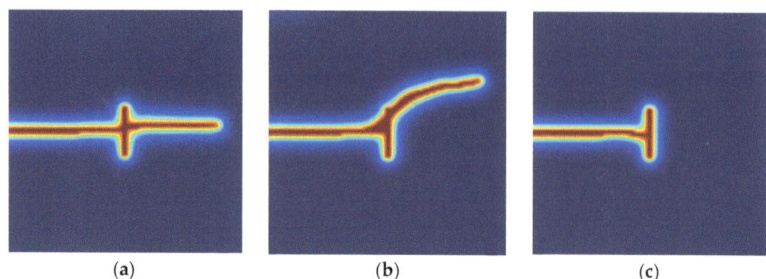

Figure 7. Interaction modes between acid fracture and natural fracture: (**a**) crossing; (**b**) diversion; (**c**) arresting.

Figure 8 presents the porosity distributions under different modes. When the hydraulic fracture contacts the natural fracture, fluid rapidly flows into the natural fracture due to its high permeability. This leads to wormhole formation around both the acid and natural fractures. The distribution of porosity is similar to the previous solution present by Zhu et al. [37]. Comparing Figure 8a,b, we observe that decreasing the stress difference causes the hydraulic fracture to divert into the natural fracture. This phenomenon can be explained by the principle of least energy dissipation. The injection pressures for these two scenarios are similar, as shown in Figure 9. In the arresting mode (Figure 8c), the reduced injection rate results in the formation of numerous wormholes propagating towards the outer boundaries. This causes rapid pressure dissipation, preventing the accumulation of high pressure necessary for the hydraulic fracture to reinitiate from the natural fracture, as evident in Figure 9.

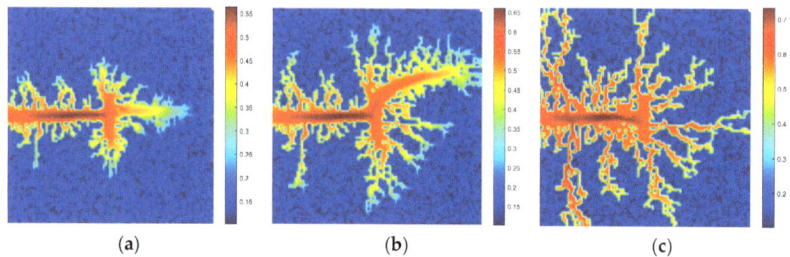

Figure 8. Porosity distributions under different modes: (**a**) crossing; (**b**) diversion; (**c**) arresting.

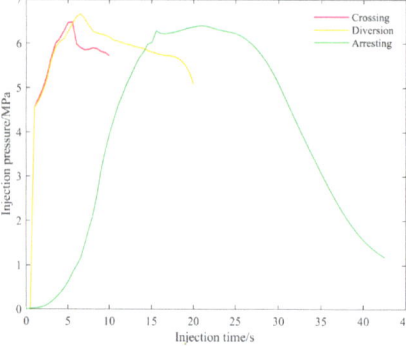

Figure 9. Injection pressure evolution for different modes.

Figure 10 displays the pressure distributions under different modes. The high-pressure area expands as the interaction mode changes from crossing to diversion and then to arresting. In the former modes, especially the crossing mode, pressure dissipation is primarily used to extend fractures. The pressure distribution is also closely related to the wormhole distribution.

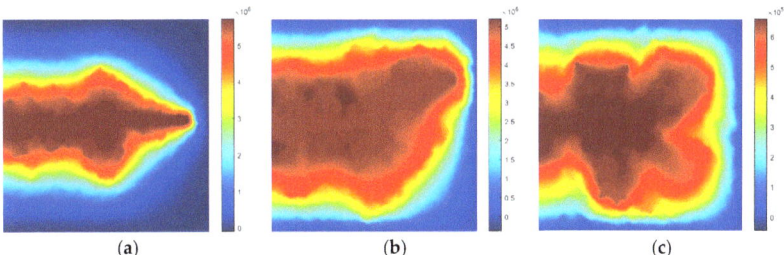

(a) (b) (c)

Figure 10. Pressure distributions under different modes: (**a**) crossing; (**b**) diversion; (**c**) arresting.

4.3. Comparison of Propagation Modes: Acid Fracture vs. Hydraulic Fracture

Acid fracturing involves acid transport and dissolution, which significantly influences the interaction process between acid fractures and natural fractures. This leads to distinct propagation modes for acid fractures compared to hydraulic fractures. To investigate the effect of acid dissolution on fracture propagation modes, we conducted several case studies.

Initially, we simulated hydraulic fracture propagations under stress differences of 0 MPa and 3 MPa, with an injection rate of 0.75×10^{-3} m^2/s. Other model parameters remained consistent with those in Table 2. These simulations resulted in diversion and crossing modes, as illustrated in Figures 10b and 11a, respectively.

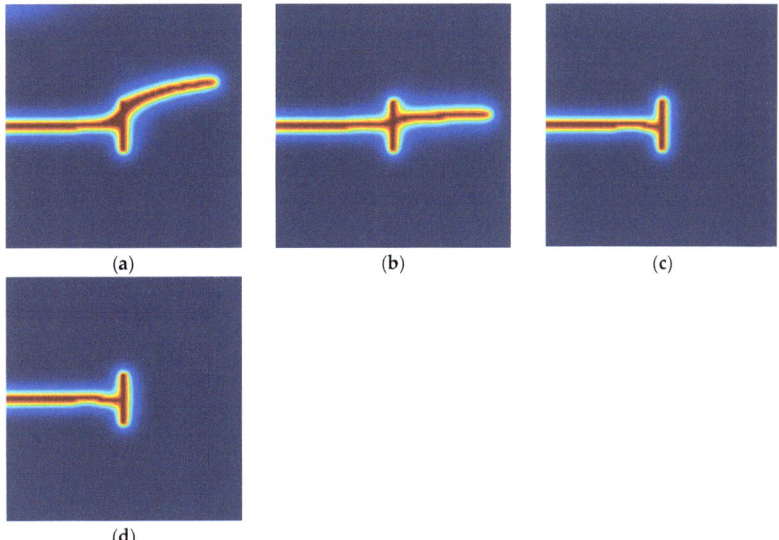

Figure 11. Fracture propagation modes for different cases: (**a**) $\Delta\sigma = 0$ MPa, hydraulic fracture; (**b**) $\Delta\sigma = 3$ MPa, hydraulic fracture; (**c**) $\Delta\sigma = 0$ MPa, acid fracture; (**d**) $\Delta\sigma = 3$ MPa, acid fracture.

Using the same injection rate, we then calculated acid fracture propagations under the two stress differences. Contrary to hydraulic fractures, acid fractures were arrested by the natural fracture in both cases, as shown in Figures 10d and 11c. The arrest mode of acid fracture is similar to the results presented by Guo et al. [34].

We compared the injection pressure evolutions for these scenarios, as depicted in Figure 12. For hydraulic fracturing, the fluid pressure rises rapidly, enabling the hydraulic fracture to either cross the natural fracture or reinitiate after diversion. In contrast, during acid fracturing, the fluid pressure increases more slowly, and the peak pressure is insufficient for fracture initiation from the natural fracture. This difference is attributed to acid dissolution leading to the formation of numerous wormholes. Fluid flows into these high-permeability channels, preventing the accumulation of high pressure necessary for acid fracture re-initiation.

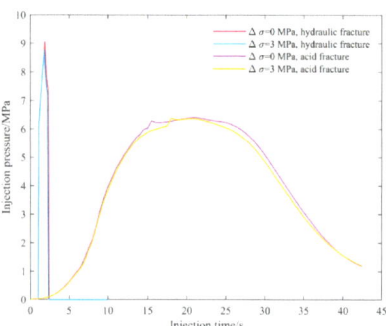

Figure 12. Injection pressure evolution for hydraulic fracture and acid fracture.

These results indicate that the hydraulic fracture is more likely to cross natural fractures, whereas acid fractures tend to become arrested or diverted. This behavior primarily results from wormhole formation due to acid dissolution, which impedes the buildup of high fluid pressure required for fracture initiation and extension.

4.4. Comprehensive Influences of Acid Dissolution with Other Factors

The interaction between acid fractures and natural fractures results from a complex interplay of acid dissolution and other factors, including stress difference, approaching angle, and treatment parameters. To investigate these effects, several scenarios were simulated for acid fracture propagation with natural fractures. Unless otherwise stated, model parameters are as provided in Table 2.

Acid fracture propagations under varying stress differences (0 MPa, 3 MPa, and 5 MPa) are illustrated in Figure 13. At a stress difference of 0 MPa, the acid fracture diverts into the natural fracture and reinitiates from the tip, extending along an arc-shaped path rather than following its initial direction. As the stress difference increases, the fracture propagation mode transitions from diversion to crossing because more strain energy is required for fracture diversion when increasing the principal stress perpendicular to the natural fracture.

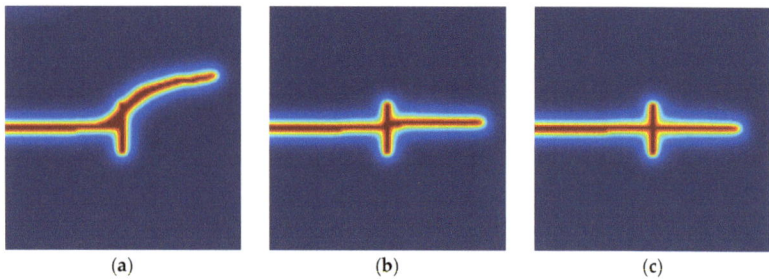

Figure 13. Fracture propagation paths under different stress differences: (a) $\Delta\sigma = 0$ MPa; (b) $\Delta\sigma = 3$ MPa; (c) $\Delta\sigma = 5$ MPa.

Acid fracture approaches to natural fractures at different angles (30°, 60°, and 90°) are illustrated in Figure 14. At angles of 30° and 60°, the acid fracture diverts into the natural fracture. However, with a 30° approach angle, the fracture alters its trajectory before contact and reinitiates at the natural fracture tip. At 60°, the direction remains unchanged until contact and then reinitiates near the tip. At lower approach angles (30° and 60°), the stress concentration at the tip of the natural fracture tends to attract and redirect the approaching acid fracture. This redirection is more pronounced at 30° due to the greater alignment with the natural fracture's orientation. The stress field perturbation becomes significant enough to alter the acid fracture's trajectory even before direct contact. In contrast, at 90°, the acid fracture crosses the natural fracture. This phenomenon is similar to hydraulic fracture propagation. These insights highlight the critical importance of understanding the geometric relationship between induced fractures and the natural fracture network for effective subsurface engineering applications.

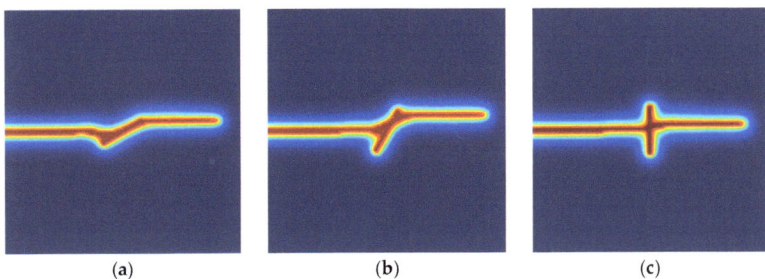

Figure 14. Fracture propagation paths with different approaching angles: (**a**) $\beta = 30°$; (**b**) $\beta = 60°$; (**c**) $\beta = 90°$.

Acid fracture propagation under varying injection rates is presented in Figure 15. At a low injection rate, the acid fracture is arrested by the natural fracture. This behavior can be attributed to insufficient pressure buildup within the fracture. The slower rate allows more time for acid–rock interaction, potentially leading to excessive etching and wormhole formation near the natural fracture interface. This increased permeability can dissipate fluid pressure, preventing the necessary stress concentration for fracture propagation beyond the natural fracture.

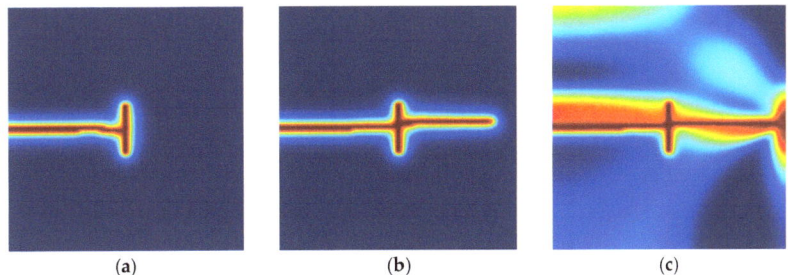

Figure 15. Fracture propagation paths under different injection rates: (**a**) $v_0 = 0.75 \times 10^{-3}$ m^2/s; (**b**) $v_0 = 1.05 \times 10^{-3}$ m^2/s; (**c**) $v_0 = 1.35 \times 10^{-3}$ m^2/s.

Increasing the injection rate leads to the acid fracture crossing the natural fracture. Higher injection rates facilitate high-pressure accumulation in the fracture, promoting fracture re-initiation from the natural fracture. This can be attributed to reduced acid–rock interaction time and enhanced pressure accumulation within the fracture. This scenario closely resembles hydraulic fracturing behavior, where the dominant mechanism shifts from chemical dissolution to mechanical breakdown of the rock formation.

Acid fracture propagations using different acid concentrations are presented in Figure 16. As acid concentration increases, the propagation mode shifts from crossing to arresting or even failure to initiate. While adding acid to the fracturing fluid can reduce breakdown pressure, excessively high concentrations can prevent fracture initiation. This phenomenon is explained by the acid dissolution effect shown in Figure 17.

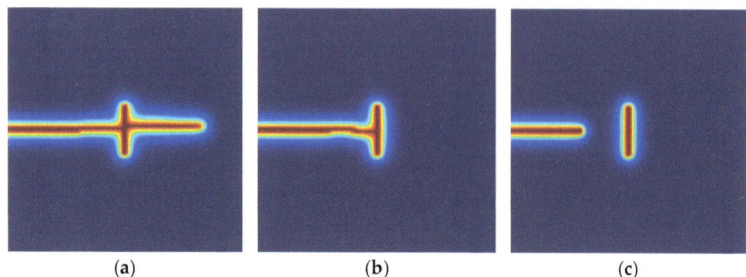

Figure 16. Fracture propagation paths with different acid concentrations: (a) $C_{f0} = 0.1$; (b) $C_{f0} = 0.15$; (c) $C_{f0} = 0.2$.

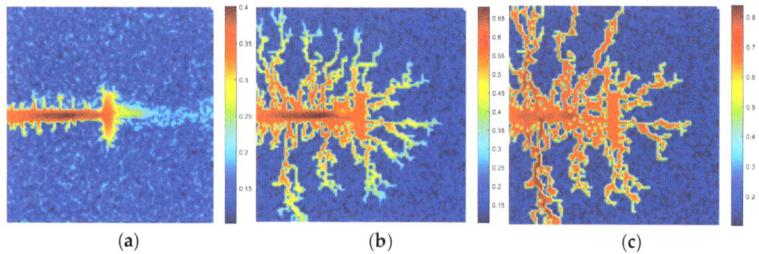

Figure 17. Porosity distributions with different acid concentrations: (a) $C_{f0} = 0.1$; (b) $C_{f0} = 0.15$; (c) $C_{f0} = 0.2$.

The process of acid fracture propagation depends on the dissolution capabilities of the acid used in fracturing fluids. At lower concentrations, the acid efficiently reduces the breakdown pressure by weakening the rock matrix, facilitating fracture initiation and propagation. However, as the concentration increases, the aggressive dissolution forms enhanced permeability channels known as wormholes. These wormholes significantly alter the fluid dynamics within the fracture, allowing hydraulic pressure to dissipate rapidly. Consequently, the pressure necessary for further fracture propagation cannot be sustained, leading to arrest or failure to initiate the fracture. This relationship highlights the dual role of acid as both an enabler and a potential inhibitor of fracture growth, depending on its concentration. To deepen the understanding of this complex interaction, future work could focus on performing experimental validation that varies acid concentration and assessing its effect on fracture propagation.

5. Conclusions

The hydro-mechano-reactive flow coupled model effectively simulates the interaction between acid fractures and natural fractures. It has been validated against established criteria for hydraulic fracturing and successfully extended to acid fracturing scenarios. The following key conclusions are obtained:

(1) Acid fractures exhibit different propagation modes compared to hydraulic fractures when interacting with natural fractures. While hydraulic fractures are more likely to cross natural fractures, acid fractures tend to be arrested due to wormhole formation and pressure dissipation.

(2) Increasing stress differences and approaching angles significantly affect fracture behavior. Higher stress differences promote fracture crossing, while smaller approaching angles favor diversion into natural fractures for both hydraulic and acid fractures.

(3) Injection rate and acid concentration play crucial roles in fracture propagation. Higher injection rates facilitate fracture crossing by increasing pressure accumulation, while excessive acid concentrations can hinder fracture initiation due to enhanced wormhole formation and pressure dissipation.

(4) The model incorporates reservoir heterogeneity, demonstrating that variations in porosity and mineral inclusions significantly influence acid dissolution, wormhole formation, and subsequent fracture propagation patterns. This highlights the importance of considering reservoir heterogeneity in fracture stimulation design.

Author Contributions: Conceptualization, Q.Z. and J.Y.; methodology, Q.Z.; validation, X.L.; formal analysis, T.L.; investigation, L.B.; writing—original draft preparation, Q.Z.; writing—review and editing, X.L.; project administration, J.Y.; funding acquisition, Q.Z. and J.Y. All authors have read and agreed to the published version of the manuscript.

Funding: This study is jointly supported by National Natural Science Foundation of China (52274038, 52034010, 51904321) and Taishan Scholars Project (tsqnz20221140).

Data Availability Statement: The data presented in this study are available on request from the corresponding author.

Conflicts of Interest: The authors declare no conflicts of interest.

References

1. Rueda Cordero, J.A.; Mejia Sanchez, E.C.; Roehl, D.; Pereira, L.C. Hydro-Mechanical Modeling of Hydraulic Fracture Propagation and Its Interactions with Frictional Natural Fractures. *Comput. Geotech.* **2019**, *111*, 290–300. [CrossRef]
2. Hu, Y.; Gan, Q.; Hurst, A.; Elsworth, D. Investigation of Coupled Hydro-Mechanical Modelling of Hydraulic Fracture Propagation and Interaction with Natural Fractures. *Int. J. Rock Mech. Min. Sci.* **2023**, *169*, 105418. [CrossRef]
3. Qin, M.; Yang, D.; Jia, Y.; Zhou, Y. Peridynamics Modeling of Hydraulic Fracture Interaction with Natural Fractures in Fractured Rock Mass. *Eng. Fract. Mech.* **2024**, *307*, 110299. [CrossRef]
4. Wang, H.; Yang, S.; Zhou, D.; Wang, Q. Influence of Non-Intersecting Cemented Natural Fractures on Hydraulic Fracture Propagation Behavior. *J. Struct. Geol.* **2024**, *181*, 105111. [CrossRef]
5. Yi, L.-P.; Li, X.-G.; Yang, Z.-Z.; Yang, C.-X. Phase Field Modeling of Hydraulic Fracturing in Porous Media Formation with Natural Fracture. *Eng. Fract. Mech.* **2020**, *236*, 107206. [CrossRef]
6. Zhou, Y.; Yang, D.; Zhang, X.; Chen, W.; Xia, X. Numerical Investigation of the Interaction between Hydraulic Fractures and Natural Fractures in Porous Media Based on an Enriched FEM. *Eng. Fract. Mech.* **2020**, *235*, 107175. [CrossRef]
7. Xiong, D.; Ma, X. Influence of Natural Fractures on Hydraulic Fracture Propagation Behaviour. *Eng. Fract. Mech.* **2022**, *276*, 108932. [CrossRef]
8. Sun, T.; Zeng, Q.; Xing, H. A Quantitative Model to Predict Hydraulic Fracture Propagating across Cemented Natural Fracture. *J. Pet. Sci. Eng.* **2022**, *208*, 109595. [CrossRef]
9. Dehghan, A.N. An Experimental Investigation into the Influence of Pre-Existing Natural Fracture on the Behavior and Length of Propagating Hydraulic Fracture. *Eng. Fract. Mech.* **2020**, *240*, 107330. [CrossRef]
10. Wang, W.; Olson, J.E.; Prodanović, M.; Schultz, R.A. Interaction between Cemented Natural Fractures and Hydraulic Fractures Assessed by Experiments and Numerical Simulations. *J. Pet. Sci. Eng.* **2018**, *167*, 506–516. [CrossRef]
11. Abe, A.; Kim, T.W.; Horne, R.N. Laboratory Hydraulic Stimulation Experiments to Investigate the Interaction between Newly Formed and Preexisting Fractures. *Int. J. Rock Mech. Min. Sci.* **2021**, *141*, 104665. [CrossRef]
12. Zhou, J.; Chen, M.; Jin, Y.; Zhang, G. Analysis of Fracture Propagation Behavior and Fracture Geometry Using a Tri-Axial Fracturing System in Naturally Fractured Reservoirs. *Int. J. Rock Mech. Min. Sci.* **2008**, *45*, 1143–1152. [CrossRef]
13. Zhang, J.; Li, Y.; Pan, Y.; Wang, X.; Yan, M.; Shi, X.; Zhou, X.; Li, H. Experiments and Analysis on the Influence of Multiple Closed Cemented Natural Fractures on Hydraulic Fracture Propagation in a Tight Sandstone Reservoir. *Eng. Geol.* **2021**, *281*, 105981. [CrossRef]
14. Qiu, G.; Chang, X.; Li, J.; Guo, Y.; Zhou, Z.; Wang, L.; Wan, Y.; Wang, X. Study on the Interaction between Hydraulic Fracture and Natural Fracture under High Stress. *Theor. Appl. Fract. Mech.* **2024**, *130*, 104259. [CrossRef]
15. Wang, X.; Chen, J.; Ren, D.; Zhu, J. Role of Natural Fractures with Topology Structure for Hydraulic Fracture Propagation in Continental Shale Reservoir. *Eng. Fract. Mech.* **2023**, *284*, 109237. [CrossRef]
16. Li, Y.; Hu, W.; Wei, S.; Zhang, Z.; Yang, P.; Song, S. Sensitivity Analysis on the Effect of Natural Fractures and Injected Fluid on Hydraulic Fracture Propagation in a Fractured Reservoir. *Eng. Fract. Mech.* **2022**, *263*, 108288. [CrossRef]

17. Tong, L.; Xiaochen, W.; Xiangjun, L.; Lixi, L.; Xuancheng, W.; Jin, C.; Hongwei, L. A Criterion for a Hydraulic Fracture Crossing a Natural Fracture in Toughness Dominant Regime and Viscosity Dominant Regime. *Eng. Fract. Mech.* **2023**, *289*, 109421. [CrossRef]
18. Dang, L.; Zhou, C.; Huang, M.; Jiang, D. Simulation of Effective Fracture Length of Prepad Acid Fracturing Considering Multiple Leak-off Effect. *Nat. Gas Ind. B* **2019**, *6*, 64–70. [CrossRef]
19. Dong, R.; Wheeler, M.F.; Ma, K.; Su, H. A 3D Acid Transport Model for Acid Fracturing Treatments with Viscous Fingering. In Proceedings of the SPE Annual Technical Conference and Exhibition, Virtual, 26–29 October 2020.
20. Ugursal, A.; Zhu, D.; Hill, A.D. Development of Acid Fracturing Model for Naturally Fractured Reservoirs. *SPE Prod. Oper.* **2019**, *34*, 735–748. [CrossRef]
21. Dai, Y.; Hou, B.; Zhou, C.; Zhang, K.; Liu, F. Interaction Law between Natural Fractures-Vugs and Acid-Etched Fracture during Steering Acid Fracturing in Carbonate Reservoirs. *Geofluids* **2021**, *2021*, 1–16. [CrossRef]
22. Chen, Y.; Ma, G.; Li, T.; Wang, Y.; Ren, F. Simulation of Wormhole Propagation in Fractured Carbonate Rocks with Unified Pipe-Network Method. *Comput. Geotech.* **2018**, *98*, 58–68. [CrossRef]
23. Zhu, T.; Wei, X.; Zhang, Z. Numerical Simulation of Hydraulic-Mechanical-Chemical Field Coupled Acid Fracturing in Complex Carbonate Reservoir. *Comput. Geotech.* **2023**, *156*, 105277. [CrossRef]
24. Zeng, Q.; Li, T.; Liu, P.; Bo, L.; Yao, C.; Yao, J. A Phase Field Framework to Model Acid Fracture Propagation with Hydro-Mechano-Reactive Flow Coupling. *Comput. Geotech.* **2024**, *174*, 106658. [CrossRef]
25. Zhao, J.J.; Shen, W.Q.; Shao, J.F.; Liu, Z.B.; Vu, M.N. A Constitutive Model for Anisotropic Clay-Rich Rocks Considering Micro-Structural Composition. *Int. J. Rock Mech. Min. Sci.* **2022**, *151*, 105029. [CrossRef]
26. Yu, Z.; Shao, J.; Sun, Y.; Wang, M.; Vu, M.; Plua, C. Numerical Analysis of Hydro-Thermal Fracturing in Saturated Rocks by Considering Material Anisotropy and Micro-Structural Heterogeneity. *Int. J. Rock Mech. Min. Sci.* **2023**, *170*, 105457. [CrossRef]
27. Giraud, A.; Huynh, Q.V.; Hoxha, D.; Kondo, D. Application of Results on Eshelby Tensor to the Determination of Effective Poroelastic Properties of Anisotropic Rocks-like Composites. *Int. J. Solids Struct.* **2007**, *44*, 3756–3772. [CrossRef]
28. Lee, S.; Wheeler, M.F.; Wick, T. Pressure and Fluid-Driven Fracture Propagation in Porous Media Using an Adaptive Finite Element Phase Field Model. *Comput. Methods Appl. Mech. Eng.* **2016**, *305*, 111–132. [CrossRef]
29. Panga, M.K.R.; Ziauddin, M.; Balakotaiah, V. Two-Scale Continuum Model for Simulation of Wormholes in Carbonate Acidization. *AIChE J.* **2005**, *51*, 3231–3248. [CrossRef]
30. Palchik, V.; Hatzor, Y.H. Crack Damage Stress as a Composite Function of Porosity and Elastic Matrix Stiffness in Dolomites and Limestones. *Eng. Geol.* **2002**, *63*, 233–245. [CrossRef]
31. Croizé, D.; Bjørlykke, K.; Jahren, J.; Renard, F. Experimental Mechanical and Chemical Compaction of Carbonate Sand. *J. Geophys. Res.* **2010**, *115*, B11204. [CrossRef]
32. Wojtacki, K.; Lewandowska, J.; Gouze, P.; Lipkowski, A. Numerical Computations of Rock Dissolution and Geomechanical Effects for CO_2 Geological Storage. *Int. J. Numer. Anal. Methods Geomech.* **2015**, *39*, 482–506. [CrossRef]
33. Schuler, L.; Ilgen, A.G.; Newell, P. Chemo-Mechanical Phase-Field Modeling of Dissolution-Assisted Fracture. *Comput. Methods Appl. Mech. Eng.* **2020**, *362*, 112838. [CrossRef]
34. Guo, Y.; Na, S. A Reactive-Transport Phase-Field Modelling Approach of Chemo-Assisted Cracking in Saturated Sandstone. *Comput. Methods Appl. Mech. Eng.* **2024**, *419*, 116645. [CrossRef]
35. Borden, M.J.; Verhoosel, C.V.; Scott, M.A.; Hughes, T.J.R.; Landis, C.M. A Phase-Field Description of Dynamic Brittle Fracture. *Comput. Methods Appl. Mech. Eng.* **2012**, *217–220*, 77–95. [CrossRef]
36. Blanton, T.L. An Experimental Study of Interaction between Hydraulically Induced and Pre-Existing Fractures. In Proceedings of the SPE Unconventional Gas Recovery Symposium, Pittsburgh, PA, USA, 16–18 May 1982; pp. 559–571.
37. Zhu, T.; Zhang, Z.; Liu, Z.; An, N.; Wei, X. Simulation of hydraulic-mechanical-chemical coupled acid fracturing of rock with lattice bonds. *Heliyon* **2024**, *10*, e26517. [CrossRef]

Disclaimer/Publisher's Note: The statements, opinions and data contained in all publications are solely those of the individual author(s) and contributor(s) and not of MDPI and/or the editor(s). MDPI and/or the editor(s) disclaim responsibility for any injury to people or property resulting from any ideas, methods, instructions or products referred to in the content.

Article

Assessment of CO_2 Sequestration Capacity in a Low-Permeability Oil Reservoir Using Machine Learning Methods

Zuochun Fan [1,2], Mei Tian [2], Man Li [2], Yidi Mi [2], Yue Jiang [2], Tao Song [3], Jinxin Cao [3] and Zheyu Liu [3,*]

1. Institute of Advanced Studies, China University of Geosciences (Wuhan), Wuhan 430074, China
2. Research Institute of Exploration and Development, Liaohe Oilfield Company, PetroChina, Panjin 124010, China
3. State Key Laboratory of Petroleum Resources and Engineering, China University of Petroleum (Beijing), Beijing 102249, China; songtao_cup@163.com (T.S.); cjxafield@163.com (J.C.)
* Correspondence: zheyu.liu@cup.edu.cn

Citation: Fan, Z.; Tian, M.; Li, M.; Mi, Y.; Jiang, Y.; Song, T.; Cao, J.; Liu, Z. Assessment of CO_2 Sequestration Capacity in a Low-Permeability Oil Reservoir Using Machine Learning Methods. *Energies* 2024, *17*, 3979. https://doi.org/10.3390/en17163979

Academic Editor: Hossein Hamidi

Received: 21 May 2024
Revised: 28 July 2024
Accepted: 9 August 2024
Published: 11 August 2024

Copyright: © 2024 by the authors. Licensee MDPI, Basel, Switzerland. This article is an open access article distributed under the terms and conditions of the Creative Commons Attribution (CC BY) license (https://creativecommons.org/licenses/by/4.0/).

Abstract: The CO_2 sequestration capacity evaluation of reservoirs is a critical procedure for carbon capture, utilization, and storage (CCUS) techniques. However, calculating the sequestration amount for CO_2 flooding in low-permeability reservoirs is challenging. Herein, a method combining numerical simulation technology with artificial intelligence is proposed. Based on the typical geological and fluid characteristics of low-permeability oil reservoirs in the Liaohe oilfield, the CMG 2020 version software GEM module is used to establish a model for CO_2 flooding and sequestration. Meanwhile, a calculation method for the effective sequestration coefficient of CO_2 is established. We systematically study the sequestration rules in low-permeability reservoirs under varying conditions of permeability, reservoir temperature, and initial reservoir pressure. The results indicate that, as the permeability and sequestration pressure of the reservoir increase, oil recovery gradually increases. The proportion of structurally bound sequestration volume increases from 55% to 60%. Reservoir temperature has minimal impact on both the recovery rate and the improvement in sequestration efficiency. Sequestration pressure primarily improves sequestration efficiency by increasing the dissolution of CO_2 in the remaining oil and water. The calculation chart for the effective sequestration coefficient, developed using artificial intelligence algorithms under multi-factor conditions, enables accurate and rapid evaluation of the sequestration potential and the identification of favorable sequestration areas in low-permeability reservoirs. This approach provides valuable technical support for CO_2 flooding and sequestration in pilot applications.

Keywords: effective burial coefficient; CCUS; numerical simulation; artificial intelligence

1. Introduction

In recent years, with the rapid development of the global economy and the improvement of living standards, global oil consumption has been rising. Since 2019, China has relied on foreign crude oil for over 70% of its needs for six consecutive years [1–3]. The main section of aging oilfields developed through water injection has reached the "double high" stage, characterized by a high water cut and high recovery rates. Stabilizing and increasing crude oil production has become challenging, making it urgent to identify new growth points for resources [4–6]. Direct emissions of CO_2, the by-product of burning fossil fuels, will pollute the atmosphere, causing the greenhouse effect [7,8]. As the world's second largest economic entity and a responsible nation, China has taken the initiative to shoulder the responsibility of tackling global climate change. It has set the ambitious targets of reaching peak carbon emissions by 2030 and achieving carbon neutrality by 2060 [9–11]. In the background of "energy independence" and "carbon peak and carbon neutrality",

CCUS (carbon capture, utilization, and storage) have received increasing attention in the field of petroleum production [12,13].

Low-permeability reservoirs face challenges such as strong reservoir heterogeneity, a complex pore structure, rapid energy depletion in depleted production formations, difficulties in replenishing energy through water injection, and low recovery rates [14,15]. In 2014, the United States was responsible for about 93% of global CO_2 flooding-enhanced oil recovery (EOR) production. About 80% of the CO_2 flooding reservoirs had permeabilities of less than 50×10^{-3} μm, demonstrating considerable economic and social benefits [16–19]. Injecting CO_2 into a reservoir can lead to the dissolution, expansion, and viscosity reduction of crude oil. Miscible flooding can significantly enhance both sweep efficiency and flooding efficiency [20,21]. However, the varying characteristics of reservoirs—including physical properties, oil quality, temperature, pressure, fluid distribution, and trap storage capacity—affect the impact of CO_2 flooding on enhanced oil recovery and burial effects [22–24]. Currently, CO_2 storage capacity assessment studies mostly focus on specific physical properties of oil reservoirs [25–27]. There is a lack of research on the changes and potential assessment of CO_2 burial amounts under varying reservoir conditions.

The calculation of CO_2 storage capacity is primarily determined by the method used to calculate effective storage capacity. Effective storage capacity considers factors such as buoyancy, overburden pressure, fluid dynamics, heterogeneity, water saturation, and others. Its value aligns more closely with actual storage capacity than theoretical estimates [28–30]. Existing method mainly use analogy or numerical simulation techniques to calculate effective CO_2 storage capacity. However, pilot test projects for CO_2 storage sites are limited, leading to restricted available parameters and reliability issues in analog-based CO_2 storage calculations [31]. The process of calculating the effective storage coefficient by numerical simulation method is complex and lacks the capability for rapid and convenient calculations, which imposes application limitations [32,33]. To address these issues, a numerical simulation model of CO_2 flooding was developed based on the geological and fluid characteristics of typical low-permeability oil reservoirs in the Liaohe oilfield. This model aims to systematically study the effects of enhanced oil recovery and CO_2 burial behavior, establish a calculation framework for the effective storage coefficient using artificial intelligence methods, and provide technical support and theoretical guidance for assessing CO_2 burial potential and identifying favorable burial areas in the study area.

2. The Establishment of a Numerical Model of CO_2 Flooding and Burial

To investigate the CO_2 flooding and burial behavior in low-permeability reservoirs, a numerical simulation component model of CO_2 flooding and burial was developed using CMG reservoir numerical simulation software. This model was calibrated based on the phase behavior of original formation fluids and the geological characteristics and production data of the study area. A mature reservoir engineering calculation model was employed to compute the theoretical CO_2 storage capacity. The solubility of CO_2 measured in the laboratory was used to adjust the theoretical storage estimates. Subsequently, the CO_2 effective storage coefficient was determined, forming the foundation for investigating CO_2 storage mechanisms and developing a predictive model for effective storage coefficients.

2.1. The Fitting of the Phase of Fluids

Based on chromatographic measurements of degassed oils and associated gasses in the study area, simulations were conducted using the Winprop module of the phase behavior simulation CMG 2020 version software under an original reservoir pressure of 30.24 MPa, a reservoir temperature of 88 °C, and a gas–oil ratio of 91.5 m^3/m^3 to configure the live oil in the study area and reconstruct the underground fluid composition. The phase equilibrium calculations primarily utilized the PR state equation, obtaining the pseudo-component composition of the model (Table 1). The experimental data of multistage degassing and reservoir fluid property parameters were obtained from oilfield reservoir fluid testing information. The saturation pressure, viscosity, gas–oil ratio, and multi-stage degassing experiments of underground crude oil were fitted, and the results are shown below (Table 2 and Figure 1), and the equation of state parameters that can reflect the reservoir fluid was obtained (Table 3).

Table 1. Pseudo-component composition of live oil in model.

Component	Molar Composition/%	Component	Molar Composition/%
N_2	0.02	C_6–C_{12}	10.46
CO_2	0.36	C_{13}–C_{21}	15.07
CH_4	48.67	C_{22}–C_{29}	12.47
C_2–C_5	9.58	C_{30}–C_{38}	3.37

Table 2. Fitting results of fluid phase features.

Saturation Pressure/MPa			Viscosity/(mPa·s)			Gas–Oil Ratio/(m³/m³)		
Experimental Value	Simulation Value	Error/%	Experimental Value	Simulation Value	Error/%	Experimental Value	Simulation Value	Error/%
15.6	16.2	3.84%	3.2	3.1	3.2%	91.5	89.2	2.5%

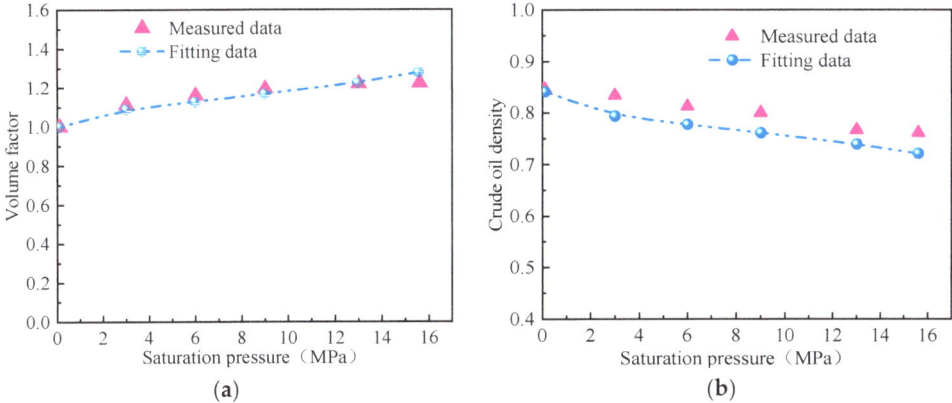

Figure 1. Fitting results of the multi-stage degassing experiment of formation fluid. (a) Fitting results of volume factor. (b) Fitting results of crude oil density.

Table 3. Characteristic parameters of the equation of state after fluid phase fitting in low-permeability thin oil reservoirs.

Component	Critical Pressure/MPa	Critical Temperature/K	Critical Volume/(L·mol⁻¹)	Acentric Factor	Molecular Weight/(g·mol⁻¹)	Ωa	Ωb
N_2	3.39	126.2	0.09	0.04	28.01	0.46	0.08
CO_2	7.38	304.2	0.094	0.23	44.01	0.46	0.08
CH_4	4.6	190.6	0.099	0.01	16.04	0.46	0.08
C_2–C_5	3.76	422.54	0.257	0.19	59.37	0.46	0.08
C_6–C_{12}	2.32	562.96	0.422	0.35	129.25	0.46	0.09
C_{13}–C_{21}	2.26	800	0.875	0.72	300.62	0.55	0.09
C_{22}–C_{29}	0.79	778.85	1.215	0.97	430.22	0.41	0.07
C_{30}–C_{38}	0.66	680.06	1.482	1.12	499.32	0.37	0.06

2.2. The Establishment of the Low-Permeability Reservoir Model

Based on the geological model of the study area, a representative well group (Figure 2) was selected, and production history matching was performed to enhance the reliability and accuracy of the model simulation results. Subsequently, numerical simulations were conducted to study CO_2 flooding and burial. The total number of grids of the model is 31,050 grids. The initial average formation pressure is 30.24 MPa, with an average permeability of 21 mD; average porosity is 16.3%, initial oil saturation is 0.55, and the rock compressibility coefficient is 4.5×10^{-6} 1/kPa. A three-dimensional schematic diagram of the model is shown below (Figure 2a). Historical production data of the study area were fitted to obtain the current distribution of the remaining oil (Figure 2b) and the relative permeability curves of the oil–water and gas–liquid phases (Figure 3).

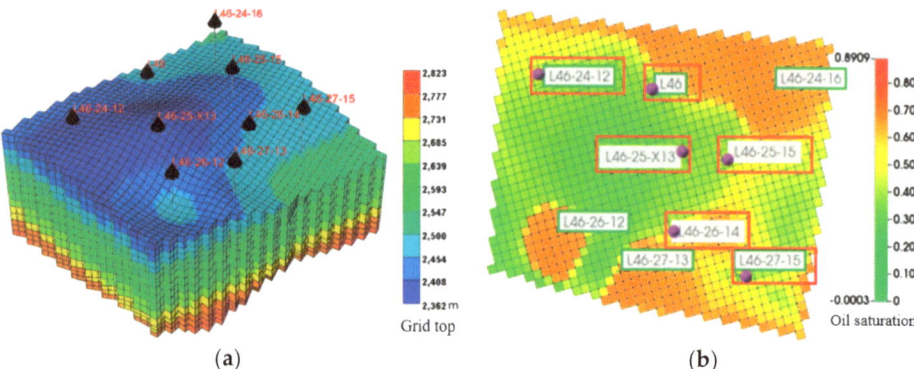

Figure 2. The established well group model and the current remaining oil distribution after historical fitting. (**a**) A typical well group model for low-permeability reservoirs in the study area. (**b**) Distribution of remaining oil after historical fitting.

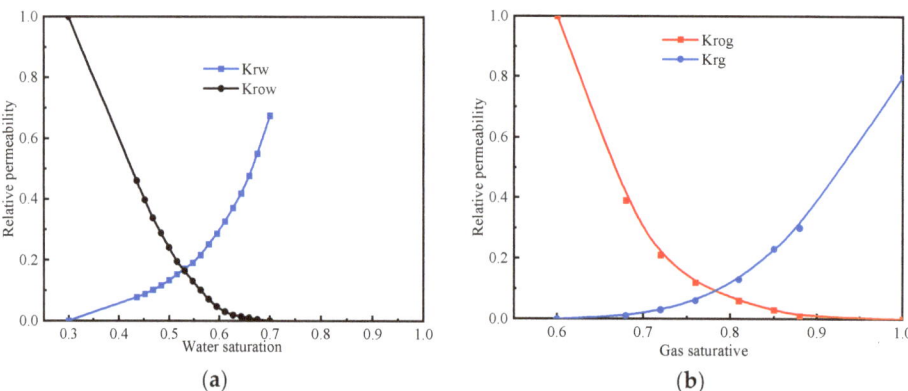

Figure 3. Relative permeability curves after historical fitting. (**a**) Oil–water relative permeability curve. (**b**) Gas–liquid relative permeability curve.

2.3. The Solution of Effective Buried Storage Coefficient

Numerical simulation is the most effective method for calculating oil recovery and effective storage coefficient [34]. CO_2 is influenced by factors such as differences in fluid viscosity and density, formation heterogeneity, water saturation, and strong water bodies. Therefore, it is more reliable to determine key parameters using "numerical simulation technology + experimental measurements" for calculating the correlation coefficient, compared to the empirical method. The solubility of CO_2 in the crude oil and water from the study area was determined through solubility measurement experiments (Figure 4).

This approach calculates the actual CO$_2$ storage considering various factors, followed by computation of the effective storage coefficient (Equation (1)) [35].

$$C_e = \frac{M_e}{M_t} \quad (1)$$

where C_e is the effective storage coefficient of the comprehensive influence of various factors, M_e is the effective burial amount of CO$_2$ in the reservoir, 10^6 t; M_t is the theoretical buried amount of CO$_2$ in the reservoir, 10^6 t.

$$M_t = \frac{\rho_r}{10^9} \left[\begin{array}{l} (0.4E_{Rb} + 0.6E_{Rh})Ah\phi(1 - S_{wi}) - V_{iw} + V_{pw} + C_{ws} \times (Ah\phi S_{wi} + V_{iw} - V_{pw}) \\ +C_{os}(1 - 0.4E_{Rb} - 0.6E_{Rh})Ah\phi(1 - S_{wi}) \end{array} \right] \quad (2)$$

where M_t is the theoretical buried amount of CO$_2$ in the reservoir, 10^6 t; ρ_r is the density of CO$_2$ in the reservoir, kg/m^3; A is the reservoir area, m^2; h is the reservoir thickness, m; Ø is the porosity of the reservoir; S_{wi} is the reservoir bound water saturation; V_{iw} is the amount of water injected into the reservoir, m^3; V_{pw} is the water produced from the reservoir, m^3; C_{ws} is the CO$_2$ solubility coefficient in water, m^3/m^3; C_{os} is the CO$_2$ solubility coefficient in oil, m^3/m^3; E_{Rb} is the oil recovery factor before CO$_2$ breakthrough; E_{Rh} is the oil recovery factor when a certain volume of CO$_2$ is injected.

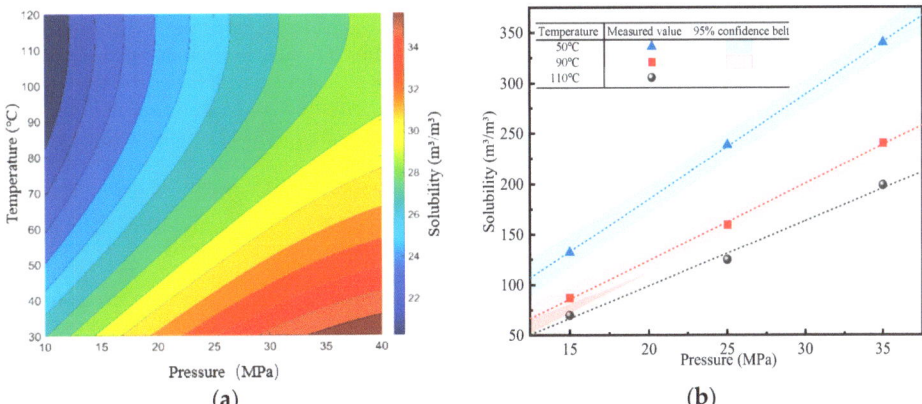

Figure 4. The results of CO$_2$ solubility determination in oil and water in the study area. (a) Solubility determination results in water. (b) Solubility determination results in oil.

The actual burial amount of CO$_2$ is obtained by numerical simulations to calculate the difference between the actual CO$_2$ injection and the CO$_2$ output, and the actual model considers the diffusion coefficient of CO$_2$ in oil and water, and the diffusion coefficient is measured by the pressure drop method experimentally [36]. The measurement method involves connecting a CO$_2$-filled container of constant volume to a core holder saturated with crude oil. At the beginning of the experiment, the valve is opened to allow communication between the container and the core holder, which proceeds at constant temperature. As CO$_2$ continues to diffuse into the core, the system pressure gradually decreases until it reaches equilibrium. Pressure changes during the experiment are recorded, and Fick's second law is applied to calculate the gas diffusion coefficient.

$$M_e = M_{inj} - M_{prd} \quad (3)$$

where: M_e is the effective burial amount of CO$_2$ in the reservoir, 10^6 t; M_{inj} is the amount of CO$_2$ injected, 10^6 t; M_{prd} is the amount of CO$_2$ produced, 10^6 t.

3. Numerical Simulation Study of CO_2 Flooding and Storage

Based on the historical fitting model, wells were arranged according to the current distribution of remaining oil. CO_2 injection was prioritized for wells with low oil saturation in the vicinity, while high-pressure gas injection was employed to maximize gravity-driven oil displacement. In Figure 2b, the injection well is highlighted by a red box, with the remaining wells designated as production wells. The single-variable method was employed to simulate CO_2 flooding and storage across varying permeability, temperature, and original reservoir pressure, based on the component model. Temperature and pressure tests are conducted via a gas injection well. The current gas injection capacity is calculated based on the principle that the bottomhole pressure should not exceed 90% of the rupture pressure, with a designed gas injection rate of 15,000 m^3/d (at standard conditions). The production wells adopt constant-pressure production, with the pressure set to the reservoir pressure in each well's grid cell after historical fitting. During the injection process, the well is shut down when the production gas–oil ratio reaches 2000 m^3/m^3. All wells are then switched to CO_2 until the formation pressure equals the original formation pressure (i.e., the final storage pressure). The CO_2 flooding and storage simulation is completed, yielding CO_2 storage results under different conditions, and calculating the contribution rate of various storage mechanisms for CO_2 storage. Several methods are employed to compute storage volumes for various sequestration mechanisms: the total storage volume is derived from the difference between the CO_2 injected at the injection well and the CO_2 produced at the production well; the dissolved storage volume in oil and water is determined using CMG 2020 version software, which calculates the CO_2 mole fraction in the oil–water phase using an equation of state. This calculation is then combined with the post-sequestration volume of oil and water in the reservoir to ascertain the dissolved storage volume. The residual storage volume is computed by subtracting the dissolved storage volume in oil and water from the total storage volume. Temperature and pressure tests are conducted via a gas injection well. The current gas injection capacity is determined based on ensuring that the bottom pressure of the injection rate does not exceed 90% of the rupture pressure. The designed injection rate is 15,000 m^3/d under standard conditions. During the injection process, wells are closed when the gas-to-oil ratio reaches 2000 m^3/m^3. Subsequently, all wells are switched to CO_2 injection until the formation pressure is restored to the original level (i.e., the final storage pressure). This process completes the simulation of CO_2-enhanced oil recovery and storage, producing calculated results for CO_2 storage under varying conditions. Consequently, the contribution rates of different storage mechanisms for CO_2 storage are calculated.

3.1. Effect of Permeability on CO_2 Flooding and Storage

Keeping all other model parameters constant, the permeability was set to 21.53 mD, 26.53 mD, and 31.53 mD, respectively, to investigate the effects of varying permeability on CO_2 flooding and storage. In low permeability oil reservoirs, with the increase in permeability, the degree of crude oil recovery improves. Figure 5 illustrates the molar fraction of CO_2 in the oil phase under varying permeability conditions. As permeability increases, the sweep range of CO_2 in the crude oil widens, facilitating contact with the crude oil and enhancing processes such as extraction, expansion, and dissolution. This effect is more beneficial for crude oil recovery in low-permeability reservoirs. Figure 6 illustrates that an increase in permeability enhances the contribution rate of structural and adsorptive storage, which represents approximately 55% to 60% of the total storage. As permeability increases, oil and water can be extracted more easily, thereby reducing the proportion of storage in the oil-water mixture. The effective storage coefficient also increases from 0.67 to 0.71, indicating that higher permeability is advantageous for CO_2 storage. This is because higher permeability facilitates greater oil extraction and provides more space in low-permeability zones for CO_2 storage. Increasing permeability significantly enhances the oil recovery and storage efficiency of CO_2 in low-permeability reservoirs.

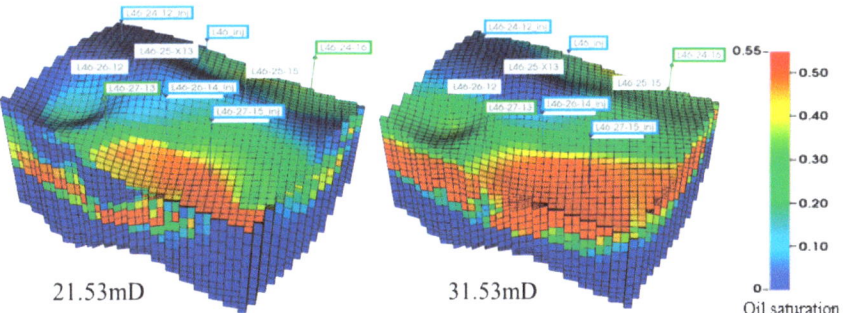

Figure 5. Mole fraction of CO_2 in the oil phase under different permeability conditions.

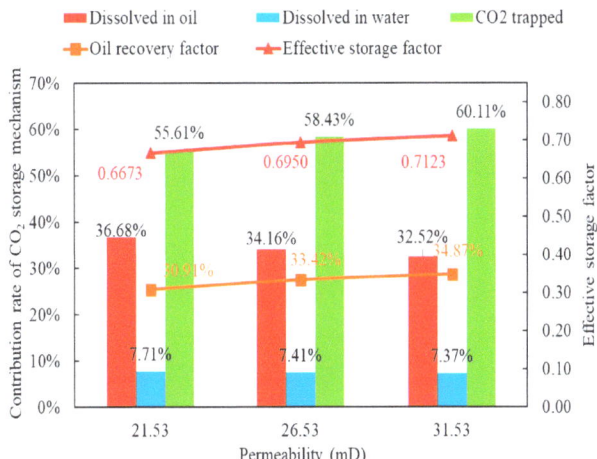

Figure 6. Effect of permeability on CO_2 storage in oil reservoirs with low permeability.

3.2. Effect of Reservoir Temperature on CO_2 Flooding and Storage

With all other model parameters held constant, reservoir temperatures were set to 83.1 °C, 88.1 °C, and 93.1 °C to investigate the effects of varying temperatures on CO_2 flooding and storage. The simulation results indicate that under temperature conditions of 83.1 °C, 88.1 °C, and 93.1 °C, the recovery rates during the depletion stage were 10.9%, 10.88%, and 10.89%, respectively. The production period during the depletion stage was 19 years. Figure 7 shows that temperature has minimal impact on the recovery rate of low-permeability oil reservoirs and the proportion of each storage mechanism. The effective storage coefficient increases with temperature because the actual amount of CO_2 stored remains relatively stable while the CO_2 density decreases with rising temperature, leading to a decrease in theoretical storage capacity and thus increasing the effective storage coefficient. The proportion of dissolved storage in oil and water decreases slightly because CO_2 becomes less soluble in oil and water at higher temperatures, reducing its dissolution. Figure 8 illustrates that as temperature increases, the average molar fraction of CO_2 in the oil phase decreases, but this decrease occurs at a slower rate.

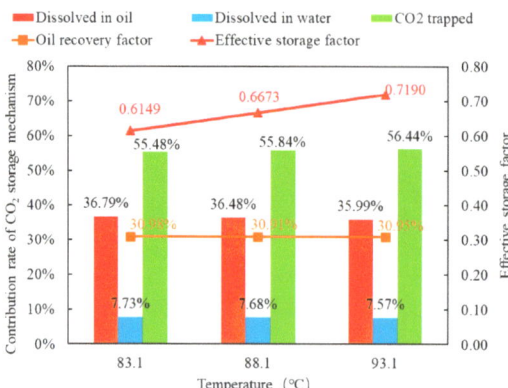

Figure 7. Effect of temperature on CO_2 storage in thin oil reservoirs with low permeability.

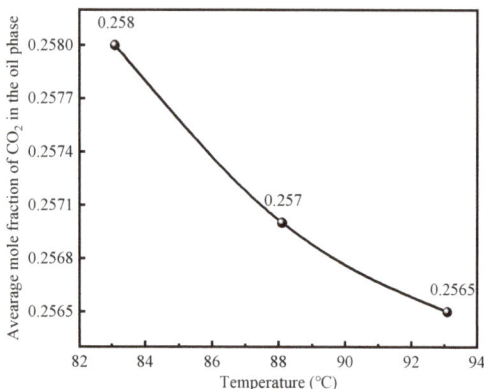

Figure 8. Effect of temperature on the average molar fraction of CO_2 in the oil phase.

3.3. Effect of Original Reservoir Pressure on CO_2 Flooding and Storage

Keeping other model parameters unchanged, the original reservoir pressures were adjusted to 28.95 MPa, 30.24 MPa, and 31.49 MPa to investigate their impacts on CO_2 flooding and storage. Simulation results indicate that as reservoir pressure increases, crude oil recovery initially increases significantly before slightly decreasing, as depicted in Figure 9. At an original formation pressure of 31.49 MPa, recovery decreases due to excessive initial formation pressure, resulting in early recovery of most crude oil from pore spaces compared to lower pressure reservoirs. Additionally, gas channeling during later stages of gas injection exacerbates this trend, reducing overall recovery efficiency. Increasing formation pressure decreases the proportion of structurally stored CO_2 while increasing the dissolved CO_2 fraction in oil and water. This relationship arises because burial upper limits are governed by original formation pressures, with higher pressures enhancing CO_2 solubility in oil and water post-burial. Consequently, dissolved CO_2 fractions rise accordingly. Figure 10 illustrates the distribution of residual oil following CO_2 flooding at various formation pressures. It is evident that as the pressure increases, the residual oil first decreases significantly before slightly increasing. Therefore, in CO_2 storage processes, a higher storage pressure is not always beneficial. An optimal storage pressure can help prevent gas channeling, thereby improving both oil recovery and storage efficiency.

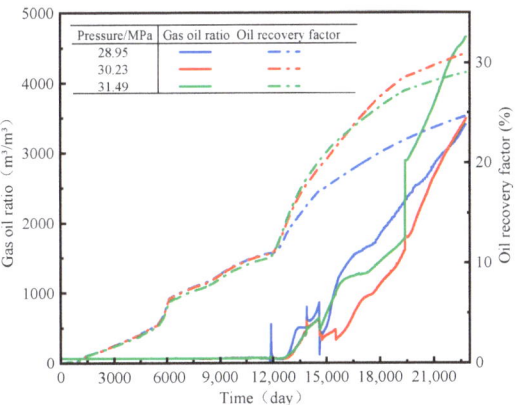

Figure 9. The curves of recovery factor and gas–oil ratio under different initial reservoir pressure conditions.

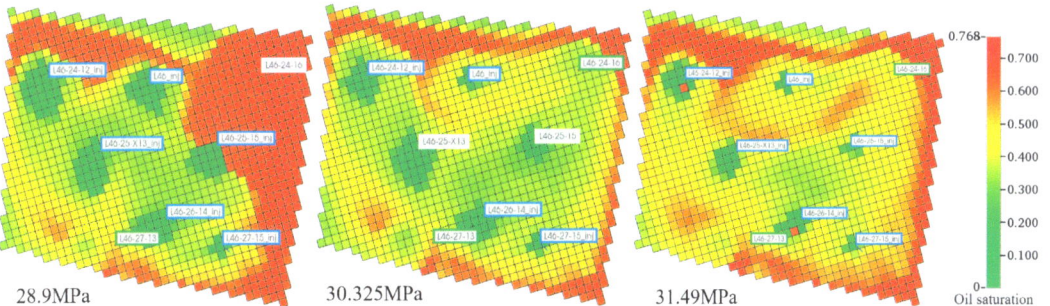

Figure 10. Distribution of residual oil after CO_2 flooding under different original formation pressures.

4. Prediction Model for the Effective Storage Coefficient Using Artificial Intelligence

4.1. Prediction Model for Effective Burial Coefficient

Building on the results of numerical simulations and utilizing the regression learner in MATLAB 2022 version software, we employed a supervised machine learning approach to develop regression models. The input variables included layer permeability, reservoir temperature, and initial formation pressure, with the effective storage coefficient as the target variable. We allocated 80% of the data to the training set and 20% to the test set for regression analysis. The regression algorithms comprised six main categories: support vector regression, Gaussian process regression (GPR), tree ensembles, neural networks, linear regression, and regression trees. The model with the lowest regression error was chosen as the surrogate model. To eliminate the influence of dimensionality, index values were standardized using the z-score method, which is based on the mean and standard deviation of the raw data. The R-square coefficient, mean square error (MSE), root mean squared error (RMSE), and mean absolute error (MAE) were used as the basis for evaluating the performance of the machine learning models. The coefficient of determination, also know as R^2, is a numerical measure that represents the relationship between a dependent variable and multiple independent variables. It reflects the reliability of the regression model in explaining variations in the dependent variable, similar to the multiple correlation coefficient.

The GPR using the quadratic rational kernel function and the exponential kernel exhibits the highest fitting accuracy for the block model of the low-permeability thin oil reservoir. The model training results are shown in Table 4. Models with a fitting accuracy greater than 0.8 include quadratic rational GPR, exponential GPR, and square exponential GPR, among others. The GPR demonstrates superior adaptability to the studied block.

Table 4. Training results of a single model in a thin oil reservoir with low permeability.

Model	Fine Model	RMSE	MSE	R^2	MAE
Gaussian process regression model	Quadratic rational GPR	0.032237	0.001039	0.919931	0.023666
	Square exponential GPR	0.036601	0.00134	0.844762	0.027608
	Matern 5/2	0.051279	0.00263	0.521953	0.039756
	Exponent GPR	0.032237	0.001039	0.919934	0.023666

4.2. Establishment and Application of Effective Storage Coefficient Plates

Using the constructed proxy model, predictions were made for the effective storage coefficient under different storage conditions (permeability, original reservoir pressure, reservoir temperature). A computational graph (Figure 11) illustrating the effective storage coefficient was established. Referring to the graph allows one to obtain the effective storage coefficient under various storage conditions and calculate the corresponding effective burial volume accordingly. Regions on the graph closer to red indicate larger effective burial coefficients, suggesting reservoir conditions more favorable for burial, facilitating the determination of favorable burial reservoir conditions. In the favorable burial area, the permeability ranges from 26 to 32 mD, and the temperature ranges from 86 °C to 98 °C. The permeability in this area increases significantly with pressure, leading to an expansion of the favorable burial zone into higher temperature and lower permeability regions. Under a pressure of 31 MPa, the effective storage coefficient increases to approximately 0.746.

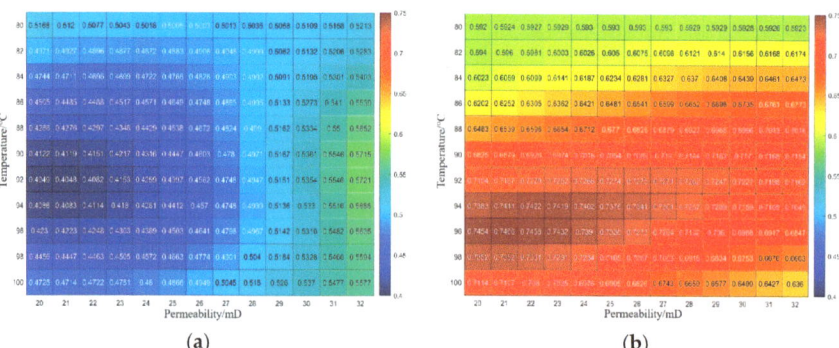

Figure 11. Calculation graph of the effective storage coefficient of a low-permeability reservoir in the study area. (a) 28 MPa. (b) 31 MPa.

Using the parameters provided, the effective buried stock was calculated. The basic reservoir parameters are listed in Table 5. The values of recovery before CO_2 breakout refers to the oil recovery from the start of CO_2 flooding to just before gas breakout occurs. The values of recovery after CO_2 breakout refers to the oil recovery from the start of CO_2 flooding to a specific moment after gas breakout occurs. The CO_2 density is 737.06 kg/m^3 at the reservoir's temperature and pressure. The solubility of CO_2 is 207.86 m^3/m^3 in crude oil and 27.63 m^3/m^3 in water under standard conditions, as measured in the study area. According to the effective storage coefficient calculation, with a permeability of 22 mD and an initial formation pressure of 31 MPa, the effective storage coefficient is 0.5981 at 82 °C. The theoretical amount of CO_2 burial, calculated using Equation (2), is 700,400 tons, while the effective CO_2 burial amount, calculated using Equation (3), is 418,900 tons.

Table 5. The basic parameters of the reservoir of the well group of the low-permeability reservoir in the block are studied.

Initial Formation Pressure/MPa	Average Permeability/mD	Reservoir Temperature/°C	Bound Water Saturation	Pore Volume/m^3	Recovery Rate/%	
					Before CO_2 Breakout	After CO_2 Breakout
31	22	88	0.3	6,108,620	2.54	20.01

5. Conclusions

(1) In low-permeability oil reservoirs, an increase in permeability results in a decrease in the contribution rate of CO_2 dissolution and sequestration in oil and water, while the proportion of structurally bound sequestration increases from 55% to 60%.

(2) Temperature has little impact on the contribution rate of different CO_2 sequestration mechanisms. The proportion of CO_2 sequestration through dissolution in oil and water decreases slightly due to the reduced solubility coefficient of CO_2 in oil and water at higher temperatures.

(3) Higher initial reservoir pressure improves the effectiveness of CO_2 enhanced oil recovery. However, when the pressure surpasses a certain threshold, gas channeling may occur during the later stages of injection, which can lead to decreased recovery and storage efficiency. During field implementation, it is crucial to ensure that the reservoir pressure exceeds the minimum miscibility pressure of CO_2 and crude oil, while also maintaining it below the maximum allowable pressure of the injection equipment and pipelines.

(4) A method was established using supervised machine learning to train regression models—with permeability, reservoir temperature, and initial reservoir pressure as the input variables, and the effective storage coefficient as the target function—to determine CO_2 effective sequestration coefficients through artificial intelligence training models. Charts depicting effective sequestration coefficients under various conditions (permeability, reservoir pressure, temperature) enable accurate and rapid calculation of effective sequestration volumes and identification of favorable sequestration areas.

Author Contributions: Conceptualization, Z.F.; methodology, T.S. and Z.L.; software, T.S. and J.C.; validation, Y.M.; formal analysis, T.S.; investigation, M.T. and M.L.; resources, Z.F.; data curation, Y.M.; writing—original draft preparation, Z.F.; writing—review and editing, T.S. and Z.L.; visualization, J.C.; supervision, Y.J. and Z.L.; project administration, Z.F. and M.T.; funding acquisition, Z.F. All authors have read and agreed to the published version of the manuscript.

Funding: This research was funded by the National Natural Science Foundation of China and Enterprise Innovation Development Joint Fund—Integrated Project, "Fundamental research on synergistic steam injection to significantly enhance heavy oil recovery through multiple chemical reactions and heat generation in the wellbore", grant number U23B6003.

Data Availability Statement: The data presented in this study are available on request from the corresponding author due to legal and privacy reasons.

Conflicts of Interest: Authors Z.F., M.T., M.L., Y.M. and Y.J. were employed by the Liaohe Oilfield Company.

References

1. Wang, Z.; Fan, Z.; Chen, X.; Fan, Z.; Wei, Q.; Wang, X.; Yue, W.; Liu, B.; Wu, Y. Global Oil and Gas Development in 2022: Situation, Trends, and Enlightenment. *Pet. Explor. Dev.* **2023**, *50*, 1167–1186. [CrossRef]
2. Wang, Q.; Li, S.; Li, R. China's Dependency on Foreign Oil Will Exceed 80% by 2030: Developing a Novel NMGM-ARIMA to Forecast China's Foreign Oil Dependence from Two Dimensions. *Energy* **2018**, *163*, 151–167. [CrossRef]
3. Tang, X.; Li, Y.; Cao, J.; Liu, Z.Y.; Chen, X.; Liu, L.; Zhang, Y.Q. Adaptability and enhanced oil recovery performance of surfactant polymer flooding in inverted seven-spot well pattern. *Phys. Fluids* **2023**, *35*, 053116. [CrossRef]
4. Wei, B.; Song, T.; Zhao, J.Z.; Valeriy, K.; Pu, W.F. Improving the Recovery Efficiency and Sensitivity of Tight Oil Reservoirs by Dissolved Gas Reinjection. *J. Southwest Pet. Univ. Sci. Technol. Ed.* **2019**, *41*, 85–95.

5. Wang, R.; Zhang, Y.; Lyu, C.; Lun, Z.M.; Cui, M.L.; Lang, D.J. Displacement characteristics of CO_2 flooding in extra-high water-cut reservoirs. *Energy Geosci.* **2024**, *5*, 100115. [CrossRef]
6. Wei, B.; Song, T.; Gao, Y.; Xiang, H.; Xu, X.G.; Valeriy, K.; Bai, J.L.; Zhai, Z.W. Effectiveness and sensitivity analysis of solution gas re-injection in Baikouquan tight formation, Mahu sag for enhanced oil recovery. *Petroleum* **2020**, *6*, 253–263. [CrossRef]
7. Zou, C.; Lin, M.; Ma, F. Development, challenges and strategies of natural gas industry under carbon neutral target in China. *Pet. Explor. Dev.* **2024**, *51*, 476–497. [CrossRef]
8. Davoodi, S.; Al-Shargabi, M.; Wood, D.A.; Mehrad, M.; Rukavishnikov, S. Carbon dioxide sequestration through enhanced oil recovery: A review of storage mechanisms and technological applications. *Fuel* **2024**, *366*, 131313. [CrossRef]
9. Shi, C.; Zhi, J.; Yao, X.; Zhang, H.; Yu, Y.; Zeng, Q.S.; Li, L.J.; Zhang, Y.X. How can China achieve the 2030 carbon peak goal—A crossover analysis based on low-carbon economics and deep learning. *Energy* **2023**, *269*, 126776. [CrossRef]
10. Zhong, Z.; Chen, Y.; Fu, M.; Li, M.Z.; Yang, K.S.; Zeng, L.P.; Liang, J.; Ma, R.P.; Xie, Q. Role of CO_2 geological storage in China's pledge to carbon peak by 2030 and carbon neutrality by 2060. *Energy* **2023**, *272*, 127165. [CrossRef]
11. Tang, X.-C.; Li, Y.-Q.; Liu, Z.-Y.; Zhang, N. Nanoparticle-reinforced foam system for enhanced oil recovery (EOR): Mechanistic review and perspective. *Pet. Sci.* **2023**, *20*, 2282–2304. [CrossRef]
12. Dou, L.; Sun, L.; Lyu, W.; Wang, M.Y.; Gao, F.; Gao, M.; Jiang, H. Trend of global carbon dioxide capture, utilization and storage industry and challenges and countermeasures in China. *Pet. Explor. Dev.* **2023**, *50*, 1246–1260. [CrossRef]
13. Tang, X.; Li, Y.; Han, X.; Zhou, Y.B.; Zhan, J.F.; Xu, M.M.; Zhou, R.; Cui, K.; Chen, X.L.; Wang, L. Dynamic characteristics and influencing factors of CO_2 huff and puff in tight oil reservoirs. *Pet. Explor. Dev.* **2021**, *48*, 946–955. [CrossRef]
14. Zhang, Z.H.; Yang, Z.M.; Liu, X.G.; Xiong, W.; Wang, X.W. A grading evaluation method for low-permeability reservoirs and its application. *Acta Pet. Sin.* **2012**, *33*, 437–441.
15. Mahdaviara, M.; Sharifi, M.; Ahmadi, M. Toward evaluation and screening of the enhanced oil recovery scenarios for low permeability reservoirs using statistical and machine learning techniques. *Fuel* **2022**, *325*, 124795. [CrossRef]
16. Wei, B.; Zhang, X.; Wu, R.; Zou, P.; Gao, K.; Xu, X.G.; Pu, W.F.; Wood, C. Pore-scale monitoring of CO_2 and N_2 flooding processes in a tight formation under reservoir conditions using nuclear magnetic resonance (NMR): A case study. *Fuel* **2019**, *246*, 34–41. [CrossRef]
17. Hill, L.B.; Li, X.; Wei, N. CO_2-EOR in China: A comparative review. *Int. J. Greenh. Gas Control* **2020**, *103*, 103173. [CrossRef]
18. Wang, L.; Wei, B.; You, J.; Pu, W.F.; Tang, J.Y.; Lu, J. Performance of a tight reservoir horizontal well induced by gas huff-n-puff integrating fracture geometry, rock stress-sensitivity and molecular diffusion: A case study using CO_2, N_2 and produced gas. *Energy* **2023**, *263*, 125696. [CrossRef]
19. Lu, L.; Liu, B. A Feasibility Research Method and Project Design on CO_2 Miscible Flooding for a Small Complex Fault Block Field. In Proceedings of the SPE International Oil and Gas Conference and Exhibition in China, Beijing, China, 2–6 November 1998; pp. SPE-50930-MS, 501–515.
20. Xu, S.; Ren, G.; Younis, R.M.; Feng, Q. Revisiting field estimates for carbon dioxide storage in depleted shale gas reservoirs: The role of geomechanics. *Int. J. Greenh. Gas Control* **2021**, *105*, 103222. [CrossRef]
21. Haishui, H.A.N.; Shiyi, Y.U.A.N.; Shi, L.I.; Xiaolei, L.; Xinglong, C. Dissolving capacity and volume expansion of carbon dioxide in chain n-alkanes. *Pet. Explor. Dev.* **2015**, *42*, 97–103.
22. He, Y.; Liu, M.; Tang, Y.; Jia, C.Q.; Wang, Y.; Rui, Z.H. CO_2 storage capacity estimation by considering CO_2 Dissolution: A case study in a depleted gas Reservoir, China. *J. Hydrol.* **2024**, *630*, 130715. [CrossRef]
23. Kutsienyo, E.J.; Ampomah, W.; Sun, Q.; Balch, R.S.; You, J.; Aggrey, W.N.; Cather, M. Evaluation of CO_2-EOR Performance and Storage Mechanisms in an Active Partially Depleted Oil Reservoir. In Proceedings of the SPE Europec featured at 81st EAGE Conference and Exhibition, London, UK, 3–6 June 2019; p. SPE-195534-MS.
24. Sedaghatinasab, R.; Kord, S.; Moghadasi, J.; Soleymanzaden, A. Relative Permeability Hysteresis and Capillary Trapping during CO_2 EOR and Sequestration. *Int. J. Greenh. Gas Control* **2021**, *106*, 103262. [CrossRef]
25. Wang, Y.Z.; Cao, R.Y.; Jia, Z.H.; Wang, B.Y.; Ma, M.; Cheng, L.S. A multi-mechanism numerical simulation model for CO_2-EOR and storage in fractured shale oil reservoirs. *Pet. Sci.* **2024**, *21*, 1814–1828. [CrossRef]
26. Zhang, R.H.; Wu, J.F.; Zhao, Y.L.; He, X.; Wang, R.H. Numerical simulation of the feasibility of supercritical CO_2 storage and enhanced shale gas recovery considering complex fracture networks. *J. Pet. Sci. Eng.* **2021**, *204*, 108671. [CrossRef]
27. Yamaguchi, A.J.; Sato, T.; Tobase, T.; Wei, X.; Huang, L.; Zhang, J.; Bian, J.; Liu, T. Multiscale numerical simulation of CO_2 hydrate storage using machine learning. *Fuel* **2023**, *334*, 126678. [CrossRef]
28. Liu, D.; Li, Y.; Agarwal, R.K. Numerical simulation of long-term storage of CO_2 in Yanchang shale reservoir of the Ordos basin in China. *Chem. Geol.* **2016**, *440*, 288–305. [CrossRef]
29. Andrić, I.; Pinaa, A.; Ferrão, P.; Fournier, J.; Lacarrière, B.; Le Corre, O. Assessing the feasibility of using the heat demand-outdoor temperature function for a long-term district heat demand forecast. *Energy Procedia* **2017**, *158*, 6079–6086. [CrossRef]
30. Thibeau, S.; Bachu, S.; Birkholzer, J.; Holloway, S.; Neele, F.; Zhou, Q. Using Pressure and Volumetric Approaches to Estimate CO_2 Storage Capacity in Deep Saline Aquifers. *Energy Procedia* **2014**, *63*, 5294–5304. [CrossRef]
31. Shen, P.P.; Liao, X.W.; Liu, Q.J. Methodology for estimation of CO_2 storage capacity in reservoirs. *Pet. Explor. Dev.* **2009**, *36*, 216–220. [CrossRef]
32. Zhao, X.; Liao, X.; Wang, W.; Chen, C.Z.; Rui, Z.H.; Wang, H. The CO_2 storage capacity evaluation: Methodology and determination of key factors. *J. Energy Inst.* **2014**, *87*, 297–305. [CrossRef]

33. Abdulwarith, A.; Ammar, M.; Dindoruk, B. Prediction/Assessment of CO_2 EOR and Storage Efficiency in Residual Oil Zones Using Machine Learning Techniques. In Proceedings of the SPE/AAPG/SEG Carbon, Capture, Utilization, and Storage Conference and Exhibition, Houston, TX, USA, 11–13 March 2024; p. SPE-CCUS-2024-4011705.
34. Ren, D.; Wang, X.; Kou, Z.; Wang, S.C.; Wang, H.; Wang, X.G.; Tang, Y.; Jiao, Z.S.; Zhou, D.S.; Zhang, R.J. Feasibility evaluation of CO_2 EOR and storage in tight oil reservoirs: A demonstration project in the Ordos Basin. *Fuel* **2023**, *331*, 125652. [CrossRef]
35. Gao, R.; Lv, C.Y.; Zhou, K.; Lun, Z.M.; Zhou, B. A CO_2 flooding dynamic storage potential calculation method based on compositional flash calculation. *Oil Drill. Prod. Technol.* **2021**, *43*, 70–75.
36. Li, S.Y.; Qiao, C.Y.; Li, Z.M.; Hui, Y.T. The effect of permeability on supercritical CO_2 diffusion coefficient and determination of diffusive tortuosity of porous media under reservoir conditions. *J. CO2 Util.* **2018**, *28*, 1–14. [CrossRef]

Disclaimer/Publisher's Note: The statements, opinions and data contained in all publications are solely those of the individual author(s) and contributor(s) and not of MDPI and/or the editor(s). MDPI and/or the editor(s) disclaim responsibility for any injury to people or property resulting from any ideas, methods, instructions or products referred to in the content.

Article

Prediction of ORF for Optimized CO$_2$ Flooding in Fractured Tight Oil Reservoirs via Machine Learning

Ming Yue [1,2,3,*], Quanqi Dai [1,2,4], Haiying Liao [1,2,4], Yunfeng Liu [1,2,4], Lin Fan [3] and Tianru Song [3]

1. State Key Laboratory of Shale Oil and Gas Enrichment Mechanisms and Effective Development, Beijing 102206, China; daiqq.syky@sinopec.com (Q.D.); liaohy.syky@sinopec.com (H.L.); liuyunfeng.syky@sinopec.com (Y.L.)
2. SINOPEC Key Laboratory of Carbon Capture, Utilization and Storage, Beijing 102206, China
3. School of Civil and Resource Engineering, University of Science and Technology Beijing, No. 30, Xueyuan Road, Beijing 100083, China; d202210001@xs.ustb.edu.cn (L.F.); songtianru910@gmail.com (T.S.)
4. Petroleum Exploration and Development Research Institute, SINOPEC, Beijing 102206, China
* Correspondence: yueming01@ustb.edu.cn

Citation: Yue, M.; Dai, Q.; Liao, H.; Liu, Y.; Fan, L.; Song, T. Prediction of ORF for Optimized CO$_2$ Flooding in Fractured Tight Oil Reservoirs via Machine Learning. *Energies* **2024**, *17*, 1303. https://doi.org/10.3390/en17061303

Academic Editor: Reza Rezaee

Received: 18 January 2024
Revised: 2 March 2024
Accepted: 5 March 2024
Published: 8 March 2024

Copyright: © 2024 by the authors. Licensee MDPI, Basel, Switzerland. This article is an open access article distributed under the terms and conditions of the Creative Commons Attribution (CC BY) license (https://creativecommons.org/licenses/by/4.0/).

Abstract: Tight reservoirs characterized by complex physical properties pose significant challenges for extraction. CO$_2$ flooding, as an EOR technique, offers both economic and environmental advantages. Accurate prediction of recovery rate plays a crucial role in the development of tight oil and gas reservoirs. But the recovery rate is influenced by a complex array of factors. Traditional methods are time-consuming and costly and cannot predict the recovery rate quickly and accurately, necessitating advanced multi-factor analysis-based prediction models. This study uses machine learning models to rapidly predict the recovery of CO$_2$ flooding for tight oil reservoir development, establishes a numerical model for CO$_2$ flooding for low-permeability tight reservoir development based on actual blocks, studies the effects of reservoir parameters, horizontal well parameters, and injection-production parameters on CO$_2$ flooding recovery rate, and constructs a prediction model based on machine learning for the recovery. Using simulated datasets, three models, random forest (RF), extreme gradient boosting (XGBoost), and light gradient boosting machine (LightGBM), were trained and tested for accuracy evaluation. Different levels of noise were added to the dataset and denoised, and the effects of data noise and denoising techniques on oil recovery factor prediction were studied. The results showed that the LightGBM model was superior to other models, with R^2 values of 0.995, 0.961, 0.921, and 0.877 for predicting EOR for the original dataset, 5% noise dataset, 10% noise dataset, and 15% noise dataset, respectively. Finally, based on the optimized model, the key control factors for CO$_2$ flooding for tight oil reservoirs to enhance oil recovery were analyzed. The novelty of this study is the development of a machine-learning-based method that can provide accurate and cost-effective ORF predictions for CO$_2$ flooding for tight oil reservoir development, optimize the development process in a timely manner, significantly reduce the required costs, and make it a more feasible carbon utilization and EOR strategy.

Keywords: CO$_2$-EOR; CO$_2$ flooding; machine learning; oil recovery prediction; tight oil reservoirs

1. Introduction

There has been a growing emphasis on exploring and developing unconventional oil and gas resources worldwide. But extracting residual oil from tight reservoirs in complex geological formations remains a significant challenge [1]. Numerous studies have demonstrated the significant impact of CO$_2$ flooding on enhancing oil recovery (EOR) in low-permeability reservoirs [2–5]. CO$_2$ flooding holds the potential for achieving high efficiency in extracting oil from reservoirs. However, the development of CO$_2$ flooding in tight reservoirs is affected by various factors, such as geology, fluid properties, CO$_2$ phase transition, and fracture structure modification, which pose challenges for predicting oil recovery factors for CO$_2$ flooding [6].

Additionally, different oil recovery factors can characterize the different development stages of the current oil and gas field [7]. Through the prediction of recovery, real-time production control can be achieved, production measures can be adjusted in a timely manner, and reservoir development can be optimized. Therefore, accurate prediction of recovery rate plays a crucial role in the development of oil and gas fields.

Currently, oil recovery factor prediction in tight oil reservoirs mainly revolves around water-driven development. The prediction methods can be broadly categorized into three main approaches: macro-equilibrium analysis, micro-experimental mechanistic analysis, and numerical simulation method [8–15]. Sun et al. [16] developed a power-function-based material balance equation for high-pressure and ultrahigh-pressure gas reservoirs and investigated the impact of reservoir pressure depletion and recovery degree on reserve estimation reliability. Cheng et al. [17] proposed a synchronization iterative oilfield oil recovery factor prediction method by combining water content curves with the exponential decline method, which is based on statistical regression experiments and field data through Buckley–Leverett theory, and these approaches have improved accuracy of oilfield recovery factor prediction. Hadia et al. [18] conducted core drive experiments to analyze the relationship between relative permeability and water saturation and predicted the recovery degree through a numerical simulation model based on the dimensionless Buckley–Leverett equation. Zhong et al. [19] studied the recovery efficiency of CO_2 flooding timing and different injection methods based on the reservoir conditions of a block in Jilin Oilfield using Eclipse 3.0. Nevertheless, the main factors affecting the recovery of CO_2 flooding in tight oil reservoirs are complex and diverse. The Macroscopic Balance Analysis and Microscopic Experimental Mechanics Analysis methods can only provide rough estimates of recovery rates, lacking precision and incurring high costs. Numerical simulation techniques require individual modeling for different reservoirs, with prediction accuracy dependent on field data, and involve lengthy simulation times. Their accuracy hinges on the availability of accurate field data, and these simulations typically require extended periods to complete. Therefore, further research is needed on the recovery prediction model for CO_2 flooding in fractured tight oil reservoirs.

In contrast, machine learning (ML) methods offer a distinct advantage. They can create unique predictive models that consider various reservoir characteristics, uncover hidden data relationships, and accurately predict production outcomes at a lower cost. In the petroleum industry, ML models have been widely applied and achieved good application results. In the petroleum industry and underground gas storage, machine learning has found application in a myriad of areas, including the evaluation of reserves in both conventional and unconventional reservoirs [20–23], the automated interpretation of well tests [24–27], forecasting production from oil and shale gas [28–31], as well as in predicting the lithology of reservoirs [32–34]. ML models have also been utilized in research for enhanced oil recovery (EOR). Van Si et al. [35] developed an artificial neural network (ANN) model designed to forecast the oil recovery factor (ORF) specific to CO_2-enhanced oil recovery (EOR) processes. Cheraghi et al. [36] suggested employing deep ANN and random forest (RF) models for identifying the most appropriate EOR techniques, leveraging data sourced from oil and gas publications. Esene et al. [37] conducted predictions of the ORF using ANN, least-squares support vector machines, and gene expression programing for carbonate water-injection processes. In another study, Pan et al. [38] constructed a machine learning model utilizing extreme gradient boosting (XGBoost) to infer reservoir porosity from well log data. They enhanced the XGBoost model's accuracy through a combination of grid search and nature-inspired optimization methods, achieving a root mean square error (RMSE) of 0.527. Further extending the exploration of machine learning applications, Huang et al. [39] evaluated the performance of ANNs, light gradient boosting machine (LightGBM), and XGBoost models in forecasting production from steam-assisted gravity drainage processes. Collectively, these investigations underscore the significant capabilities of machine learning models in forecasting the oil recovery factor and enhancing oil recovery methodologies. Compared to traditional methods of predicting recovery rates, ML can deeply mine the relationship between complex data and recovery, extract data features to identify the main controlling factors affecting recovery rates, and efficiently, accurately, and cost-effectively predict the recovery rates of reservoirs under different geological conditions. While previous research has explored machine learning (ML) models, their application in the rapid prediction of CO_2 flooding systems in tight oil reservoirs has not been extensively studied. Given the difficulty of accurately simulating underground fracturing conditions in laboratory settings and the associated high costs, the majority of recent studies have turned to numerical simulations to gather data. However, these studies frequently neglect the effect of data noise on their outcomes, potentially leading to variances between the research conclusions and real-world scenarios.

Therefore, the study is dedicated to crafting and evaluating a range of ML models to find the optimal one for application. The goal is to identify a model that significantly reduces both the time and financial costs associated with experiments while ensuring the precision of predictions regarding the ORF in the context of CO_2 flooding through horizontal wells in tight oil reservoirs, thereby providing valuable insights for future gas injection strategies in these reservoirs. For testing these models, we considered a wide array of production and geological parameters, compiling a comprehensive dataset. To more accurately reflect real-world conditions, we introduced noise into the dataset and then applied denoising techniques. This approach allows us to assess the impact of noise and denoising on our research outcomes. The findings of our study present an effective solution for swiftly predicting the ORF of CO_2 flooding in tight oil reservoirs and have potential applications in other EOR methods.

2. Methodology

This section outlines the core workflow of a novel prediction method for CO_2 Enhanced Oil Recovery (CO_2-EOR) rates. Initially, a numerical model is developed, drawing on real-world development scenarios. Key factors that influence CO_2-EOR rates are determined from prior studies. Then, using Latin hypercube sampling (LHS), a dataset for numerical simulation is created. To enhance the dataset's realism and quality, it is further processed through noise addition and denoising techniques. A general workflow for ML-based prediction of recovery degree is illustrated in Figure 1. The specific steps of the work are described in detail in the following subsections.

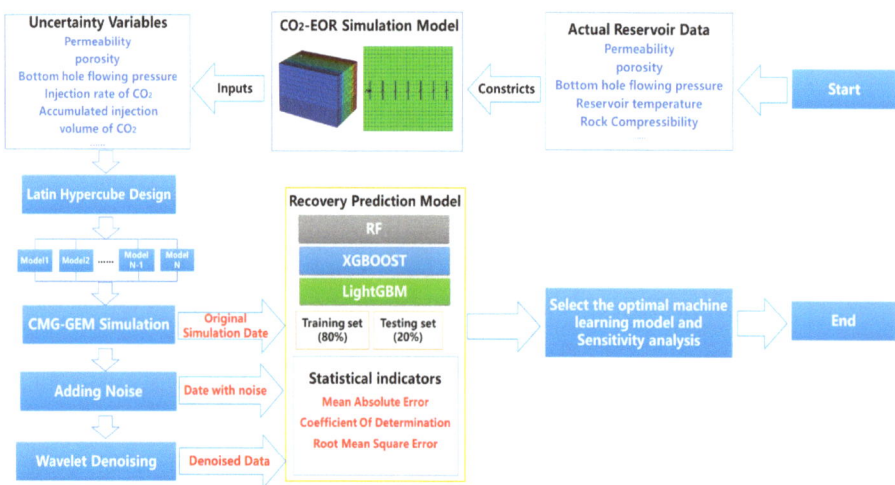

Figure 1. Workflow of ORF prediction using three ML models.

2.1. Data Preparation

2.1.1. Reservoir Model Description

Changqing tight reservoir, ideal for CO_2 miscible flooding due to its vast area and access to substantial gas resources, is the chosen site for CO_2 injection. The project is further supported by favorable on-site road conditions. To model the CO_2 injection process accurately without the influence of reservoir boundaries, we employed CMG-GEM numerical simulation software to create a simulation model. This model features a single-well radial grid layout measuring 2440 m × 1640 m × 26 m, covering 4 km^2. Utilizing the Cartesian grid system, the formation is divided into regular grids: 61 in the I direction, 41 in the J direction, and 13 in the K direction, with standard grid sizes of 40 m × 40 m × 2 m. The central encrypted grid is finer, with dimensions of 8 m × 8 m × 2 m. Figure 2 showcases the model's 3D distribution and grid layout.

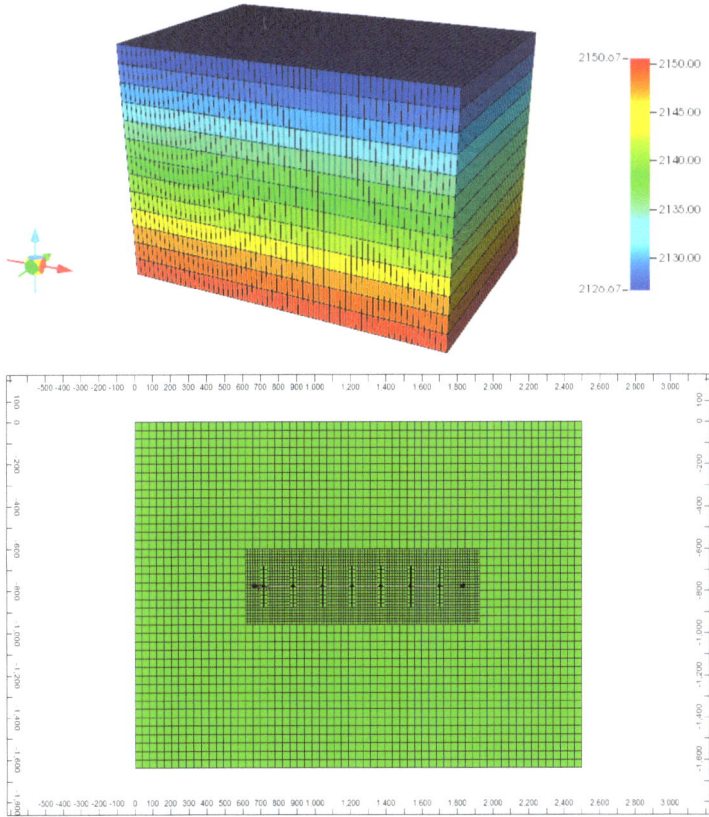

Figure 2. CMG reservoir geological model.

The original reservoir pressure is 20.9 MPa, the saturation pressure is 10.18 MPa, and the reservoir temperature is 84 °C. The porosity and permeability of the matrix are assumed to be uniformly distributed in this model. The boundary conditions, initial conditions, and specific parameters are presented in Table 1.

Table 1. Reservoir parameter settings.

Parameters	Value	Units
Reservoir depth	2126	m
Reservoir pressure	20.9	MPa
Saturation pressure	10.18	MPa
Reservoir temperature	84	°C
Rock Compressibility	1×10^{-8}	1/kPa
Permeability	0.39	mD
Porosity	0.071	-
Fracture conductivity	30	mD·m
Horizontal well length	1020	m
Fracture half-length	120	m
The maximum CO_2 injection volume	1500	t
Injection rate of CO_2	50	t/day
Minimum bottomhole flow pressure	11	MPa
Maximum surface oil rate	50	m^3/d
Soaking time	20	d

The fluid phase data were fitted using the results of the formation fluid phase simulation and the fluid phase permeation curves were taken from phase permeation data derived from laboratory long-core testing experiments, as shown in Figure 3.

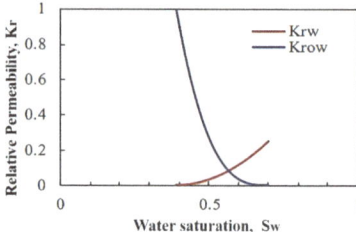 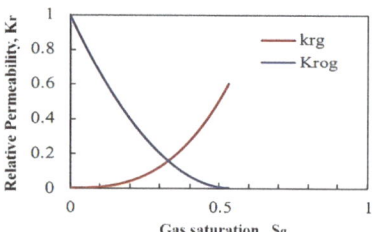

Figure 3. Oil–water relative permeability curves.

2.1.2. Obtaining Numerical Simulation Data

In the simulation process, continuous CO_2 injection into fractured horizontal wells was modeled over an 18-year period, with daily oil production rates varying between 1 m^3/d and 2 m^3/d. Following the screening criteria for CO_2 flooding as outlined by Carcoana et al. [21,40,41], this study aimed to refine ORF prediction accuracy and model applicability by considering a broader spectrum of factors and incorporating more detailed characteristic parameters.

To achieve this, the study gathered a large dataset through the definition of uncertainty variables and the application of Latin hypercube sampling, guided by previous sensitivity analyses that highlighted key factors in the EOR-CO_2 process [1,42–45]. Consequently, nine parameters were selected for detailed analysis: porosity (Por), permeability (Perm), reservoir thickness (Thickness), fracture half-length (FHL), bottom hole flowing pressure (BHP), injection rate of CO_2 (CO_2-INJ), cumulative injected CO_2 mass (CO_2-CMASS), soaking time (SOAK-T), and number of fractures (Numfrac). Based on the nine selected influential factors and using the parameter ranges provided in Table 2, Latin hypercube sampling (LHS) was applied to sample these nine parameters, resulting in 4090 data samples. And a new reservoir model was generated based on these 9 parameters. The CMOST optimization tool facilitated parallel computing to calculate the reservoir recovery rate 10 years later. It will take 16,360 min to obtain the calculation results of these 4090 models in this study. The integration of Builder and CMOST allows for the simulation of different geological implementations, as illustrated in Figure 4.

Table 2. The range of values for the model parameters in the Latin hypercube experimental design.

Parameter	Symbol	Minimum	Maximum	Base Case	Units
Porosity	Por	0.03	0.12	0.071	-
Permeability	Per	0.05	1.05	0.39	mD
Reservoir thickness	Thickness	6.5	35	26	m
Fracture half-length	FHL	60	120	100	m
Bottom hole flowing pressure	BHP	11	14	12	MPa
Injection rate of CO_2	CO_2-INJR	30	150	100	t/day
Accumulated injection mass of CO_2	CO_2-Mass	750	3500	1500	t
Soaking time	SOAK-T	5	50	20	day
Number of fractures	Numfrac	5	10	7	-

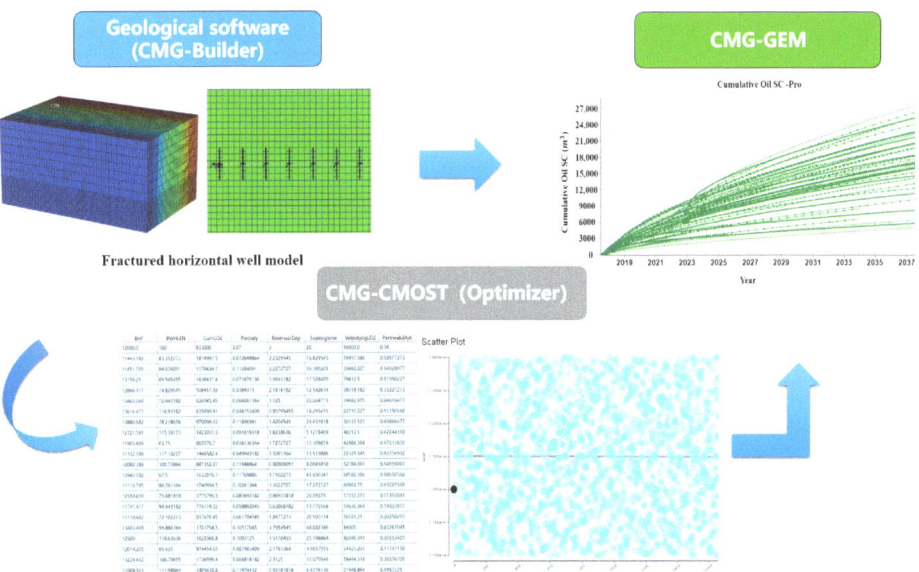

Figure 4. The integrated process Petrel and CMOST optimizer for considering geological realizations to generate the training samples.

2.1.3. Data Preprocessing

In this study, the impact of noise addition and denoising on the dataset's predictive results was investigated. Adding noise to the dataset aimed to improve the machine learning model's generalization capacity, mitigating the risk of overfitting and accommodating wider data variability, thereby aligning the simulation more closely with real-world data. To further enhance the model's performance and the precision of CO_2-EOR rate predictions, the study employed wavelet denoising techniques on the dataset with added noise, followed by a standardization process.

Obtaining Data with Noise

The dataset used in this section is derived from the reservoir numerical simulation model constructed in Section 2.1.2. It consists of a total of 4090 groups of data. Each group of models calculates the ORF for the corresponding model. The dataset includes the ORF and nine parameters mentioned in the previous section, namely Por, Per, Thickness, FHL, BHP, CO_2-MASS, CO_2-INJR, SOAK-T, and Numfrac, forming a set of data for each model.

In order to enhance the resemblance of the simulated data to the actual data collected in the field, we introduced different levels of noise to the simulated data. We added noise with the same noise ratio to all 4090 datasets, creating a noise dataset with the same noise level. Subsequently, we will assess the impact of noise corruption on the data.

The formula to add noise is represented by the following:

$$D_{noise} = D + \alpha \cdot D \cdot \varepsilon \tag{1}$$

where D is the original numerical simulation data, α is the noise level, and ε is the random number.

Three datasets were generated, each containing 4090 data points, with noise levels set at 0.05, 0.1, and 0.15, respectively. This study then examined how the predictive accuracy of machine learning models was impacted by these varying degrees of noise.

Obtaining Denoised Data

In practical applications, noise can interfere with the accurate analysis and processing of signals, leading to challenges in making precise judgments. In our previous section, we intentionally introduced random noise to analog data to simulate real-world conditions. Therefore, it becomes crucial to denoise the signal in order to enhance the quality of analysis and facilitate subsequent processing at various levels. To boost the model's accuracy and refine our dataset, we employed an efficient and widely applicable wavelet denoising technique. This method was used to clean the datasets that had noise ratios of 0.05, 0.10, and 0.15, as identified in the earlier section of our study, The principle of wavelet denoising is as follows:

Assuming there is a noisy signal of length N:

$$D_{noise}(n) = D(n) + \alpha \cdot e(n) \tag{2}$$

where $D(n)$ is the truth data and $e(n)$ is the noise.

The WT involves concentrating the energy of a noisy signal in some of the larger wavelet coefficients after wavelet decomposition. In contrast, noise energy is spread throughout the wavelet domain, leading to smaller wavelet coefficients being predominantly influenced by noise. This property allows us to consider larger wavelet coefficients as the signal and smaller ones as the noise. Wavelets, with their decorrelation feature, play a crucial role in signal processing, image processing, data analysis, and prediction [46–48].

The continuous WT of a one-dimensional continuous function $D(n)$ is given by:

$$W_r(a,b) := \int_{-\infty}^{+\infty} D(n)\overline{\psi_{a,b}(n)}dn = \frac{1}{\sqrt{|a|}}\int_{-\infty}^{+\infty} D(n)\psi\left(\frac{n-b}{a}\right)dn \tag{3}$$

where $W_r(a,b)$ is the corresponding wavelet coefficient, $\psi_{a,b}(n)$ is the wavelet function, $\psi(n)$ is the fundamental wavelet, a is the scaling factor, and b is the translation factor.

On the other hand, the wavelet inversion is given by:

$$D(n) := C_\psi^{-1}\int_{-\infty}^{+\infty}\int_{-\infty}^{+\infty} W_r(a,b)\psi_{a,b}(n)\frac{da}{a^2}db \tag{4}$$

$$C_\psi = \int_{-\infty}^{+\infty} \frac{|\widehat{\psi(\omega)}|}{|\omega|}d\omega < \infty \tag{5}$$

$\widehat{\psi(\omega)}$ is the Fourier transform of $\psi(n)$.

In the experiment, we utilized WT technology to filter the analog datasets with four different noise levels. Taking the example of cumulative injected CO_2 data with 15% noise, the comparison before and after filtering is depicted in Figure 5.

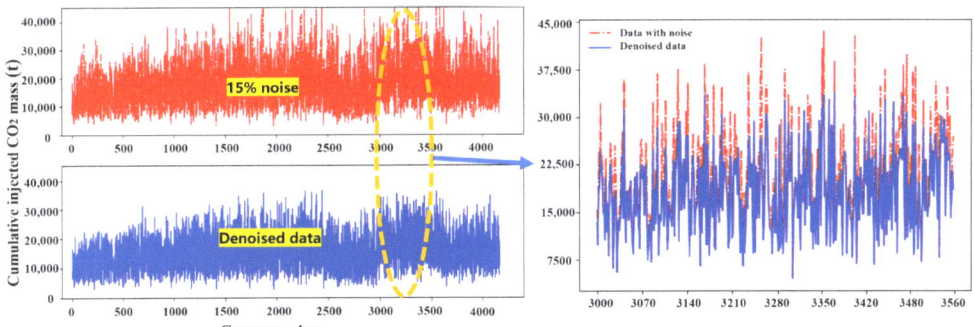

Figure 5. The comparison of data with 15% noise before and after denoising.

Data Normalization

To improve model generalization and accuracy, the original dataset from the simulation, the noisy dataset with added noise at different ratios, and the denoised dataset using WT are all normalized. This normalization removes the influence of scale and reduces data fluctuation interference, facilitating more reliable and meaningful comparisons and predictions. The normalization equation is as follows:

$$X = \frac{x - x_{\min}}{x_{\max} - x_{\min}} \qquad (6)$$

where X is the normalized data, $x_{\min}$ is the minimum value of this type of data, and $x_{\max}$ is the maximum value of this type of data.

2.2. Theory of Machine Learning Techniques

2.2.1. Random Forest

Random Forest (RF) serves as a multifunctional algorithm for both classification and regression, employing an ensemble approach to enhance prediction accuracy and stability. It constructs numerous regression trees from randomly selected subsets of the training data and predictors. Training each tree with bootstrap samples and applying binary splits on a chosen subset of predictors at every node, RF effectively selects features and grows trees. This methodology ensures the RF model's effectiveness in diverse prediction scenarios by leveraging the collective strength of multiple trees for more reliable outcomes [49].

2.2.2. XGBoost

XGBoost is an advanced boosting ensemble method applied to both regression and classification, aimed at reducing training error by assembling weak learners into a robust combined model [50–53]. It begins with training an initial model on a randomly chosen data sample and employs incremental boosting to correct previous models' errors. XGBoost's distinctiveness lies in its objective function, which blends a loss function—to minimize the gap between predicted and actual values—with a regularization term to deter overfitting, ensuring a balance between accuracy and model simplicity.

2.2.3. Light Gradient Boosting Machine (LightGBM)

The LightGBM model, a recent advancement leveraging the gradient boosting tree technique, was selected for this study for its precision and scalability [54]. Its effectiveness is largely owed to its enhanced loss function, which builds upon the Taylor objective function with a second-order extension. This method captures more detailed information about the objective function, significantly improving model performance. The following is the mathematical form of the loss function:

$$L_t = \sum_{j=1}^{J}\left[G_{tj}w_{tj} + \frac{1}{2}(H_{tj} + \lambda)w_{tj}^2\right] + \gamma J \tag{7}$$

$$G_{tj} = \sum_{x_i \in R_{tj}} g_{ti}, H_{tj} = \sum_{x_i \in R_{tj}} h_{ti} \tag{8}$$

where G_{tj} and H_{tj} represent the first and second derivatives of the objective function for each sample within a leaf-node area, respectively, w_{tj} is the optimal value assigned to the Jth leaf node of each decision tree, J refers to the total count of leaf nodes, and γ and λ are user-defined values.

The information gain employed in the segmentation of each leaf node is:

$$\text{Gain}' = \frac{1}{2}\left[\frac{G_J^2}{H_L + \lambda} + \frac{G_R^2}{H_R + \lambda} - \frac{(G_L + G_R)^2}{H_L + H_R + \lambda}\right] - \gamma \tag{9}$$

Additionally, LightGBM shifts away from XGBoost's level-wise approach to adopt a leaf-wise growth strategy with depth limitations, significantly boosting its efficiency. It selects the leaf with the highest splitting gain from all the existing leaves and performs splitting and cycling, achieving higher accuracy. However, it is important to note that this approach may occasionally result in overfitting. To mitigate this issue, the max_depth parameter can be set to control the depth of the tree and prevent excessive complexity.

Figure 6 illustrates the architecture of LightGBM. The LightGBM network model is built on the gradient-boosted decision tree (GBDT) algorithm framework and incorporates several techniques to enhance efficiency and accuracy. It utilizes Gradient-Based One-Side Sampling (GOSS) for sampling, reducing computational and time costs by focusing on relevant samples. The model also employs a histogram algorithm to find the best data segmentation points, reducing memory usage and segmentation complexity. Additionally, it uses a leaf node growth algorithm with a depth limit to improve accuracy and prevent overfitting. By leveraging these techniques, LightGBM achieves a balance between efficiency and accuracy, making it well suited for handling large datasets and delivering high-performance results.

Compared to XGBoost's presorting algorithm, LightGBM optimizes time complexity from O (Data * features) to O (Bins * features). Additionally, the histogram-based algorithm consumes approximately seven times less memory than the presorting algorithm.

The EFB algorithm plays a role in reducing feature dimensions by converting numerous mutually exclusive features into low-dimensional dense features. This effectively avoids unnecessary calculations involving redundant features with zero values.

Overall, LightGBM offers the benefits of scalability and high accuracy. With the continuous expansion of oilfield datasets, LightGBM holds potential for applications in predicting the ORF for CO_2-EOR and even in practical field operations within the petroleum industry.

2.3. Workflow

The ML models were trained using the input variables: Por, Perm, Thickness, FHL, BHP, CO_2-CMASS, CO_2-INJR, SOAK-T, and Numfrac. Figure 1 illustrates the key processes involved in the proposed methodology.

2.3.1. Dataset Partitioning

In this study, as outlined in Section 2.1, we generated three datasets: original, noise-added, and denoised. We allocated 80% of each dataset for training the models, with the balance 20% reserved for performance evaluation. To ensure robust model validation, we employed 10-fold cross-validation, dividing the training segment into ten parts—nine for training and one for validation in turn. This technique allowed for the comprehensive utilization of data for training while preserving the integrity of the test set, thus yielding a more reliable measure of the model's true accuracy.

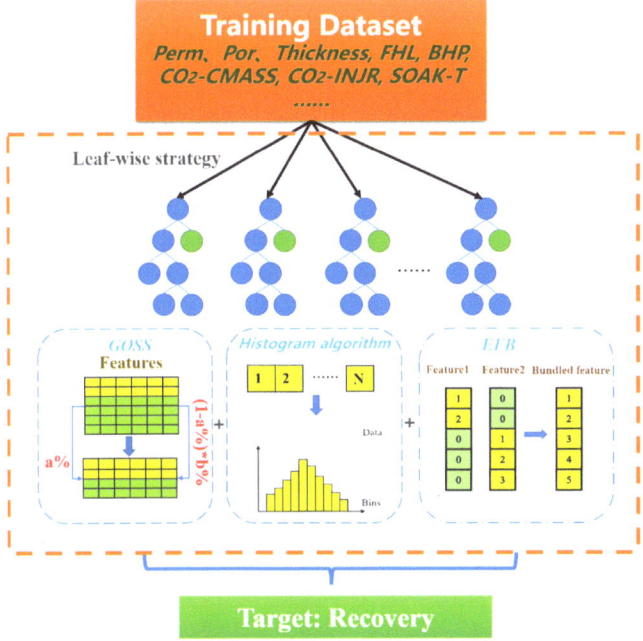

Figure 6. The architecture of LightGBM.

2.3.2. ML Model Development

The random search method (Figure 7) is employed to identify hyperparameters using RMSE as the evaluation metric, aiming to enhance the model's accuracy. Table 3 shows the search range of selected hyperparameters of the three regression models based on RF, XGboost, and LightGBM at different noise levels.

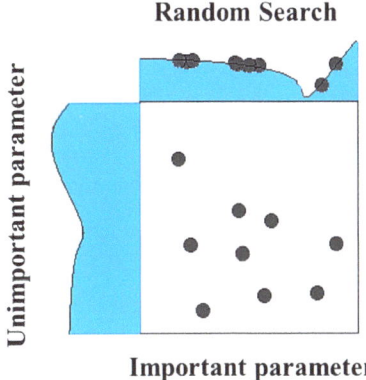

Figure 7. Schematic diagram of random search method.

Table 3. Search range of selected hyperparameters.

Model	Hyperparameter	Range
RF	n_estimators	10, 50, 100, 300, 500
	max_depth	10, 20, 40, 70, 100
	min_samples_leaf	1, 2, 4, 6, 8
	max_features	0.2, 0.4, 0.7, 0.8, 1
	learning_rate	0.001, 0.01, 0.05, 0.1, 1
	min_samples_split	1, 2, 4, 6, 8
XGboost	n_estimators	10, 50, 80, 100, 200
	max_depth	1, 2, 4, 6, 8
	num_leaves	8, 16, 32, 64, 128
	learning_rate	0.001, 0.01, 0.05, 0.1, 1
	randam_state	0, 6, 12, 20, 30
	min_child_weight	0.1, 0.2, 0.4, 0.6, 0.8
	subsample	0.5, 0.6, 0.7, 0.8, 1
	colsample_bytree	0.5, 0.6, 0.7, 0.8, 1
LightGBM	n_estimators	50, 100, 300, 500, 800
	max_depth	3, 4, 5, 6, 7
	num_leaves	8, 16, 32, 64, 128
	learning_rate	0.01, 0.05, 0.1, 0.5, 1
	max_bin	10, 30, 50, 60, 70
	bagging_fraction	0, 0.1, 0.4, 0.7, 1
	bagging_freg	10, 40, 50, 60, 80
	bagging_seed	10, 20, 40, 60, 80
	Feature_fraction	0.5, 0.6, 0.7, 0.8, 0.9

2.3.3. Model Performance Evaluation

The evaluation indicators of the ORF prediction regression model were set as follows [55]: correlation factor (R^2), root mean square error (RMSE), and mean absolute percentage error (MAE).

$$R^2 = 1 - \frac{\sum_{i=1}^{N}(y_{pre} - y_{tru})^2}{\sum_{i=1}^{N}(\overline{y_{tru}} - y_{tru})^2} \qquad (10)$$

$$RMSE = \sqrt{\frac{1}{m}\sum_{i=1}^{m}(y_{tru} - y_{pre})^2} \qquad (11)$$

$$MAE = \frac{1}{N}\sum_{i=1}^{N}|y_{tru} - y_{pre}| \qquad (12)$$

3. Results and Discussion

This section focuses on assessing the proposed RF, XGBoost, and LightGBM models' effectiveness in forecasting CO_2-EOR. We also examine how data noise and subsequent denoising actions affect the accuracy of model predictions. By analyzing data through these models, we have pinpointed critical factors that impact the CO_2 recovery in tight oil reservoirs, providing valuable insights for optimizing CO_2-EOR strategies in oilfields.

3.1. Evaluation of Model Performance

Hyperparameter tuning plays a crucial role in achieving optimal ML model performance. Consequently, for all types of ML models, the tuning process should be prioritized to guarantee the precision of the prediction model. As illustrated in Table 4, we identified optimal parameters for RF, XGBoost, and LightGBM models across different noise levels by the random search method outlined in Section 2.3.2.

Table 4. Optimal parameters for different models.

Model	Hyperparameter	Optimal Value (Original Data)	Optimal Value (5% Noise)	Optimal Value (10% Noise)	Optimal Value (15% Noise)
RF	n_estimators	100	200	200	100
	max_depth	70	70	70	20
	min_samples_leaf	2	2	2	1
	max_features	0.7	0.7	0.8	0.8
	learning_rate	0.1	0.1	0.1	0.05
	min_samples_split	4	4	2	5
XGboost	n_estimators	80	80	100	100
	max_depth	4	4	6	6
	num_leaves	16	32	32	16
	learning_rate	0.1	0.1	0.1	0.05
	randam_state	9	12	20	20
	min_child_weight	0.6	0.8	0.8	0.8
	subsample	1	1	1	0.8
	colsample_bytree	1	1	1	0.8
Lightbgm	n_estimators	300	300	500	300
	max_depth	5	5	5	5
	num_leaves	32	32	32	32
	learning_rate	0.05	0.01	0.01	0.05
	max_bin	50	50	60	60
	bagging_fraction	0.6	0.7	0.6	0.4
	bagging_freg	40	40	50	80
	bagging_seed	40	40	60	60
	Feature_fraction	0.8	0.8	0.8	0.8

Table 5 illustrates the performance metrics (R^2, RMSE, and MAE) of each ML model based on the aforementioned hyperparameters and in predicting ORF using the original dataset. Generally, a higher R^2 and lower values of MAE and RMSE indicate better predictive accuracy. In the training phase, all models showed excellent results, with R^2 values exceeding 0.99. LightGBM was the standout, achieving an R^2 of 0.996, RMSE of 0.008, and MAE of 0.009. Its dominance extended to the testing phase, where it maintained high accuracy (R^2 = 0.995, RMSE = 0.009, and MAE = 0.010).

Table 5. Prediction accuracy of training and testing sets.

Data	Indicator	RF	XGboost	LightGBM
Training	R^2	0.992	0.995	0.996
	RMSE	0.017	0.013	0.008
	MAE	0.011	0.010	0.009
Testing	R^2	0.959	0.985	0.995
	RMSE	0.031	0.023	0.009
	MAE	0.018	0.014	0.010

The data obtained from numerical simulations are typically free from noise interference but, in real-world measurements, data noise is unavoidable. Previous studies, Sun and Thanh et al. [7,52], have used numerical simulation data for machine learning models to evaluate CO_2 storage capacity and effectiveness. However, they did not consider the presence of noise in on-site data. To simulate the presence of noise in on-site data and enhance the generalization of the trained machine learning model in this study, we introduced different levels of noise using the method described in Section 2.1.3. Subsequently, we performed the denoising processing (Section 2.1.3) to investigate the impact of noisy data and denoised data on the prediction results of ORF.

3.2. Effect of Noise on the ML Model Oil Recovery Factor Predictions

After adjusting the hyperparameters of the three machine learning models for ORF prediction (as presented in Table 4), we evaluated each model's performance across diverse noise levels. Figure 8 shows that an increase in the noise ratio is associated with a discernible decline in the accuracy of the machine learning model's predictions for recovery. Figure 8a demonstrates that, at a 5% noise level, the correlation coefficient between predicted and measured ORFs from test data predominantly aligns with the fitted line (slope = 1), indicating accurate predictions by RF, XGBoost, and LightGBM ($R^2 > 0.95$). In Figure 8b, the RF model's R^2 significantly drops to 0.891 at a 10% noise level from 0.954 at 5% noise. However, XGBoost and LightGBM maintain strong accuracy ($R^2 > 0.91$). Figure 8c depicts that, at a 15% noise level, all models exhibit R^2 values below 0.87, RMSE values exceeding 0.055, and MAE values surpassing 0.043.

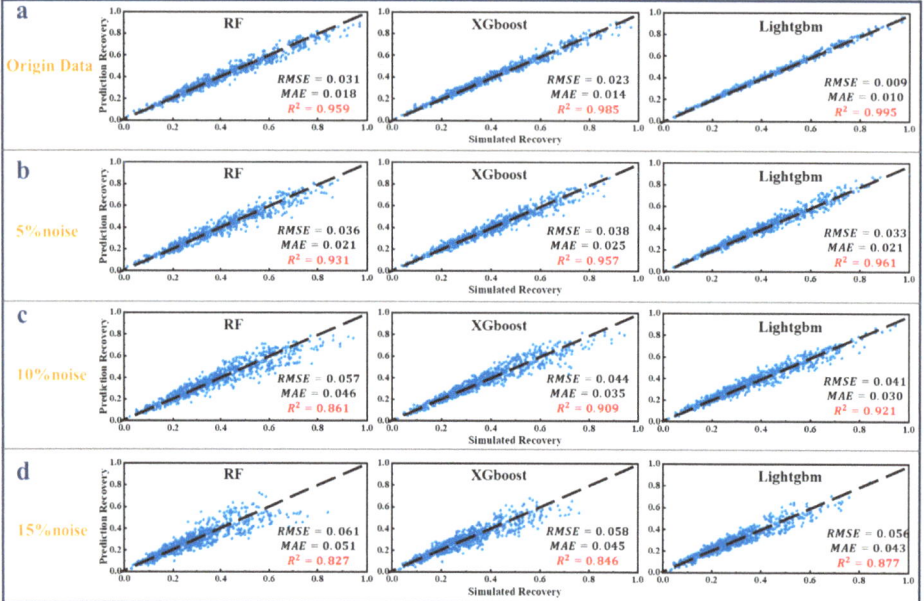

Figure 8. Cross correlation between ORF predicted by LightGBM model and ORF obtained from numerical simulation under different noise levels (**a**) prediction result of original dataset (**b**) prediction result at the 5% noise level (**c**) prediction result at the 10% noise level (**d**) prediction result at the 15% noise level.

Figure 9 presents the relationship between predicted and simulated ORFs for CO_2 flooding in tight oil reservoir and offers a comparative view of R^2, RMSE, and MAE among the three ML models. LightGBM excels in training and testing, while the RF model performs best in training with added noise but yields the poorest test results, potentially indicating overfitting in noisy scenarios.

To summarize, all three ML models exhibit commendable ORF prediction capabilities. Nevertheless, the LightGBM model stands out due to its enhanced robustness, stability, and resistance to interference. It consistently delivers superior results across various conditions. As a result, this paper conducts an in-depth analysis of the LightGBM model, aiming to assess its potential applicability in CO_2-EOR scenarios.

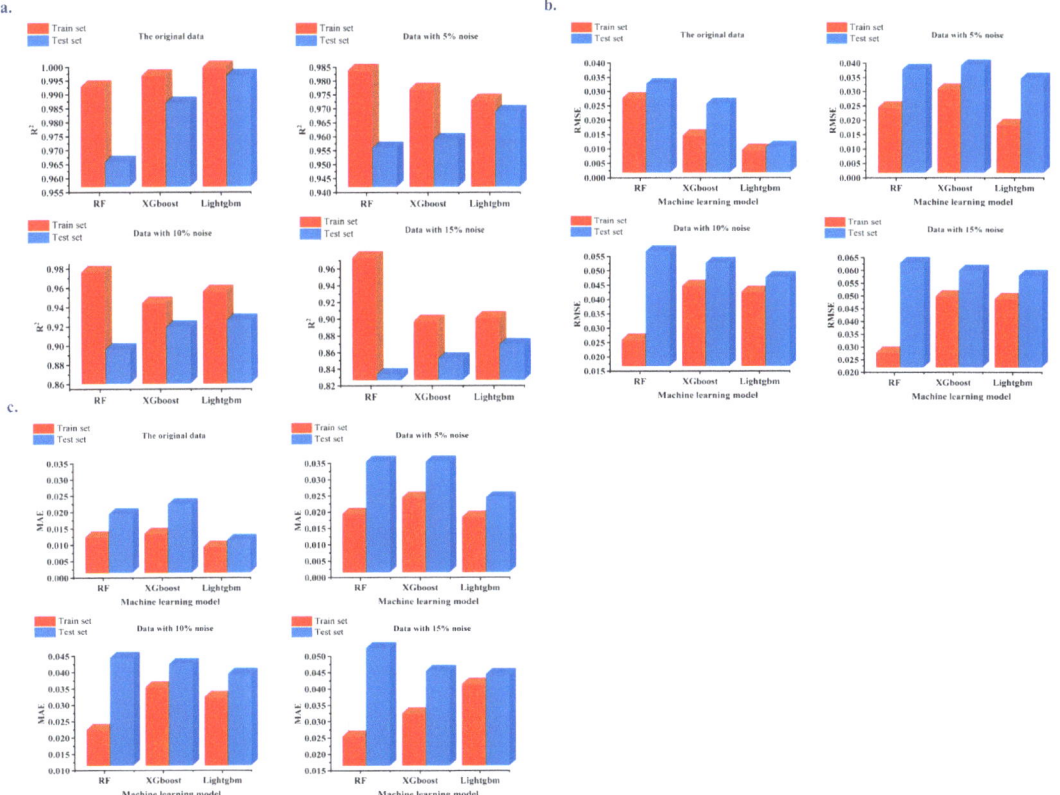

Figure 9. The statistical performance of the ML models under different noise levels: (**a**) R^2, (**b**) RMSE, and (**c**) MAE.

3.3. Model Analysis after Data Denoising

To enhance the prediction accuracy of the LightGBM model for oil recovery, we employed the WT method to denoise datasets with varying noise levels. Initially, we identified the optimal decomposition level for wavelet threshold denoising.

We opted for bdN and symN wavelet bases due to their robust orthogonality, precise positioning, and superior localization capabilities. Specifically, we randomly chose the bd6 and sym10 wavelet bases for denoising, ensuring parameter consistency. The threshold was determined using a unified global threshold, heuristic principles, and a soft threshold function. After denoising, the datasets were used to train the LightGBM model. The optimal decomposition level was assessed using RMSE, MAE, and R^2 metrics. The test set's denoising quality evaluation results are presented in Table 6.

From the denoising results using the two wavelet bases, the noisy datasets achieved the lowest RMSE, lowest MAE, and highest R^2 at a decomposition level of 1. Over-decomposition can occur with too many filtering layers, leading to a loss of signal details. Thus, the optimal decomposition level for wavelet threshold denoising of noisy data is 1. After setting this level, we used 13 wavelet basis functions from four wavelet families to decompose the noisy data. We evaluated the model training outcomes using the same metrics, and the test set's denoising quality results are presented in Table 7.

Table 6. Prediction results after denoising of the test set.

Type of Wavelet Bases	Level	5% Noise			10% Noise			15% Noise		
		RMSE	MAE	R^2	RMSE	MAE	R^2	RMSE	MAE	R^2
Bd6	J = 1	0.032	0.019	0.966	0.044	0.032	0.921	0.055	0.046	0.864
	J = 2	0.069	0.054	0.787	0.076	0.062	0.725	0.083	0.068	0.682
	J = 3	0.104	0.077	0.570	0.106	0.084	0.529	0.109	0.091	0.500
Sym10	J = 1	0.039	0.029	0.924	0.049	0.039	0.904	0.063	0.051	0.827
	J = 2	0.072	0.056	0.771	0.079	0.064	0.712	0.086	0.068	0.698
	J = 3	0.102	0.078	0.512	0.107	0.086	0.493	0.110	0.092	0.461

Table 7. Prediction results after denoising of the test set.

Type of Wavelet Bases	5% Noise			10% Noise			15% Noise		
	RMSE	MAE	R^2	RMSE	MAE	R^2	RMSE	MAE	R^2
Haar	0.054	0.043	0.891	0.059	0.045	0.865	0.077	0.064	0.783
Bd4	0.49	0.038	0.905	0.056	0.046	0.848	0.068	0.057	0.791
Bd6	0.032	0.019	0.956	0.044	0.032	0.921	0.055	0.046	0.864
Bd8	**0.027**	**0.015**	**0.969**	**0.037**	**0.027**	**0.939**	**0.045**	**0.033**	**0.912**
Bd9	0.033	0.021	0.955	0.050	0.041	0.896	0.064	0.051	0.831
Sym7	0.059	0.045	0.855	0.063	0.051	0.827	0.069	0.060	0.797
Sym8	0.044	0.033	0.915	0.058	0.046	0.853	0.065	0.052	0.836
Sym9	**0.037**	**0.028**	**0.931**	**0.045**	**0.33**	**0.911**	**0.061**	**0.050**	**0.843**
Sym10	0.039	0.029	0.924	0.049	0.039	0.904	0.063	0.051	0.827
Coif1	0.108	0.081	0.471	0.117	0.095	0.431	0.124	0.101	0.327
Coif2	0.087	0.069	0.685	0.101	0.085	0.507	0.108	0.088	0.513
Coif3	0.072	0.061	0.803	0.077	0.065	0.765	0.083	0.069	0.692
Coif4	0.069	0.058	0.793	0.073	0.064	0.727	0.079	0.068	0.688

Table 7 shows that using the Bd8 wavelet base for broadband denoising on datasets with varying noise levels yields the lowest RMSE, minimum MAE, and highest correlation coefficient. The DB8 wavelet base has been chosen for denoising the noisy dataset.

Figure 10 provides a comparative analysis of test set prediction results before and after denoising the dataset. Utilizing the Bd8 wavelet for denoising brought the predicted and simulated ORF data points in the cross-plot closer to the fit line (slope = 1), signifying an improvement in the model's predictive accuracy.

In Figure 10a, the dataset with a 5% noise level displays a slight enhancement in prediction accuracy post-denoising. The R^2 value increases by a mere 0.08, while both RMSE and MAE decrease by 0.06. In contrast, Figure 10c highlights that the dataset with 15% noise sees a notable uptick in prediction accuracy after denoising: R^2 rises by 0.35 and RMSE and MAE drop by 0.011 and 0.010, respectively. A key observation from Figure 10 is that wavelet denoising appears more beneficial for datasets with pronounced noise levels. For datasets with minimal noise, the impact of denoising is subdued. This phenomenon can be linked to the LightGBM model's inherent resilience to noise, as it retains high predictive accuracy ($R^2 > 0.96$), even when faced with an added 5% noise. However, for datasets with low noise, denoising could inadvertently strip away valuable information that might seem noisy, potentially compromising the model's predictive capability.

3.4. Screening and Evaluation of Main Control Factors

Figure 11 presents the ranking results based on the feature selection method of the LightGBM model. LightGBM ranks each feature based on both average information gain and total information gain, resulting in a comprehensive ranking of influential factors. As evident from Figure 11, permeability stands out as the most influential factor, ranking first. Porosity and reservoir thickness are also significantly affected, ranking second and third, respectively. Following these, the factors of fracture count, CO_2 mass, BHP, half-length of fracture, soak time, and carbon dioxide injection rate are less influential.

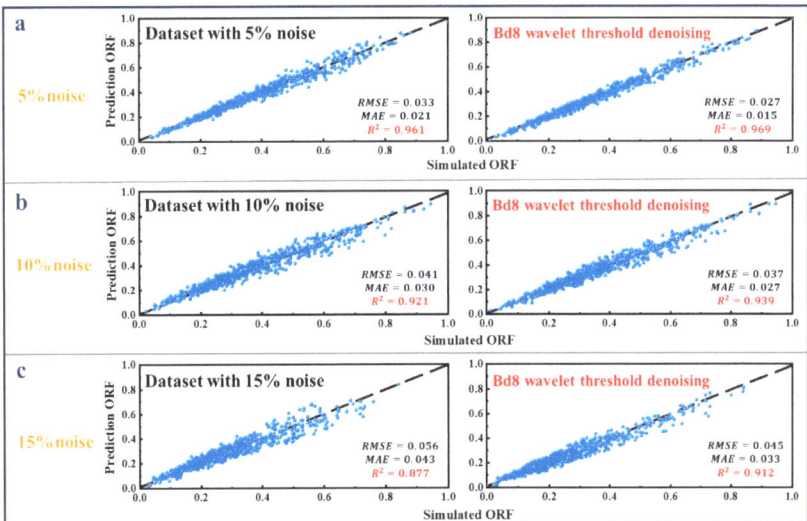

Figure 10. The statistical performance of the LightGBM model under different noise levels: (**a**) R^2, (**b**) RMSE, and (**c**) MAE.

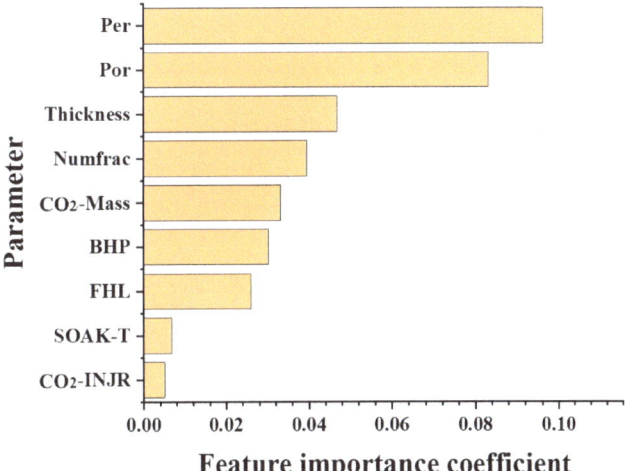

Figure 11. Ranking of feature importance.

Permeability, porosity, reservoir pressure, and permeability have long been used as screening criteria for evaluating CO_2-EOR. In this study, we incorporated CO_2 accumulation, injection rate, and soak time to investigate the impact of these factors on the CO_2 flooding efficiency. Although CO_2 has a significant diluting effect and can theoretically enhance oil recovery to a greater extent, as seen in Figure 11, the influence of CO_2 accumulation on the injection volume only ranks fifth. This suggests that the effectiveness of CO_2 flooding is significantly influenced by permeability and porosity. For low-permeability and tight reservoirs, conducting CO_2-EOR operations may require a screening of the reservoir conditions.

4. Conclusions

This article introduces a novel approach for the rapid and precise prediction of recovery in CO_2 flooding operations within tight oil reservoirs through the use of ML models. By conducting thorough data mining on the collected data, this study develops an ML model specifically tailored for assessing CO_2 flooding efficiency in such reservoirs. The key findings are summarized as follows:

(1) By considering actual blocks as examples, a numerical simulation model for CO_2 flooding in low-permeability tight oil reservoirs has been developed. Utilizing the Latin hypercube design method, a comprehensive dataset comprising 4090 numerical simulations is generated, providing a robust foundation for the ML model to analyze ORF.

(2) The study examines the impact of introducing varying levels of noise (5%, 10%, and 15%) to the simulation data on the predictive accuracy of LightGBM, XGBoost, and RF models regarding ORF. Findings reveal that the LightGBM model outperforms the others, demonstrating superior predictive capabilities for CO_2 flooding recovery efficiency in tight oil reservoirs, with R^2 values of 0.995, 0.961, 0.921, and 0.877 for the original, 5% noise, 10% noise, and 15% noise datasets, respectively.

(3) This research identifies the primary factors influencing CO_2-enhanced oil recovery, ranked as follows: permeability, porosity, reservoir thickness, number of fracturing fractures, CO_2 mass, BHP, fracture half-length, soak time, and CO_2 injection rate.

(4) The method proposed here stands as a promising alternative to conventional CO_2-ORF prediction techniques. Embracing ML for supplementary decision making offers a more adaptable and accurate framework for evaluations, reducing the risk of misjudgments associated with static indicator ranges.

Employing ML as proxies for predicting recovery presents distinct challenges. To guarantee the universality of the models, extensive and high-quality geological and production data from diverse reservoirs are essential for training. Moreover, the increased volume and complexity of data necessitate substantial investment in rapidly optimizing model parameters to boost accuracy.

Moving forward, our focus will shift to analyzing the impact of various petroleum component parameters on CO_2 flooding. It aims to refine our model's adaptability and to elevate the precision of CO_2-EOR predictions across diverse reservoir conditions. Furthermore, the model will be applied to some actual reservoirs. This expansion entails blending geological and production data from actual reservoirs with simulated datasets, then conducting preprocessing on this amalgamated dataset. Training the model with this refined data will verify its feasibility in real-world conditions.

Author Contributions: Methodology, Y.L.; Software, H.L.; Investigation, T.S.; Resources, Q.D.; Writing—original draft, M.Y.; Writing—review & editing, L.F. All authors have read and agreed to the published version of the manuscript.

Funding: This study was supported by the SINOPEC Key Laboratory of Carbon Capture, Utilization and Storage.

Data Availability Statement: Data are contained within the article.

Conflicts of Interest: Authors Ming Yue, Quanqi Dai, Haiying Liao and Yunfeng Liu were employed by the company SINOPEC. The remaining authors declare that the research was conducted in the absence of any commercial or financial relationships that could be construed as a potential conflict of interest.

Nomenclature

EOR	enhancing oil recovery
ORF	oil recovery factor
R^2	correlation factor
RMSE	root mean square error
MAE	mean absolute percentage error
ML	machine learning
RF	random forest
XGBoost	extreme gradient boosting
LightGBM	light gradient boosting machine
ANN	artificial neural network
Por	porosity
Perm	permeability
Thickness	reservoir thickness
FHL	fracture half-length
BHP	bottom hole flowing pressure
CO_2-INJ	injection rate of CO_2
CO_2-CMASS	cumulative injected CO_2 mass
SOAK-T	soaking time
Numfrac	number of fractures
D	original numerical simulation data
α	noise level
ε	random number
$D(n)$	truth data
$e(n)$	noise data
$W_r(a,b)$	corresponding wavelet coefficient
$\psi_{a,b}(n)$	wavelet function
$\psi(n)$	fundamental wavelet
a	scaling factor
b	translation factor
$\hat{\psi}(\omega)$	Fourier transform of $\psi(n)$
X	normalized data
x_{min}	minimum value of this type of data
x_{max}	maximum value of this type of data
G_{tj}	the first derivatives of the objective function for each sample within a leaf-node area
H_{tj}	the second derivatives of the objective function for each sample within a leaf-node area
J	the total count of leaf nodes
w_{tj}	the optimal value assigned to the Jth leaf node of each decision tree
γ	user-defined values
λ	user-defined values

References

1. Vo Thanh, H.; Sheini Dashtgoli, D.; Zhang, H.; Min, B. Machine-learning-based prediction of oil recovery factor for experimental CO_2-Foam chemical EOR: Implications for carbon utilization projects. *Energy* **2023**, *278*, 127860. [CrossRef]
2. Farajzadeh, R.; Eftekhari, A.A.; Dafnomilis, G.; Lake, L.W.; Bruining, J. On the sustainability of CO_2 storage through CO_2—Enhanced oil recovery. *Appl. Energy* **2020**, *261*, 114467. [CrossRef]
3. Zuloaga-Molero, P.; Yu, W.; Xu, Y.; Sepehrnoori, K.; Li, B. Simulation Study of CO_2-EOR in Tight Oil Reservoirs with Complex Fracture Geometries. *Sci. Rep.* **2016**, *6*, 33445. [CrossRef] [PubMed]
4. Chen, T.; Song, L.; Zhang, X.; Yang, Y.; Fan, H.; Pan, B. A Review of Mineral and Rock Wettability Changes Induced by Reaction: Implications for CO_2 Storage in Saline Reservoirs. *Energies* **2023**, *16*, 3484. [CrossRef]
5. Bello, A.; Ivanova, A.; Cheremisin, A. A Comprehensive Review of the Role of CO_2 Foam EOR in the Reduction of Carbon Footprint in the Petroleum Industry. *Energies* **2023**, *16*, 1167. [CrossRef]
6. Zhang, J.; Tian, L. Tight oil recovery prediction based on extreme gradient boosting algorithm and support vector regression algorithm variable weight combination model. *Sci. Technol. Eng.* **2022**, *22*, 4778–4787.

7. Sun, R.; Pu, H.; Yu, W.; Miao, J.; Zhao, J.X. Simulation-based enhanced oil recovery predictions from wettability alteration in the Middle Bakken tight reservoir with hydraulic fractures. *Fuel* **2019**, *253*, 229–237. [CrossRef]
8. Miura, K.; Wang, J. An Analytical Model to Predict Cumulative Steam Oil Ratio (CSOR) in Thermal Recovery SAGD Process. In Proceedings of the Canadian Unconventional Resources and International Petroleum Conference, Calgary, AB, Canada, 19–21 October 2010; p. 137604.
9. Teng, L.; Zhang, D.; Li, Y.; Wang, W.; Wang, L.; Hu, Q.; Ye, X.; Bian, J.; Teng, W. Multiphase mixture model to predict temperature drop in highly choked conditions in CO_2 enhanced oil recovery. *Appl. Therm. Eng.* **2016**, *108*, 670–679. [CrossRef]
10. Guo, C.; Li, H.; Tao, Y.; Lang, L.; Niu, Z. Water invasion and remaining gas distribution in carbonate gas reservoirs using core displacement and NMR. *J. Cent. South Univ.* **2020**, *27*, 531–541. [CrossRef]
11. Al-Jifri, M.; Al-Attar, H.; Boukadi, F. New proxy models for predicting oil recovery factor in waterflooded heterogeneous reservoirs. *J. Pet. Explor. Prod.* **2021**, *11*, 1443–1459. [CrossRef]
12. Fathaddin, M.T.; Thomas, M.M.; Pasarai, U. Predicting oil recovery through CO_2 flooding simulation using methods of continuous and water alternating gas. *J. Phys. Conf. Ser.* **2019**, *1402*, 55015. [CrossRef]
13. Yuan, Z.; Wang, J.; Li, S.; Ren, J.; Zhou, M. A new approach to estimating recovery factor for extra-low permeability water-flooding sandstone reservoirs. *Pet. Explor. Dev.* **2014**, *41*, 377–386. [CrossRef]
14. Yue, M.; Song, T.; Chen, Q.; Yu, M.; Wang, Y.; Wang, J.; Du, S.; Song, H. Prediction of effective stimulated reservoir volume after hydraulic fracturing utilizing deep learning. *Pet. Sci. Technol.* **2023**, *41*, 1934–1956. [CrossRef]
15. Huang, J.; Wang, H. Pore-Scale Simulation of Confined Phase Behavior with Pore Size Distribution and Its Effects on Shale Oil Production. *Energies* **2021**, *14*, 1315. [CrossRef]
16. Sun, H.; Wang, H.; Zhu, S.; Nie, H.; Liu, Y.; Li, Y.; Li, S.; Cao, W.; Chang, B. Reserve evaluation of high pressure and ultra-high-pressure reservoirs with power function material balance method. *Nat. Gas Ind. B* **2019**, *6*, 509–516. [CrossRef]
17. Cheng, M.; Lei, G.; Gao, J.; Xia, T.; Wang, H. Laboratory Experiment, Production Performance Prediction Model, and Field Application of Multi-slug Microbial Enhanced Oil Recovery. *Energy Fuels* **2014**, *28*, 6655–6665. [CrossRef]
18. Hadia, N.; Chaudhari, L.; Aggarwal, A.; Mitra, S.K.; Vinjamur, M.; Singh, R. Experimental and numerical investigation of one-dimensional waterflood in porous reservoir. *Exp. Therm. Fluid Sci.* **2007**, *32*, 355–361. [CrossRef]
19. Zhong, Q.; Shi, Y.; Liu, P.; Peng, B.; Zhuang, Y. Study on injecting time of CO_2 flooding in low permeability reservoir. *Fault-Block Oil Gas Field* **2012**, *19*, 346–349.
20. Al-qaness, M.A.A.; Ewees, A.A.; Thanh, H.V.; AlRassas, A.M.; Dahou, A.; Elaziz, M.A. Predicting CO2 trapping in deep saline aquifers using optimized long short-term memory. *Environ. Sci. Pollut. Res.* **2023**, *30*, 33780–33794. [CrossRef]
21. Esmaili, S.; Mohaghegh, S.D. Full field reservoir modeling of shale assets using advanced data-driven analytics. *Geosci. Front.* **2016**, *7*, 11–20. [CrossRef]
22. Miah, M.I.; Ahmed, S.; Zendehboudi, S. Connectionist and mutual information tools to determine water saturation and rank input log variables. *J. Pet. Sci. Eng.* **2020**, *190*, 106741. [CrossRef]
23. Yasin, Q.; Sohail, G.M.; Ding, Y.; Ismail, A.; Du, Q. Estimation of Petrophysical Parameters from Seismic Inversion by Combining Particle Swarm Optimization and Multilayer Linear Calculator. *Nat. Resour. Res.* **2020**, *29*, 3291–3317. [CrossRef]
24. Muojeke, S.; Venkatesan, R.; Khan, F. Supervised data-driven approach to early kick detection during drilling operation. *J. Pet. Sci. Eng.* **2020**, *192*, 107324. [CrossRef]
25. Hegde, C.; Pyrcz, M.; Millwater, H.; Daigle, H.; Gray, K. Fully coupled end-to-end drilling optimization model using machine learning. *J. Pet. Sci. Eng.* **2020**, *186*, 106681. [CrossRef]
26. Gurina, E.; Klyuchnikov, N.; Zaytsev, A.; Romanenkova, E.; Antipova, K.; Simon, I.; Makarov, V.; Koroteev, D. Application of machine learning to accidents detection at directional drilling. *J. Pet. Sci. Eng.* **2020**, *184*, 106519. [CrossRef]
27. Zhu, W.; Song, T.; Wang, M.; Jin, W.; Song, H.; Yue, M. Stratigraphic subdivision-based logging curves generation using neural random forests. *J. Pet. Sci. Eng.* **2022**, *219*, 111086. [CrossRef]
28. Gupta, S.; Fuehrer, F.; Jeyachandra, B.C. In Production Forecasting in Unconventional Resources using Data Mining and Time Series Analysis. In Proceedings of the SPE/CSUR Unconventional Resources Conference, Calgary, AB, Canada, 30 September–2 October 2014.
29. Lala, A.M.S.; Lala, H.M.S. Study on the improving method for gas production prediction in tight clastic reservoir. *Arab. J. Geosci.* **2017**, *10*, 70. [CrossRef]
30. Lin, B.; Guo, J.; Liu, X.; Xiang, J.; Zhong, H. Prediction of flowback ratio and production in Sichuan shale gas reservoirs and their relationships with stimulated reservoir volume. *J. Pet. Sci. Eng.* **2020**, *184*, 106529. [CrossRef]
31. Liu, W.; Yang, Y.; Qiao, C.; Liu, C.; Lian, B.; Yuan, Q. Progress of Seepage Law and Development Technologies for Shale Condensate Gas Reservoirs. *Energies* **2023**, *16*, 2446. [CrossRef]
32. Al-Mudhafar, W.J. Integrating lithofacies and well logging data into smooth generalized additive model for improved permeability estimation: Zubair formation, South Rumaila oil field. *Mar. Geophys. Res.* **2019**, *40*, 315–332. [CrossRef]
33. Al-Mudhafar, W.J. Integrating machine learning and data analytics for geostatistical characterization of clastic reservoirs. *J. Pet. Sci. Eng.* **2020**, *195*, 107837. [CrossRef]
34. Pan, B.; Song, T.; Yue, M.; Chen, S.; Zhang, L.; Edlmann, K.; Neil, C.W.; Zhu, W.; Iglauer, S. Machine learning–based shale wettability prediction: Implications for H2, CH4 and CO_2 geo-storage. *Int. J. Hydrogen Energy* **2024**, *56*, 1384–1390. [CrossRef]

35. Van, S.L.; Chon, B.H. Effective Prediction and Management of a CO_2 Flooding Process for Enhancing Oil Recovery Using Artificial Neural Networks. *J. Energy Resour. Technol.* **2017**, *140*, 032906. [CrossRef]
36. Cheraghi, Y.; Kord, S.; Mashayekhizadeh, V. Application of machine learning techniques for selecting the most suitable enhanced oil recovery method; challenges and opportunities. *J. Pet. Sci. Eng.* **2021**, *205*, 108761. [CrossRef]
37. Esene, C.; Zendehboudi, S.; Shiri, H.; Aborig, A. Deterministic tools to predict recovery performance of carbonated water injection. *J. Mol. Liq.* **2020**, *301*, 111911. [CrossRef]
38. Pan, S.; Zheng, Z.; Guo, Z.; Luo, H. An optimized XGBoost method for predicting reservoir porosity using petrophysical logs. *J. Pet. Sci. Eng.* **2022**, *208*, 109520. [CrossRef]
39. Huang, Z.; Chen, Z. Comparison of different machine learning algorithms for predicting the SAGD production performance. *J. Pet. Sci. Eng.* **2021**, *202*, 108559. [CrossRef]
40. Shen, B.; Yang, S.; Gao, X.; Li, S.; Ren, S.; Chen, H. A novel CO2-EOR potential evaluation method based on BO-LightGBM algorithms using hybrid feature mining. *Geoenergy Sci. Eng.* **2023**, *222*, 211427. [CrossRef]
41. Taber, J.J.; Martin, F.D.; Seright, R.S. EOR Screening Criteria Revisited—Part 1: Introduction to Screening Criteria and Enhanced Recovery Field Projects. *Spe Reserv. Eng.* **1997**, *12*, 189–198. [CrossRef]
42. Lee, J.H.; Park, Y.C.; Sung, W.M.; Lee, Y.S. A Simulation of a Trap Mechanism for the Sequestration of CO_2 into Gorae V Aquifer, Korea. *Energy Sources Part A Recovery Util. Environ. Eff.* **2010**, *32*, 796–808.
43. Liu, B.; Zhang, Y. CO2 Modeling in a Deep Saline Aquifer: A Predictive Uncertainty Analysis Using Design of Experiment. *Environ. Sci. Technol.* **2011**, *45*, 3504–3510. [CrossRef] [PubMed]
44. Abbaszadeh, M.; Shariatipour, S.M. Investigating the Impact of Reservoir Properties and Injection Parameters on Carbon Dioxide Dissolution in Saline Aquifers. *Fluids* **2018**, *3*, 76. [CrossRef]
45. Gao, M.; Liu, Z.; Qian, S.; Liu, W.; Li, W.; Yin, H.; Cao, J. Machine-Learning-Based Approach to Optimize CO_2-WAG Flooding in Low Permeability Oil Reservoirs. *Energies* **2023**, *16*, 6149. [CrossRef]
46. Li, S.; Wang, Z.; Kang, Y.; Hou, J. Noise reduction of a safety valve pressure relief signal based on an improved wavelet threshold function. *J. Vib. Shock.* **2021**, *40*, 143–150.
47. Li, W.; Xu, W.; Zhang, T. Improvement of Threshold Denoising Method Based on Wavelet Transform. *Comput. Simul.* **2021**, *38*, 348–351.
48. Song, T.; Zhu, W.; Chen, Z.; Jin, W.; Song, H.; Fan, L.; Yue, M. A novel well-logging data generation model integrated with random forests and adaptive domain clustering algorithms. *Geoenergy Sci. Eng.* **2023**, *231*, 212381. [CrossRef]
49. Liaw, A.; Wiener, M. Classification and regression by randomForest. *R News* **2002**, *2*, 18–22.
50. Vo Thanh, H.; Lee, K. Application of machine learning to predict CO_2 trapping performance in deep saline aquifers. *Energy* **2022**, *239*, 122457. [CrossRef]
51. Vo Thanh, H.; Yasin, Q.; Al-Mudhafar, W.J.; Lee, K. Knowledge-based machine learning techniques for accurate prediction of CO_2 storage performance in underground saline aquifers. *Appl. Energy* **2022**, *314*, 118985. [CrossRef]
52. Meng, M.; Zhong, R.; Wei, Z. Prediction of methane adsorption in shale: Classical models and machine learning based models. *Fuel* **2020**, *278*, 118358. [CrossRef]
53. Gholami, H.; Mohamadifar, A.; Collins, A.L. Spatial mapping of the provenance of storm dust: Application of data mining and ensemble modelling. *Atmos. Res.* **2020**, *233*, 104716. [CrossRef]
54. Ke, G.; Meng, Q.; Finley, T.; Wang, T.; Chen, W.; Ma, W.; Ye, Q.; Liu, T.Y. Lightgbm: A highly efficient gradient boosting decision tree. *Adv. Neural Inf. Process. Syst.* **2017**, *30*, 1–2.
55. Stazio, A.; Victores, J.G.; Estevez, D.; Balaguer, C. A Study on Machine Vision Techniques for the Inspection of Health Personnels' Protective Suits for the Treatment of Patients in Extreme Isolation. *Electronics* **2019**, *8*, 743. [CrossRef]

Disclaimer/Publisher's Note: The statements, opinions and data contained in all publications are solely those of the individual author(s) and contributor(s) and not of MDPI and/or the editor(s). MDPI and/or the editor(s) disclaim responsibility for any injury to people or property resulting from any ideas, methods, instructions or products referred to in the content.

Article

A Viscoplasticity Model for Shale Creep Behavior and Its Application on Fracture Closure and Conductivity

Shiyuan Li [1,2,*], Jingya Zhao [2], Haipeng Guo [1,3], Haigang Wang [1,3], Muzi Li [1,3], Mengjie Li [2], Jinquan Li [2] and Junwang Fu [2]

1. Hebei Cangzhou Groundwater and Land Subsidence National Observation and Research Station, Cangzhou 061000, China; guohaipeng@mail.cgs.gov.cn (H.G.); wanghaigang@mail.cgs.gov.cn (H.W.); limuzi@mail.cgs.gov.cn (M.L.)
2. School of Petroleum Engineering, China University of Petroleum-Beijing, Beijing 102249, China; zjy08310@163.com (J.Z.); limengjie0330@163.com (M.L.); liaucho@outlook.com (J.L.); fujunwang2001@163.com (J.F.)
3. China Institute of Geo-Environment Monitoring, Beijing 100081, China
* Correspondence: lishiyuan1983@cup.edu.cn

Citation: Li, S.; Zhao, J.; Guo, H.; Wang, H.; Li, M.; Li, M.; Li, J.; Fu, J. A Viscoplasticity Model for Shale Creep Behavior and Its Application on Fracture Closure and Conductivity. *Energies* **2024**, *17*, 1122. https://doi.org/10.3390/en17051122

Academic Editors: Reza Rezaee, Hai Sun, Wenchao Liu and Daobing Wang

Received: 24 January 2024
Revised: 19 February 2024
Accepted: 23 February 2024
Published: 27 February 2024

Copyright: © 2024 by the authors. Licensee MDPI, Basel, Switzerland. This article is an open access article distributed under the terms and conditions of the Creative Commons Attribution (CC BY) license (https:// creativecommons.org/licenses/by/ 4.0/).

Abstract: Hydraulic fracturing is the main means for developing low-permeability shale reservoirs. Whether to produce artificial fractures with sufficient conductivity is an important criterion for hydraulic fracturing evaluation. The presence of clay and organic matter in the shale gives the shale creep, which makes the shale reservoir deform with time and reduces the conductivity of the fracture. In the past, the influence of shale creep was ignored in the study of artificial fracture conductivity, or the viscoelastic model was used to predict the conductivity, which represents an inaccuracy compared to the actual situation. Based on the classical Perzyna viscoplastic model, the elasto-viscoplastic constitutive model was obtained by introducing isotropic hardening, and the model parameters were obtained by fitting the triaxial compression creep experimental data under different differential stresses. Then, the constitutive model was programmed in a software platform using the return mapping algorithm, and the model was verified through the numerical simulation of the triaxial creep experiment. Then, the creep calculation results of the viscoplastic constitutive model and the power law model were compared. Finally, the viscoplastic constitutive model was applied to the simulation of the long-term conductivity of the fracture to study the influence of creep on the fracture width, and sensitivity analysis of the influencing factors of the fracture width was carried out. The results show that the numerical calculation results of the viscoplastic model were in agreement with the experimental data. The decrease in fracture width caused by pore pressure dissipation and reservoir creep after 72 h accounts for 32.07% of the total fracture width decrease.

Keywords: viscoplasticity; creep behavior; fracture closure

1. Introduction

In the current context of "carbon peak" and "carbon neutrality", the demand for natural gas as a clean and low-carbon fossil fuel is gradually increasing [1]. As an important source of natural gas, shale has attracted increasing attention [2], and pores and natural micro-fractures of shale are the main reservoir space of shale gas. However, shale has extremely low porosity and permeability [3], usually ranging from millidarcy to nanodarcy, and must be fractured to create complex artificial fracture networks to form effective productivity. The key to hydraulic fracturing is determining whether a fracture with high conductivity can be formed. The presence of clay and organic matter in the shale gives the shale creep [4,5]. The proppants in the fracture are embedded in the shale reservoirs due to shale elastic and creep deformation during the production of shale gas, which causes fracture closure and long-term conductivity loss [6–8].

A large number of creep experiment results using shale show that the majority of the creep deformation is unrecoverable plastic deformation [9–11]. For example, Sone and Zoback conducted a creep experiment with unloading/reloading differential stress paths using a Haynesville shale sample, and the results showed that there is a significant plastic component in the strain response [11]. Chang and Zoback conducted a laboratory experiment on room-dried unconsolidated GOM shale under hydrostatic pressure and triaxial compressive stress, and the results showed that shale exhibits negligible creep strain accompanying unloading [9]. Therefore, it is appropriate to use the viscoplastic model to describe shale creep.

There are two commonly used viscoplastic theories. One is the Perzyna model, and the other is the Duvaut–Lions model. In 2020, Borja pointed out that the two models can obtain similar creep results when the two models' material parameters are selected appropriately [12]. Figure 1 shows the creep strain responses calculated by the Perzyna and Duvaut–Lions model when adjusting the ratio η/τ, η is the viscosity of the shale rock and τ is the relaxation time. Since Perzyna proposed the elastic–viscoplastic theoretical framework in 1963, many scholars have extended it to form their constitutive models and applied them to engineering practice. In 2010, Chang combined the Perzyna viscoplastic constitutive law with a modified Cambridge clay plastic yield model to describe the viscoplastic behavior of room-dried shale [13]. In 2012, Darabi used the classical Perzyna viscoplastic model to predict the mechanical response of asphalt concrete under cyclic loading and unloading creep test and found that with the increase in the number of cycles, the predicted value of the model significantly deviated from the experimental data, so the viscoplastic-softening model was proposed [14]. In 2020, Kabwe replaced the Newtonian element with a spring-pot in the Maxwell and VP components and obtained the fraction-order derivative viscoelastic viscoplastic (FDVP) model to estimate the delayed deformation characterized by squeezing [15]. In 2020, Haghighat developed a viscoplastic model to reproduce creep behavior and inelastic deformation by combining the Perzyna-type viscoplastic model and the modified Cam-clay model [16].

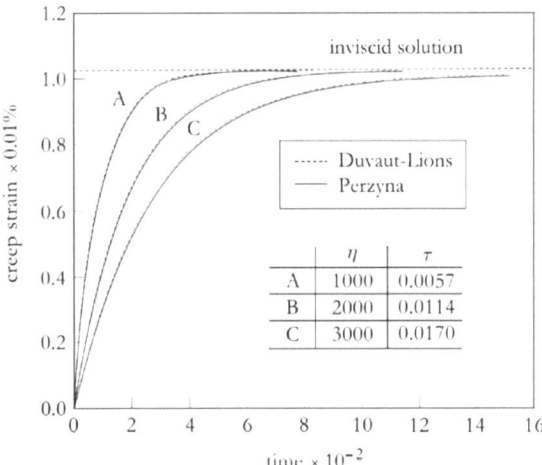

Figure 1. Comparison of axial creep strain with time—two viscoplastic models. Units: η is in MPa^3·h; τ and time are in hours [12].

However, the effect of reservoir creep is rarely considered in fracture conductivity modeling. Meanwhile, several scholars have applied viscoelastic models to account for the creep embedments [17–20]. For example, [19] introduced the modified Burgers creep model into the discontinuous embedment model for multiple-particle-size proppants. They combined it with the KC equation to develop a conductivity prediction model. The results show that with an increase in rock viscosity, the fracture conductivity decreased to the same level at approximately 300 days. Ref. [18] applied the Burgers model to investigate the role of shale creep on proppant embedment and fracture conductivity. Ref. [17] utilized the fractional Maxwell model to characterize the viscoelastic deformation of tight sandstones. Combining the fractional Maxwell model with Hertz contact theory, an analytical model of fracture width was established. Therefore, based on the classical Perzyna viscoplastic model, the elasto-viscoplastic constitutive model is obtained by introducing isotropic hardening. The constitutive model is programmed in a software platform using the return mapping algorithm, and the model is verified. Then, the elasto-viscoplastic constitutive model is applied to the simulation of the long-term conductivity of the fracture to study the influence of creep on the fracture width.

2. Constitutive Formulation

2.1. Elastic–Viscoplastic Constitutive Formulation

The bedding planes and the fractures within them are the main sources of creep anisotropy [10], and the results of many shale creep experiments show that the strain of shale in the horizontal bedding sample is larger than that in the vertical bedding sample [4,21]. Shale comprises hard materials (quartz, feldspar, and pyrite) and soft materials (clay, kerogen, etc.). The creep is mainly caused by clay and organics. Ref. [22] conducted nanoindentation tests on hard frames (quartz and pyrite), clay, and kerogen in shale. Experimental results showed that the hard frame exhibited anisotropic responses, whereas the responses of clay and kerogen were isotropic [23–25]. Therefore, we assume that the creep behavior of shale is isotropic, and that the anisotropy of shale is mainly reflected in elastic deformation. Since we focus on the creep properties of shale, the anisotropy of shale elasticity is not considered, and a linear elastic constitutive formulation is adopted.

We assumed that the total strain ε_{ij} can be decomposed into elastic strain ε_{ij}^e and viscoplastic strain ε_{ij}^{vp}

$$\varepsilon_{ij} = \varepsilon_{ij}^e + \varepsilon_{ij}^{vp} \tag{1}$$

The linear elastic constitutive equation is as follows [26]:

$$\sigma_{ij} = C_{ijkl}\varepsilon_{kl}^e \tag{2}$$

where σ_{ij} is the stress tensor, and C_{ijkl} is the fourth-order elasticity tensor. C_{ijkl} can be expressed as

$$C_{ijkl} = \lambda \delta_{ij}\delta_{kl} + 2\mu \delta_{ik}\delta_{jl} \tag{3}$$

where λ and μ are Lame constants, which can be expressed by the elastic modulus E and Poisson's ratio ν. δ_{ij}, δ_{kl}, δ_{ik}, and δ_{jl} are Kronicker symbols.

According to the Perzyna-type viscoplastic model [27–29], the viscoplastic strain rate $\dot{\varepsilon}_{ij}^{vp}$ can be expressed as

$$\dot{\varepsilon}_{ij}^{vp} = \dot{p}\frac{\partial g}{\partial \sigma_{ij}} \quad \dot{p} = \frac{\langle f \rangle}{\eta} \tag{4}$$

where $\dot{p}$ is the viscoplastic multiplier, which determines the magnitude of the viscoplastic strain rate, and the unit is s^{-1}. $\frac{\partial g}{\partial \sigma_{ij}}$ determines the direction of the viscoplastic strain rate. η is the viscosity constant, and the unit is MPa·s. f and g are the yield function and plastic potential function, respectively. The associated viscoplastic flow rule is adopted, Therefore, f equals g. $\langle f \rangle$ is defined as follows:

$$\langle f \rangle = \begin{cases} 0, f \leq 0 \\ f, f > 0 \end{cases} \tag{5}$$

The von Mises yield function considering isotropic hardening can be expressed as follows:

$$f = \sigma_e - \sigma_{y0} - hp \tag{6}$$

where $\sigma_e(\sigma_e = \sqrt{\frac{3s_{ij}s_{ij}}{2}})$ is the von Mises equivalent stress, $s_{ij}(s_{ij} = \sigma_{ij} - \frac{\sigma_{kk}}{3}\delta_{ij})$ is the stress deviator tensor, σ_{y0} is the initial yield stress, and h is the hardening material parameter. σ_{y0} determines the size of the initial yield surface.

When the von Mises equivalent stress is less than or equal to the initial yield stress ($\sigma_e \leq \sigma_{y0}$, $f \leq 0$), the viscoplastic strain rate $\dot{p}$ is 0, and accordingly, the viscoplastic strain p is also 0. When the von Mises equivalent stress is greater than the initial yield stress ($\sigma_e > \sigma_{y0}$, $f > 0$), viscoplastic strain rate $\dot{p}$ is generated, and viscoplastic strain p accumulates accordingly (the yield surface expands due to isotropic hardening). The magnitude of the viscoplastic strain rate is $\frac{f}{\eta}$, substituting (6) into

$$\dot{p} = \frac{f}{\eta} = \frac{\sigma_e - hp - \sigma_{y0}}{\eta} \tag{7}$$

By solving this differential Equation (7), the viscoplastic strain p can be written as

$$p = Ce^{-\frac{h}{\eta}t} + \frac{\sigma_e - \sigma_{y0}}{h} \tag{8}$$

When $t = 0$, the viscoplastic strain p is p_0.

$$p = C + \frac{\sigma_e - \sigma_{y0}}{h} = p_0 \tag{9}$$

It can be seen that C can be obtained from Equation (9). Therefore, C is not an independent material constant but depends on von Mises equivalent stress σ_e, the hardening material parameter h, and the initial yield stress σ_{y0}.

When $t \to \infty$, the viscoplastic strain p tends to a fixed value. Figure 2 shows a schematic of the elastic–viscoplastic model.

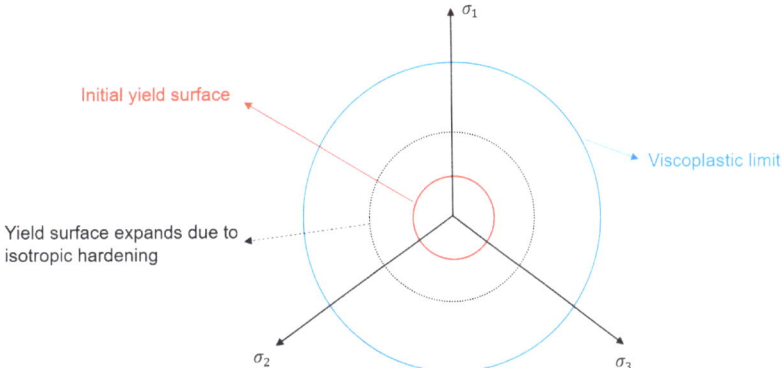

Figure 2. Schematic illustration of elastic–viscoplastic model.

2.2. Power Law Model

The power law model is introduced here only for comparison creep calculation results with the elastic–viscoplastic model.

The creep strain rate of the power law model of time hardening form can be expressed in Equation (10):

$$\dot{\varepsilon} = A\sigma_e^m t^n \tag{10}$$

where σ_e^m is the von Mises equivalent stress, and t is the time. A, m, and n are material constants. A and m must be positive and $-1 < n \leq 0$.

3. Modeling Creep in Shale

3.1. Triaxial Creep Experiment

The shale samples used in the experiment come from the Chang 7 reservoir in Ordos Basin, with a depth of about 3000–4000 m. We conducted three creep experiments on three cylindrical samples with a diameter of 25 mm and a height of 50 mm in a servo-controlled triaxial apparatus at a temperature of 110 °C. The specimen wrapped in the heat-shrink jacket was placed in the confining cell. Hydrostatic confining pressure was applied to the specimen to 5 MPa and then held constant. Next, the axial differential stress was increased to a fixed value and kept constant for 8 h. Figure 3 shows the axial strain data of three shale samples under a confining pressure of 5 MPa and differential stresses of 10 MPa, 15 MPa, and 20 MPa.

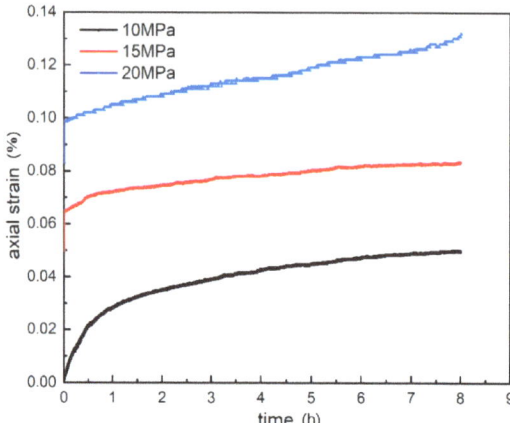

Figure 3. Axial strain data when the differential stress is 10MPa, 15 MPa, and 20 MPa.

The core samples do not enter the tertiary creep stage when differential stress is 10 MPa or 15 MPa, and only show primary and secondary creep, in which the creep strain rate decreases with time and remains constant. When differential stress is 20 MPa, the shale sample exhibits all three stages of creep. A comparison of axial strain data under different differential stresses shows that the shale sample enters the steady-state creep phase faster with increase in differential stress. For example, when the differential stress is 15 MPa, the time from primary creep to steady-state creep is about half that when the differential stress is 10 MPa. When the differential stress is 20 MPa, the time from primary creep to steady-state creep is shortened, and primary creep is hardly observed.

3.2. Parameter Identification

The elastic parameters in the elastic–viscoplastic constitutive model are calculated as follows. We take the ratio of stress to strain of the elastic stage of the axial strain data as the elastic modulus E, and Poisson's ratio ν is assumed to be 0.3. Equation (8) is used to fit the creep strain experimental data to obtain viscoplastic model parameters. Uniaxial or triaxial compression tests must be performed to determine the initial yield stress σ_{y0}, and 1 MPa is assumed here. The fitting results are shown in Figure 4.

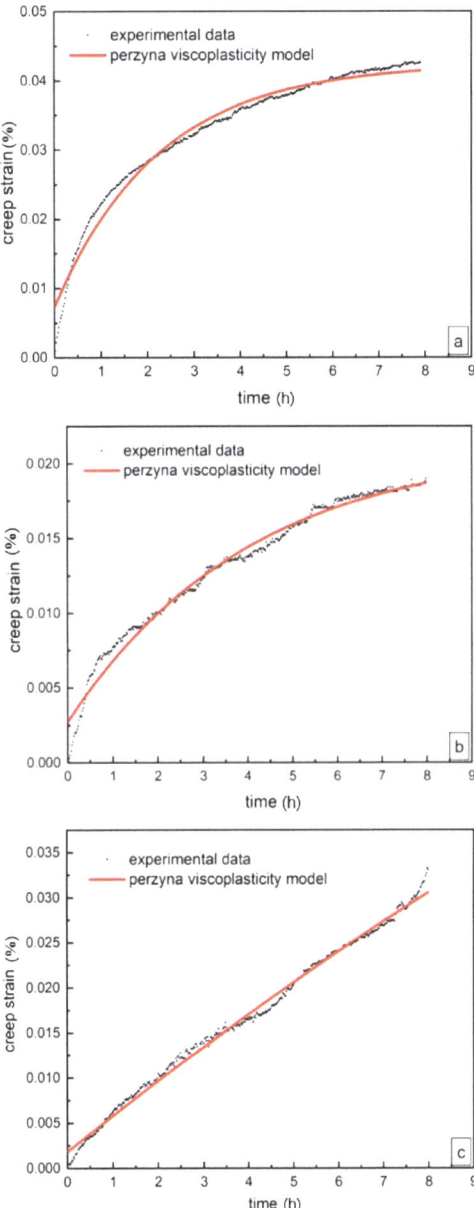

Figure 4. Viscoplastic model fitting experimental data: (**a**) differential stress is 10 MPa; (**b**) differential stress is 15 MPa; (**c**) differential stress is 20 MPa.

Table 1 lists the viscoplastic model parameters obtained by fitting experimental data.

Table 1. Material parameters obtained by fitting.

Differential Stress (MPa)	C	η (MPa·h)	h (MPa)
10	-3.51×10^{-4}	47,598.67	21,174.86
15	-1.85×10^{-4}	264,287.20	65,884.27
20	-0.00137	464,192	13,708

3.3. Algorithm Implementation

The elastic–viscoplastic constitutive model is programmed into the software platform using the return mapping algorithm [30]. The algorithm is divided into two steps: the first step is elastic prediction, and the second step is inelastic correction [26].

Elastic prediction refers to the update of stress calculated according to Equation (11). The stress obtained at this point is called the probing stress $\sigma_{ij}{}^{tr}$.

$$\sigma_{ij}{}^{tr} = \sigma_{ij}{}^{n} + \Delta\sigma = \sigma_{ij}{}^{n} + C_{ijkl} : \Delta\varepsilon \quad (11)$$

In the formula, $\Delta\sigma$ is the stress increment, C_{ijkl} represents the fourth-order elastic tensor, and $\Delta\varepsilon$ is the strain increment. We set the magnitude of the viscoplastic strain p as the state variable and keep it constant at this step, i.e., $p^{n+1} = p^{n}$, with the initial $p = 0$.

We can substitute Equation (11) into the expression of the yield function to obtain Equation (12) and check if it is greater than 0.

$$f(\sigma_{ij}{}^{tr}) = \sigma_e{}^{tr} - \sigma_{y0} - hp^{n+1} \quad (12)$$

In the formula, $\sigma_e{}^{tr}$ is the von Mises equivalent stress calculated from $\sigma_{ij}{}^{tr}$. If the yield function $f \leq 0$ is used, the material exhibits elasticity, and the stress tensor $\sigma_{ij}{}^{n+1}$ in step $n + 1$ is the probing stress.

If the yield function $f > 0$, the material has entered the viscoplastic stage; subsequently, it enters the second step of inelastic correction. The stress derivation for step $n + 1$ is as follows:

$$\sigma_{ij}{}^{n+1} = \sigma_{ij}{}^{n} + \Delta\sigma = \sigma_{ij}{}^{n} + C_{ijkl} : (\Delta\varepsilon - \Delta\varepsilon_{vp}) = \sigma_{ij}{}^{n} + C_{ijkl} : \Delta\varepsilon - C_{ijkl} : \Delta\varepsilon_{vp} \quad (13)$$

In the formula, $\Delta\varepsilon_{vp}$ is the increment in viscoplastic strain. Substituting Equation (11) into Equation (13) yields

$$\sigma_{ij}{}^{n+1} = \sigma_{ij}{}^{tr} - C_{ijkl} : \Delta\varepsilon_{vp} \quad (14)$$

Substituting the expression $\Delta\varepsilon_{vp}$ into Equation (14) yields

$$\sigma_{ij}{}^{n+1} = \sigma_{ij}{}^{tr} - C_{ijkl} : \dot{p}\frac{\partial f}{\partial \sigma_{ij}}\Delta t \quad (15)$$

In the formula, Δt is the time increment. Substituting $\frac{\partial f}{\partial \sigma_{ij}} = \frac{3s_{ij}{}^{tr}}{2\sigma_e{}^{tr}}$ into Equation (15) yields

$$\sigma_{ij}{}^{n+1} = \sigma_{ij}{}^{tr} - C_{ijkl} : \dot{p}\frac{3s_{ij}{}^{tr}}{2\sigma_e{}^{tr}}\Delta t \quad (16)$$

Considering $C_{ijkl} : s_{ij}{}^{tr} = 2Gs_{ij}{}^{tr} + \lambda s_{ii}\delta_{ij} = 2Gs_{ij}{}^{tr}$, substituting it into Equation (16) yields

$$\sigma_{ij}{}^{n+1} = \sigma_{ij}{}^{tr} - 2G\dot{p}\frac{3s_{ij}{}^{tr}}{2\sigma_e{}^{tr}}\Delta t \quad (17)$$

According to Equation (17), the stress $\sigma_{ij}{}^{n+1}$ at step $n + 1$ depends on the value of $\dot{p}$. The following solution process of $\dot{p}$ is provided.

We can satisfy the equation $\dot{p} = \frac{f}{\eta}$; for ease of calculation, we can rewrite it as $\Delta p = \dot{p}\Delta t = \frac{f}{\eta}\Delta t$ and solve it through Newton's iteration. The specific derivation process is as follows:

$$\Delta p - \frac{f}{\eta}\Delta t + (1 - \frac{f'_{\Delta p}}{\eta}\Delta t)d\Delta p = 0 \tag{18}$$

In the formula, $f'_{\Delta p}$ is the derivative of the yield function f to Δp. Substituting the expression of the yield function into Equation (18) yields the following:

$$\Delta p - \frac{\sigma_e^{n+1} - \sigma_{y0} - hp^{n+1}}{\eta}\Delta t + (1 - \frac{\partial(\sigma_e^{n+1} - \sigma_{y0} - hp^{n+1})}{\partial \Delta p}\frac{\Delta t}{\eta})d\Delta p = 0 \tag{19}$$

Considering that $\sigma_e^{n+1} = \sigma_e^{tr} - 3G\Delta p$, substituting it into Equation (19) yields the following:

$$\Delta p - \frac{\sigma_e^{n+1} - \sigma_{y0} - hp^{n+1}}{\eta}\Delta t + (1 - \frac{\partial(\sigma_e^{tr} - 3G\Delta p - \sigma_{y0} - hp^{n+1})}{\partial \Delta p}\frac{\Delta t}{\eta})d\Delta p = 0 \tag{20}$$

Solution:

$$(\Delta p)^{n+1} = (\Delta p)^n + \frac{\frac{f}{\eta}\Delta t - (\Delta p)^n}{1 + (3G + h)\frac{\Delta t}{\eta}} \tag{21}$$

The increment in viscoplastic strain is calculated based on Δp, i.e.,

$$\Delta \varepsilon_{vp} = \Delta p \frac{\partial f}{\partial \sigma_{ij}} \tag{22}$$

The increment in elastic strain is

$$\Delta \varepsilon_e = \Delta \varepsilon - \Delta \varepsilon_{vp} \tag{23}$$

The stress at step $n + 1$ is

$$\sigma_{ij}^{n+1} = \sigma_{ij}^n + \Delta \sigma = \sigma_{ij}^n + C_{ijkl} : \Delta \varepsilon_e \tag{24}$$

The initial elastic stiffness matrix is still used here. The flowchart of the entire algorithm is shown in Figure 5.

3.4. Elastic–Viscoplastic Model Validation

In this study, we simulate numerical creep experiments under triaxial loading to verify the correctness of the viscoplastic constitutive relationship. Using the axisymmetric model shown in Figure 6, the leftmost dashed line in the figure represents the axis of symmetry. In order to compare with the experimental data, the geometric dimensions, loads, and boundary conditions were set the same as those of the shale sample creep experiment. The elastic parameters are taken as Young's modulus E = 25 GPa and Poisson's ratio ν = 0.3. The parameters of the viscoplastic material are determined by fitting the viscoplastic constitutive model to the first set of experimental data and the third set of experimental data.

In order to improve the accuracy of the calculation, the unit type used is CAX4. We use structured grid partitioning technology to divide the model into 1000 units. We set two static analysis steps. In the first analysis step, confining pressure and axial pressure are applied, and the load increases linearly with time, with a loading time of 90 s. In the second analysis step, the load remains unchanged, with a total time of 7.995 h. In order to obtain the same number of data points as the experiment, a fixed step size of 10 s is used.

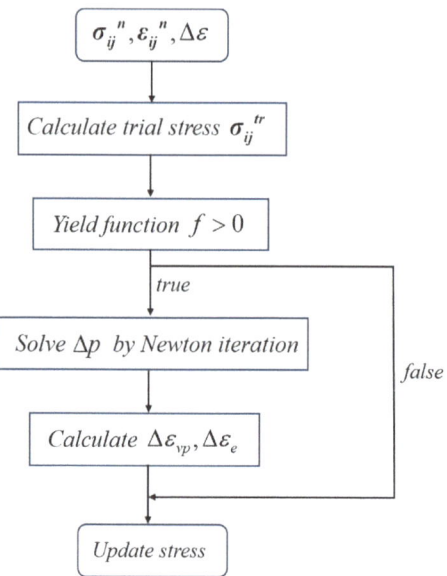

Figure 5. Return mapping algorithm for elasto-viscoplastic constitutive model.

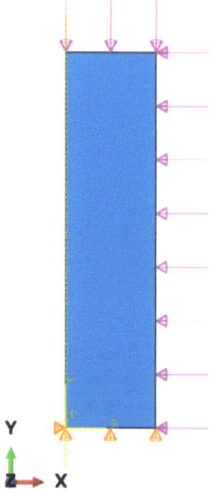

Figure 6. Numerical creep experimental model under triaxial compression (The arrow represents pressure loading and the triangle represents support).

Figure 7 shows a comparison between simulation results and experimental data.

As shown in Figure 7a, during the decay creep stage, there is a gap between the numerical calculation results of the viscoplastic model and the experimental data. After entering the steady-state creep stage (about 3 h), the simulation results agree with the experimental data. As shown in Figure 7b, the numerical calculation results of the viscoplastic model are in good agreement with the experimental data.

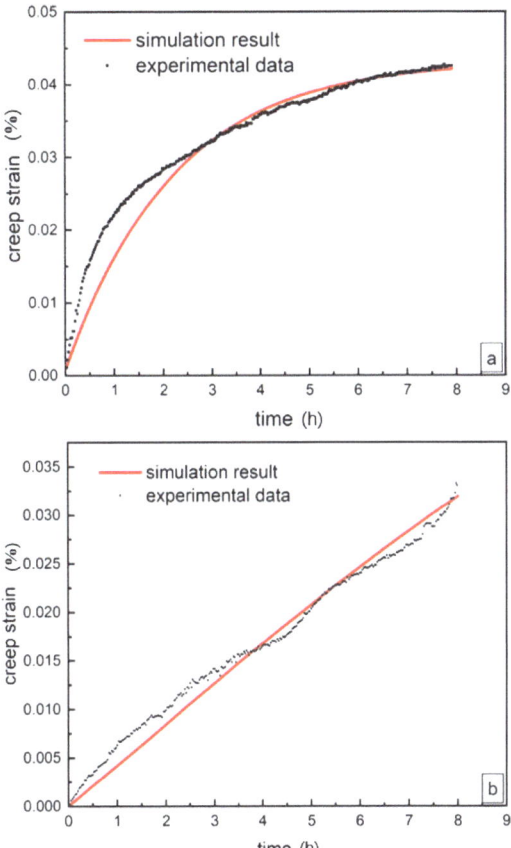

Figure 7. Comparison between simulation results and experimental data. (**a**) differential stress is 10 MPa; (**b**) differential stress is 20 MPa.

There may be two possible reasons for the discrepancy between the numerical calculation results of the viscoplastic model during the attenuation creep stage in Figure 7a and the experimental data. The first is the influence of shale anisotropy: compared to sandstone, shale exhibits significant anisotropy due to bedding planes. The viscoplastic model itself is an ideal model that does not consider the presence of internal pores, fractures, and defects in the rock core. However, there are many natural original fractures within the actual shale core itself. These original fractures may close in the early stage of creep, and over time, they may expand and even cause creep fractures (Wang, 2012) [31]. Therefore, the rate of decay creep is relatively high in the early creep stage.

Figure 8 compares the simulation results of the viscoplastic model and the power law model inherent in the software platform and the third set of experimental data. As shown in the figure, both the viscoplastic model and the power law model are in good agreement with the creep experimental results.

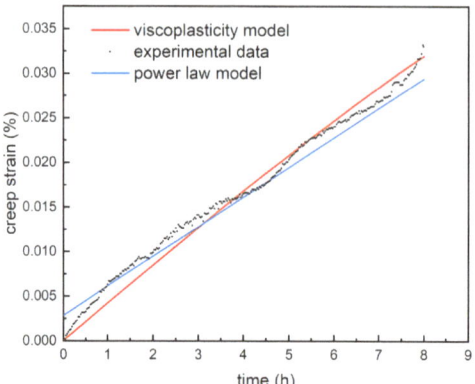

Figure 8. Comparison of simulation results of viscoplastic model and power law model with experimental data.

3.5. Parameter Analysis of Elastic–Viscoplastic Model

This section conducts sensitivity analysis on the viscosity parameters, initial yield stress, and strengthening material parameters in the elasto-viscoplastic constitutive model to understand their impact on creep strain.

Figure 9 shows the variation curve of creep strain over time under different viscosity parameters. Figure 10 shows the time-dependent curves of creep strain at initial yield stresses of 0.3 MPa, 0.8 MPa, and 1.3 MPa. Figure 11 shows the time-dependent creep strain curves of reinforced materials with parameters of 11,210.76 MPa, 21,210.76 MPa, and 31,210.76 MPa.

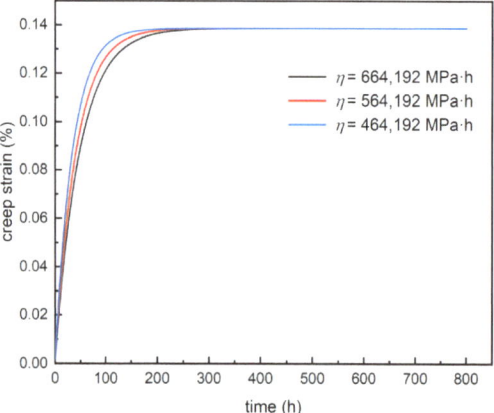

Figure 9. Influence of different viscous parameters on creep curve.

In Figure 9, under certain other conditions, as the viscosity parameter increases, the creep strain rate decreases, and the time to reach the "limit" of creep strain extends. That is, the viscosity parameter only affects the time to reach the maximum creep strain and does not affect the final creep strain. When the viscosity parameter is set, the creep strain limit is closer to the initial yield surface—that is, the viscoplastic solution is closer to the "pure" plastic solution [29]. As shown in Figure 10, under certain other conditions, the larger the initial yield stress, the smaller the creep strain rate and creep strain. For every 0.5 MPa increase in initial stress, the creep strain decreases by approximately 0.23×10^{-4}. As shown in Figure 11, under certain other conditions, as the strengthening material parameters increase, the creep strain decreases. Furthermore, there is a non-linear relationship between creep strain and strengthening material parameters.

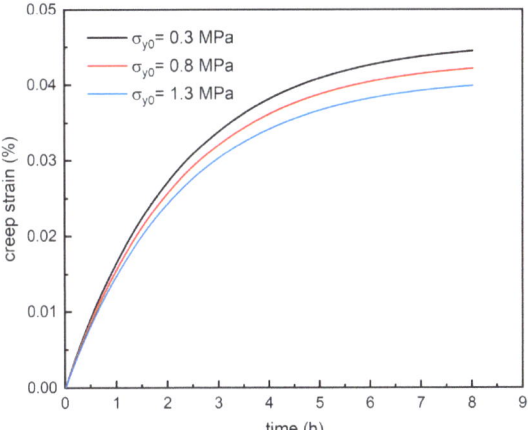

Figure 10. Effect of initial yield stress on creep strain.

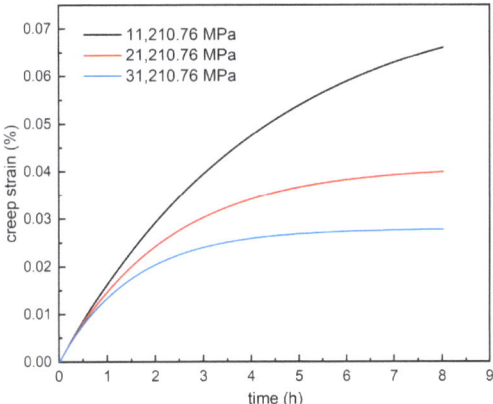

Figure 11. Creep strain versus time for different strengthening material parameters.

4. Fracture Conductivity Simulation

We use finite element analysis software to apply the elasto-viscoplastic constitutive model to the analysis of fracture conductivity, study the influence of creep on fracture width, and analyze the factors influencing fracture width.

4.1. Fracture Conductivity Model Setup

We can simplify the plane strain problem of the rock fracture system and adopt the following assumptions in modeling:

1. We do not consider the poro-elastic characteristics of shale;
2. The permeability of the fracture remains constant; it does not take into account the changes in fracture permeability caused by the decrease in pore pressure, the increase in effective stress on the proppant, and the detachment of the reservoir caused by the embedding of the proppant, leading to the migration of debris and proppant;
3. There is an intermediate layer in the middle of the fracture, in addition to the relatively large size of the proppant, which is believed to be composed of relatively small

proppant particles, pores, and rock debris, and crystallization and clay-like minerals formed by pressure solution diagenesis in the proppant [20];
4. We can ignore the influence of proppant gravity and fracture surface roughness.

Figure 12 shows the geometric model of the rock fracture system. Shale reservoirs are located above and below, with proppant-filled layers between the reservoirs. The circular particles in the middle are relatively larger proppant particles (double-layer proppant is used here). The intermediate layer is in contact with the relatively larger proppants of the upper and lower reservoirs (the following proppants refer to larger proppants). The reservoir is 1.5 cm thick, with a fracture length of 5 cm, a width of 2 cm, and a proppant radius of 5 mm. Except for the CPE4P element, which is a seepage displacement coupling element with pore pressure degrees of freedom, all other components use the CPE4 element.

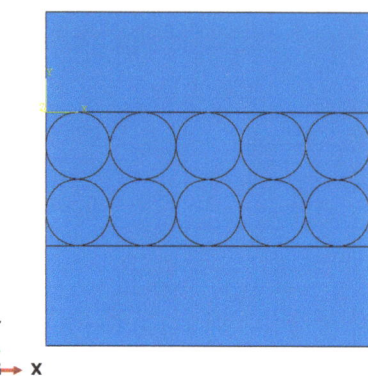

Figure 12. Geometry model of rock fracture system.

The material parameters are shown in Table 2.

Table 2. Material parameters.

Model Material Parameter	Specific Value
Elastic modulus of proppant (GPa)	30
Proppant Poisson's ratio	0.3
Elastic modulus of shale reservoirs (GPa)	20
Poisson's ratio of shale	0.25
Initial yield strength (MPa)	1
Strengthening material parameters (MPa)	14,276.18
Viscosity parameter (MPa·h)	461,543.53
Elastic modulus of intermediate layer (MPa)	100
Intermediate Poisson's ratio	0.3
Liquid gravity (kN/m^3)	10
Permeability coefficient (m/s)	10^{-10}
Fluid bulk modulus (GPa)	2

Figure 13 shows the setting of boundary conditions for the model. We can fix the horizontal and vertical displacement of the bottom surface. The horizontal displacement of all components is limited, and only a vertical pressure of 85 MPa is applied on the top surface. Considering that the pore fluid in the middle layer flows toward the wellbore under closed pressure, we set the right boundary of the middle layer to have a pore pressure of 0. The initial porosity ratio of the intermediate layer is set to 0.67. The decrease in pore fluid pressure causes an increase in the effective stress on the proppant, and the compaction process of the proppant is achieved through the contact between the larger proppant and the upper and lower reservoirs, as well as between the intermediate layer and the upper and lower reservoirs.

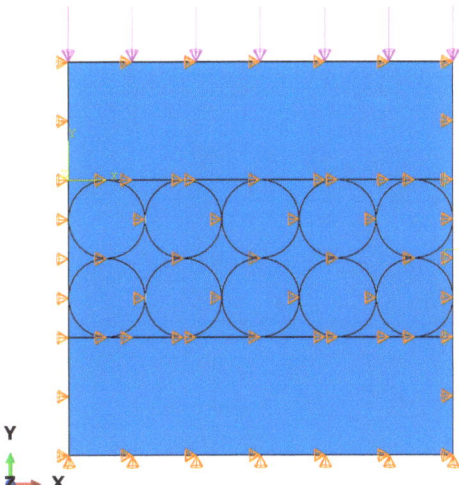

Figure 13. Load and boundary condition settings (The arrow represents pressure loading and the triangle represents support).

The specific settings for contact are as follows. Each proppant is set as the contact surface, the upper and lower surfaces of the reservoir are set as the contact surface, the upper and lower surfaces of the intermediate layer are set as the contact surface, and the entire intermediate layer is set as the contact surface. A contact pair is set between each proppant in the first layer and the lower surface of the reservoir. Usually, the surface with high stiffness is chosen as the main surface, so the proppant is the main surface, and the lower surface of the reservoir is the secondary surface. A contact pair is set between each proppant in the second layer and the upper surface of the reservoir, with the proppant as the main surface and the upper surface of the reservoir as the secondary surface. A contact pair is set between the upper surface of the intermediate layer and the lower surface of the reservoir, with the lower surface of the reservoir being the main surface and the upper surface of the intermediate layer being the secondary surface. A contact pair is set between the lower surface of the intermediate layer and the upper surface of the reservoir, with the upper surface of the reservoir being the main surface and the lower surface of the intermediate layer being the secondary surface. The contact properties are set to normal hard contact, and the penalty friction formula with a friction coefficient of 0.5 is used for tangential contact. We can set binding constraints between the entire intermediate layer and each proppant, with the proppant as the main surface and the intermediate layer as the secondary surface.

4.2. Results

Figures 14 and 15, respectively, show the overall stress distribution cloud maps of the model at 0.3 h and 72 h, as well as the stress distribution cloud maps of the support agent. Figure 16 shows the overall vertical displacement cloud maps of the model at 0.3 h and 72 h.

Based on Figures 14 and 15, it can be seen that the von Mises stress in the contact area between the upper and lower layers of the proppant is the highest, with a maximum von Mises stress of 1151.19 MPa. The von Mises stress in the contact area between the proppant and the upper and lower reservoir rocks is secondary, with a maximum von Mises stress of 854.54 MPa. This is mainly because the elastic moduli of the upper and lower layers of proppants are equal, and they are less prone to deformation when in contact. The elastic modulus of the proppant is slightly higher than that of shale, and the contact between the two is relatively prone to deformation. Moreover, as the closing pressure increases and time passes, the proppant is more likely to be embedded into shale reservoirs.

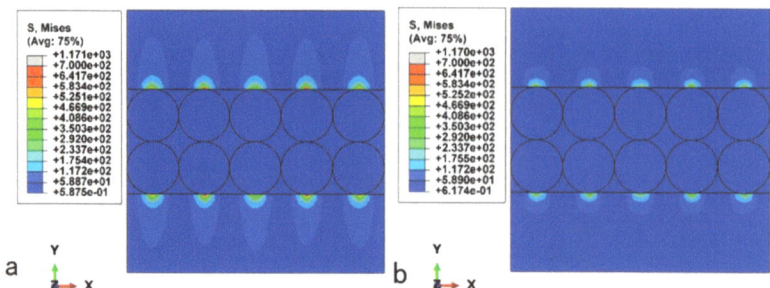

Figure 14. von Mises stress of conductivity model (**a**) at 0.3 h; (**b**) at 72 h. (The scientific notation e+03 means × 10^3).

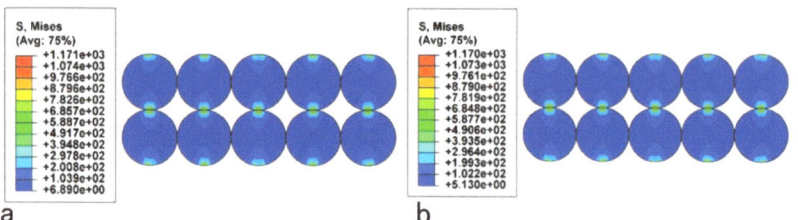

Figure 15. von Mises stress of proppants (**a**) at 0.3 h; (**b**) at 72 h. (The scientific notation e+03 means × 10^3).

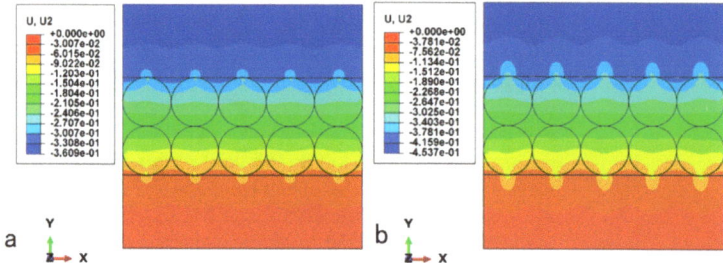

Figure 16. Vertical displacement of conductivity model (**a**) at 0.3 h; (**b**) at 72 h. (The scientific notation e+03 means × 10^3).

Comparing the overall von Mises stress cloud maps of the 0.3 h and 72 h models with the von Mises stress cloud maps of the proppant, it is easy to find that the von Mises stress at 0.3 h is lower than that at 72 h, especially in the contact area between the proppant and the upper and lower shale reservoir rocks, indicating stress relaxation at the fracture surface, which is caused by the dissipation of pore pressure in the intermediate layer.

As shown in Figure 16, the vertical displacement decreases sequentially from the top to the bottom. When the calculation is terminated, the proppant is slightly embedded in the formation, and the amount of proppant embedded in the middle is about 0.0488 mm. The variation in fracture width (referring to the minimum fracture width here) is 0.352 mm. The embedding amount accounts for 27.7% of the half fracture width variation, and the seam width variation rate, which is the ratio of fracture width variation to the original fracture width, is 1.76%.

We defined a path along the exit direction of the fracture surface. Figure 17 shows the distribution pattern of pore pressure in the middle layer along the outlet direction of the fracture surface at different times. Obviously, the pore pressure gradually decreases along the direction of the fracture surface outlet, reflecting the process of pore pressure dissipation. Figure 18 shows the distribution pattern of flow velocity along the outlet direction of the fracture surface at the end of the calculation. It is easy to see that the flow velocity gradually increases along the direction of the fracture surface outlet, and the flow velocity at the insertion point of the support agent significantly decreases, which indirectly reflects the decrease in permeability in the embedding area and the decrease in diversion capacity.

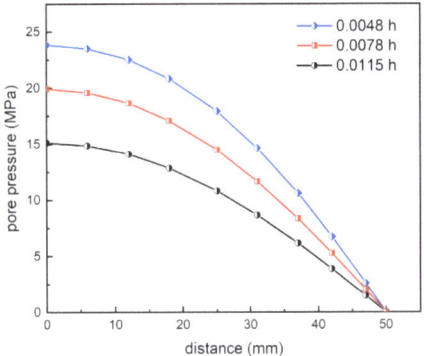

Figure 17. Pore pressure versus distance.

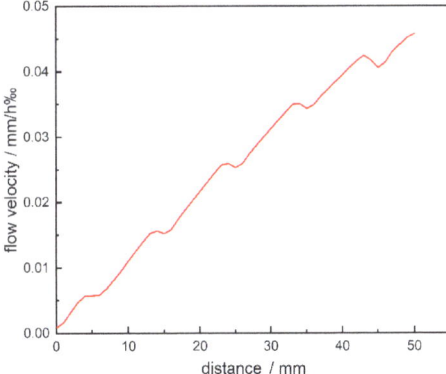

Figure 18. Flow velocity versus distance.

Figure 19 shows the variation in fracture width over time. As shown in the figure, the instantaneous change in fracture width caused by closed stress is about 0.239 mm, accounting for 67.93% of the total fracture width change. Subsequently, the change in fracture width caused by pore pressure dissipation and rock creep accounts for about 32.07%. It can be seen that the changes in fracture width caused by the decrease in pore pressure of the proppant filling layer and the changes in fracture width caused by rock creep cannot be ignored.

4.3. Analysis of Factors Influencing Fracture Width

This section investigates the effects of physical quantities such as the closure pressure, elastic modulus and Poisson's ratio of proppants, the elastic modulus and Poisson's ratio of rocks, viscosity parameters, the strengthening material parameters, and the initial yield stress on the fracture width.

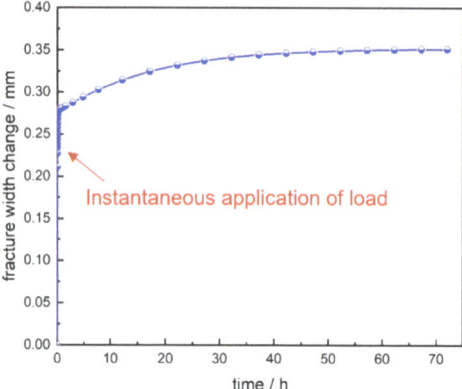

Figure 19. Change in fracture width versus time.

Figure 20 shows the variation in fracture width over time at closure pressures of 45 MPa, 65 MPa, 85 MPa, and 105 MPa. As shown in Figure 20, under certain other conditions, as the closure stress increases, the fracture width gradually decreases. For every 20 MPa increase in closure stress, the fracture width decreases by approximately 0.07 mm. As the closure pressure increases, the time for the fracture width to reach stability is extended. The variation in fracture width can be divided into three stages. At the moment of loading, due to the lack of time for pore pressure to dissipate, the reduction in fracture width is mainly caused by the elastic deformation of the reservoir and proppant. Subsequently, as the pore pressure dissipates, the effective stress on the proppant increases, causing further deformation of the proppant and reservoir, resulting in a decrease in fracture width. After the dissipation of pore pressure, due to the creep properties of the reservoir rock, the reservoir continues to deform, and the fracture width further decreases. As the viscoplastic model of the reservoir is an "upper limit" model, the fracture width will gradually approach a fixed value over time. When the closure stress is increased, it impacts all three stages.

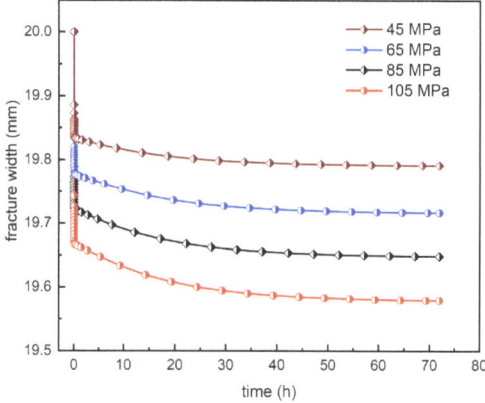

Figure 20. Fracture width versus time for different closing pressures.

Figure 21 shows the variation in fracture width with time when the elastic moduli of the proppant are 30 GPa, 40 GPa, 50 GPa, and 60 GPa. When other conditions are constant, the larger the elastic modulus of the proppant, the wider the fracture width. This is because the larger the elastic modulus of the proppant, the more effectively it can play its role; that is, the proppant is less likely to deform and embed, and fractures are less likely to close.

When the elastic modulus of the proppant increases from 30 GPa to 40 GPa, the decrease in fracture width is greater than when the elastic modulus of the proppant increases from 40 GPa to 50 GPa. This indicates that when the elastic modulus of the proppant increases to a value of around 40 GPa, it can better support fractures. If the elastic modulus of the proppant continues to increase, it cannot significantly reduce the fracture width. When comparing the time-varying curves of the fracture width with different elastic moduli of proppants, the curves are parallel to each other, indicating that the elastic modulus of proppants mainly affects the instantaneous application of closing stress and the early stage of pore pressure dissipation. For the time period after small changes in pore pressure, changes in the elastic modulus of proppants have almost no effect on the fracture width. This is because the change in elastic deformation caused by the change in elastic modulus is not related to the length of time but only to stress.

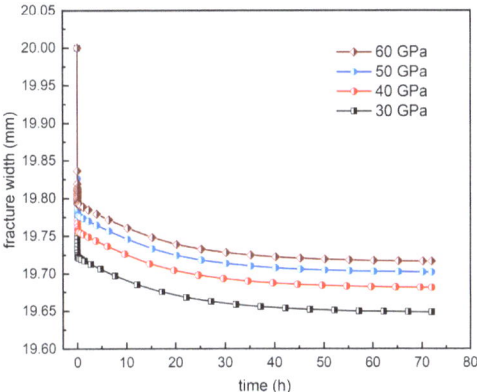

Figure 21. Fracture width versus time for different elastic moduli of proppant.

Figure 22 shows the variation in fracture width with time when the Poisson's ratio of the proppant is taken as 0.2, 0.3, 0.4, and 0.45. As shown in the figure, when other conditions are constant, the larger the Poisson's ratio, the wider the gap. Due to the limitation of all lateral displacements in the diversion capacity evaluation model, deformation can only occur in the vertical direction. The larger the Poisson's ratio of the proppant, the stronger its ability to resist vertical deformation, and the less likely it is to reduce the fracture width. Similarly to Figure 21, the time-dependent curves of fracture width under the influence of different proppant Poisson's ratios are parallel to each other, indicating that the proppant Poisson's ratio mainly affects the instantaneous application of closure stress and the early stage of pore pressure dissipation. For the later stage of pore pressure dissipation, the proppant Poisson's ratio has almost no effect on fracture width. This is because the change in elastic deformation caused by Poisson's ratio is not related to the length of time but only to the change in stress.

Figure 23 shows the variation in fracture width over time for shale reservoirs with elastic moduli of 10 GPa, 15 GPa, 20 GPa, and 30 GPa. As shown in the figure, when other conditions are constant, the larger the elastic modulus of shale reservoirs, the wider the fracture width. The elastic modulus reflects the ability of a reservoir to resist elastic deformation. The larger the elastic modulus, the stronger the resistance to deformation, and the less likely the fracture width is to decrease. The reduction in fracture width caused by the change in elastic modulus from 10 GPa to 20 GPa is 2.61 times that caused by the change in elastic modulus from 20 GPa to 30 GPa. This is because the elastic modulus of the proppant is 30 GPa. As the elastic modulus of the reservoir approaches 30 GPa, the two are evenly matched, making the proppant less likely to deform and the fracture width less likely to decrease. When the elastic modulus of the reservoir and the elastic modulus of the proppant differ significantly, the proppant is prone to embedding and cannot effectively support the fractures.

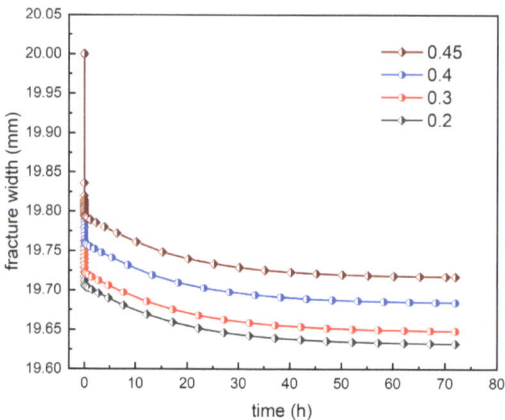

Figure 22. Fracture width versus time for different Poisson's ratios of proppant.

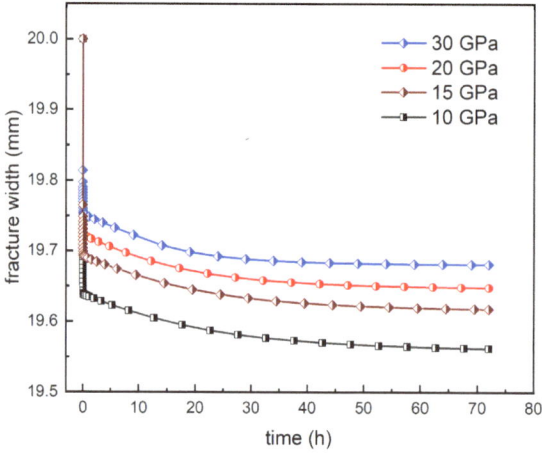

Figure 23. Fracture width versus time for different elastic moduli of reservoir.

Figure 24 shows the variation in fracture width over time for shale reservoirs with Poisson's ratios of 0.2, 0.25, 0.3, and 0.35. As shown in the figure, when other conditions are constant, the larger the Poisson's ratio of the reservoir, the wider the fracture width. The reason for this is the same as the effect of the Poisson's ratio of the proppant on the fracture width, so it will not be repeated here. The width of the fracture is not linearly proportional to the Poisson's ratio of the reservoir. When the reservoir Poisson's ratio increased from 0.2 to 0.25, the fracture width increased by 0.0087 mm; when the reservoir Poisson's ratio increased from 0.25 to 0.3, the fracture width increased by 0.0111 mm; and when the reservoir Poisson's ratio increased from 0.3 to 0.35, the fracture width increased by 0.0144 mm. Similarly to Figure 22, the Poisson's ratio of the reservoir mainly affects the instantaneous application of closure pressure and the early stage of pore pressure dissipation. For the later stage of pore pressure dissipation, the Poisson's ratio of the reservoir has almost no effect on fracture width.

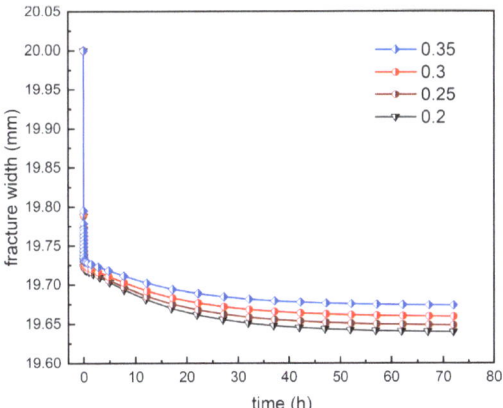

Figure 24. Fracture width versus time for different Poisson's ratios of reservoir.

Figure 25 shows the viscosity parameters of shale reservoirs, which are η = 661,543.53 MPa·h, η = 561,543.53 MPa·h, and η = 461,543.53 MPa·h. The variation law of fracture width with time at 361,543.53 MPa·h. As shown in the figure, for the 72 h diversion capacity evaluation model, changes in viscosity parameters have almost no effect on the final fracture width.

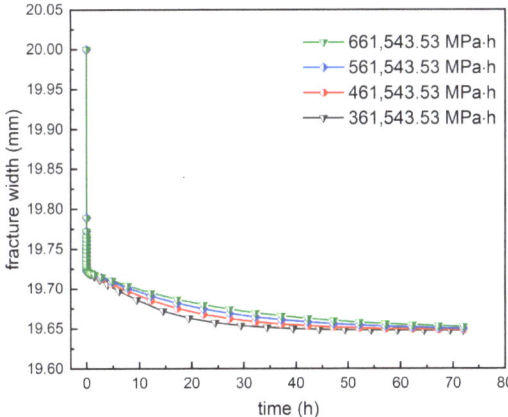

Figure 25. Fracture width versus time for different viscous parameters of reservoir.

Figure 26 shows the variation in fracture width over time at initial yield stresses of 1 MPa, 21 MPa, 41 MPa, and 61 MPa. As shown in the figure, under certain other conditions, as the initial yield stress of the reservoir increases, the fracture width increases. For every 10 MPa increase in initial yield stress, the fracture width increases by approximately 0.01 mm. Although this value is small, it cannot be ignored when measuring the effect of initial yield stress on fracture width on a longer time scale, such as 10 years, given a creep time of 72 h.

Figure 27 shows the variation in fracture width over time when the strengthening material parameters are h = 14,276.18 MPa, h = 24,276.18 MPa, h = 34,276.18 MPa, and h = 44,276.18 MPa. As shown in the figure, when other conditions are constant, as the strengthening material parameters increase, the fracture width increases. The parameters of strengthening materials mainly affect the process of stress, which no longer changes after the dissipation of pore pressure. Increasing the parameters of strengthening materials is equivalent to increasing the subsequent yield stress, resulting in a decrease in creep strain rate, a decrease in creep strain, and an increase in fracture width. As the variation in pore pressure is small, the variation curves of fracture width with time corresponding to different strengthening materials are not parallel, indicating that the strengthening material parameters are not linearly related to fracture width. This is consistent with the influence of strengthening material parameters on creep strain in Figure 11. Comparing Figures 26 and 27, the influence of strengthening material parameters on seam width is greater than that of initial yield stress on seam width. This is because the initial yield stress is only the "threshold value" for entering viscoplasticity, while the strengthening material parameters affect the entire creep process.

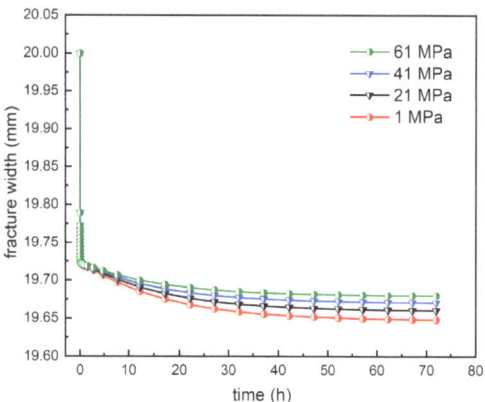

Figure 26. Fracture width versus time for different initial yield stresses of reservoir.

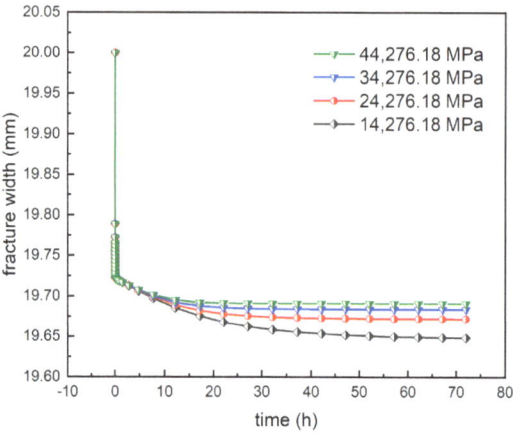

Figure 27. Fracture width versus time for different hardening material parameters of reservoir.

5. Conclusions

This study introduces isotropic strengthening based on the classic Perzyna viscoplastic model to obtain an elastic–viscoplastic constitutive model. The model was used to fit triaxial compression creep experimental data under different differential stresses, obtain model parameters, and program the model in the UMAT subroutine interface for model validation. Finally, the viscoplastic model was applied to the calculation and simulation of long-term fracture conductivity to study the effect of creep on fracture width and to analyze the factors affecting fracture width. The conclusions drawn are as follows:

(1) According to the 8 h high-temperature creep test results of shale with confining pressure of 5 MPa and differential stresses of 10 MPa, 15 MPa, and 20 MPa, shale mainly exhibits characteristics of attenuation creep and steady-state creep.
(2) The verification results of the viscoplastic model indicate that, during the attenuation creep stage at a differential stress of 10 MPa, there is a gap between the numerical calculation results of the viscoplastic model and the experimental data. After entering the steady-state creep stage, the simulation results agree with the experimental data. When the differential stress is 20 MPa, the overall agreement between the viscoplastic model and experimental data is good.
(3) The sensitivity analysis of material parameters in the viscoplastic model shows that the viscosity parameters only affect the time to reach the final creep strain and do not affect the final creep. There is a linear inverse relationship between creep strain and initial yield stress. There is a non-linear inverse relationship between creep strain and strengthening material parameters.
(4) The simulation results of the artificial fracture diversion capacity evaluation model show that, after 72 h, the reduction in fracture width caused by pore pressure dissipation and reservoir creep accounts for 32.07% of the total reduction in fracture width. Further, the viscoplastic deformation of reservoirs cannot be ignored in predicting the hydraulic conductivity of artificial fractures.
(5) The process of fracture width variation can be divided into three stages. The instantaneous application of closed stress, due to the lack of time for pore pressure to dissipate, results in a decrease in fracture width mainly caused by the elastic deformation of the reservoir and proppant. Then, as the pore pressure dissipates, the effective stress on the proppant increases and is transmitted to the reservoir through contact, and the fracture width continues to decrease. When the pore pressure has dissipated and the effective stress on the proppant no longer changes due to the creep properties of the reservoir, the strain will still gradually increase over time until it reaches the maximum value of creep strain, and the fracture width will no longer change.

Author Contributions: Conceptualization, S.L.; Methodology, S.L.; Software, J.Z.; Validation, M.L. (Mengjie Li) and J.L.; Formal analysis, M.L. (Mengjie Li); Investigation, J.L.; Writing—original draft, J.Z.; Writing—review & editing, J.F.; Supervision, S.L.; Project administration, H.G., H.W. and M.L. (Muzi Li); Funding acquisition, H.G., H.W. and M.L. (Muzi Li). All authors have read and agreed to the published version of the manuscript.

Funding: This work was supported by the Open Fund of Hebei Cangzhou Groundwater and Land Subsidence National Observation and Research Station (CGLOS-2023-09) and the National Natural Science Foundation of China (No. 52174011). We also thank the following partners at China University of Petroleum (Beijing) for their contributions: Min Zhang and Houze Chen.

Data Availability Statement: All relevant data are within the paper.

Conflicts of Interest: The authors declare no conflict of interest.

References

1. Zhong, C.; Hou, D.; Liu, B.; Zhu, S.; Wei, T.; Gehman, J.; Alessi, D.S.; Qian, P.-Y. Water footprint of shale gas development in china in the carbon neutral era. *J. Environ. Manag.* **2023**, *331*, 117238. [CrossRef]
2. Ma, Z.; Pi, G.; Dong, X.; Chen, C. The situation analysis of shale gas development in China-based on Structural Equation Modeling. *Renew. Sustain. Energy Rev.* **2017**, *67*, 1300–1307. [CrossRef]
3. Masłowski, M.; Labus, M. Preliminary Studies on the Proppant Embedment in Baltic Basin Shale Rock. *Rock Mech. Rock Eng.* **2021**, *54*, 2233–2248. [CrossRef]
4. Sone, H.; Zoback, M.D. Mechanical properties of shale-gas reservoir rocks—Part 2: Ductile creep, brittle strength, and their relation to the elastic modulus. *Geophysics* **2013**, *78*, D390–D399. [CrossRef]
5. Zheng, D.; Miska, S.; Ozbayoglu, E. The Influence of Formation Creeping on Wellbore Integrity. In Proceedings of the SPE 2021 Symposium Compilation, Virtual, 26 November 2021. [CrossRef]
6. Guo, J.; Liu, Y. Modeling of Proppant Embedment: Elastic Deformation and Creep Deformation. In Proceedings of the SPE International Production and Operations Conference & Exhibition, SPE-157449-MS, Doha, Qatar, 14–16 May 2012.

7. Katende, A.; Allen, C.; Rutqvist, J.; Nakagawa, S.; Radonjic, M. Experimental and numerical investigation of proppant embedment and conductivity reduction within a fracture in the Caney Shale, Southern Oklahoma, USA. *Fuel* **2023**, *341*, 127571. [CrossRef]
8. Katende, A.; O'Connell, L.; Rich, A.; Rutqvist, J.; Radonjic, M. A comprehensive review of proppant embedment in shale reservoirs: Experimentation, modeling and future prospects. *J. Nat. Gas Sci. Eng.* **2021**, *95*, 104143. [CrossRef]
9. Chang, C.; Zoback, M.D. Viscous creep in room-dried unconsolidated Gulf of Mexico shale (I): Experimental results. *J. Pet. Sci. Eng.* **2009**, *69*, 239–246. [CrossRef]
10. Rassouli, F.S.; Zoback, M.D. Comparison of Short-Term and Long-Term Creep Experiments in Shales and Carbonates from Unconventional Gas Reservoirs. *Rock Mech. Rock Eng.* **2018**, *51*, 1995–2014. [CrossRef]
11. Sone, H.; Zoback, M.D. Time-dependent deformation of shale gas reservoir rocks and its long-term effect on the in situ state of stress. *Int. J. Rock Mech. Min. Sci.* **2014**, *69*, 120–132. [CrossRef]
12. Borja, R.I.; Yin, Q.; Zhao, Y. Cam-Clay plasticity. Part IX: On the anisotropy, heterogeneity, and viscoplasticity of shale. *Comput. Methods Appl. Mech. Eng.* **2020**, *360*, 112695. [CrossRef]
13. Chang, C.; Zoback, M.D. Viscous creep in room-dried unconsolidated Gulf of Mexico shale (II): Development of a viscoplasticity model. *J. Pet. Sci. Eng.* **2010**, *72*, 50–55. [CrossRef]
14. Darabi, M.K.; Abu Al-Rub, R.K.; Masad, E.A.; Huang, C.-W.; Little, D.N. A modified viscoplastic model to predict the permanent deformation of asphaltic materials under cyclic-compression loading at high temperatures. *Int. J. Plast.* **2012**, *35*, 100–134. [CrossRef]
15. Kabwe, E.; Karakus, M.; Chanda, E.K. Creep constitutive model considering the overstress theory with an associative viscoplastic flow rule. *Comput. Geotech.* **2020**, *124*, 103629. [CrossRef]
16. Haghighat, E.; Rassouli, F.S.; Zoback, M.D.; Juanes, R. A viscoplastic model of creep in shale. *Geophysics* **2020**, *85*, MR155–MR166. [CrossRef]
17. Ding, X.; Zhang, F.; Zhang, G. Modelling of time-dependent proppant embedment and its influence on tight gas production. *J. Nat. Gas Sci. Eng.* **2020**, *82*, 103519. [CrossRef]
18. Fan, M.; Han, Y.; Chen, C. Thermal–Mechanical Modeling of a Rock/Proppant System to Investigate the Role of Shale Creep on Proppant Embedment and Fracture Conductivity. *Rock Mech. Rock Eng.* **2021**, *54*, 6495–6510. [CrossRef]
19. Liu, Y.; Mu, S.; Guo, J.; Yang, X.; Chen, C.; Liu, H. Analytical model for fracture conductivity with multiple particle sizes and creep deformation. *J. Nat. Gas Sci. Eng.* **2022**, *102*, 104607. [CrossRef]
20. Luo, Z.; Zhang, N.; Zhao, L.; Liu, F.; Liu, P.; Li, N. Modeling of pressure dissolution, proppant embedment, and the impact on long-term conductivity of propped fractures. *J. Pet. Sci. Eng.* **2020**, *186*, 106693. [CrossRef]
21. Geng, Z.; Bonnelye, A.; David, C.; Dick, P.; Wang, Y.; Schubnel, A. Pressure Solution Compaction during Creep Deformation of Tournemire Shale: Implications for Temporal Sealing in Shales. *J. Geophys. Res. Solid Earth* **2021**, *126*, e2020JB021370. [CrossRef]
22. Yin, Q.; Liu, Y.; Borja, R.I. Mechanisms of creep in shale from nanoscale to specimen scale. *Comput. Geotech.* **2021**, *136*, 104138. [CrossRef]
23. Rafieepour, S.; Zheng, D.; Miska, S.; Ozbayoglu, E.; Takach, N.; Yu, M.; Zhang, J. Combined Experimental and Well Log Evaluation of Anisotropic Mechanical Properties of Shales: An Application to Wellbore Stability in Bakken Formation. In Proceedings of the SPE Annual Technical Conference and Exhibition, Virtual, 21–22 October 2020. [CrossRef]
24. Zheng, D.; Ozbayoglu, E.; Miska, S.; Zhang, J. Combined Experimental and Well Log Study of Anisotropic Strength of Shale. In Proceedings of the SPE Annual Technical Conference and Exhibition, San Antonio, TX, USA, 16–18 October 2023. [CrossRef]
25. Zheng, D.; Ozbayoglu, E.; Miska, S.; Zhang, J. Experimental Study of Anisotropic Strength Properties of Shale. In Proceedings of the 57th U.S. Rock Mechanics/Geomechanics Symposium, Atlanta, GA, USA, 15–28 June 2023. [CrossRef]
26. Hartley, P. Introduction to Computational Plasticity. *J. Phys. A Math. Gen.* **2006**, *39*, 3850. [CrossRef]
27. Lazari, M.; Sanavia, L.; di Prisco, C.; Pisanò, F. Predictive potential of Perzyna viscoplastic modelling for granular geomaterials. *Int. J. Numer. Anal. Methods Geomech.* **2019**, *43*, 544–567. [CrossRef]
28. Perzyna, P. Fundamental Problems in Viscoplasticity. *Adv. Appl. Mech.* **1966**, *9*, 243–377. [CrossRef]
29. Song, F.; Rodriguez-Dono, A.; Olivella, S. Hydro-mechanical modelling and analysis of multi-stage tunnel excavations using a smoothed excavation method. *Comput. Geotech.* **2021**, *135*, 104150. [CrossRef]
30. Wang, X.; Wang, L.B.; Xu, L.M. Formulation of the return mapping algorithm for elastoplastic soil models. *Comput. Geotech.* **2004**, *31*, 315–338. [CrossRef]
31. Wang, J. Mechanical Properties and Researches of Roadway Supporting Technology of Oil Shale Under the Action of Water. Ph.D. Thsis, Liaoning University of Technology, Jinzhou, China, 2012.

Disclaimer/Publisher's Note: The statements, opinions and data contained in all publications are solely those of the individual author(s) and contributor(s) and not of MDPI and/or the editor(s). MDPI and/or the editor(s) disclaim responsibility for any injury to people or property resulting from any ideas, methods, instructions or products referred to in the content.

Article

Experimental Study of the Fluid Contents and Organic/Inorganic Hydrocarbon Saturations, Porosities, and Permeabilities of Clay-Rich Shale

Fenglan Wang [1,2], Binhui Li [2,3,4], Sheng Cao [2,3,4], Jiang Zhang [2,3,4], Quan Xu [2,3,4] and Qian Sang [5,*]

1. Daqing Oilfield Co., Ltd., Daqing 163002, China
2. National Key Laboratory for Multi-Resources Collaborative Green Production of Continental Shale Oil, Daqing 163712, China; libinhui@petrochina.com.cn (B.L.); zhangjiang1@petrochina.com.cn (J.Z.); xuquan1@petrochina.com.cn (Q.X.)
3. Exploration and Development Research Institute of Daqing Oilfield Co., Ltd., Daqing 163712, China
4. Heilongjiang Provincial Key Laboratory of Reservoir Physics & Fluid Mechanics in Porous Medium, Daqing 163712, China
5. School of Petroleum Engineering, China University of Petroleum (East China), Qingdao 266580, China
* Correspondence: 20190005@upc.edu.cn

Abstract: Unlike conventional reservoirs, shale is particularly complex in its mineral composition. As typical components in shale reservoirs, clay and organic matter have different pore structures and strong interactions with fluids, resulting in complex fluid occurrence-states in shale. For example, there are both free water and adsorbed water in clay, and both free oil and ad/absorbed oil in organic matter. Key properties such as fluid content, organic/inorganic porosity, and permeability in clay-rich shale have been poorly characterized in previous studies. In this paper, we used a vacuum-imbibition experimental method combined with nuclear magnetic resonance technique and mathematical modeling to characterize the fluid content, organic/inorganic porosity, saturation, and permeability of clay-rich shale. We conducted vacuum-imbibition experiments on both shale samples and pure clay samples to distinguish the adsorbed oil and water in clay and organic matter. The effects of clay content and total organic matter content (TOC) on porosity and adsorbed-fluid content are then discussed. Our results show that, for the tested samples, organic porosity accounts for 26–76% of total porosity. The oil content in organic matter ranges from 29% to 69% of the total oil content, and 2% to 58% of the organic oil content is ad/absorbed in kerogen. The inorganic porosity has a weak positive correlation with clay content, and organic porosity increases with rising levels of organic matter content. The organic permeability is 1–3 orders of magnitude lower than the inorganic permeability.

Keywords: vacuum-imbibition; clay; shale; organic saturation; inorganic saturation; organic permeability; inorganic permeability

1. Introduction

With the maturity of horizontal well drilling technology and large-scale volume fracturing technology, the commercial development of shale oil has gradually been realized. Compared with conventional reservoirs, shale oil reservoirs are rich in organic matter and have high clay mineral content, diverse pore types, and complex spatial structures. Shale oil reservoirs have the characteristics of low porosity, low permeability, and high specific surface area. The presence of clay minerals can lead to severe water sensitivity effects, while significant capillary phenomena can cause water-locking phenomena. Depositional environments can significantly impact the components, contents, structures, and distribution of clay in shale oil layers. These factors of clay minerals in shale reservoirs have two aspects in their influence on the reservoir's physical properties and rock-mechanical properties: one is the influence on the mechanical properties of shale rocks; the increased amounts of clay minerals will reduce natural fractures and reduce the "fracturability" of rock layers. The second is the influence on fluid continuity and flow characteristics, because the distribution of clay in shale will influence the continuity of organic matter, fluid distributions, and relationship between the two phases of permeability. The interactions between clay minerals and fluids can significantly affect the flow capacity of fluids [1,2]. Clay minerals will swell when they encounter water. The mechanism of clay swelling has been well studied, and the main understanding achieved describes the formation of hydrogen bonds on the surface of clay minerals which results in the adsorption of water [3,4], and the hydration of ions in clay minerals to form a diffuse double layer, which increases the inter-crystal distance of clay crystals [5]. The swelled clay will disperse and produce precipitation, blocking flow channels, resulting in reduced permeability [6,7]. The interaction between oil and clay minerals mainly focuses on the characterization of the occurrence-state of oil. The polar part of oil adsorbs to clay minerals due to hydrogen bonding [8]. When the water content increases, the proportion of adsorbed oil decreases and the proportion of free oil increases [9], thereby increasing the mobility of oil. However, due to the effect of clay swelling, the increase in flow ability is limited [10]. The above studies reveal the storage and flow characteristics of oil and water in clay minerals, but they do not systematically characterize the content and distribution characteristics of fluids in shale samples with complex mineral compositions.

The shale oil reservoir is rich in organic matter. The oil in shale reservoirs with high oil content and oil of medium to high maturity is of good quality, has low viscosity, and is mainly stored in the pores and kerogen. The occurrence-states of continental shale oil are diverse, mainly including free oil, oil adsorbed on the surface of minerals, and oil absorbed in the organic matter [11,12]. The free oil is mainly composed of small molecular components and is mainly found in micro-fractures, fractures between layers, and large-sized pores in the mudstone matrix [13,14]. The adsorbed oil is mainly composed of medium to large molecular components and is mainly adsorbed onto the surface of rock minerals and kerogen macromolecule skeletons through physical adsorption and non-covalent chemical adsorption [15,16]. The absorbed oil is mainly composed of medium to small molecular components, which are mainly small molecules entrapped within the internal network structure of kerogen or dissolved in asphaltene and residual water [17,18]. Two types of experimental characterization methods have been developed for studying the occurrence-state of shale oil: one is the solvent extraction method [18], and the other is the pyrolysis method [19]. Some scholars have also used molecular simulation technology to quantitatively evaluate the adsorption and free-oil content of either single-component hydrocarbons or mixed hydrocarbons in pores with different properties (type, size, shape, wettability, etc.) and under different conditions (temperature, pressure) [20]. The thickness and density-distribution of the adsorption layer were described in their studies.

Based on the wettability and occurrence characteristics of oil/water in organic/inorganic pores, through the fluid imbibition experimental method, Sang et al. [21] determined the inorganic porosity, organic porosity, maximum organic saturation, maximum inorganic saturation, and adsorption-relative miscible-fluid saturation of continental shale rock samples from the Jiyang Depression in eastern China. The oil content in organic matter (including free oil, adsorbed oil, and adsorbed oil) accounted for 6% to 55% of the total saturated oil content, of which 50% to 90% was ad- or absorbed oil. The inorganic and organic permeability of the measured shale samples were obtained by fitting the imbibition curves of the oil and water. The organic permeability was approximately one to two orders of magnitude lower than the inorganic permeability [22]. The above experiments and fitting processes did not consider the influence of clay on the contents of ad- and adsorbed fluid. The content levels of ad- and adsorbed fluid in the organic matter and clay minerals were not separated.

The objectives of this study are to characterize the fluids in different occurrence-states of shale rock samples rich in organic matter and clay minerals, and to obtain key parameters that are important in fluid-flow studies by considering the adsorbed fluid in organic matter and clay minerals. In this paper, a total of eight shale samples and two types of clay minerals were selected to conduct vacuum-imbibition tests in order to separately consider the organic porosity, inorganic porosity, organic saturation, and inorganic saturation of the shale samples. The experimental results were fitted to obtain the organic permeability and inorganic permeability of the tested samples. The clay contents and organic-matter contents of the samples were characterized to analyze the factors affecting the organic/inorganic saturation and the fluid content in different occurrence-states.

2. Experimental Methods

2.1. Rock Samples

Shale rock samples were collected from Well GY 7, Well GY 10HC, Well GY 4HC, Well GY 9HC, and Well GY 16 in the Songliao Basin, which is located in Daqing, Heilongjiang Province, in northern China. The mineralogy of the core samples was characterized by X-ray diffraction, and the TOC of core samples was characterized by the high-temperature pyrolysis method. The mineralogy and TOC of all samples are shown in Table 1. The shale samples have high content levels of clay minerals, mostly ranging from 30% to 50%. The TOC of samples are in the range of 1–3%.

Table 1. Mineralogy data and TOC of samples (wt.%).

Number	Samples	Clay	Quartz	Feldspar	Calcite	Dolomite	Pyrite	Ankerite	Barite	TOC
#1	GY 7	50.1	32.7	9.0	7.4	6.2	1.9	5.4	-	2.5
#2	GY 10HC-1	42.6	37.0	9.4	-	8.5	2.5	-	-	2.0
#3	GY 10HC-2	8.8	6.1	-	-	81.3	-	-	3.8	0.4
#4	GY 4HC-189	16.4	19.6	2.7	1.7	58.2	1.4	-	-	2.4
#5	GY 9HC-208	37.4	35.4	16.5	0.7	6.4	3.6	-	-	2.1
#6	GY 9HC-210	41.5	37.8	16.7	-	-	4.0	-	-	2.9
#7	GY 16-24	41.3	31.7	14.8	-	6.1	6.1	-	-	1.3
#8	GY 16-32	39.3	33.6	18.9	4.3	-	3.9	-	-	1.0

2.2. Pore-Structure Characterization

Shale matrix has a complex pore structure, with pore sizes ranging from nanometers to micrometers. In this work, the morphology of the sample surface was observed using an SEM with argon ion-beam milling. Pore-size distribution was measured by the N_2 adsorption method and the high-pressure mercury injection method. The N_2 adsorption method was conducted at 77 K, using a Micromeritics ASAP 2020 surface-area and porosity analyzer. Pore-size distribution was calculated using the Barret–Joyner–Halenda (BJH) method [23]. An AutoPore IV 9500 was used to measure the pore throat sizes in the high-pressure mercury injection test.

2.3. Vacuum-Imbibition Tests

In this study, vacuum-imbibition tests were conducted to determine the organic and inorganic porosity, organic and inorganic saturation, and content of ad/absorbed oil of the shale rock samples. The helium was saturated to obtain the volume of the total pore space, and the oil and water imbibition tests were performed to obtain the volumes of the organic and inorganic pores.

An apparatus for vacuum-imbibition tests is shown in Figure 1. The experimental set-up consists of a gas cylinder, vacuum pump, pressure sensors, sealed test cells, and high-precision burettes. Each burette was sealed with plastic wrap and aluminum foil to prevent evaporation of the liquid. Due to the low amount of imbibed liquid in this experiment, we chose burettes with graduations at every 0.01 mL to ensure accurate measurement of the imbibed contents. The apparatus for the oil and water vacuum-imbibition test is divided into two parts; the upper part includes the valve V-2 and a high-precision burette, and the lower part includes the valve V-1 and the sealed cell. The role of the V-1 is to connect to the vacuum pump for vacuuming, and after the vacuuming is completed, the V-1 is closed to maintain the vacuum of the cell. The V-2 is the valve for the high-precision burette, preventing the liquid from leaking out from the bottom when the high-precision burette is filled with liquid. The upper and lower parts are assembled together for the imbibition experiment.

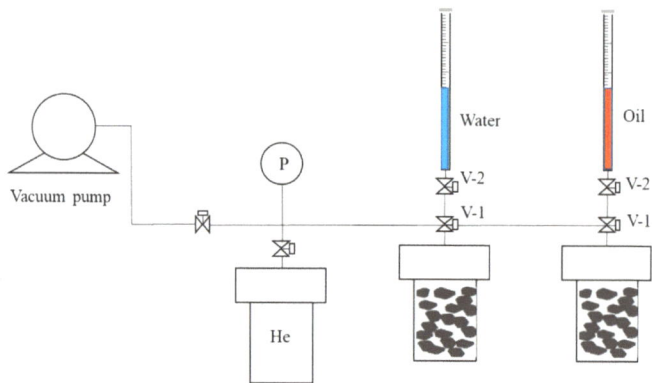

Figure 1. Schematic of the vacuum-imbibition test.

Due to the significant impact on volume accuracy made by the container's volume in the experimental results, it is necessary to accurately measure the internal-space volume of the sealed cell. The internal volume of the sealed cell refers to all the volume below the valve V-1. First, the sealed cell was vacuumed for at least 4 h; then, it was connected to the burette filled with distilled water, and the volume of water imbibed was recorded. The internal volume was calibrated, at least three times, until the volume error of the three calibrations was less than 0.1 mL. The calibrated volumes were then averaged to calculate the accurate volume of the sealed cell.

The sealed cell was connected to the high-precision burette filled with oil or water. The valve V-2 was opened, and the initial position of the liquid was recorded. Then, the valve V-1 was opened, and the level of the liquid in the burette at first rapidly dropped to fill the space in the sealed cell not occupied by the rock sample. Then, the level of the liquid dropped slowly. The liquid position V_i and the corresponding time t were recorded until the liquid level no longer dropped.

The procedure of the vacuum-imbibition test is as follows:

(1) The shale samples were broken into blocks of approximately 1 cm × 1 cm × 0.5 cm and extracted with CH_2Cl_2 for 15 days, and then dried for 48 h at 100 °C.
(2) The processed rock-samples were divided into two parts of equal weight and placed in two sealed cells for water and oil imbibition tests. An air tightness test was conducted.
(3) The samples were vacuumed, and then saturated with helium gas. The saturated gas content was obtained according to Boyle's law to calculate the total pore volumes and porosities of the shale rock samples.
(4) The samples were then vacuumed again, and oil and water were imbibed into the 2 cells. The imbibed volumes of oil and water were recorded, along with the time. The entire experimental process was conducted at a constant temperature of 25 °C. The n-dodecane (n-C_{12}) was used as the oil phase, and the KCl solution with 8% mass concentration was used as the water phase.

2.4. Two-Dimensional Nuclear Magnetic Resonance Characterization

In this study, the nuclear magnetic resonance (NMR) technique is used to characterize the oil and water in different occurrence-states. In order to distinguish different hydrogen-bearing components in shale, we obtained the transverse relaxation (T_2) and its relationship with the longitudinal relaxation time (T_1) for the shale samples. The dry samples and the samples after the oil and water had been imbibed were subjected to NMR tests, and the components of kerogen, structural water, free water, adsorbed water, free oil, and adsorbed oil were identified from the T_1–T_2 maps.

The NMR tests were carried out on a MacroMR12-150H-VTHP instrument (Shanghai Niumag, Shanghai, China) operated at 12 MHz. The measurement parameters of the NMR were set as follows: waiting time, 4000 ms; number of echoes, 10,000; regulate first data, 0.08 ms; sampling frequency, 250 kHz; pulse 90, 20 µs.

3. Results and Discussion

3.1. Microstructure of Pore-Space and Pore Size Distribution

Combined with the argon ion-beam milling technique, scanning electron microscopy (SEM) was used to characterize the pores at nanometer resolution. Nano- and micro-scale pores were identified and analyzed based on SEM images. Some representative images are shown in Figure 2. Various minerals such as quartz, clay, kerogen, dolomite, and pyrite were observed from the SEM images. The pore type includes intergranular pores, intragranular dissolution pores, clay mineral pores, organic matter pores, and micro-fractures. It can be seen from the images that the organic matter and clay minerals are mixed, and there are large pores or micro-fractures on the edge of the kerogen. Some of the kerogen has a certain degree of pore development inside.

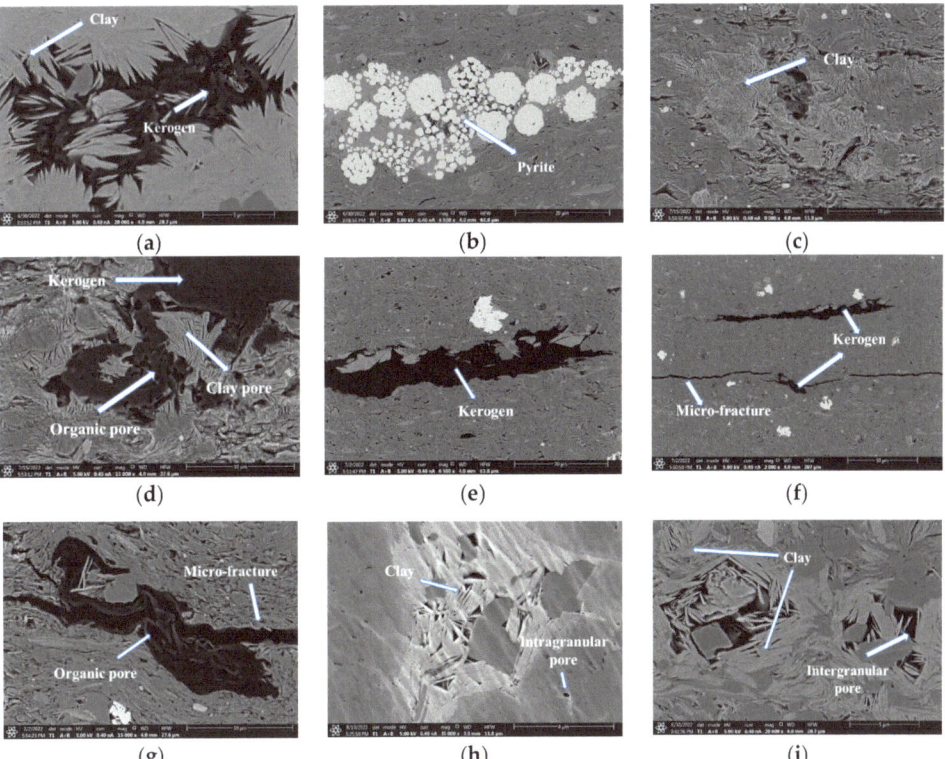

Figure 2. SEM images of shale samples: (**a**,**b**) sample #1, Well GY 7, 2406.02 m; (**c**,**d**) sample #2, Well GY 10HC, 2588.58 m; (**e**) sample #4, Well GY 4HC, 2478.2 m; (**f**,**g**) sample #5, Well GY 9HC, 2353.9 m; (**h**,**i**) sample #7, Well GY 16, 2350.17 m.

Due to the inconsistencies between resolution and field range, the SEM technique can only provide local pore structure information of rock samples, and cannot fully characterize the pore-size distribution of rock samples. Combined with high-pressure mercury injection and N_2 adsorption testing, the full pore-size range of shale samples was characterized.

As shown in Figure 3a, the shale samples have a large number of pores with sizes of less than 100 nanometers. The pores with sizes of 2–100 nanometers and 10–200 μm account for the largest proportion of the pore volume and contribute to the main storage capacity. The pores with sizes of 2–100 nanometers mainly exist in the shale matrix, and are manifested as intergranular pores, intragranular pores, and intercrystalline pores. The pores with sizes of 10–200 μm are mainly micro-fractures.

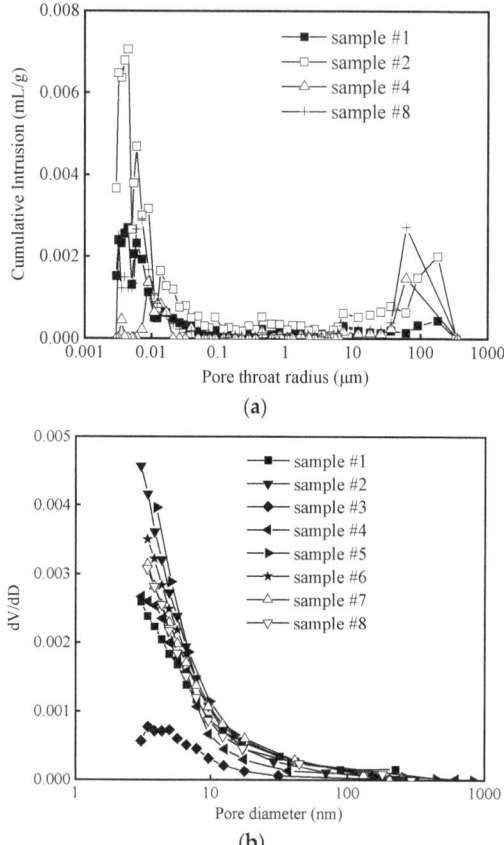

Figure 3. Pore-size distribution of tested samples: (**a**) Pore throat radius distribution measured by the mercury injection test. (**b**) Pore diameter distribution measured by the N_2 adsorption test.

Figure 3b shows that the nanopores with pore diameters of less than 10 nanometers occupy the majority of the pore space. Its characteristics are related to the high clay content, which develops a large number of nanopores, mainly in the form of sheet pores or slit pores. Combined with high-pressure mercury porosimetry results, we can conclude that for shale samples, the pore size of nanoscale pores is generally less than 100 nanometers.

3.2. Inorganic and Organic Saturations and Porosities

In shale reservoirs, shale oil exists in various forms of occurrence, including free oil in inorganic pores and fractures, adsorbed oil in clay minerals, organic free-oil in organic pores, and ad- and adsorbed oil in the kerogen matrix. The inorganic minerals in shale reservoirs are mostly water-wet surfaces with relatively large pore sizes compared to organic pores. Organic pores have hydrophobic surfaces and are mostly nano-scale pores. For the vacuum-imbibition process, the single-phase flow occurs, and the surface energy of the oil is greater than that of water; therefore, oil can enter both the organic and inorganic pores, while water can only enter the hydrophilic inorganic pores and cannot enter the hydrophobic organic matter [21].

The vacuum-imbibition results of the shale samples are shown in Figure 4. The vacuum-imbibition curves of the oil and water show the following characteristics: (1) The imbibition of oil and water can be divided into two stages. At the early stage, the imbibition rate of oil and water is fast, and the imbibition volume rises rapidly. At the later stage, the imbibition rate gradually slows down, and the imbibition volume slowly rises until reaching a state of equilibrium at which the imbibition volume no longer changes. (2) The final imbibed volume of water is smaller than the final imbibed volume of oil. (3) It takes longer for oil to reach equilibrium than for water. For example, sample #6 took about 100 h to imbibe 0.029 mL/g of water, while the imbibition time and volume of oil were 730 h and 0.054 mL/g, respectively. The oil imbibition volume was 0.025 mL/g more than that of water. The final imbibed water volume of sample #1 was 0.024 mL/g, reaching equilibrium in approximately 4.2 days, while the final imbibed oil volume was 0.10 mL/g (4.17 times of the imbibed water volume), and took 66.2 days to reach equilibrium (15.8 times of the equilibrium time for water imbibition).

The imbibition time of oil is longer than that of water, which can be explained in following aspects. First, the overall pore size of organic pores is smaller than that of inorganic pores, resulting in lower permeability and poorer flow capacity of organic pores. Therefore, the flow of oil in organic pores is slower (as shown in Figure 5). Second, the viscosity of oil is higher than that of water, resulting in greater flow resistance, slowing down the flow velocity of oil. Furthermore, oil migrates in the organic matter by molecular diffusion, and the speed of diffusion movement is very slow, resulting in the continuous imbibition of oil into the sample at a lower speed in the later stage.

Based on the imbibed volumes of oil and water, we can distinguish organic and inorganic porosity and organic and inorganic saturation in shale and calculate the amount of ad/absorbed oil in kerogen. For example, the inorganic pore volume is equal to the imbibed water volume, because the imbibed water occupies all the inorganic pores. The gas-measured pore volume is the sum of the inorganic and organic pore volumes, so the organic pore volume is the gas-measured volume minus the imbibed-water volume. The imbibed oil is composed of the oil occupying pore space and the oil adsorbing or dissolving in kerogen, so the amount of ad/absorbed oil in kerogen is the imbibed oil volume minus the gas-measured pore volume.

In shale, besides organic matter, clay minerals also have adsorption effects on fluids due to the large specific surface area of the mineral. For shale samples with low levels of clay content, the analysis of vacuum-imbibition experimental data can ignore the influence of clay [21]. However, in this study, the samples from the Gulong shale have high levels of clay content, and clay has adsorption effects on both water and oil, so both water and oil can exist in an adsorbed state in the clay. The water imbibed into shale pores includes two parts: free water in inorganic pores, and adsorbed water in clay. The oil imbibed into shale pores includes four parts: free oil in inorganic pores, adsorbed oil in clay, free oil in organic pores, and ad/absorbed oil in kerogen.

The adsorption effect of clay on oil and water cannot be ignored for the determination of inorganic/organic porosity and saturation. For example, due to the large amount of adsorbed water in clay, if the influence of clay is not considered, the imbibed water volume is equal to the inorganic pore volume, resulting in an overestimation of the inorganic pore volume. Therefore, when analyzing the vacuum-imbibition data, it is necessary to further consider the adsorption effects of clay on oil and water.

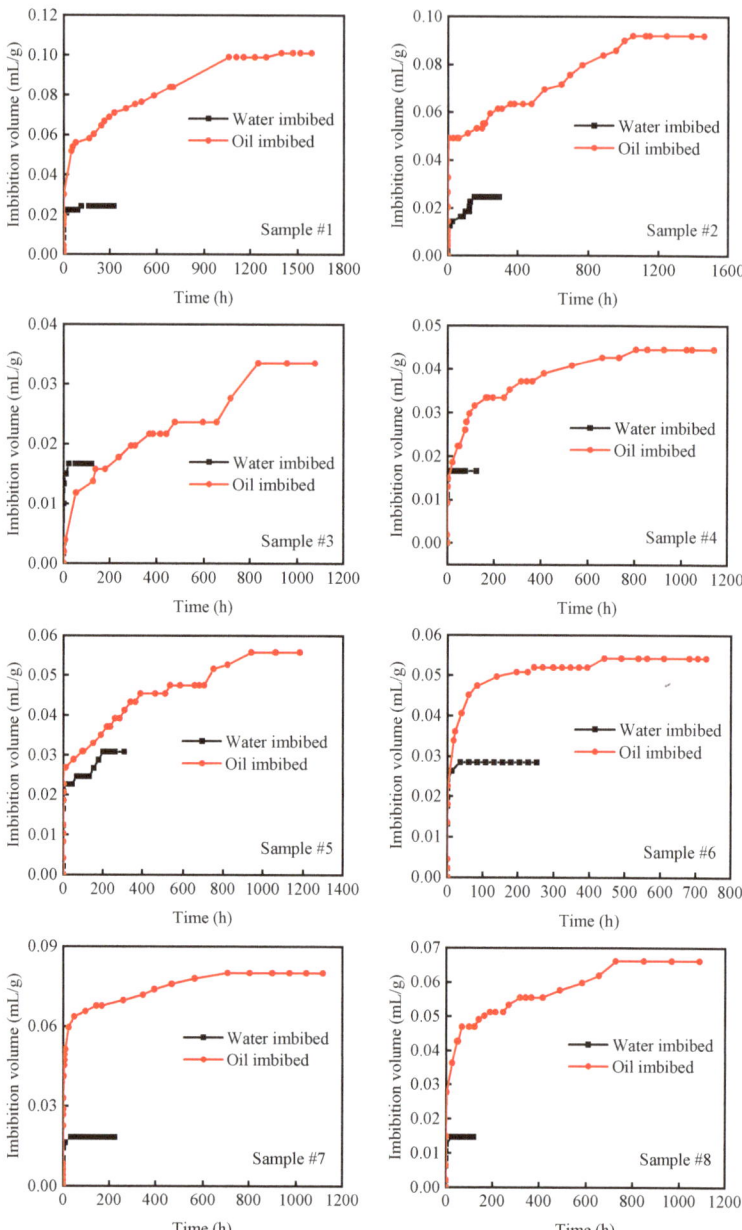

Figure 4. Vacuum-imbibition curves of water and oil in eight shale samples. The characteristics of the oil and water vacuum-imbibition curves show significant differences, indicating that the flow space and flow capacity of the oil and water are different. Water can only enter inorganic pores, while oil can enter both inorganic pores and organic pores, so oil eventually occupies more pore space than does water. In addition, due to the strong interaction between organic matter and oil, oil molecules can diffuse into the organic matter and combine with the molecular skeleton of kerogen to form ad/absorbed oil. This further increases the imbibition volume of oil, resulting in a significantly higher imbibed volume of oil than water.

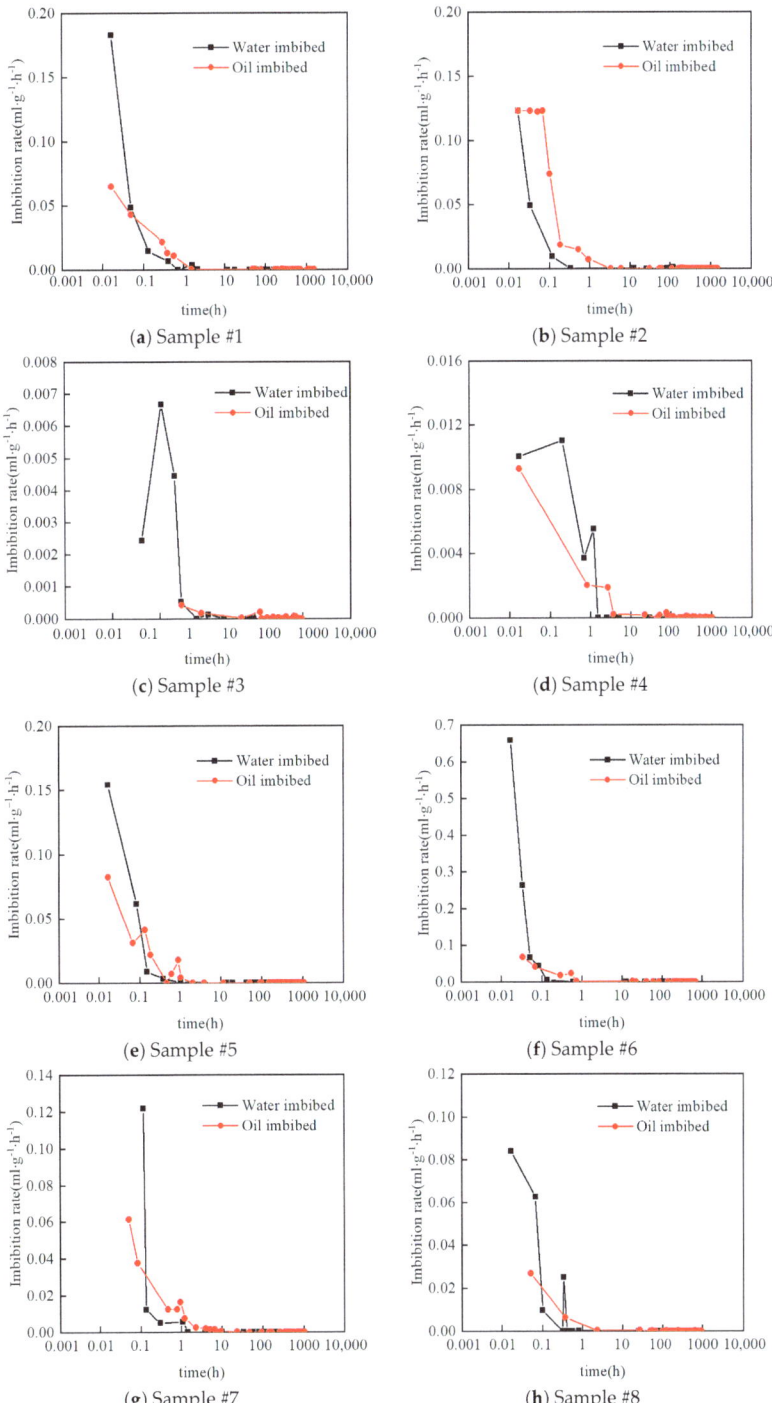

Figure 5. Vacuum-imbibition rate curves of water and oil in eight shale samples.

The clay composition of the selected shale samples is shown in Table 2. The clay components are dominated by illite, chlorite, and illite–chlorite mixed layers, with illite and chlorite accounting for the majority. Therefore, we selected pure samples of illite and chlorite to analyze the adsorption of clay on oil and water. The samples of illite and chlorite are shown in Figure 6. The experimental method and process were the same as those for the shale samples.

Figure 7 shows the vacuum-imbibition curves of water and oil in chlorite and illite samples. The vacuum-imbibition curve characteristics of illite and chlorite are similar to those of shale samples, with higher imbibition rates of oil and water in the early stage, followed by a decrease in the imbibition rate in the later stage. The final level of imbibed oil is higher than that of water, and the time for oil imbibition is longer than that for water. The time for oil and water imbibition to reach equilibrium in clay samples is shorter than that in shale samples, which is due to the lower mass of clay samples in the experiment, resulting in a reduced time for oil and water imbibition to reach equilibrium.

Table 2. Clay compositions of shale samples (wt.%).

Sample Number	Illite	Kaolinite	Chlorite	Illite/Smectite
#1	41	-	20	39
#2	66	-	7	27
#3	42	31	27	-
#4	39	-	37	24
#5	64	-	10	26
#6	68	-	9	23
#7	44	-	18	38
#8	60	-	7	33
#1	41	-	20	39

Figure 6. Samples of chlorite and illite.

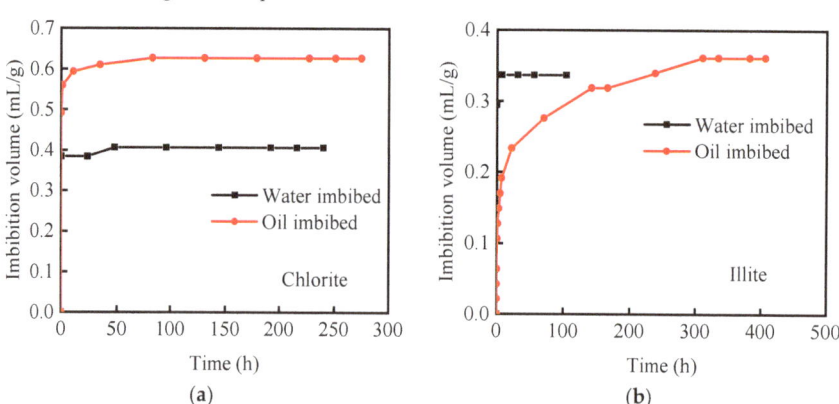

Figure 7. Vacuum-imbibition curves of water and oil in chlorite and illite samples: (a) Chlorite; (b) Illite.

The imbibition curves of illite and chlorite show a difference in the imbibed volumes of oil and water. The final imbibed oil and imbibed water volumes of chlorite are higher than those of illite, indicating that the adsorption capacity of chlorite is stronger than that of illite. Through the vacuum-imbibition experiment, we determined that the adsorbed water volume and adsorbed oil volume in chlorite are 0.41 mL/g and 0.63 mL/g, respectively, and the adsorbed water volume and adsorbed oil volume in illite are 0.34 mL/g and 0.36 mL/g, respectively.

The oil and water imbibed volumes of the shale rock samples were calculated by removing the volumes adsorbed in chlorite and illite; the inorganic/organic porosity and saturation can then be calculated, and is shown in Table 3:

(1) Inorganic porosity: The ratio of the inorganic pore volume to the apparent volume of the crushed sample. The inorganic pore volume is calculated by subtracting the adsorbed water volume in clay from the gas-measured pore volume.
(2) Organic porosity: The ratio of the organic pore volume to the apparent volume of the crushed sample. The organic pore volume is the gas-measured pore volume minus the inorganic pore volume.
(3) Inorganic saturation: The ratio of the volume of oil imbibed into the inorganic material to the total volume of imbibed oil. The oil imbibed into the inorganic material is the sum of the free oil in the inorganic pores (inorganic pore volume) and the clay-adsorbed oil.
(4) Organic saturation: The ratio of the volume of oil imbibed into the organic matter to the total volume of imbibed oil. The oil imbibed into the organic matter is the sum of the free oil in organic pores and the ad/absorbed oil in the organic matter.
(5) The percentage of adsorbed water in clay: The ratio of the clay-adsorbed water volume to the water volume within the inorganic matter.
(6) The percentages of ad/absorbed oil in organic matter: The ratio of the volume of ad/absorbed oil in the organic matter to the volume of oil within the organic matter.

Table 3. Organic oil saturation, inorganic oil saturation, organic porosity, and inorganic porosity of different shale samples.

Sample Number	Bulk Volume (cm³)	Oil Volume in Organic Pores (mL/g)	Oil Volume in Inorganic Pores (mL/g)	Helium Saturation Volume (mL/g)	Organic Saturation (%)	Inorganic Saturation (%)	Organic Porosity (%)	Inorganic Porosity (%)	Percentage of Organic Pores (%)
#1	20.434	0.101	0.024	0.048	47.59	52.41	8.86	2.82	75.87
#2	18.502	0.092	0.025	0.048	66.09	33.91	6.81	5.72	54.36
#3	21.818	0.034	0.017	0.019	34.62	65.38	1.41	3.94	26.27
#4	20.708	0.045	0.017	0.025	32.80	67.20	3.72	2.79	57.13
#5	19.689	0.056	0.031	0.040	29.85	70.15	3.11	6.72	31.63
#6	18.220	0.054	0.029	0.037	32.16	67.84	3.06	6.22	33.00
#7	19.591	0.080	0.018	0.040	56.75	43.25	7.33	2.77	72.59
#8	18.984	0.066	0.015	0.035	68.65	31.35	5.71	3.01	65.46

As typical components in shale, clay and organic matter control the pore spaces. Additionally, clay and organic matter have adsorption and dissolution effects on the oil and water, affecting the fluid content of shale. In this study, by combining mineral composition analysis and vacuum-imbibition experimental data, the effects of clay content and TOC on the inorganic/organic porosity and saturation were investigated.

Figure 8 shows the trends of inorganic porosity (a), inorganic saturation (b), and adsorbed water in clay (c), together with the clay content levels of the shale samples. Figure 8a shows that the inorganic porosity has a weak growth trend with increases in clay content, which is related to the pore structure of clay in the shale samples. According to previous SEM scanning results, the pores in clay minerals are relatively developed, while the number of pores in other inorganic minerals is relatively low. In this case, as the clay content increases, the number and volume of inorganic pores increase, resulting in an increase in inorganic porosity. The relationship between inorganic saturation and clay content is not clear (as shown in Figure 8b). This is because the rock samples also contain other types of inorganic minerals besides clay, and the degree of their pore space development also affects inorganic saturation. Figure 8c shows that the content of adsorbed water in clay has a weak correlation with the clay content levels. The higher the level of clay content, the greater the specific surface area of pores, resulting in a stronger adsorption capacity.

Figure 9 shows the trends of organic porosity (a), organic saturation (b), and ad/absorbed oil (c) with the TOC of the shale samples. There are a large number of pores in organic matter, and the higher the organic matter content level, the more the organic pore space should be. However, the results showed that the organic porosity has a weak growth trend with increases in the TOC. More test results are still needed to obtain a more obvious correlation. Figure 9b,c indicate that the relationships between both organic saturation and the amount of ad/absorbed oil in kerogen and the TOC are not clear. The adsorption capacity of organic matter is not only related to its content, but also affected by thermal maturity and kerogen type. The organic-matter properties of the shale samples tested are complex, so the organic saturation and the ad/absorbed oil content in kerogen do not show strong correlations with organic matter content.

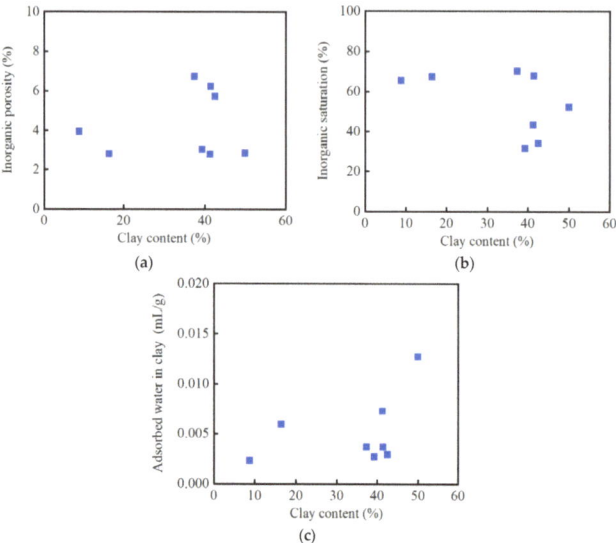

Figure 8. Relationships between (**a**) inorganic porosity and clay content; (**b**) inorganic saturation and clay content; (**c**) adsorbed water in clay and clay content.

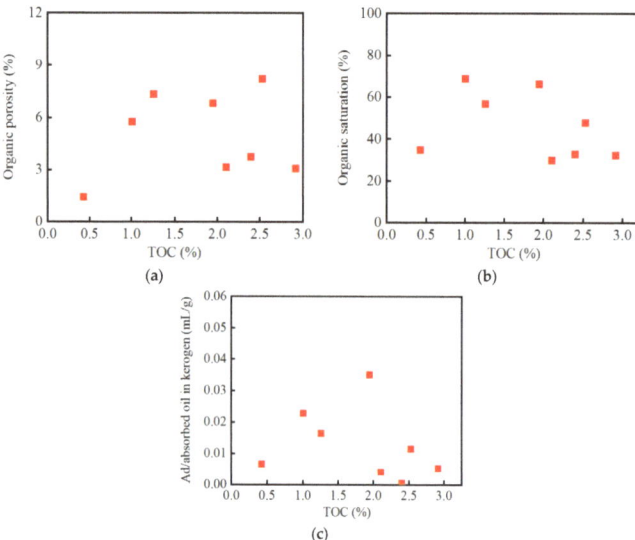

Figure 9. Relationship between (**a**) organic porosity and TOC; (**b**) organic saturation and TOC; (**c**) ad/absorbed oil in kerogen and TOC.

3.3. NMR Spectrum of Fluid in Different States

Li et al. [24] established an NMR T_1-T_2 map division method for distinguishing each hydrogen-bearing component. According to the T_1–T_2 map proposed by Li et al. [24], we identify the signals corresponding to different components. Figure 10 shows the T_1–T_2 map of shale samples before and after imbibition experiments. It can be seen that water is dominated by free water and structural water, while oil is dominated by free oil and adsorbed oil. The signal intensity of oil is significantly greater than that of water, which is consistent with the vacuum-imbibition experimental results.

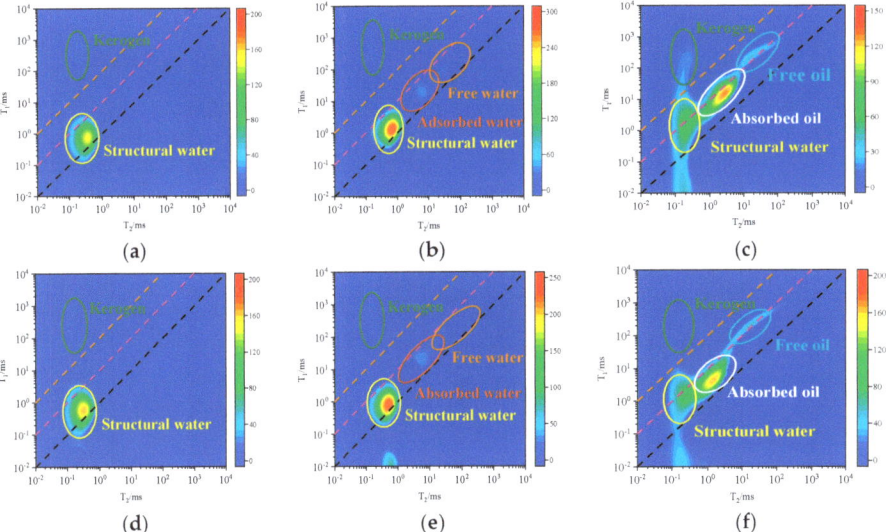

Figure 10. T_1–T_2 map of shale samples: (**a**) sample #2, dry sample; (**b**) sample #2, after imbibition of water; (**c**) sample #2, after imbibition of water; (**d**) sample #8, dry sample; (**e**) sample #8, after imbibition of water; (**f**) sample #8, after imbibition of water.

3.4. Mathematical Model Coupling Organic and Inorganic Pores Considering Ad/Absorbed Fluid in Organic Matter and Clay

The continuous dual-porosity method is applied to simulate the oil–water imbibition process, in which the water phase flows only into the inorganic pores, while the oil phase flows into both inorganic and organic pores. The oil–water imbibition curve is fitted to solve for the permeabilities of inorganic and organic pores.

Shale reservoirs are composed of two continuous media systems: inorganic pores and organic pores. The two systems are coupled, and there is fluid exchange, forming a complex multi-scale coupled-flow system. Due to the different mineral compositions and pore-size distributions of inorganic and organic pores, there are differences in the permeabilities of the two systems. Considering the above factors, mathematical models for oil and water vacuum-imbibition are established. The basic assumptions of the model are as follows:

(1) Oil exists in the form of a free state in inorganic pores, in the form of a free state and an adsorbed state in clay pores, and in the forms of an adsorbed state and a free state in organic pores.
(2) Water only flows into inorganic pores. Oil can enter both organic and inorganic pores.
(3) The matrix and the fluid are slightly compressible, and the influence of gravity is neglected.
(4) The shale sample is spherical and isotropic.
(5) During the experiment, the temperature and boundary pressure remained constant.

(6) The adsorption of clay and kerogen on oil is considered. The adsorption amount is not a function of time, which means that the adsorption process is instantaneous.

The organic–inorganic dual-medium model is shown in Figure 11, and the flow equations for inorganic pores and organic pores are shown in Equations (1) and (2).

$$\nabla \cdot \left(\rho_l \frac{k_{im}}{\mu_l} \nabla p_{im} \right) - q = \rho_l c_{t\_im} \frac{\partial p_{im}}{\partial t} + R_m \frac{\partial C_s}{\partial t} (l = o, w) \tag{1}$$

$$\nabla \cdot \left(\rho_l \frac{k_{om}}{\mu_l} \nabla p_{om} \right) + q + \beta q_m = \rho_l c_{t\_om} \frac{\partial p_{om}}{\partial t} (l = o) \tag{2}$$

$$q = \frac{\alpha \rho_l k_{im}}{\mu_l} (p_{im} - p_{om}) \tag{3}$$

where q is the cross-flow between inorganic and organic pores, kg/(m^3·s); C_s is adsorption mass per unit volume of clay, kg/m^3; R_m is the percentage of clay mass, Fraction; r_l is the density of liquid, kg/m^3; k_{im} is the permeability of inorganic pores, m^2; µL is the viscosity of liquid, Pa·s; P_{im} is the pressure of inorganic pores, Pa; c_{t_im} is the compression coefficient of inorganic pores, 1/Pa; t is the time, s; k_{om} is the permeability of organic pores, m^2; P_{om} is the pressure of organic pores, Pa; c_{t_om} is the compression coefficient of organic pores, 1/Pa; q_m is the flow of ad/absorbed oil per unit volume, kg/(m^3·s); β is proportion of organic matter, Fraction; α is cross flow coefficient, 1/m^2.

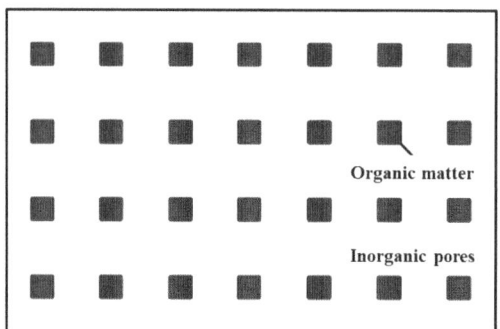

Figure 11. Organic-inorganic dual medium model.

The finite-difference method is used to solve the mathematical model of water imbibition, and the numerical solution is fitted to the experimental results. The fitting accuracy is improved by adjusting the inorganic permeability, and the cumulative flow curve with the best fitting-effect is obtained, thus determining the inorganic permeability. The determined inorganic permeability is then substituted into the oil imbibition equation, and the organic permeability is obtained by changing the organic permeability under the best-fitting accuracy.

The fitted results of the shale samples are shown in Figure 12. The oil vacuum-imbibition process of all eight samples can be well fitted. With the increase of time, the fitting values have a slight deviation. This is because the slow diffusion of oil molecules into organic matter and clay minerals is the main process at this stage. In the tests, the oil variation caused by diffusion is very small and the measurement error is relatively large, which may lead to a slight deviation between the fitting results and the test results. For the water vacuum-imbibition process of samples #2, #5, and #6, the fitting results are quite different from the test results. The discrepancies between the water imbibition equation and the behavior of Samples #2, #5, and #6 may be due to the following two reasons: (1) The samples used in the vacuum imbibition tests and the mineral tests were selected from different positions of the same layer sample. The heterogeneity of the samples caused differences between the mineral composition of the vacuum imbibition test samples and the mineralogy test samples. The mineral content (R_m) in the model was input based on the test results, and the inaccuracy of the input parameters may have led to large differences between the fitting results and the test results. (2) The water adsorption capacity of clay minerals (C_s) input in the model is derived from the average value of vacuum imbibition test results (as shown in Figure 6). The types of clay minerals in rock samples are not limited to chlorite and illite, which may cause the input water absorption capacity parameters (C_s) to be inconsistent with the actual ones. This may also lead to a large difference between the fitting results and the test results. The inorganic and organic permeabilities fitted from oil- and water-imbibition experiments are shown in Table 4. The order of magnitude of the inorganic permeability is 10^{-3} to 10^{-2} mD, and the order of magnitude of the organic permeability is 10^{-5} to 10^{-4} mD. The organic permeability is one to three orders of magnitude lower than the inorganic permeability.

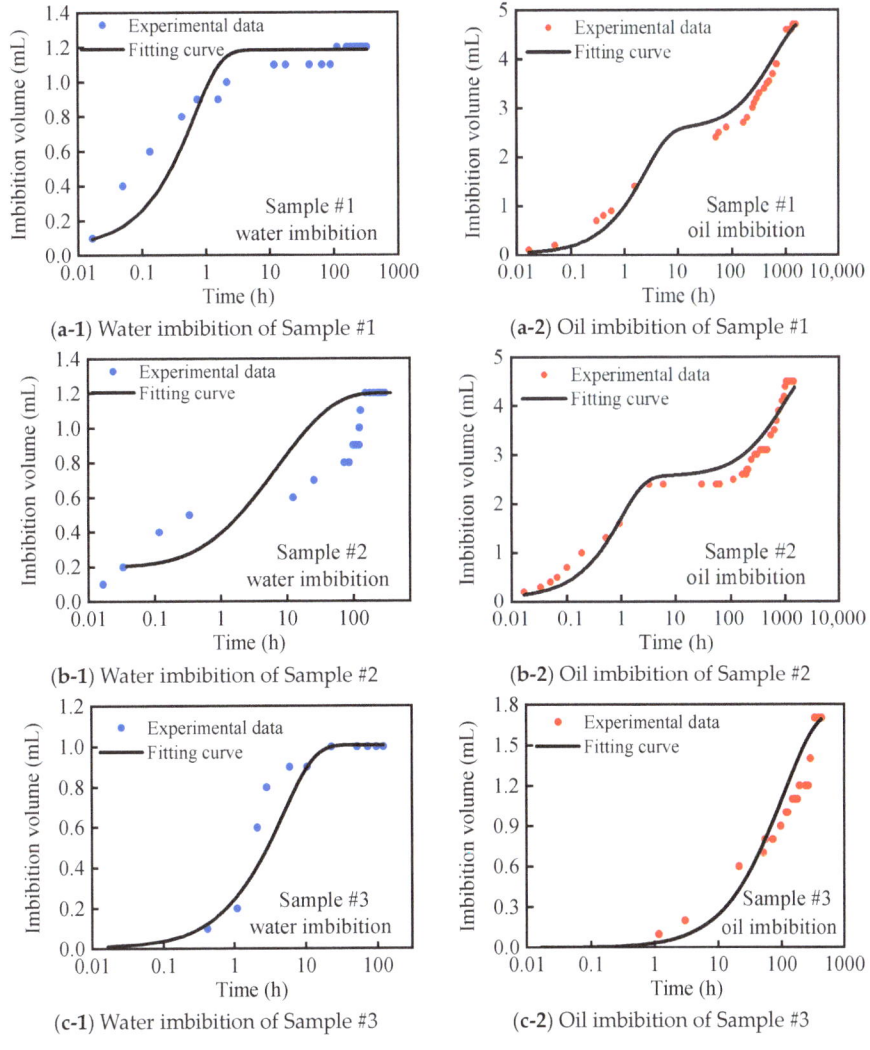

Figure 12. *Cont.*

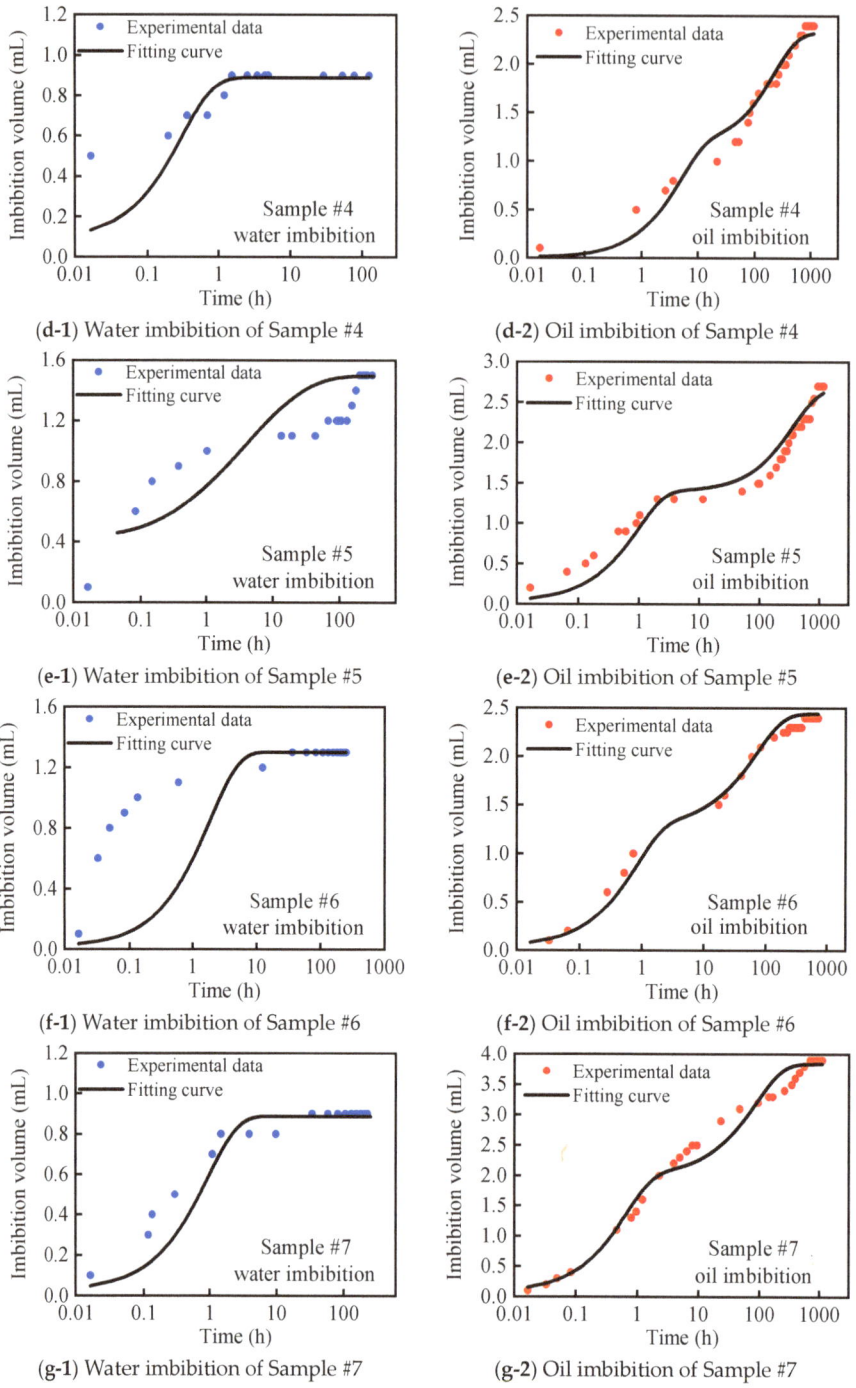

Figure 12. *Cont.*

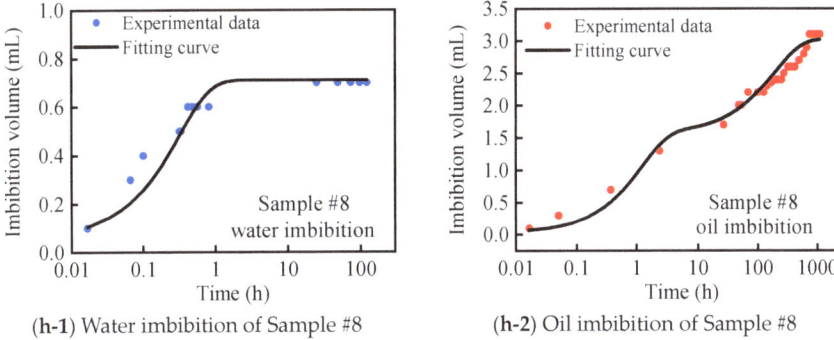

(h-1) Water imbibition of Sample #8   (h-2) Oil imbibition of Sample #8

Figure 12. Fitting results of oil- and water-imbibition curves for the shale samples.

Table 4. Inorganic and organic permeability of the shale samples.

Sample Number	Inorganic Permeability (mD)	Organic Permeability (mD)
#1	4.00×10^{-2}	6.00×10^{-5}
#2	2.38×10^{-2}	1.19×10^{-5}
#3	1.33×10^{-3}	1.33×10^{-4}
#4	4.00×10^{-2}	1.00×10^{-5}
#5	4.35×10^{-2}	1.74×10^{-5}
#6	5.71×10^{-2}	4.76×10^{-4}
#7	4.55×10^{-2}	2.12×10^{-4}
#8	3.08×10^{-2}	1.15×10^{-4}

4. Conclusions

In this study, eight sets of shale samples and two types of clay minerals were selected to conduct oil and water vacuum-imbibition experiments. The organic and inorganic saturations, porosities, and permeabilities, as well as the content levels of fluids in different occurrence-states were determined.

(1) Based on the different interactions between oil, water, and gas and the organic and inorganic pores in shale, a fluid saturation method was established to distinguish the contents of each fluid phase. For the shale samples tested, the organic porosities are 1–9%, the inorganic porosities are 2–7%, the organic saturations are 29–69%, and the inorganic saturations are 31–71%. The oil in shale can be divided into inorganic free-oil, organic free-oil, and oil ad/absorbed in the kerogen. A total of 2% to 58% of the organic oil content is ad/absorbed in the kerogen.

(2) A two-dimensional nuclear magnetic resonance method was used to characterize the occurrence-states of the oil and water in the shale samples. Water was mainly in the forms of free water and structural water, while oil was mainly in the forms of free oil and adsorbed oil. The signal intensity of oil is significantly higher than that of water, indicating that the imbibed oil volume is higher than the imbibed water volume.

(3) Based on the inorganic–organic coupling mathematical model, by considering ad/absorbed fluid, the organic and inorganic permeabilities of the shale samples were determined. The organic permeability is one to three orders of magnitude lower than that of the inorganic permeability.

Author Contributions: Methodology, J.Z. and Q.X.; resources, F.W. and J.Z.; software, J.Z.; validation, Q.X.; investigation, B.L., S.C., J.Z., Q.X. and Q.S.; writing—original draft preparation, Q.S.; writing—review and editing, F.W. and S.C.; supervision, Q.S.; project administration, B.L.; funding acquisition, Q.S. and B.L. All authors have read and agreed to the published version of the manuscript.

Funding: The authors declare that this study received funding from the Major Project of the China National Petroleum Corporation (2021ZZ10-02), and the Natural Science Foundation of Shandong Province, China (No. ZR2020ME091). The funders were not involved in the study design; the collection, analysis, or interpretation of data; the writing of this article; or the decision to submit it for publication.

Data Availability Statement: The original contributions presented in the study are included in the article; further inquiries can be directed to the corresponding author.

Conflicts of Interest: Author Fenglan Wang was employed by the Daqing Oilfield Co., Ltd. Authors Binhui Li, Sheng Cao, Jiang Zhang, and Quan Xu were employed by the Exploration and Development Research Institute of Daqing Oilfield Co., Ltd. The remaining authors declare that the research was conducted in the absence of any commercial or financial relationships that could be construed as a potential conflict of interest.

Nomenclature

Abbreviations

TOC	Total organic matter content
GY	Guye well
BJH	Barret–Joyner–Halenda method
NMR	Nuclear magnetic resonance
T_2	The transverse relaxation
T_1	The longitudinal relaxation
SEM	Scanning electron microscopy
XRD	X-ray diffraction
PDI	Polydispersity Index

Symbols

q	The cross-flow between inorganic and organic pores, kg/(m^3·s)
C_s	Adsorption mass per unit volume of clay, kg/m^3
R_m	The percentage of clay mass, Fraction
r_l	The density of liquid, kg/m^3
k_{im}	The permeability of inorganic pores, m^2
μ_l	The viscosity of liquid, Pa·s
P_{im}	The pressure of inorganic pores, Pa
c_{t_im}	The compression coefficient of inorganic pores, 1/Pa
t	The time, s
k_{om}	The permeability of organic pores, m^2
P_{om}	The pressure of organic pores, Pa
c_{t_om}	The compression coefficient of organic pores, 1/Pa
q_m	The flow of ad/absorbed oil per unit volume, kg/(m^3·s)
β	Proportion of organic matter, Fraction
α	Cross-flow coefficient, 1/m^2

References

1. Liu, X.; Jian, X.; Liang, L. Investigation of pore structure and fractal characteristics of organic-rich Yanchang formation shale in central China by nitrogen adsorption/desorption analysis. *Plateau Meteorol.* **2015**, *22*, 62–72. [CrossRef]
2. Jian, X.; Liu, X.; Liang, L. Experimental study on the pore structure characteristics of the Upper Ordovician Wufeng Formation shale in the southwest portion of the Sichuan Basin, China. *J. Nat. Gas Sci. Eng.* **2015**, *22*, 530–539.
3. Hensen, E.; Smit, B. Why clays swell. *J. Phys. Chem. B* **2002**, *106*, 12664–12667. [CrossRef]
4. Tunega, D.; Gerzabek, M.H.; Lischka, H. Ab Initio Molecular Dynamics Study of a Monomolecular Water Layer on Octahedral and Tetrahedral Kaolinite Surfaces. *J. Phys. Chem. B* **2004**, *108*, 5930–5936. [CrossRef]
5. Norrish, K.; Quirk, J. Crystalline Swelling of Montmorillonite: Manner of Swelling of Montmorillonite. *Nature* **1954**, *173*, 256–257. [CrossRef]
6. Wang, Z.; Yue, X.; Han, D. Effect of clay mineral and fluid on flow characteristics of the low-permeability cores. *Pet. Geol. Recovery Effic.* **2007**, *14*, 89–92.
7. Liang, J.; Fang, Y. Experimenal study of seepage characteristics of tiny-particle clay. *Chin. J. Rock Mech. Eng.* **2010**, *29*, 1222–1230.
8. Skipper, N.; Lock, P.; Titiloye, J.; Swenson, Z.; Howells, W.; Fernandez-Alonso, F. The structure and dynamics of 2-dimensional fluids in swelling clays. *Chem. Geol.* **2006**, *230*, 182–196. [CrossRef]

9. Wei, J.; Yang, H.; Zhu, J.; Shen, W.; Yuan, P.; He, H.; Chen, M. Molecular dynamic simulation of interlayer micro-structure in orgnaic montmorilionite. *Mineral. Petrol.* **2009**, *29*, 33–37.
10. Jiang, J.; Yu, C.; Zhou, Y. The study and aplication of clay swelling model considering wettability. *Pet. Geol. Eng.* **2016**, *30*, 87–89, 144.
11. Zou, C.; Yang, Z.; Cui, J.; Zhu, R.; Hou, L.; Tao, S.; Yuan, X.; Wu, S.; Lin, S.; Wang, L.; et al. Formation mechanism, geological characteristics and development strategy of nonmarine shale oil in China. *Pet. Explor. Dev.* **2013**, *40*, 14–26. [CrossRef]
12. Zhang, J.; Lin, L.; Li, Y.; Tang, X.; Zhu, L.; Xing, Y.; Jiang, S.; Jing, T.; Yang, S. Classification and evaluation of shale oil. *Earth Sci. Front.* **2012**, *19*, 322–331.
13. Wu, X.; Gao, B.; Ye, X.; Bian, R.; Nie, H.; Lu, F. Shale oil accumulation conditions and exploration potential of faulted basins in the east of China. *Oil Gas Geol.* **2013**, *34*, 455–462.
14. Fangxing, N.; Xuejun, W.; Xuefeng, H. An analysis on occurrence state and mobility of shale oil in Jiyang depression. *Xinjiang Oil Gas* **2015**, *11*, 1–5.
15. Yang, C.; Sheng, G.; Dang, Z. Sorption mechanism of polycyclic aromatic hydrocarbons (PAHs) on kerogen. *Environ. Chem.* **2007**, *26*, 472–475.
16. Cai, J. *Organo-Clay Complexes in Muddy Sediments and Mudstones*; Science Press: Beijing, China, 2004; pp. 53–144.
17. Juanhong, G.; Yanrong, Z.; Yonghe, Y. Evolutional characteristics of the kerogen molecular structure during the low-mature stage: An infrared spectra analysis. *Geochimica* **2014**, *43*, 529–537.
18. Qian, M.; Jiang, Q.; Li, M.; Li, Z.; Liu, P.; Ma, Y.; Cao, T. Quantitative characterization of extractable organic matter in lacustrine shale with different occurrences. *Pet. Geol. Exp.* **2017**, *39*, 278–286.
19. Jiang, Q.; Li, M.; Qian, M.; Li, Z.; Li, Z.; Huang, Z.; Zhang, C.; Ma, Y. Quantitative characterization of shale oil in different occurrence states and its application. *Pet. Geol. Exp.* **2016**, *38*, 842–849.
20. Wang, S.; Feng, Q.; Zha, M.; Lu, S.; Qin, Y.; Xia, T.; Zhang, C. Molecular dynamics simulation of liquid alkane occurrence state in pores and fractures of shale organic matter. *Pet. Explor. Dev.* **2015**, *42*, 772–778. [CrossRef]
21. Sang, Q.; Zhang, S.; Li, Y.; Dong, M.; Bryant, S. Determination of organic and inorganic hydrocarbon saturations and effective porosities in shale using vacuum-imbibition method. *Int. J. Coal Geol.* **2018**, *200*, 123–134. [CrossRef]
22. Li, S.; Sang, Q.; Dong, M.; Luo, P. Determination of inorganic and organic permeabilities of shale. *Int. J. Coal Geol.* **2019**, *215*, 103296. [CrossRef]
23. Barrett, E.P.; Joyner, L.G.; Halenda, P.P. The determination of pore volume and area distributions in porous substances. I. Computations from nitrogen isotherms. *J. Am. Chem. Soc.* **1951**, *73*, 373–380. [CrossRef]
24. Li, J.; Huang, W.; Lu, S.; Wang, M.; Chen, G.; Tian, W.; Guo, Z. Nuclear Magnetic Resonance T_1-T_2 Map Division Method for Hydrogen-Bearing Components in Continental Shale. *Energy Fuels* **2018**, *32*, 9043–9054. [CrossRef]

Disclaimer/Publisher's Note: The statements, opinions and data contained in all publications are solely those of the individual author(s) and contributor(s) and not of MDPI and/or the editor(s). MDPI and/or the editor(s) disclaim responsibility for any injury to people or property resulting from any ideas, methods, instructions or products referred to in the content.

Article

CO_2–Water–Rock Interaction and Its Influence on the Physical Properties of Continental Shale Oil Reservoirs

Sheng Cao [1,2,3], Qian Sang [4,*], Guozhong Zhao [1,2,3], Yubo Lan [1,2,3], Dapeng Dong [1,2,3] and Qingzhen Wang [1,2,3]

1. National Key Laboratory for Multi-Resources Collaborative Green Production of Continental Shale Oil, Daqing 163712, China
2. Exploration and Development Research Institute of Daqing Oilfield Co., Ltd., Daqing 163712, China
3. Heilongjiang Provincial Key Laboratory of Reservoir Physics & Fluid Mechanics in Porous Medium, Daqing 163712, China
4. School of Petroleum Engineering, China University of Petroleum (East China), Qingdao 266580, China
* Correspondence: 20190005@upc.edu.cn

Citation: Cao, S.; Sang, Q.; Zhao, G.; Lan, Y.; Dong, D.; Wang, Q. CO_2–Water–Rock Interaction and Its Influence on the Physical Properties of Continental Shale Oil Reservoirs. Energies 2024, 17, 477. https://doi.org/10.3390/en17020477

Academic Editor: Reza Rezaee

Received: 15 December 2023
Revised: 11 January 2024
Accepted: 16 January 2024
Published: 18 January 2024

Copyright: © 2024 by the authors. Licensee MDPI, Basel, Switzerland. This article is an open access article distributed under the terms and conditions of the Creative Commons Attribution (CC BY) license (https://creativecommons.org/licenses/by/4.0/).

Abstract: Shale oil resources are abundant, but reservoirs exhibit strong heterogeneity with extremely low porosity and permeability, and their development is challenging. Carbon dioxide (CO_2) injection technology is crucial for efficient shale oil development. When CO_2 is dissolved in reservoir formation water, it undergoes a series of physical and chemical reactions with various rock minerals present in the reservoir. These reactions not only modify the reservoir environment but also lead to precipitation that impacts the development of the oil reservoir. In this paper, the effects of water–rock interaction on core porosity and permeability during CO_2 displacement are investigated by combining static and dynamic tests. The results reveal that the injection of CO_2 into the core leads to reactions between CO_2 and rock minerals upon dissolution in formation water. These reactions result in the formation of new minerals and the obstruction of clastic particles, thereby reducing core permeability. However, the generation of fine fractures through carbonic acid corrosion yields an increase in core permeability. The CO_2–water–rock reaction is significantly influenced by the PV number, pressure, and temperature. As the injected PV number increases, the degree of pore throat plugging gradually increases. As the pressure increases, the volume of larger pore spaces gradually decreases, resulting in an increase in the degree of pore blockage. However, when the pressure exceeds 20 MPa, the degree of carbonic acid dissolution will be enhanced, resulting in the formation of small cracks and an increase in the volume of small pores. As the temperature reaches the critical point, the degree of blockage of macropores gradually increases, and the blockage of small pores also occurs, which eventually leads to a decrease in core porosity.

Keywords: shale formation; CO_2 flooding; CO_2–water–rock reaction; blocking action; dissolution reaction

1. Introduction

Although shale oil resources are abundant, the heterogeneity of reservoirs is significant, with low porosity and permeability, making their development challenging and resulting in a low degree of primary exploitation. Because of the obvious water sensitivity, waterflood cannot be adopted. Therefore, carbon dioxide (CO_2) injection technology serves as a crucial approach to achieving the efficient development of the resource. CO_2 flooding has been widely used in conventional reservoirs as an efficient oil displacement technology and a way to reduce greenhouse gases. In recent years, field tests for CO_2 injection development in shale oil reservoirs have been gradually initiated. Compared to tight oil reservoirs, shale reservoirs typically contain abundant organic matter called kerogen. Kerogen possesses a strong ability to adsorb and dissolve crude oil, whereas CO_2 exhibits a potent capability to extract hydrocarbons from shale formations [1,2]. Laboratory experiments on CO_2 huff and puff for various types of shale oil samples (Mancos and Eagle Ford core; diameter:

1.5 inches; length: 2 inches; huff and puff pressure: 850–3500 psi) have demonstrated ultimate recovery rates ranging from 33% to 85%. The ultimate recovery is found to be correlated with shale properties and the operational parameters (soaking time, huff and puff times) of multiple huff and puff cycles [3]. Simultaneously, the interaction between CO_2 and kerogen can induce alterations in the specific surface area, porosity, and microstructure of shale [4,5], while CO_2 can undergo adsorption and dissolution within kerogen [6].

Once CO_2 infiltrates the formation, it initially undergoes a reaction with the reservoir fluid, followed by a series of diverse reactions involving CO_2, reservoir fluid, and reservoir rock minerals. CO_2 is dissolved in formation water under high-temperature and high-pressure conditions to form carbonic acid, which is mainly divided into two steps of ionization, and the main ionization equations are as follows:

$$H_2CO_3 \leftrightarrow H^+ + HCO_3^- \tag{1}$$

$$HCO_3^- \leftrightarrow H^+ + CO_3^{2-} \tag{2}$$

Shale reservoirs are rich in carbonate and silicate minerals. Carbonate minerals mainly include dolomite, calcite, etc., which are easy to react with carbonic acid solutions and form new secondary minerals under the conditions of high temperatures and high pressure in the formation [7], and the reaction equations are as follows:

$$CaMg(CO_3)_2 + 2H^+ = Ca^{2+} + Mg^{2+} + 2HCO_3^- \tag{3}$$

$$CaMg(CO_3)_2 + 2H_2CO_3 = Ca^{2+} + Mg^{2+} + 4HCO_3^- \tag{4}$$

$$CaMg(CO_3)_2 + 2H_2O = Ca^{2+} + Mg^{2+} + 2HCO_3^- + 2OH^- \tag{5}$$

Most silicate minerals (albite, potassium feldspar, clay minerals other than quartz) are extremely unstable under acidic environmental conditions. They are readily soluble in water and generate secondary minerals [8]. The reaction equations are as follows:

$$2KAlSi_3O_8 + 2H^+ + 9H_2O = Al_2Si_2O_5(OH)_4 + 2K^+ + 4H_4SiO_4 \tag{6}$$

$$2NaAlSi_3O_8 + 3H_2O + 2CO_2 = Al_2Si_2O_5(OH)_4 + 4SiO_2 + 2Na^+ + 2HCO_3^- \tag{7}$$

$$CaAl_2Si_2O_8 + H_2CO_3 + H_2O = CaCO_3 + Al_2SiO_5(OH)_4 \tag{8}$$

$$(Fe/Mg)_5Al_2Si_3O_{10}(OH)_8 + 5CaCO_3 + 5CO_2 = 5Ca[Fe/Mg](CO_3)_2 + Al_2Si_2O_5(OH)_4 + 2H_2O \tag{9}$$

The effects of CO_2 reaction with different minerals on reservoir properties have been widely reported. Most of these studies focus on siliciclastic and carbonate formations [9–21]. Ross et al. [9] observed that CO_2 reacted with limestone and dolomite in the core, and the core's permeability increased after the dissolution of the dolomite and limestone. Knet et al. [10] observed that the carbonate minerals and clay minerals in sandstone were dissolved, and the fine particles were deposited into the pore throats through fluid migration. Sayegh et al. [11] observed that detrital particles such as illite and calcite in sandstone dissolved in large quantities and clogged in pores and throats as reservoir fluids migrated. Minerals such as carbonate cements are dissolved, creating a large number of micropores. Qu [12] conducted experiments on the reaction between different minerals and CO_2, and the dissolution effect of reservoir minerals (calcite, dolomite, carbonate rock, etc.) gradually increased with the increase in temperature when the temperature conditions changed. Shi et al. [13] analyzed the mineral composition of a sandstone core after CO_2 displacement, and the results showed that the carbonate mineral composition in the core increased significantly after CO_2 injection. Yu et al. [14] carried out CO_2 displacement experiments on a core of saturated formation water under the temperature and pressure conditions of the reservoir, and the experimental results showed that the dissolution reaction of carbonate minerals was the most violent, and the most obvious reaction was calcite among carbonate minerals, followed by flake aluminite minerals and iron dolomite with the lowest degree of dissolution. Wang et al. [15] found that after the temperature rises to a critical temperature, CO_2 changes from a gaseous state to a supercritical state, and the minerals in the core are violently dissolved. Secondary minerals are formed on the pore surface inside the core. Xiao et al. [16] evaluated the effect of CO_2–water–rock interactions on the characteristics of a carbonate reservoir at high pressure and temperature. With the increase in CO_2 pressure, the surface dissolution of calcite appeared more obvious. With the increase in the reaction temperature, the surface dissolution of calcite also appeared more obvious. Liu and Cheng [19] revealed the possible geochemical effects of cement mineral variations on water–rock–CO_2 interactions at 180 °C and 18 MPa. The sensitive orders of cement mineral variations due to water–rock–CO_2 interactions are carbonates, argillaceous, and siliceous minerals.

The reported results [16–21] show that the effects of temperature and pressure on the dissolution reaction and clogging are important. Different temperature and pressure conditions will result in different changes in porosity and permeability. At present, there is no systematic study on the influence of CO_2–water–rock reaction on core permeability and porosity under different temperature and pressure conditions. Shale reservoirs are more complex than sandstone and carbonate reservoirs due to their diverse and complex mineral types and strong heterogeneity. Shale reservoirs often comprise organic-rich mudstone layers, carbonate rocks, and sandstone interlayers, which all have diverse mineral types and a higher mineral content compared to sandstone reservoirs. In shale reservoirs, feldspar minerals and carbonate minerals, which are prone to dissolution reactions, coexist with detrital minerals that are prone to migration and clogging. While dissolution reactions can enhance the porosity and permeability of shale reservoirs to some extent, the sediments produced during the reaction are more likely to further clog the already fine pores. Therefore, it is urgent to study the dissolution and scaling laws of CO_2 on shale reservoirs and evaluate the effects of dissolution and scaling on the pore structure and porosity parameters of reservoirs, so as to provide reference data for formulating CO_2 development plans.

The aim of this paper is to study the impact of CO_2–water–rock reactions on the physical properties of shale cores during CO_2 displacement. Firstly, a static experiment of CO_2–water–rock reaction was carried out by using a high-temperature and high-pressure reactor. By analyzing the changes of different ion concentrations in formation water before and after CO_2 injection, the effects of temperature and pressure on precipitation formation were obtained. Additionally, scanning electron microscopy (SEM) was used to observe the dissolution of the pore structure of the rock samples after static experiments. Secondly, dynamic CO_2 displacement experiments were carried out. The T_2 spectra of the core before and after displacement was obtained using nuclear magnetic resonance technology, and the change in porosity was analyzed. By measuring the changes in permeability before and after core displacement, the influence of inorganic salt precipitation generated by CO_2–water–rock reaction on the physical properties of shale cores under different conditions was comprehensively evaluated.

2. Experimental
2.1. Experimental Material

The main mineral composition of the shale rock samples is shown in Table 1. The porosity and permeability of all samples are shown in Table 2. The ion contents of the formation water samples are shown in Table 3. The purity of CO_2 used in the tests is greater than 99.8%.

Table 1. Main mineral types and contents of shale cores.

Number	Plagioclase	Calcite	Quartz	Clay	Feldspar	Dolomite
#1	24.0	2.1	44.7	10.7	11.9	6.6
#2	15.0	6.0	48.0	9.7	12.3	9.0
#3	5.0	13.0	33.0	34.0	9.0	6.0
#4	24.0	6.0	24.0	18.0	16.0	12.0
#5	26.7	2.3	44.0	9.0	10.0	8.0
#6	19.0	6.8	48.1	12.0	9.1	5.0
#7	23.2	3.0	32.0	15.0	6.8	20.0
#8	18.6	9.2	42.7	14.0	8.7	7.0
#9	10.4	7.4	44.2	18.0	4.0	16.0
#10	21.5	7.7	41.8	16.0	4.0	9.0
#11	21.9	8.7	37.4	15.0	9.0	8.0
#12	28.0	6.1	45.6	8.2	6.1	6.0
#13	16.9	13.0	46.1	5.0	12.0	7.0
#14	23.3	5.4	38.4	10.0	13.9	9.0
#15	22.5	5.9	43.4	9.0	11.2	8.0

Table 2. Porosity and permeability parameters of cores.

Number	Length/cm	Diameter/cm	Porosity/%	Permeability/mD
#1	4.78	2.48	13.55	5.40×10^{-2}
#2	4.74	2.48	7.73	1.60×10^{-2}
#3	4.77	2.48	10.25	2.30×10^{-2}
#4	4.79	2.46	9.81	3.30×10^{-2}
#5	4.86	2.49	15.34	1.17×10^{-1}
#6	4.76	2.48	12.34	1.12×10^{-1}
#7	4.75	2.47	12.66	5.40×10^{-2}
#8	5.75	2.47	8.14	3.30×10^{-2}
#9	5.29	2.48	5.53	3.00×10^{-3}
#10	5.14	2.48	11.34	1.17×10^{-1}
#11	3.81	2.48	11.67	1.12×10^{-1}
#12	4.77	2.47	12.89	8.82×10^{-1}
#13	3.56	2.48	5.53	3.00×10^{-3}
#14	5.04	2.47	11.34	1.70×10^{-2}
#15	4.32	2.47	12.89	8.82×10^{-1}

Table 3. Ion contents of formation water.

Ion Species	Ion Content/(mg/L)
Ca^{2+}	5683.46
Mg^{2+}	421.803
Ba^{2+}	169.06
Na^+	15,030.60
Sr^{2+}	443.25
K^+	292.26
CO_3^{2-}	-
HCO_3^-	252

2.2. CO_2–Water–Rock Static Reaction

The experimental equipment is shown in Figure 1, including the CO_2 cylinder, the CO_2–water reaction cylinder, the vacuum pump, the hand pump, and the oven.

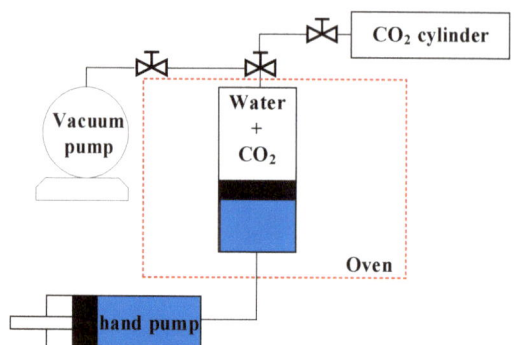

Figure 1. CO_2–water reaction: installation diagram of high-temperature and -pressure cylinder.

Prior to conducting the experiment, the leakage of the reactor was tested and the main ions present in the formation water were determined through an ICP-MS (inductively coupled plasma–mass spectrometry) analysis. Formation water and core samples were added to the reactor. CO_2 was injected after evacuation and a hand hump was used to raise the pressure of the entire system to a specified level. The reactor was subsequently placed inside a constant-temperature oven for a defined duration. After the experiment, the gas was slowly released and left undisturbed for a specified duration. Subsequently, the pH meter was employed to measure the variation in the pH value of the formation water prior to and after the CO_2–formation water reaction. The concentrations of cations, HCO_3^-, and CO_3^{2-} in the formation water were determined using the titration and ICP-MS methods. Moreover, the alterations observed in the mineral composition of the core, and the extent of core porosity dissolution subsequent to the reaction, were investigated utilizing the ICP-MS and electron microscopy scanning techniques. We studied the influence of different factors on the amount of precipitation by changing the experimental temperature and pressure.

2.3. CO_2–Water–Rock Dynamic Displacement Experiment

The experimental equipment is shown in Figure 2, including the high-pressure micro-metering pump, core holder, pressure gauge, electronic balance, thermostat, and nuclear magnetic resonance (NMR) instrument.

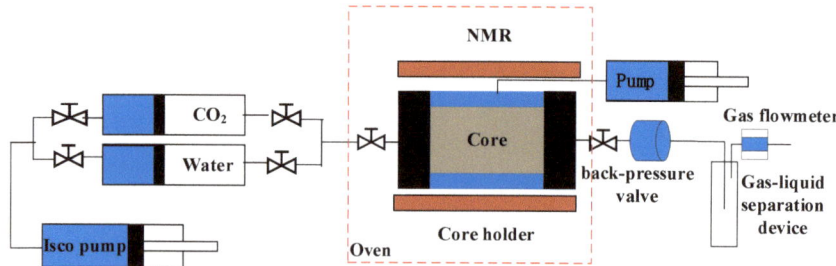

Figure 2. Image of CO_2 displacement experimental device based on NMR technology.

The air tightness of the core holder was tested. Through the analysis of the main ions in the formation water before the experiment using the ICP-MS method, it was clear that the cations which are easy to react with CO_2 form precipitated cations. After 48 h of drying, the dry weight was weighed and the formation water was saturated through pressure saturation. The porosity was calculated by comparing the mass difference before and after saturation with the total volume of the core, and transverse relaxation time (T_2) spectrum sampling was carried out using NMR. The saturated core was placed into the core holder, and the displacement pump conducted CO_2 displacement of the formation water in constant-pressure mode, with the pressure difference controlled by the back pressure valve set to 1 MPa. Based on the experimental plan, various experimental conditions were established for displacement, aiming to investigate the influence of CO_2 flooding water on reservoir pore physical properties, under varying temperatures (30 °C, 40 °C, 50 °C, 60 °C), displacement pressures (5 MPa, 10 MPa, 15 MPa, 20 MPa), and displacement PV values (25 PV, 50 PV, 75 PV, 100 PV). After completing the displacement, the T_2 spectrum of the core's re-saturated formation water was sampled and the core permeability was tested. The T_2 spectrum difference of the saturated water before and after displacement was compared to analyze the changes in the core's physical properties following the CO_2–water–rock reaction.

3. Results and Discussion

3.1. Influence of Pressure and Temperature Conditions on Formation Water pH Value

Figure 3 shows the influence of different temperatures on the pH value of the produced liquid at 20 MPa. As the solubility of CO_2 in the formation water decreases due to the rise in temperature, the CO_2 dissolved in the formation water also decreases. The reduction in CO_2 results in a decrease in carbonic acid and a subsequent reduction in H^+ in the formation water. Simultaneously, elevated temperatures promote the ionization reaction of carbonic acid toward the product, leading to an increase in H^+ in the formation water. Consequently, the pH of the resulting liquid increases after the CO_2–formation water reaction takes place.

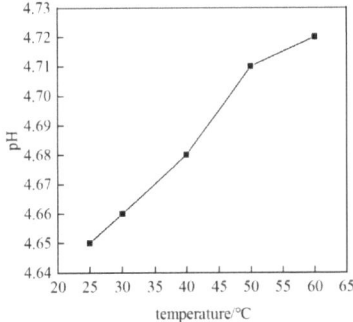

Figure 3. The pH of the system after the reaction of 25% CO_2 at 20 MPa.

Figure 4 demonstrates the impact of varying pressures on the pH value of the resulting liquid at 60 °C. When pressurizing the high-temperature and high-pressure reactor at a specific temperature, the solubility of CO_2 in formation water increases correspondingly, resulting in an elevation in carbonic acid due to the increased dissolution of CO_2 in the formation water. The rise in CO_2 levels facilitates the ionization reaction towards the product direction, resulting in an increase in the ionized H^+ concentration and consequently leading to a decrease in the pH of the formation water. Additionally, the increase in pressure accelerates the rate of the ionization reaction, thus promoting the continuous ionization of H_2CO_3 and HCO_3^- into H^+ ions and leading to a decrease in the pH value of the formation water. The H^+ concentration of the formation water increases under the combined action of the two effects, and the pH value of the formation water enhances towards acidity, gradually decreasing with the increase in pressure.

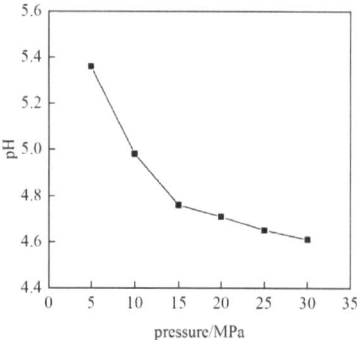

Figure 4. The pH of the system after 5% CO_2 reaction at 60 °C under different pressures.

3.2. Change in Ion Concentration in Formation Water during Static Reaction

Based on the X-ray diffraction (XRD) analysis of the core, it was found that the core contains a higher concentration of minerals such as plagioclase, potassium feldspar, and dolomite, which are more susceptible to reacting with CO_2 during injection. In order to verify the reaction of the core under the influence of carbonic acid, the formation water in the original static experiment was replaced with distilled water, the core fragments were placed in a high-temperature and high-pressure reactor, the distilled water was saturated after a vacuum, and the CO_2 was injected and left to stand for 40 h. Subsequent to the completion of the experiment, the resulting liquid was collected for ion detection. The changes in ion content within the produced liquid were then analyzed under different influencing factors, as presented in Table 4.

Table 4. Ion content of produced liquid.

Experiment Number	Pressure/MPa	Temperature/°C	$\rho(K^+)/(mg \cdot L^{-1})$	$\rho(Na^+)/(mg \cdot L^{-1})$	$\rho(Ca^{2+})/(mg \cdot L^{-1})$
1	5	60	2.34	17.64	80.32
2	10	60	4.67	21.56	115.68
3	15	60	6.23	27.83	174.36
4	20	60	7.68	37.97	220.71
5	20	30	3.31	16.51	100.03
6	20	40	4.37	24.88	167.28
7	20	50	5.96	30.45	190.33
8	20	60	7.68	37.97	220.71

Following the completion of the experiment, the concentration of common cations within the resulting solution was determined utilizing the ICP-MS method. K^+, Na^+ and Ca^{2+} could be detected in the produced liquid after the reaction by passing CO_2 into the high-temperature and high-pressure reactor. Since the distilled water in the reactor did not contain the above ions, the ions detected in the produced liquid were all generated through the dissolution of reservoir minerals. The presence of Na^+ and K^+ can primarily be attributed to the dissolution of plagioclase and potassium feldspar, respectively, indicating varying degrees of dissolution for these minerals within the core. Because the mineral content of plagioclase in the core is relatively high, the Na^+ concentration in the produced solution increases significantly, while the K^+ concentration in the produced solution increases due to the dissolution of potassium feldspar under the action of the acid solution. However, due to the low content of potassium feldspar in the core, the concentration only increases slightly. The presence of Ca^{2+} in the solution primarily originates from the dissolution of carbonate minerals like dolomite and calcite. Due to the reactivity of carbonate minerals with H^+, the concentration of Ca^{2+} in the resulting solution tends to increase more significantly compared to other ions. Simultaneously, it can be observed that as the pressure increases, the concentrations of Na^+, K^+, and Ca^{2+} in the resulting liquid gradually elevate. Under the influence of pressure, the dissolution capacity of rocks gradually amplifies as the pressure increases. When the pressure exceeds the critical level, the dissolution degree further increases. The elevation of temperature additionally serves to stimulate the dissolution of core minerals, particularly after surpassing the critical pressure. High temperatures continue to advance the dissolution reaction within the core. Therefore, the formation's high-temperature and -pressure conditions are conducive to the dissolution of rock minerals.

The alterations in particle size and PDI (polydispersity index) of formation water before and after the CO_2–water–rock static tests were analyzed using a Malvern laser particle size analyzer. The PDI represents whether the particle size distribution is uniform, and the smaller the PDI value, the more uniform the particle size distribution and the more uniform the particle size. On the contrary, the larger the PDI value, the wider the particle size distribution and the more uneven the particle size. The alterations in the particle size and PDI of the formation water were compared before adding core fragment samples and after the formation water reacted with the core samples and supercritical CO_2 at a pressure of 20 MPa and temperature of 60 °C. The results are presented in Table 5.

Table 5. Particle size and PDI before and after reaction.

	Particle Size/nm	PDI
Pre-reaction	1377	0.296
	1081	0.284
	1130	0.296
Post-reaction	2500	0.362
	2854	0.423
	2971	0.534

As depicted in Table 5, the average particle size of the formation water prior to the reaction is recorded as 1196 nm, with a PDI of 0.292. Conversely, after the reaction, the average particle size of the formation water increases to 2775 nm, accompanied by a raised PDI of 0.439. Following the reaction, the average particle size of the formation water experienced a significant increase of 1579 nm, while the PDI witnessed a noticeable elevation of 0.147. The uniformity of mineral particles on the surface of rock samples is rather inadequate, rendering them susceptible to dissolution and reaction when exposed to high-temperature and -pressure conditions. Carbonate minerals such as dolomite and calcite exhibit instability under acidic conditions, leading to the release of secondary minerals and crystalline substances from reacting salts into the formation water. Consequently, this process contributes to the increase in particle size in the formation water. After the dissolution of clay cements and other minerals, the cementation weakens and dislodges into the formation water, contributing to an increase in the average particle size and enhancing the particle size heterogeneity in the water.

After conducting the CO_2–water–rock static experiment, SEM was utilized to observe the microscopic changes on the core's surface and the dissolution state. The surface of the core sample was smoothed and then scanned with an electron microscope. The specific results are shown in Figures 5 and 6. Figure 5 clearly illustrates that, prior to the reaction, the surface of the core was smooth and even, devoid of any fine particles. Once the pressure increased to 5 MPa, the rock surface underwent a transformation, becoming rough and experiencing slight acidic corrosion from the carbonic acid. This lead to the emergence of minute pores at the marked locations in the figure. As the pressure increases, mineral dissolution becomes more intense and results in the emergence of more and more micro pores. In addition, the dissolution of minerals causes cemented clay minerals to disintegrate, leading to an increase in debris, as depicted in Figure 5c. When CO_2 reaches the supercritical state, the reaction becomes highly efficient. In Figure 5d,e, numerous small irregular particles are observed adhering to the surface of the rock. These particles may be derived from newly formed mineral particles resulting from the reaction between CO_2, water, and minerals, or they could be fine particles produced after the dissolution of initially larger minerals. Under intensified dissolution, the small pores formed due to the initial dissolution process also enlarge, forming solution pores with a larger size. However, the surfaces of these solution pores may be partially obstructed by clay particles and newly formed fine particles. At varying temperatures, Figure 6 demonstrates that as the temperature increases, the dissolution process becomes more pronounced. This leads to a gradual augmentation in the number of small pores. After the critical condition of CO_2 is reached, the particles on the core surface increase in number, the small pores become larger, large pores such as solution pits appear, and the dissolution effect on the minerals on the rock surface is strong.

3.3. Changes in Core Physical Properties after CO_2 Displacement
3.3.1. Effect of Injection PV Number

Table 6 shows core flooding information and displacement conditions. Samples of saturated water, both before and after displacement, were collected and compared by analyzing their respective T_2 spectra.

Table 6. CO_2 flooding conditions.

Experiment Number	Displacement Pressure/MPa	PV Number	Displacement Pressure Difference/MPa	Experimental Temperature/°C
1	10	25	1	60
2	10	50	1	60
3	10	75	1	60
4	10	100	1	60

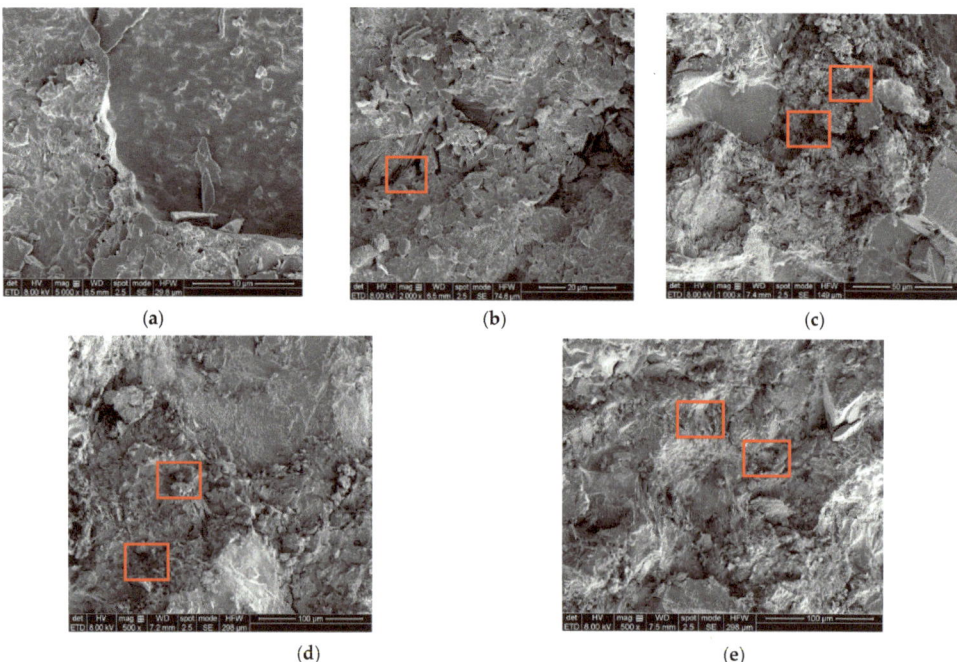

Figure 5. SEM diagram of core dissolution under different pressures: (**a**) 0 MPa; (**b**) 5 MPa; (**c**) 10 MPa; (**d**) 15 MPa; (**e**) 20 MPa. Red box denotes the change on the sample.

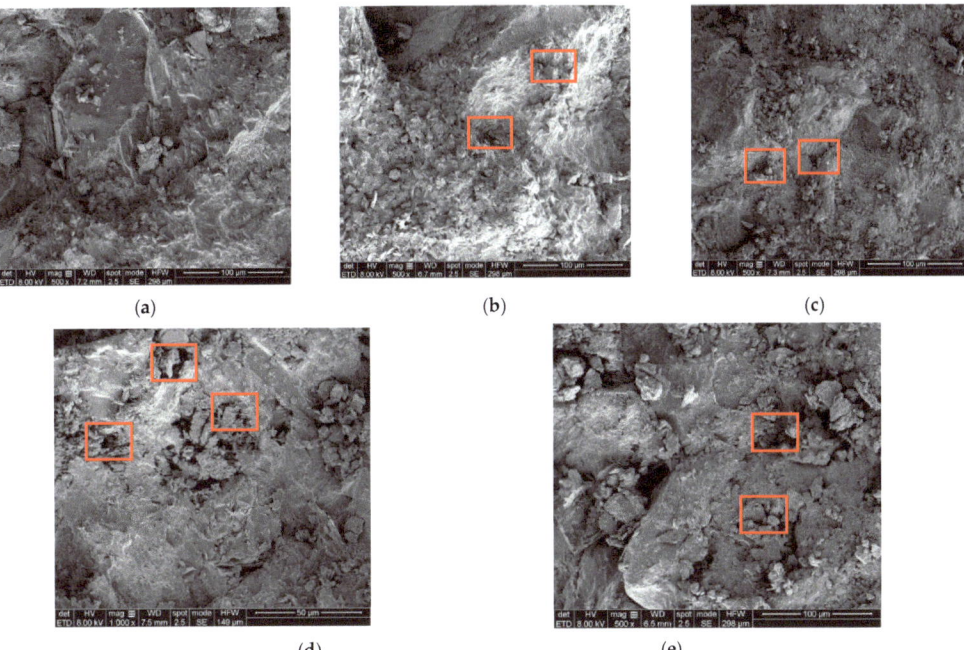

Figure 6. SEM diagram of core dissolution under different temperatures: (**a**) 20 °C; (**b**) 30 °C; (**c**) 40 °C; (**d**) 50 °C; (**e**) 60 °C. Red box denotes the change on the sample.

Figure 7 shows the T_2 spectra of saturated water before and after 25 PV flooding. From Figure 7, it can be observed that after CO_2 flooding of 25 PV in saturated water–rock cores, the quantity of secondary saturated water is significantly lower than that of the initial saturated water. This indicates that, following the 25 PV flooding, the pore throat becomes obstructed by the newly formed minerals resulting from the water–rock reaction of CO_2 and the debris generated by the mineral corrosion from carbonic acid formation. The saturation of water before and after CO_2 flooding undergoes a significant change between 1 ms to 100 ms, implying that the water–rock reaction of CO_2 primarily impacts the intermediate and larger-sized pores during the initial stages of CO_2 flooding.

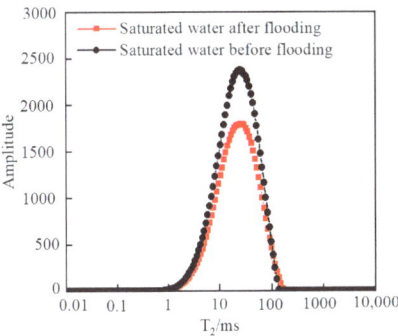

Figure 7. Saturated water T_2 spectra before and after 25 PV flooding.

Figure 8 shows the T_2 spectra of saturated water before and after 50 PV flooding. After the 50 PV injection, it can be seen from the figure that the pore water saturation at 0.1–1 ms is reduced to a certain extent, which indicates that blockage occurs in the small hole. The saturation of water also exhibited a certain level of change before and after displacement between 10 and 100 ms. This implies that, as the reaction time increased, the debris created through the dissolution process following CO_2 injection gradually accumulated in the larger-sized pores, leading to the formation of partial blockages.

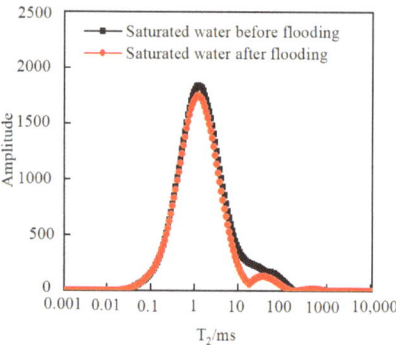

Figure 8. Saturated water T_2 spectra before and after 50 PV flooding.

Figure 9 shows the T_2 spectra of saturated water before and after 75 PV flooding. After injecting 75 PV of fluid, the saturated water volume in the small pores (represented by 1–10 ms) and the saturated water volume in the larger pores (represented by 10–100 ms) decreased significantly. This suggests a more severe pore blockage and a reduction in the pore throat volume. However, the saturated water volume of the small pores (represented by 0.1–1 ms) only decreased slightly, indicating that, with the increasing reaction time, the plugging at mesopores and larger pores decreased, resulting in an increase in pore volume. Nonetheless, CO_2 still reacted with water in the small pores, resulting in a certain degree of plugging.

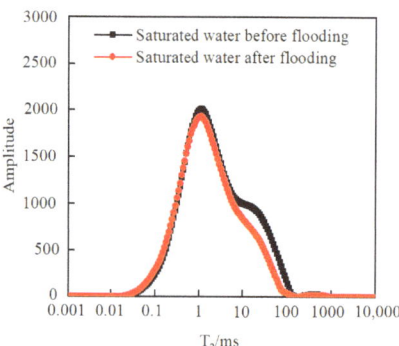

Figure 9. Saturated water T_2 spectra before and after 75 PV flooding.

Figure 10 shows the T_2 spectra of saturated water before and after 100 PV flooding. As the CO_2 flooding time increased, the signal volume after CO_2 flooding decreased significantly, revealing a more severe degree of plugging. The most significant change occurred in the region of mesopores and larger pores (represented by 10–100 ms), indicating that rock debris particles generated by CO_2–water–rock dissolution were concentrated in the area of larger pores. Conversely, the signal volume in the small and medium pores (represented by 0.1–10 ms) increased, indicating an increase in the saturated water volume. Therefore, it can be inferred that the dissolution caused small cracks in the pore channels, leading to an increase in pore volume in the small and medium pores.

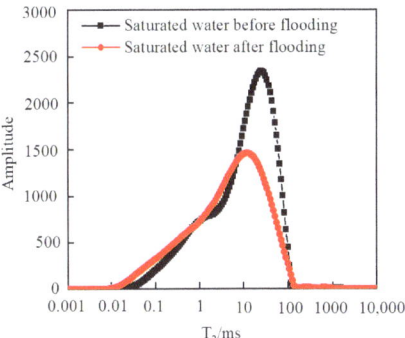

Figure 10. Saturated water T_2 spectra before and after 100 PV flooding.

After the reaction between CO_2, water, and rock, secondary minerals and stripped particles are formed and migrate with the fluid, blocking the pores and reducing the overall pore volume of the core. Utilizing the principle of nuclear magnetic resonance, the peak area of the T_2 curve reflects the signal quantity emitted by hydrogen in the entire core, which corresponds to the amount of saturated water in the core. The T_2 spectra of saturated water are measured before and after CO_2 flooding to characterize these changes. Figure 11 shows the schematic diagram for calculating the degree of pore plugging.

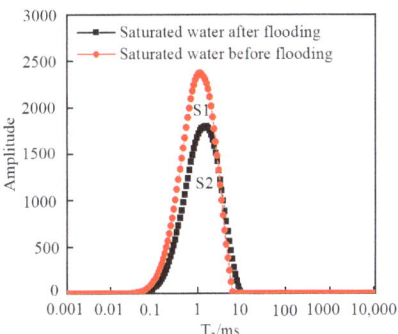

Figure 11. Schematic diagram of core pore-plugging calculation.

The figure illustrates the schematic diagram of the degree of core pore plugging. Assuming that the peak area of the T_2 spectrum of saturated water before CO_2 flooding is represented by S_1, and the peak area of the T_2 spectrum of saturated water after CO_2 flooding is represented by S_2; the pore plugging rate (B) can be obtained by comparing the difference in the T_2 spectrum before and after saturation, as shown in Equation (10).

$$B = \frac{S_1 - S_2}{S_1} \tag{10}$$

where B is the pore plugging rate; S_1 is the peak area value of saturated water before CO_2 flooding; and S_2 is the peak area value of saturated water after flooding. Table 7 shows the pore permeability changes of the core before and after CO_2 displacement. From the table, it is evident that the permeability of the core decreases after CO_2 flooding compared to its initial value, with a reduction ranging from 9% to 18%. Similarly, the porosity of the core also decreases after CO_2 flooding, with a reduction ranging from 6% to 17%. With the increase in the PV number, the pore permeability change rate and pore plugging rate still increased significantly, indicating that with the continuous injection of CO_2, the precipitation plugging effect generated by the CO_2–water–rock reaction was greater than the dissolution effect of CO_2 on different minerals.

Table 7. Table of changes of porosity and permeability before and after CO_2 flooding.

Core Number	Permeability (before) /mD	Permeability (after) /mD	Rate of Change /%	Porosity (before) /%	Porosity (after) /%	Rate of Change /%	Pore Throat Blockage Rate/%
#1	3.30×10^{-2}	4.90×10^{-2}	9.26	13.55	12.66	6.56	7.21
#2	3.30×10^{-2}	1.40×10^{-2}	12.50	7.73	7.12	7.89	8.42
#3	2.30×10^{-2}	2.00×10^{-2}	13.04	10.25	8.84	13.75	11.66
#4	3.30×10^{-2}	2.70×10^{-2}	18.18	9.81	8.14	17.06	17.94

3.3.2. Effect of Pressure

In this experiment, the CO_2 flooding pressure conditions were modified while keeping other conditions constant, as depicted in Table 8. Figure 12 shows the T_2 spectra of saturated formation water before and after CO_2 flooding under different pressures. It is apparent in Figure 12a that the larger pores are the first to experience blockage, leading to a significant decrease in secondary saturation flag; the smaller and medium-sized pores of the core (0.1~10 ms) show little change in their secondary saturation flag. Figure 12b shows the T_2 spectra of saturated formation water before and after CO_2 flooding at a pressure of 10 MPa. As the CO_2 flooding pressure increased, both the temperature and pressure were higher than the critical conditions of CO_2 (31.6 °C, 7.39 MPa). CO_2 transitioned from a gaseous state to a supercritical state. In the larger pores, represented by 10–100 ms, the semaphore of the T_2 spectrum exhibited a significant reduction following the CO_2 flooding. This suggests that the blockage resulting from the water–rock reaction was intensified in these larger pores. As the CO_2 flooding pressure increased, the amount of debris generated by dissolution also increased and began to gradually accumulate within the micropores. At the same time, the change rate of the core porosity and permeability also indicates that the CO_2–water–rock reaction intensifies with the increase in the CO_2 flooding pressure. Figure 12c shows the T_2 spectra of saturated formation water before and after CO_2 flooding under a pressure of 15 MPa. As the CO_2 flooding pressure increases, the peak value of the T_2 spectrum measured with saturated water after CO_2 flooding shifts towards the left. This shift indicates a reduction in the overall pore diameter of the core, in which the pores become blocked due to the CO_2–water–rock reaction. The amount of saturated water in larger pores decreases, which suggests the gradual accumulation of debris generated by dissolution or the formation of new minerals. This accumulation ultimately leads to a decrease in secondary saturated water. The semaphore of small and medium pores increased slightly after displacement, indicating that small fractures were produced by dissolution, the pore volume of small pores increased, and the change rate of the pore permeability did not continue to increase. After the injection pressure rises to 20 MPa, it is obvious from Figure 12d that the CO_2–water–rock reaction becomes more and more intense under the influence of the rising pressure, and the signal volume of the T_2 spectrum of the core after CO_2 flooding becomes less and less, but the secondary saturated water in small and medium pores increases. The observed trend indicates that as the pressure rises, the dissolution process becomes more pronounced, leading to the formation and connection of small fractures with larger pores. Dissolution plays a predominant role in driving the water–rock reaction.

Table 8. Core displacement conditions.

Experiment Number	Displacement Pressure/MPa	PV Number	Displacement Pressure Difference/MPa	Experimental Temperature/°C
1	5	50	1	60
2	10	50	1	60
3	15	50	1	60
4	20	50	1	60

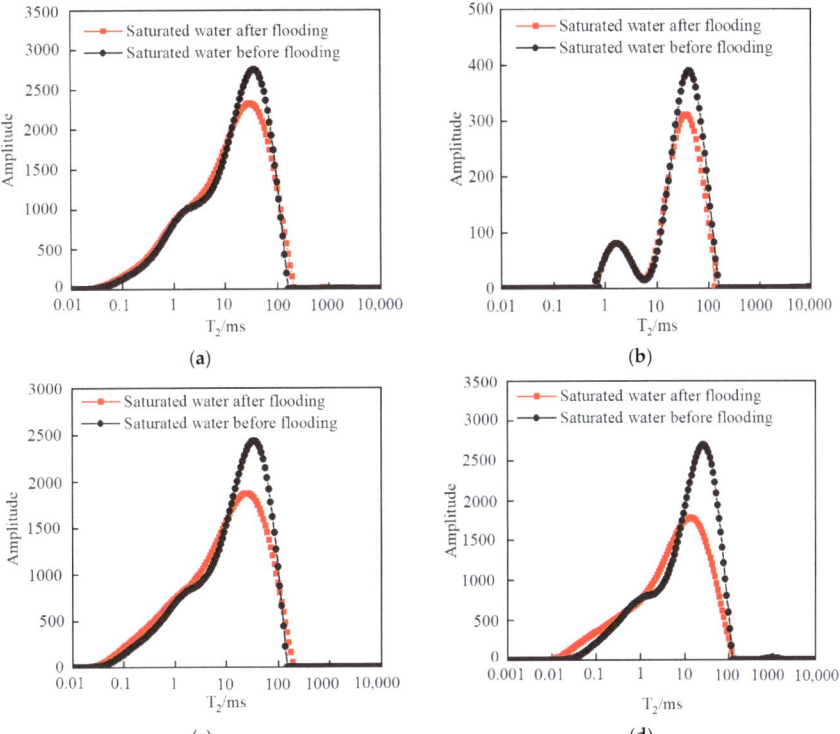

Figure 12. Saturated water T_2 spectra before and after CO_2 flooding at different pressures: (**a**) 5 MPa; (**b**) 10 MPa; (**c**) 15 MPa; (**d**) 20 MPa.

Table 9 shows the pore permeability changes of the core before and after CO_2 flooding under different pressure conditions. The table illustrates that the permeability of the core after CO_2 flooding was lower than that of the core before CO_2 flooding, with a reduction of approximately 10–15%. The porosity of the core after CO_2 flooding also exhibited a decrease of around 11–17% when compared to the initial porosity before CO_2 flooding. When CO_2 was injected into the core, the CO_2–water–rock reaction caused the dissolution of some minerals, resulting in the generation of new small pores. During the continuous flooding, the dissolved minerals were transported to the macropores, and some of the pores were blocked. With the increase in the injection pressure, the amount of injected CO_2 increases, and the CO_2–water–rock reaction becomes more violent, which will lead to more minerals shedding, and there is a greater probability that the macropores will be blocked during the migration process. As the CO_2 flooding pressure increases, the rate of change in pore permeability and pore throat plugging initially rises and then declines. This trend suggests that at low pressures, the plugging effect of the CO_2–water–rock reaction in larger pores is predominant. However, at high pressures, the CO_2–water–rock reaction leads to the generation of new small pores, indicating that dissolution becomes the dominant mechanism.

Table 9. Changes in porosity and permeability before and after CO_2 flooding under different pressures.

Core Number	Displacement Pressure /MPa	Permeability (before) /mD	Permeability (after) /mD	Rate of Change /%	Porosity (before) /%	Porosity (after) /%	Rate of Change /%	Pore Throat Blockage Rate/%
#5	5	1.17×10^{-1}	1.05×10^{-1}	10.26	15.34	13.51	11.93	8.21
#6	10	1.12×10^{-1}	0.95×10^{-1}	15.20	12.34	10.16	17.67	8.74
#7	15	5.40×10^{-2}	4.60×10^{-2}	14.81	12.66	10.46	17.37	13.41
#8	20	3.30×10^{-2}	2.90×10^{-2}	12.12	8.14	6.78	16.71	13.27

3.3.3. Effect of Temperature

In this group of experiments, temperature conditions were changed, and other conditions remained the same, as shown in Table 10.

Table 10. Core information and displacement conditions.

Experiment Number	Displacement Pressure/Mpa	PV Number	Displacement Pressure Difference/Mpa	Experimental Temperature/°C
1	10	50	1	30
2	10	50	1	40
3	10	50	1	50
4	10	50	1	60

Figure 13 shows the T_2 spectra of saturated formation water before and after CO_2 flooding at different temperatures. Table 11 shows the pore permeability changes of the core before and after CO_2 flooding under different temperature conditions. Figure 13a shows the T_2 spectra of saturated formation water before and after CO_2 flooding at 30 °C (before reaching the critical temperature of CO_2). The reaction between CO_2 and formation water rocks in the gas state is still weak, and the water quantity before and after saturation is only slightly decreased, with the change rate of porosity decreasing at 4.58% and permeability decreasing at 5.22%. Only a slight blockage occurs in the larger pores represented by 10–100 ms, resulting in a decrease in the secondary saturated water quantity after displacement. Figure 13b shows the T_2 spectra measured for the saturated formation water before and after the core reaction at 40 °C. Compared with 30 °C, the critical temperature of CO_2 was reached when the experimental temperature reached 40 °C, and CO_2 was in a supercritical state, which intensified the water–rock reaction of CO_2. The larger pores represented by 10–100 ms have a certain degree of blockage. Figure 13c shows the T_2 spectra measured for the saturated formation water before and after the core reaction at 50 °C. As the temperature increases, the interaction between CO_2 and the water–rock reaction is further intensified. Consequently, the secondary saturation of water in small and medium pores decreases significantly, indicating a more severe blockage compared to that at 40 °C. The overall porosity of the core shows a change rate of 12.86%, with a corresponding decrease in permeability of 10.48%. Figure 13d shows the T_2 spectra measured for the saturated formation water before and after the core reaction at 60 °C. At 60 °C, the degree of the CO_2–water–rock reaction further intensifies, leading to the dissolution of minerals within the core. During the migration, the dissolved particles become lodged in various pore throats, causing a reduction in the secondary water saturation after displacement. This blockage affects both large and small pores to varying degrees. As a result, the overall porosity of the core decreases by 15.2%, with a corresponding decrease in permeability of 15.88%. It is evident that an increase in temperature facilitates and enhances the CO_2–water–rock reaction.

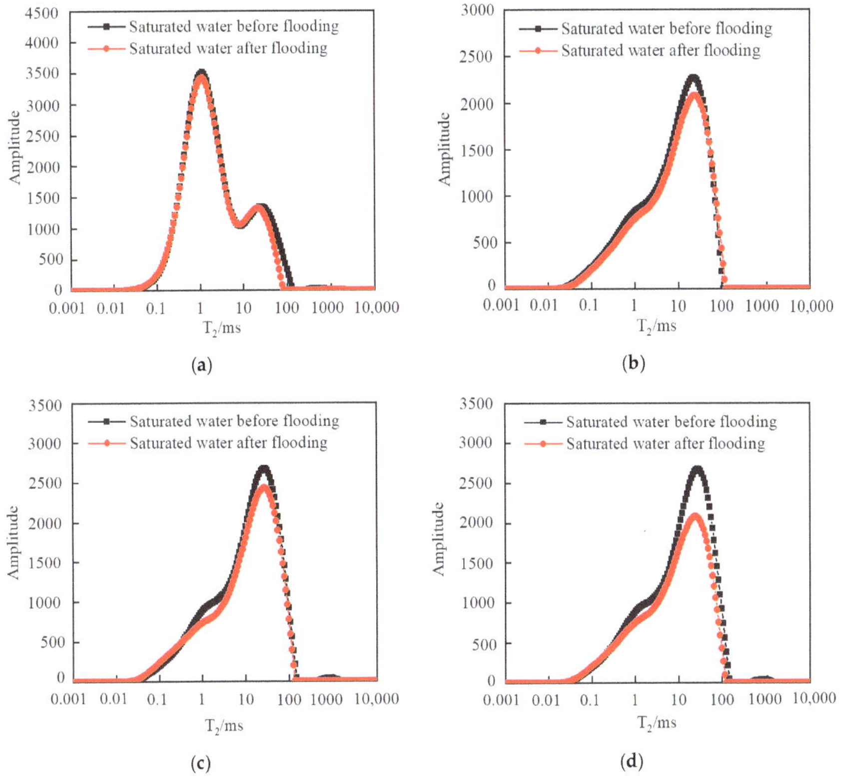

Figure 13. Saturated water T_2 spectra before and after CO_2 flooding at different temperatures: (a) 30 °C; (b) 40 °C; (c) 50 °C; (d) 60 °C.

Table 11. Changes in porosity and permeability before and after CO_2 flooding under different temperatures.

Core Number	Experimental Temperature /°C	Permeability (before) /mD	Permeability (after) /mD	Rate of Change/%	Porosity (before) /%	Porosity (after) /%	Rate of Change /%	Pore Throat Blockage Rate/%
#9	30	3.00×10^{-3}	2.90×10^{-3}	5.22	5.53	5.28	4.58	2.78
#10	40	1.17×10^{-1}	1.05×10^{-1}	10.21	11.34	10.28	9.34	7.84
#11	50	1.12×10^{-1}	1.00×10^{-1}	10.48	11.67	9.94	12.86	10.22
#12	60	8.82×10^{-1}	7.42×10^{-1}	15.88	12.89	12.12	15.20	14.78

3.3.4. Effect of Core Permeability

For cores with different permeability, the experimental temperature is 60 °C, the CO_2 flooding pressure is 10 MPa, the pressure difference is controlled at 1 MPa, and the injection PV number is 50. Figure 14 shows the T_2 spectra of saturated formation water before and after CO_2 flooding with different permeability. It can be seen from the comparison that the core with low permeability is greatly affected by the water–rock reaction, and the secondary saturated water volume decreases more than before the experiment. Considering the small pore radius of cores with low permeability, the migration of secondary minerals formed by the CO_2–water–rock reaction and shed clay particles tends to accumulate in these tiny pores, resulting in significantly stronger plugging effects compared to the other two cores.

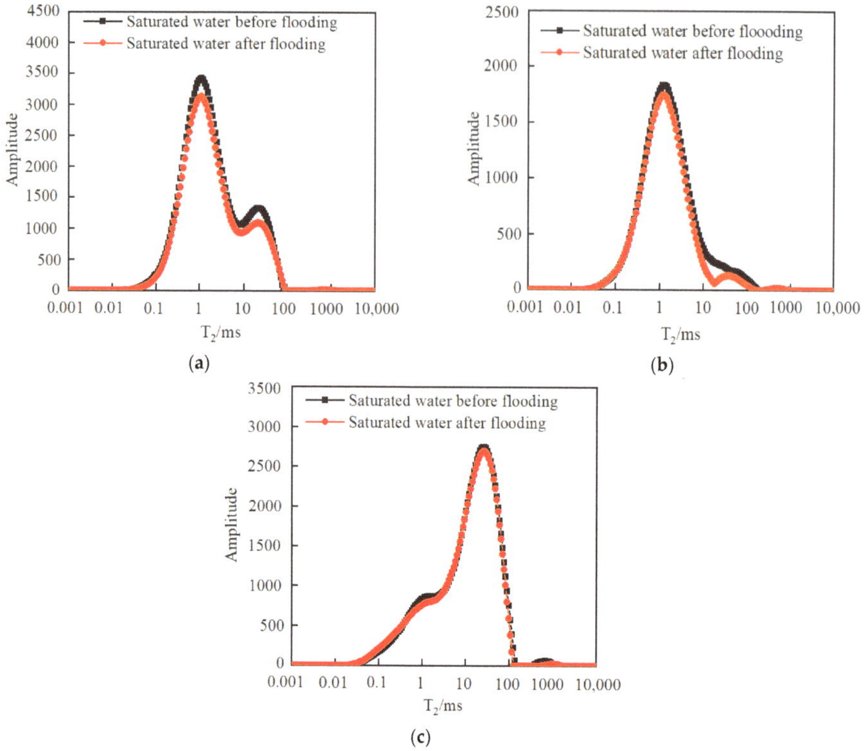

Figure 14. Saturated water T_2 spectra before and after CO_2 flooding with different permeability: (a) $k = 10^{-3}$ mD; (b) $k = 10^{-2}$ mD; (c) $k = 10^{-1}$ mD.

Table 12 shows the pore permeability changes of the cores with different permeability before and after CO_2 flooding. The data presented in the table illustrate that the permeability of the cores decreases after displacement, with a reduction ranging from 2.11% to 10.78%. Similarly, the porosity of the cores also experiences a decrease after CO_2 flooding, ranging from 2.74% to 8.84%. As the core permeability decreases, there is a gradual increase in both the pore permeability change rate and pore throat plugging rate, indicating that the plugging effects resulting from the CO_2–water–rock reaction's precipitation have a more significant impact on cores with lower permeability.

Table 12. Changes in porosity and permeability of different core samples before and after CO_2 flooding.

Core Number	Permeability (before) /mD	Permeability (after) /mD	Rate of Change /%	Porosity (before) /%	Porosity (after) /%	Rate of Change /%	Pore Throat Blockage Rate /%
#13	3.00×10^{-3}	2.60×10^{-3}	10.78	5.53	4.79	8.84	9.02
#14	1.70×10^{-2}	1.60×10^{-2}	5.66	11.34	10.77	5.02	6.84
#15	8.82×10^{-1}	8.64×10^{-1}	2.11	2.89	12.56	2.74	2.07

4. Conclusions

This paper explores the water–rock reaction conditions and their influence on the core physical properties of continental shale reservoirs through static CO_2–water–rock reactions and dynamic CO_2 flooding experiments. The main conclusions are as follows:

(1) CO_2 has a notable effect on the pH value of the formation water. To mitigate this influence during CO_2 injection for reservoir development purposes, it is necessary to increase the injection pressure to reduce the pH value of the reservoir. By creating an acidic environment, some minerals in the reservoir can be dissolved, and the reservoir's permeability can be enhanced to some extent.

(2) As the temperature increased, the solubility of CO_2 in the formation water decreased. Consequently, the concentration of CO_3^{2-} ions decreased, and the amount of carbonate precipitation gradually decreased. Conversely, increasing the pressure promoted the ionization of carbonic acid and bicarbonate, leading to an increase in CO_2 solubility. Upon pressure release, the CO_2 concentration decreased, resulting in an increase in formation water precipitation.

(3) The average particle size of the produced liquid increased by 1579 nm, and the PDI increased by 0.147. It is important to note that the core is susceptible to corrosion under high-temperature and -pressure conditions. As the temperature and pressure increased, the degree of dissolution gradually intensified. The concentration of Na^+ and K^+ ions primarily increased due to the dissolution of plagioclase and potassium feldspar minerals. Furthermore, the concentration of Ca^{2+} ions primarily increased as a result of the dissolution of dolomite and calcite carbonate minerals.

(4) As the injected PV increased, CO_2 first entered the macropores, resulting in a decrease in the amount of secondary saturated water within the macropores and an increasing degree of pore throat blockage. As the injected PV continued to increase, the dissolution process led to the formation of small cracks, resulting in an increased amount of secondary saturated water compared to the initial stage, and the volume of small pores increased. Pressure plays a significant role in the CO_2–water–rock reaction. When the pressure reaches the supercritical state, the dissolution process intensifies at the pore throat. The resulting debris from dissolution can then block the flow path, leading to a significant reduction in signal within larger pores and causing severe blockage. Temperature also has an effect on the water–rock reaction. As the temperature increases to the critical temperature, the macropores are the first to become blocked, and the degree of blockage in the macropores gradually increases. The small pores are also blocked, leading to a decrease in the porosity of the core.

In this study, the influence of CO_2–water–rock reaction on the porosity and permeability of shale cores under different temperature and pressure conditions was quantitatively characterized. The results show that the blockage caused by CO_2–water–rock reflection mainly occurs in macropores, and the degree of blockage is higher than that of small pores. Under high-temperature and -pressure conditions, due to the intensification of dissolution, some new small pores can also be generated. These results provide a basic understanding of the development plan and clarify the degree of formation damage caused by CO_2 injection in shale reservoirs at different stages of development, which is helpful to determine the CO_2 injection pressure and temperature. In addition, the results can also be used to predict the content and stability of CO_2 stored in such reservoirs. Shale reservoirs have complex lithologies, and not all samples of lithology are tested due to the long experimental period. Other types of shale core testing require further conduct. The results do not take into account the CO_2–crude oil interaction and its effect on porosity and permeability. The above contents will be systematically studied in future studies.

Author Contributions: Methodology, D.D.; resources, D.D.; investigation, S.C., Q.S., G.Z., D.D. and Q.W.; writing—original draft preparation, S.C. and Q.S.; writing—review and editing, G.Z., Y.L., and Q.W.; supervision, Q.S.; project administration, Y.L.; funding acquisition, Q.S. and G.Z. All authors have read and agreed to the published version of the manuscript.

Funding: The authors declare that this study received funding from the Natural Science Foundation of Shandong Province of China (No. ZR2020ME091), and the Major Project of the China National Petroleum Corporation (2021ZZ10-02). The funders were not involved in the study design, collection, analysis, interpretation of data, the writing of this article or the decision to submit it for publication.

Data Availability Statement: The original contributions presented in the study are included in the article, further inquiries can be directed to the corresponding author.

Conflicts of Interest: Authors Sheng Cao, Guozhong Zhao, Yubo Lan, Dapeng Dong and Qingzhen Wang were employed by the company Exploration and Development Research Institute of Daqing Oilfield Co., Ltd. The remaining authors declare that the research was conducted in the absence of any commercial or financial relationships that could be construed as a potential conflict of interest.

Nomenclature

Abbreviations

SEM	scanning electron microscopy
ICP-MS	inductively coupled plasma–mass spectrometry
T_2	transverse relaxation time
NMR	nuclear magnetic resonance
XRD	X-ray diffraction
PDI	polydispersity index

Symbols

PV	the pore volume, dimensionless
B	the pore plugging rate, fraction
S_1	the peak area value of saturated water before flooding
S_2	the peak area value of saturated water after flooding

References

1. Jarboe, P.; Candela, P.; Zhu, W.; Kaufman, A. Extraction of hydrocarbons from high-maturity Marcellus shale using supercritical carbon dioxide. *Energy Fuels* **2015**, *29*, 7897–7909. [CrossRef]
2. Jin, L.; Hawthorne, S.; Sorensen, J.; Pekot, L.; Kurz, B.; Smith, S.; Heebink, L.; Herdegen, V.; Bosshart, N.; Dalkhaa, C.; et al. Advancing CO_2 enhanced oil recovery and storage in unconventional oil play—Experimental studies on Bakken shales. *Appl. Energy* **2017**, *208*, 171–183. [CrossRef]
3. Gamadi, T.; Sheng, J.; Soliman, M.; Menouar, H.; Watson, M.; Emadibaladehi, H. *An Experimental Study of Cyclic CO_2 Injection to Improve Shale Oil Recovery, SPE Improved Oil Recovery Symposium*; Society of Petroleum Engineers: Tulsa, OK, USA, 2014; pp. 1–9.
4. Jiang, Y.; Luo, Y.; Lu, Y.; Qin, C.; Liu, H. Effects of supercritical CO_2 treatment time, pressure, and temperature on microstructure of shale. *Energy* **2016**, *97*, 173–181. [CrossRef]
5. Yin, H.; Zhou, J.; Jiang, Y.; Xian, X.; Liu, Q. Physical and structural changes in shale associated with supercritical CO_2 exposure. *Fuel* **2016**, *184*, 289–303. [CrossRef]
6. Psarras, P.; Holmes, R.; Vishal, V.; Wilcox, J. Methane and CO_2 adsorption capacities of kerogen in the Eagle Ford shale from molecular simulation. *Acc. Chem. Res.* **2017**, *50*, 1818–1828. [CrossRef] [PubMed]
7. Zhu, Z.; Li, M.; Lin, M.; Peng, B.; Sun, L. Review of the CO_2-Water-Rock Interaction in Reservoir. *Bulletin of Mineralogy. Petrol. Geochem.* **2011**, *30*, 104–112.
8. Dove, P.M.; Crerar, D.A. Kinetics of quartz dissolution in electrolyte solutions using a hydrothermal mixed flow reactor. *Geochim. Cosmochim. Acta* **1990**, *54*, 955–969. [CrossRef]
9. Ross, G.D.; Todd, A.C.; Tweedie, J.A. *The Effect of Simulated CO_2 Flooding on the Permeability of Reservoir Rocks*; Elsevier: Amsterdam, The Netherlands, 1981; Volume 37, pp. 169–174.
10. Bowker, K.A.; Shuler, P.J. Carbon dioxide injection and resultant alteration of the Weber Sandstone, Rangely Field, Colorado. *AAPG Bull.* **1991**, *39*, 129–136.
11. Sayegh, S.G.; Krause, F.F.; Girard, M.; DeBree, C. Rock/fluid interactions of carbonated brines in a sandstone reservoir: Pembina Cardium, Alberta, Canada. *SPE Form. Eval.* **1990**, *5*, 399–405. [CrossRef]
12. Qu, X. *The Experiment Research of CO_2-Sandstone Interaction and Application CO_2 Gas Reservoir*; Jilin University: Changchun, China, 2007.
13. Shi, M. Experiment research of compatibility between CO_2 and formation water, reservoir respectively. *J. Chongqing Univ. Sci. Technol.* **2011**, *13*, 55–57.
14. Yu, Z.; Yang, S.; Liu, L.; Li, S.; Yang, Y. An experimental study on water-rock interaction during water flooding in formations saturated with CO_2. *Acta Pet. Sin.* **2012**, *33*, 1032–1042.
15. Wang, G.; Zhao, J.; Zhang, F.; Tao, Y.; Yang, X.; Wang, H. Interactions of CO_2-brine-rock in sandstone reservoir. *J. Cent. South Univ.* **2013**, *44*, 1167–1173.

16. Cui, G.; Zhang, L.; Tan, C.; Ren, S.; Zhuang, Y.; Enechukwu, C. Injection of supercritical CO_2 for geothermal exploitation from sandstone and carbonate reservoirs: CO_2–water–rock interactions and their effects. *J. CO2 Util.* **2017**, *20*, 113–128. [CrossRef]
17. Siqueira, A.T.; Iglesias, S.R.; Ketzer, M.J. Carbon dioxide injection in carbonate reservoirs—A review of CO_2-water-rock interaction studies. *Greenh. Gases* **2017**, *7*, 802–816. [CrossRef]
18. Xiao, N.; Li, S.; Lin, M. Experimental Investigation of CO_2-Water-Rock Interactions during CO_2 Flooding in Carbonate Reservoir. *Open J. Yangtze Oil Gas* **2017**, *2*, 108–124. [CrossRef]
19. Liu, N.; Cheng, J. Geochemical effects of cement mineral variations on water–rock–CO_2 interactions in a sandstone reservoir as an experiment and modeling study. *Greenh. Gases* **2019**, *9*, 789–810. [CrossRef]
20. Ma, B.; Cao, Y.; Zhang, Y.; Eriksson, A.K. Role of CO_2-water-rock interactions and implications for CO_2 sequestration in Eocene deeply buried sandstones in the Bonan Sag, eastern Bohai Bay Basin, China. *Chem. Geol.* **2020**, *541*, 119585. [CrossRef]
21. Ahmat, K.; Cheng, J.; Yu, Y.; Zhao, R.; Li, J. CO_2-Water-Rock Interactions in Carbonate Formations at the Tazhong Uplift, Tarim Basin, China. *Minerals* **2022**, *12*, 635. [CrossRef]

Disclaimer/Publisher's Note: The statements, opinions and data contained in all publications are solely those of the individual author(s) and contributor(s) and not of MDPI and/or the editor(s). MDPI and/or the editor(s) disclaim responsibility for any injury to people or property resulting from any ideas, methods, instructions or products referred to in the content.

Article

High Permeability Streak Identification and Modelling Approach for Carbonate Reef Reservoir

Dmitriy Shirinkin, Alexander Kochnev, Sergey Krivoshchekov *, Ivan Putilov, Andrey Botalov, Nikita Kozyrev and Evgeny Ozhgibesov

Petroleum Geology Department, Perm National Research Polytechnic University, Komsomolsky Prospect, 29, 614990 Perm, Russia; shirinkindo.40@mail.ru (D.S.); sashakoch93@gmail.com (A.K.); ivan.putilov@pstu.ru (I.P.)
* Correspondence: krivoshchekov@gmail.com

Abstract: Reef reservoirs are characterised by a complex structure of void space, which is a combination of intergranular porosity, fractures, and vuggy voids distributed chaotically in the carbonate body in different proportions. This causes great uncertainty in the distribution of porosity and permeability properties in the reservoir volume, making field development a complex and unpredictable process associated with many risks. High densities of carbonate secondary alterations can lead to the formation of zones with abnormally high porosity and permeability—high permeability streaks or super-reservoirs. Taking into account super-reservoirs in the bulk of the deposit is necessary in the dynamic modelling of complex-structure reservoirs because it affects the redistribution of filtration flows and is crucial for reservoir management. This paper proposes a method for identifying super-reservoirs by identifying enormously high values of porosity and permeability from different-scale study results, followed by the combination and construction of probabilistic curves of superreservoirs. Based on the obtained curves, three probabilistic models of the existence of a superreservoir were identified: P10, P50, and P90, which were further distributed in the volume of the reservoir and on the basis of which new permeability arrays were calculated. Permeability arrays were simulated in a dynamic model of the Alpha field. The P50 probabilistic model showed the best history matching after one iteration.

Keywords: porosity; permeability; reef; complex carbonate reservoir; super-reservoir; dynamic modelling

Citation: Shirinkin, D.; Kochnev, A.; Krivoshchekov, S.; Putilov, I.; Botalov, A.; Kozyrev, N.; Ozhgibesov, E. High Permeability Streak Identification and Modelling Approach for Carbonate Reef Reservoir. *Energies* **2024**, *17*, 236. https://doi.org/10.3390/ en17010236

Academic Editors: Hai Sun, Wenchao Liu and Daobing Wang

Received: 24 November 2023
Revised: 20 December 2023
Accepted: 27 December 2023
Published: 2 January 2024

Copyright: © 2024 by the authors. Licensee MDPI, Basel, Switzerland. This article is an open access article distributed under the terms and conditions of the Creative Commons Attribution (CC BY) license (https:// creativecommons.org/licenses/by/ 4.0/).

1. Introduction

Management of carbonate reservoirs with complex structures is complicated by many reasons associated with high heterogeneity and anisotropy of properties, cyclic sedimentation, facies variability, and extensive propagation of secondary processes. These factors combine to introduce a high degree of uncertainty in the geological structure, which must be taken into account during static and dynamic reservoir model preparation [1,2]. Distinguishing zones with different secondary changes, and hence different reservoir properties, is an important task for understanding the reservoir structure. Reservoir properties can change dramatically both vertically and horizontally; therefore, the task of predicting properties in the interwell space of the reservoir has to be addressed [3].

Rational management of carbonate reservoirs requires preparation models that take into account all features of the reservoir structure. Description and modelling of carbonate reservoirs have become the focus of various studies [4–7]. There are different approaches to creating static and dynamic models of carbonate reservoirs. Rock typing and facies modelling approaches, in conjunction with seismic trends, are often used to model reservoir properties. When creating a dynamic model, studies at various scales are taken into account, such as thin sections, core, well logging, well tests, and seismic attributes, to best reflect the complex structure of the reservoir [8–12].

Reservoir permeability is the main property that determines fluid filtration in the reservoir rocks. For carbonate reservoirs, standard petrophysical dependence (permeability-porosity) often does not reflect the heterogeneity of properties [13,14]. A number of methods describe the separation of rock types in the reservoir volume and set separate dependencies for each rock type [15]. Another classical method for creating a fractured reservoir model is dual media modelling, which reflects both the matrix and the fractured components of the rock [16–18].

Modern approaches to permeability prediction in the interwell space are based on consideration of seismic trends using machine learning and statistical methods [19–22].

Approaches taking into account uncertainty and diversity are common in studying highly heterogeneous reservoirs [23,24]. Multivariate modelling allows one to go through a range of basic reservoir parameters (porosity, permeability, aquifer, etc.), perform more justified history matching, obtain several model implementations, and make optimistic, pessimistic, and realistic forecasts of technological performance. The methodology is well-proven in uncertainty assessment, but it has its own shortcomings. As a rule, the criteria of geological realism and consistency of parameters are not established, resulting in physically unfeasible model implementations [25,26].

An important factor influencing permeability heterogeneity is the presence of secondary reservoir processes. Calcitization, dolomitization, recrystallization, leaching, and fracturing are widespread in carbonate sediments of different strata [27,28].

Secondary processes can also have a major influence on changing the structure of the reservoir void space. Intervals with abnormally high permeability may form during rock leaching. Modelling of highly permeable streaks in carbonate reservoirs is an existing problem that should be taken into account when designing a reservoir management system, mostly in organising a waterflooding system, as there is a high risk of premature flooding and fingering water breakthrough [29–31].

Different conditions are associated with different genesis of highly permeable streaks in the volume of the reservoir. Some works describe the influence of karst formation processes on the occurrence of highly permeable streaks [32–35]. Other works describe the genesis of highly permeable streaks in the process of primary sedimentation caused by the peculiarities of the reef structure and the presence of spill channels [36].

The aim of this study is to develop a methodology for the identification and modelling of highly permeable streaks to create a dynamic reservoir model and optimise the history matching process. The resulting model will ensure reliable forecasting of oil production and injection control.

In the scope of this study, the Geological Settings section will describe the geological structure of the field and the main features of the area and present the concept of karst formation. The Materials and Methods section reflects the main methods for identifying highly permeable streaks along the wellbore and modelling them in the volume of the deposit, as well as the materials on the basis of which this study was conducted—the results of core studies; well logging; well testing; etc. The Results and Discussion section presents the main results of modelling highly permeable streaks in the volume of the reservoir, as well as the results of verification of the obtained models through comparison with actual field data. The Conclusion presents the main conclusions and recommendations on modelling highly permeable streaks.

2. Geological Settings

The object of this study is the Alpha field located within the Denisov depression (Figure 1). The sediments of the Alpha field are composed of reefs formed in the Early Famennian age. The process of reef structure formation in the territory under study includes four consecutive reef-building cycles: one Zadonian and three Yeletsian. The reservoir is composed of sediments from the reef itself and the backreef shelf.

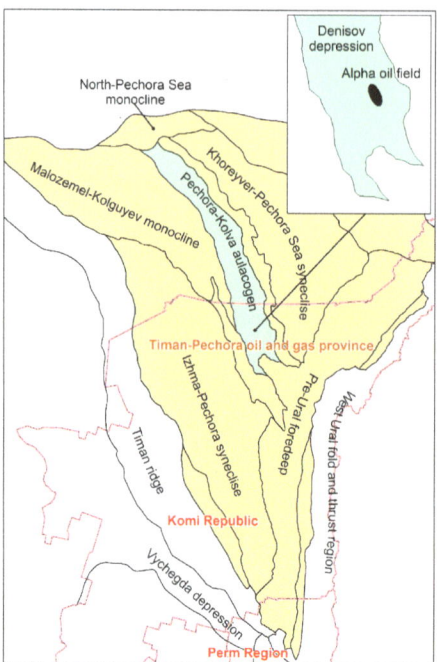

Figure 1. Extract from the tectonic map of this study area.

The lithological description of oil-saturated rocks is based on the results of core sampling taken during the drilling of nine wells. The sediments are represented by detrital-algal, spherical-patterned, organogenic-clastic, grey, grey-brown, irregularly dolomitized, and recrystallized limestones. Carbonates formed in the backreef shelf conditions are characterised by a more uniform distribution of porosity and permeability (standard deviation-349 mD), while rocks of the reef facies are characterised by a higher dispersion (standard deviation-846 mD) caused by an increased extent of secondary alterations of carbonates (Figure 2).

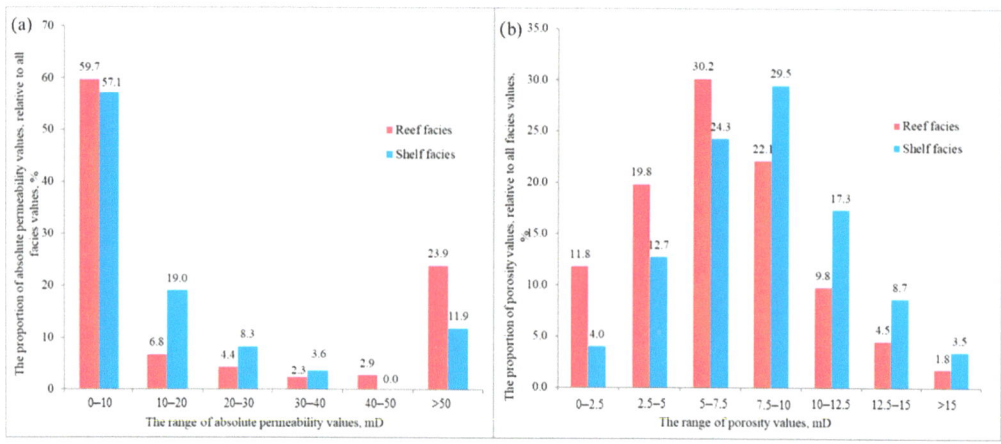

Figure 2. Histogram of the distribution of absolute permeability (**a**) and porosity (**b**) values based on core test data.

The carbonate reservoir is characterised by a complex type of void space. It is a complex configuration consisting of fractures, vuggy pores, and intergranular porosity unevenly distributed in the rock body. Increased zonal concentration of fractures and vuggy pores leads to the formation of zones with abnormally high porosity and permeability, which, in turn, leads to the formation of super-reservoirs. Core porosity and permeability characteristics of sediments in the Alpha field are presented in Table 1.

Table 1. Porosity and permeability characterisation of the Alpha deposit.

Number of Plugs Core Samples, pcs.	Number of Full-Size Core Samples, pcs.	Porosity, Reef Facies, %		Permeability, Reef Facies, mD		Porosity, Shelf Facies, %		Permeability, Shelf Facies, mD	
		Range	Average	Range	Average	Range	Average	Range	Average
2547	300	0.1–24.2	6.77	0.001–18,143	124.29	0.1–29.6	8.43	0.001–4439	49.87

In the Famennian sediments of the field under study, there are both open and closed fractures, as well as mineralised fractures (with calcite, dolomite, and sulphate group minerals) and fractures filled with bituminous and clayey matter (Figure 3). According to the core, the fractures are scant and/or occur as series-inclined, vertical, subhorizontal, and multidirectional rectilinear, curved, and branched, with lengths ranging from a few centimetres to 1.2 m. According to the classification of fracture openness, the rocks of the Famennian age contain fractures ranging from very narrow (from a few fractions of a millimetre) to very wide (up to 7.00 mm and more) (Figure 4). The fractures cut through and go around formational elements, fenestrae, and fragments of stylolites in the rocks, often connecting fenestral cavities, etc. The fractures are especially developed in lithotypes of limestones (boundstones, rudstones, packstones, and grainstones) and are less present in secondary dolomites.

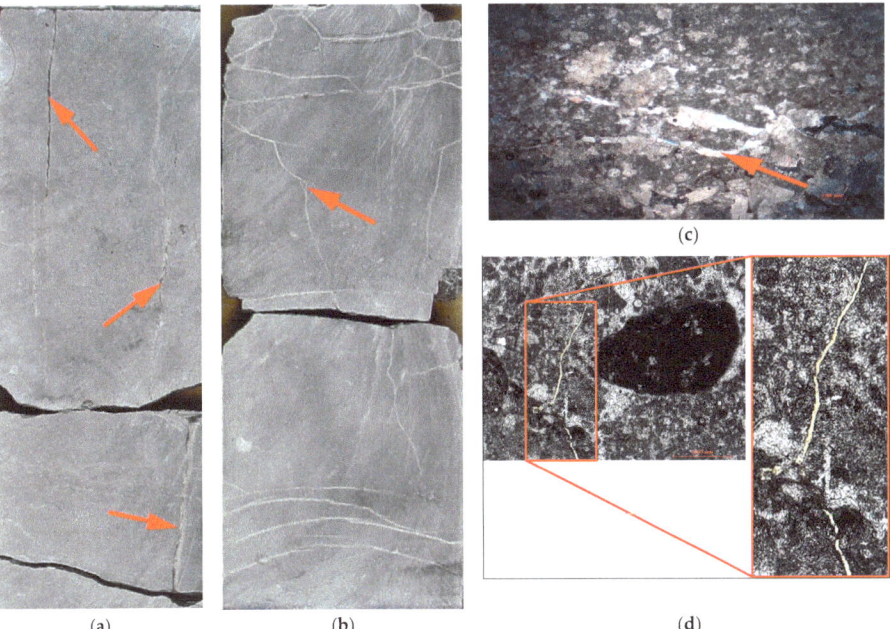

Figure 3. Fractures by openness and filling: (**a**)—open (core photo, arrows indicate open fractures); (**b**)—closed (core photo, arrow indicates closed fracture); (**c**)—mineralized (thin section photo, arrow indicates fracture mineralized with dolomite and sulphates); (**d**)—filled with bituminous and clayey matter (thin section photo, arrow indicates fracture filled with bituminous organic matter).

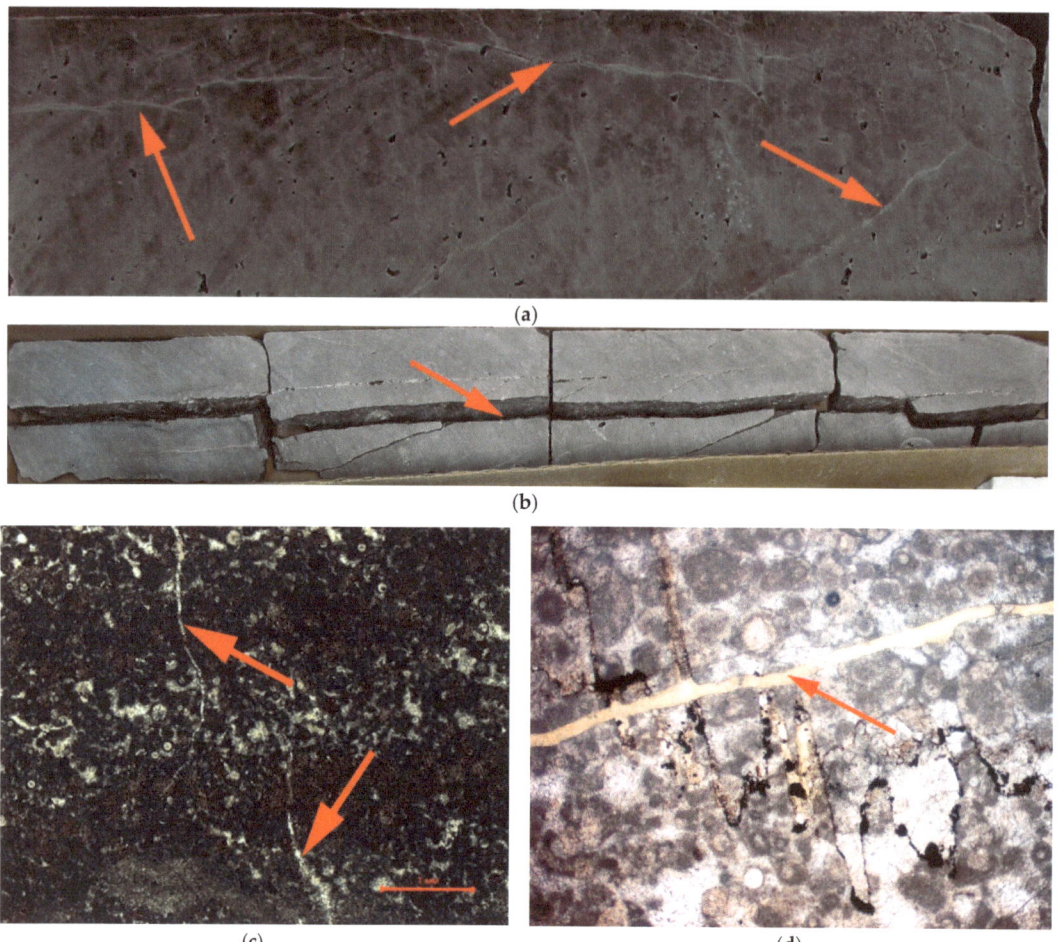

Figure 4. Fractures by size: (**a**)—short (core photo, arrows indicate fractures up to 10 cm long); (**b**)—long (core photo, arrow indicates fracture over 70 cm long); (**c**)—very small (thin section photo, arrows indicate fractures 0.02 mm wide); (**d**)—macro-fracture (thin section photo, arrow indicates fracture 2 mm wide).

Based on the geological peculiarities of the structure and formation conditions of the Timan-Pechora region, there can be two hypotheses of karst formation in the reef sediments of the Famennian age of the Denisov trough. The first mechanism of karst formation in the sediments is related to sea level fluctuations. During the periods of regressive phases of sedimentation, the reefs rose above sea level and were thus exposed to karst formation under the influence of the sun, wind, and fresh water (Figure 5a). The second concept of karst formation in reefs is related to the active tectonic regime of the basin, where younger rocks undergo erosion under the influence of inversion movements. Together, active tectonics and erosion events bring older rocks to the earth's surface, where they are exposed to atmogenic waters, causing karst formation (Figure 5b).

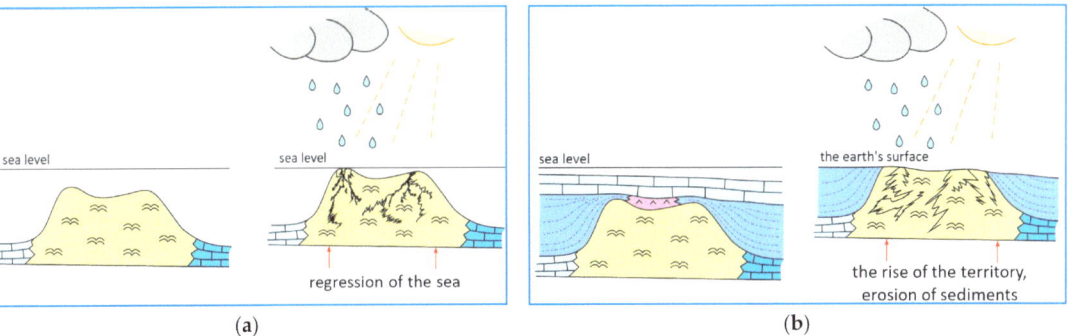

Figure 5. Mechanisms of karst formation in reef sediments: (**a**)—first stage, regression of the sea; (**b**)—second stage, erosion of sediments on the earth's surface.

The study of the core material revealed that reef rocks of the Zadonian-Yeletsian age have a complex structure of void space in both lateral and vertical directions. The complexity is caused by a significant propagation of vuggy pores (Figure 6) and karst fractures (Figure 7) in the rocks. In general, the core contains either a small or a large number of isometric or elongated vuggy pores, isolated or communicating, in sizes ranging from very small to very large (up to 80 mm).

Figure 6. Vuggy pores in Famennian reef reservoirs: (**a**)—small vuggy pores; (**b**)—medium-sized vuggy pores; (**c**)—large vuggy pores.

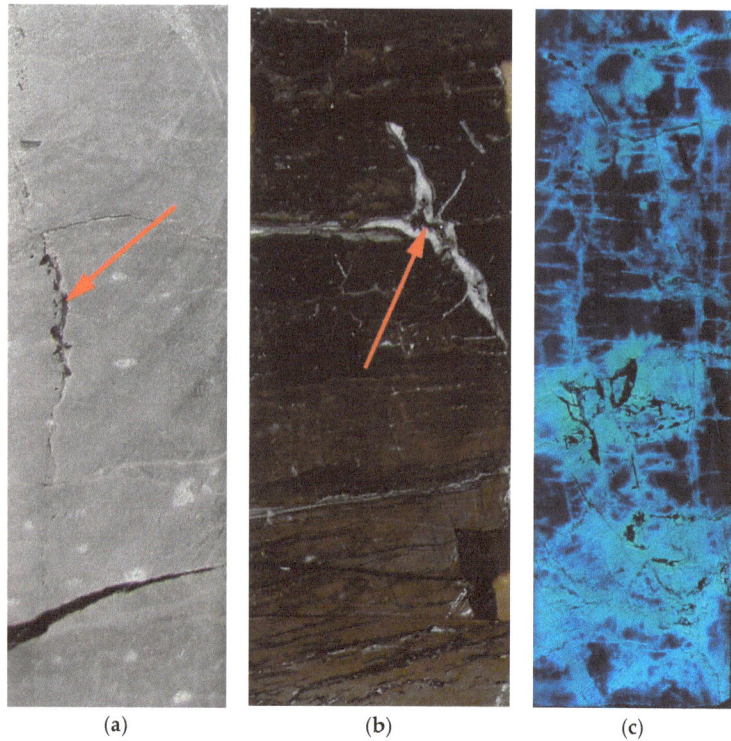

Figure 7. Karst fractures in Famennian reef reservoirs: (**a**)—karst fracture; (**b**)—karst fracture healed by calcite; (**c**)—karst fracture system.

3. Materials and Methods

In this work, the identification of highly permeable streaks along the wellbore is achieved through probabilistic methods using a combination of 10 different studies that characterise the porosity and permeability of the reservoir. The essence of the method is to identify intervals with abnormally high porosity and permeability for each of the studies; hence, the more anomalies of different geological and geophysical properties are observed in one interval, the higher the probability of the presence of a super-reservoir.

The first and foremost property that can be used to identify a super-reservoir is the absolute permeability value determined on 2547 plug core samples. Absolute permeability directly characterises the filtration capacity of the rock, so zones with abnormally high permeability can, with a high degree of certainty, indicate the presence of a super-reservoir in this interval. The approach of analysing the accumulated correlation between permeability and porosity values was used to assess the degree of their relationship and to identify anomalies [37,38].

Further, the values of filtration flow unit–FZI (1), which were first described by Amaefule et al. in 1993 [39] and are based on the Kozeny–Carman equation, were used to identify the super-reservoir.

$$\text{FZI} = \frac{\text{RQI}}{\varphi_z} \quad (1)$$

where "RQI" is the reservoir quality index, mD;

"φ_z" is the indicator of normalised porosity, unit fractions.

RQI is calculated according to Formula (2):

$$RQI = 0.0314\sqrt{\frac{k}{\varphi}} \qquad (2)$$

where "k" is the permeability coefficient, mD;
"φ" is the porosity coefficient, unit fractions.
"φ_z" characterises the void ratio—the ratio of pore volume to grain volume ratio—and is determined according to Formula (3):

$$\varphi_z = \frac{\varphi}{1-\varphi} \qquad (3)$$

Thus, the calculation of the FZI coefficient is reduced to Formula (4):

$$FZI = \frac{0.0314\sqrt{\frac{k}{\varphi}}}{\frac{\varphi}{1-\varphi}} \qquad (4)$$

Hence, as the FZI value increases, the permeability will increase and the porosity will decrease as well, and it is thus possible to identify samples exposed to secondary changes that are typical of a super-reservoir.

Differentiation of core samples into different classes according to the FZI parameter was carried out using the DRT technique [40]. The formula for determining the DRT class is given below (5):

$$DRT = 2\ln(FZI) + 10.6 \qquad (5)$$

At the next stage, a number of well-logging curves that characterise the reservoir properties of the rock were used to identify zones with abnormally high porosity and permeability, which may indicate the presence of a super-reservoir. A correlation matrix was built to evaluate the relationship between well logging curves and porosity and permeability based on the results of core studies. Well, logging curves that best characterise reservoir properties were selected. These curves were:

1. Porosity coefficient curves determined by acoustic (KPA), density (KPD), and neutron (KPN) methods, and the effective porosity coefficient determined by nuclear magnetic logging (CMFF);
2. Permeability coefficient curves that are determined by nuclear magnetic logging using the SDR (KSDR) model and the permeability coefficient calculated using the Timur-Coates (KTIM) model;
3. Fraction of oil in total void volume (FOIL) curve.

The presence of super-reservoir zones can be indicated by increased fracture density in a particular wellbore interval, indicating the high activity of secondary rock transformations. Formation Micro-Imager Logs (FMI) are used to identify these fractures, and, accordingly, abnormally high fracture densities derived from FMI can indicate the presence of a super-reservoir in a particular wellbore interval.

4. Results

The first criterion for high-permeable streak identification was the absolute value of permeability determined on 2547 plug core samples. Relationships between cumulative values of effective porosity and absolute permeability for the core samples were used to estimate the reservoir property relationships. Porosity and permeability data were sorted from minimum to maximum permeability values, and correlation coefficients were calculated for cumulative plugs n = 3, 4,... 2547. These correlations enabled us to estimate the relationship of parameters over the whole range of porosity values, which made it possible to identify ranges of different types of porous space (Figure 8).

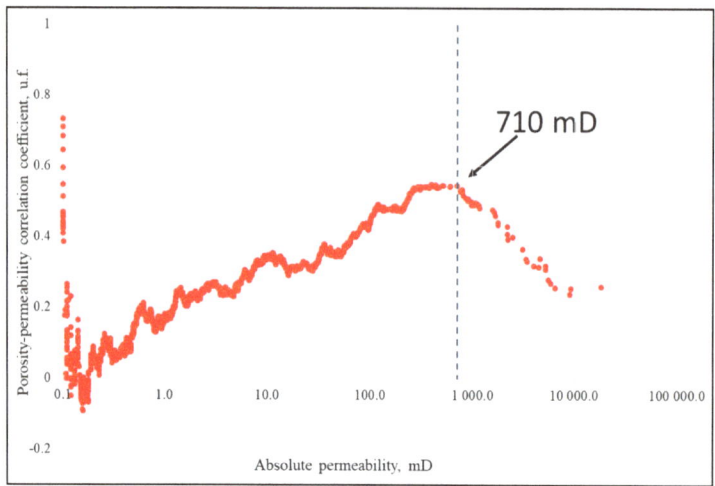

Figure 8. The dependence of the cumulative correlation coefficient between porosity and permeability on permeability values.

Figure 8 shows how the correlation coefficient between porosity and permeability gradually increases once a sufficient value is reached; that is, when permeability values increase, the relationship between porosity and permeability also becomes stronger. In other words, the increase in rock permeability is due to the increase in effective void space.

At the level of 710 mD, an inflection point is observed, after which the curve goes sharply down. This means that these core samples with abnormally high permeability are outliers relative to the total sample, and the high permeability values are not due to high values of classical intergranular porosity but to other factors, which, first of all, are fractures and communicating vuggy pores. Consequently, the value at the inflection point, equal to 710 mD, can be chosen as a boundary value to identify a super-reservoir.

The boundary value for the super-reservoir equal to 710 mD can also be seen in the absolute permeability distribution plot (Figure 9).

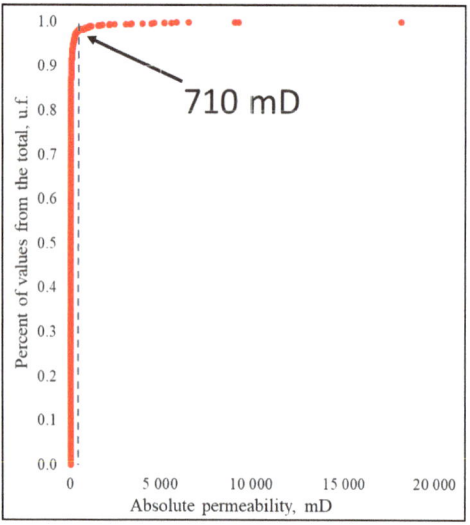

Figure 9. Absolute permeability distribution plot.

Figure 9 shows that most of the samples have relatively low permeability, forming a subvertical curve. Further, there is a sharp bend, and the curve goes to the subhorizontal position. This part of the sample characterises abnormally high values of permeability, which is not typical for the total number. Thus, 1.6% of samples are attributed to a super-reservoir by the absolute permeability value.

The next method to identify the super-reservoir was the hydraulic flow unit (FZI) calculation technique. The FZI value was calculated for each of the core samples. Further, the FZI distribution was plotted (Figure 10), and the point of inflection and exit of the distribution curve to a subhorizontal position was determined, reflecting abnormally high values that are not typical for the total sample.

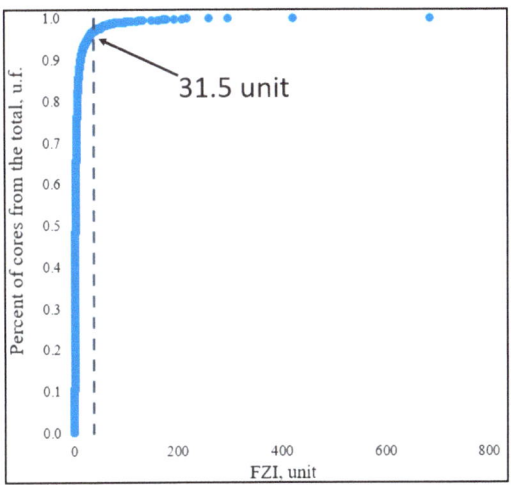

Figure 10. FZI distribution plot.

The FZI value at the inflection point is 31.5 units, which corresponds to class 18 according to the DRT classification (Figure 11).

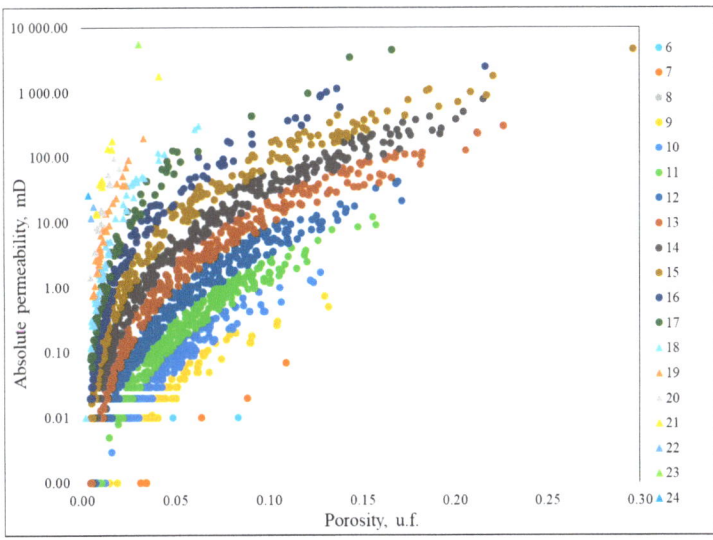

Figure 11. Classification of DRT core samples.

Core samples with abnormally high FZI values (>31.5), falling into DRT classes 18–24, are characterized by filtration typical of samples with an increased degree of secondary alterations (vuggy pores, fractures), which suggests the presence of a super-reservoir in this part of the section. Thus, only 4.5% of samples are attributed to a super-reservoir.

Next, a database of 40 well logging curves was compiled and correlated with the porosity and permeability determined in the core studies. The correlation matrix is presented in Table 2.

Table 2. Results of the correlation between core sampling and well-logging data.

	Depth	AF90	AF30	AF60	AF20	AF10	AMF	BIT	CFTC	CNTC
Permeability	0.007	0.019	0.021	0.031	0.037	−0.013	0.046 [1]	−0.036	−0.122	−0.091
Porosity	0.080	0.041	0.142	0.181	0.052	0.068	0.237	−0.109	−0.487	−0.317
	GR	HCAL	HDRA	HMIN	HMNO	HTEM	PEFZ	RHOZ	RLA1	RLA2
Permeability	−0.056	−0.070	−0.039	−0.038	−0.038	−0.052	−0.024	−0.005	−0.021	−0.028
Porosity	−0.219	−0.328	−0.194	−0.153	−0.152	−0.088	−0.189	−0.067	−0.140	−0.173
	RLA3	RLA4	RLA5	RT_HRLT	RXOZ	SP	**KPA**	DOLM	Kpob	**KPD**
Permeability	−0.024	−0.022	−0.014	−0.018	−0.045	−0.003	0.211 [2]	0.019	0.161	0.188
Porosity	0.159	0.156	−0.093	0.121	−0.137	−0.162	0.670	0.007	0.343	0.622
	KPkv	**KPN**	**KSDR**	**KTIM**	LIME	**FOIL**	PORW	SHALE	TCMR	**CMFF**
Permeability	0.089	0.176	0.371	0.194	−0.035	0.199	−0.029	−0.015	0.042	0.254
Porosity	0.128	0.695	0.302	0.363	−0.102	0.474	0.021	0.104	0.000	0.705

[1] Statistically significant correlations are highlighted in red (p-value < 0.05 u.f.). [2] Methods selected for further research are highlighted in bold font.

Out of 40 well logging curves, seven (KPA, KPD, KPN, CMFF, KSDR, KTIM, and FOIL) were selected for further work, which inherently reflect reservoir quality and show the highest correlation with core permeability.

To identify abnormally high values that may indicate the presence of a super-reservoir, distribution plots were built for each of the well logging methods (Figure 12). The super-reservoir boundary values are identified at the point of inflection and the changing direction of the distribution curve to the sub-horizontal position that characterises the abnormal properties.

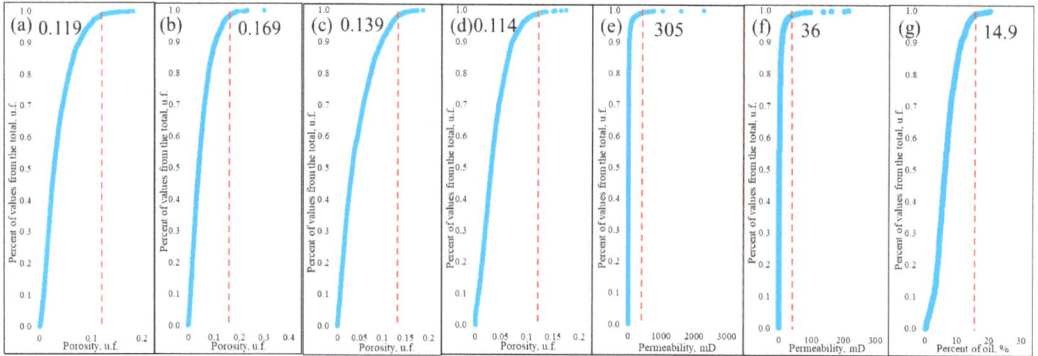

Figure 12. Distribution plots of well logging curves: (**a**) KPA, (**b**) KPD, (**c**) KPN, (**d**) CMFF, (**e**) KSDR, (**f**) KTIM, and (**g**) FOIL.

As a result, only 2–2.5% of the values in each of the well logging curves are attributed to the super-reservoir.

The next criterion for super-reservoir identification was FMI results. Using a moving average along the wellbore, synthetic fracture density curves according to FMI were built (Figure 13), and intervals with fracture densities above 1 fracture per metre were identified as probable high permeable streaks.

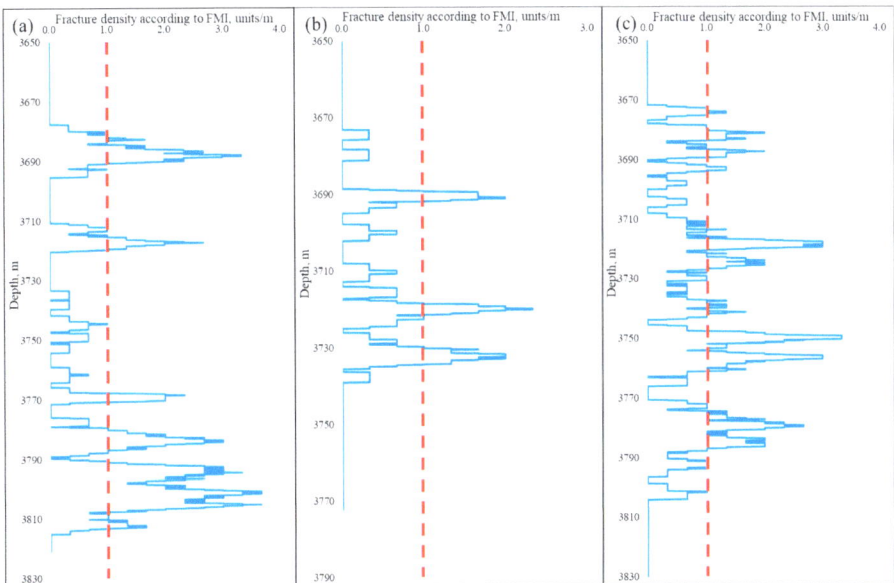

Figure 13. Synthetic fracture density curves identified by FMI: (**a**) well No. 1; (**b**) well No. 2; (**c**) well No. 3.

At the next stage, the super-reservoir intervals identified in the set of studies of various scales were summarised along the wellbore of each well, and synthetic curves of probability of super-reservoir presence were obtained (Figure 14).

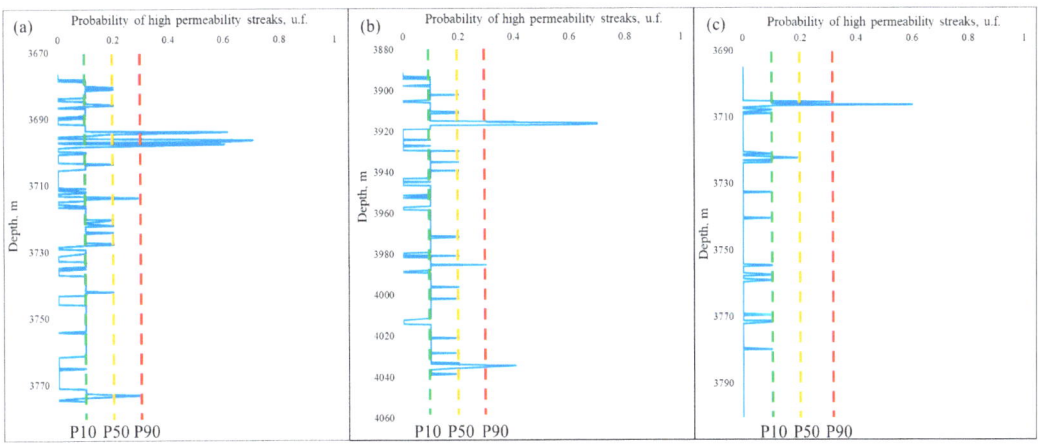

Figure 14. Synthetic curves of probability of super-reservoir presence: (**a**) well No. 1; (**b**) well No. 2; (**c**) well No. 3.

At the next stage, the supercollector intervals for each study are given the same weighting coefficient, which equals 0.1 units. Afterwards, all weighting coefficients were summed up for each wellbore study. Consequently, the more studies that reveal anormous reservoir properties intersect in one interval, the higher the probability of the presence of a superreservoir in this interval is expected. In this way, sums of weighting coefficients for intervals were obtained, and, as a result, synthetic curves of the probability of the presence of a superreservoir were generated (Figure 14).

As described earlier, each of the criteria has extremely strict requirements for super-reservoir identification (only 1–4% of maximum values), so the coincidence of at least two various scale studies of different natures in one scale interval of a plug core sample indicates the presence of a super-reservoir with a high degree of certainty.

To account for the uncertainties associated with the distribution of highly permeable streaks, the obtained synthetic curves were assigned criteria with different significance levels (P10, P50, and P90). Identification conditions at the P10 level reflect the maximum possible number of highly permeable streaks; in this case, all intervals where at least one of the studies indicates the presence of a highly productive interval fit the criteria for super-reservoir identification. The P50 level reflects an intermediate distribution of highly permeable reservoirs; only those intervals where two or more studies indicate the presence of a super-reservoir are eligible for super-reservoir identification, which significantly reduces the total thickness of highly productive intervals. The third level of super-reservoir identification, P90, shows the most conservative result of all. In this case, only those intervals where the presence of a super-reservoir is confirmed by three or more studies are identified as a super-reservoir.

At the next stage, LAS-files were generated for the wells and loaded into the project of the existing geological model, where the values 0 and 1 alternate along the wellbore opposite to the depth marker, 0–absence of a super-reservoir, 1–presence of a super-reservoir.

The basic parameters of the geologic model are described below:

Number of cells—45,638,389; grid type—corner points; average layer size by thickness—0.38 m, cell size—50 × 50 m; number of layers—1743; formations—D_3el_3, D_3el_2, D_3el_1, D_3el_trans, D_3zd.

Then, using stochastic indicator modelling, the super-reservoir was distributed in the model volume (Figure 15).

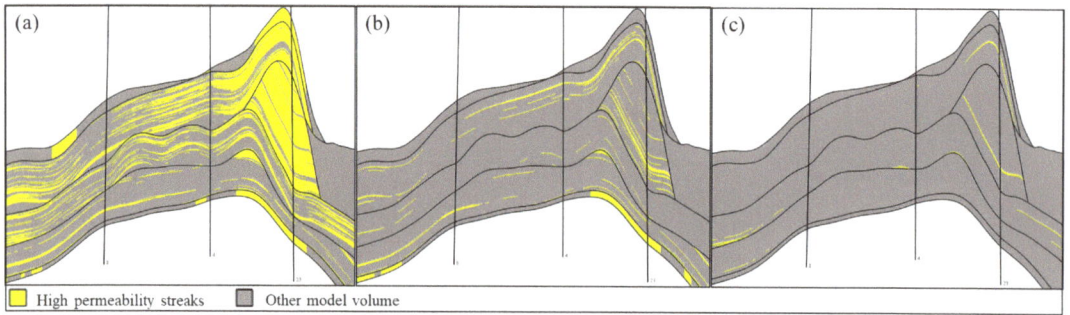

Figure 15. Distribution of the super-reservoir in the model volume: (**a**) P10; (**b**) P50; (**c**) P90.

Permeability arrays were built based on the distribution of highly permeable intervals. The permeability array calculated by standard petrophysical dependence is taken as a basis (Figures 16 and 17); further, for the intervals attributed to the super-reservoir with probability P10, P50, and P90, the permeability is calculated on the basis of petrophysical dependence obtained for core samples, which according to FZI are attributed to the super-reservoir (Figures 11, 16 and 17).

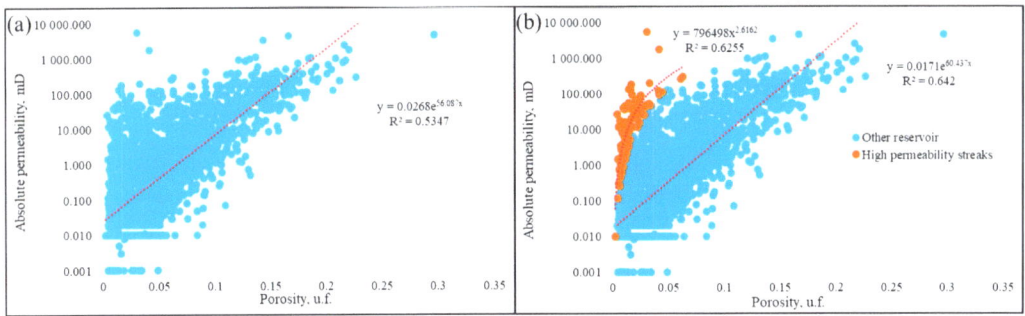

Figure 16. Permeability-porosity dependence plot: (**a**) standard petrophysical dependence; (**b**) petrophysical dependence for a super-reservoir.

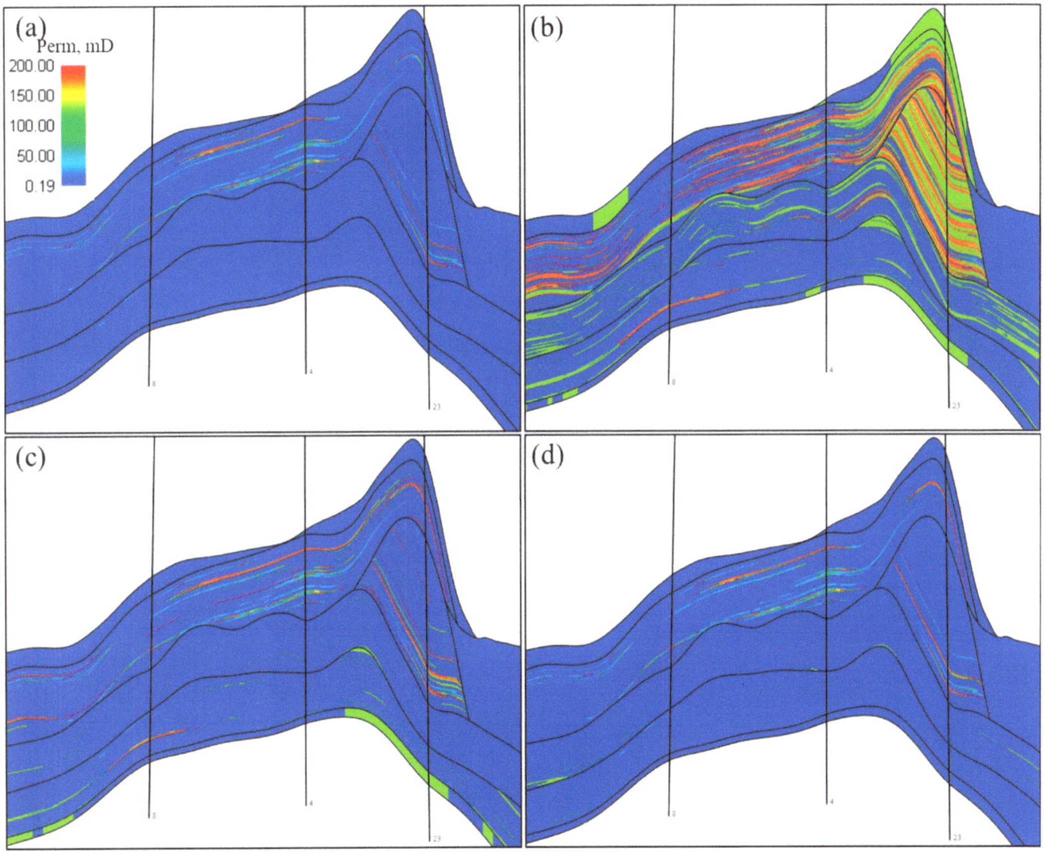

Figure 17. Comparison of obtained permeability arrays: (**a**) standard petrophysical dependence; (**b**) P10; (**c**) P50; (**d**) P90.

Comparisons of average permeability for each permeability array are shown in Table 3.

Table 3. Comparison of average permeability for the obtained permeability arrays.

	Standard Petrophysical Dependence	P10	P50	P90	Well Flow Test	Core
Average permeability, mD	28.5	523.9	53.6	49.8	93.5	24.5

After that, the obtained permeability arrays were loaded into the existing dynamic model. In the dynamic model, a PVT model was implemented with black oil and relative curves according to core studies (Figure 18). A high trend is used for supercollectors; a base trend is used for D_3el_3, D_3el_2, D_3el_1, and D_3zd; a low trend is used for D_3el_trans and D_3el_2.

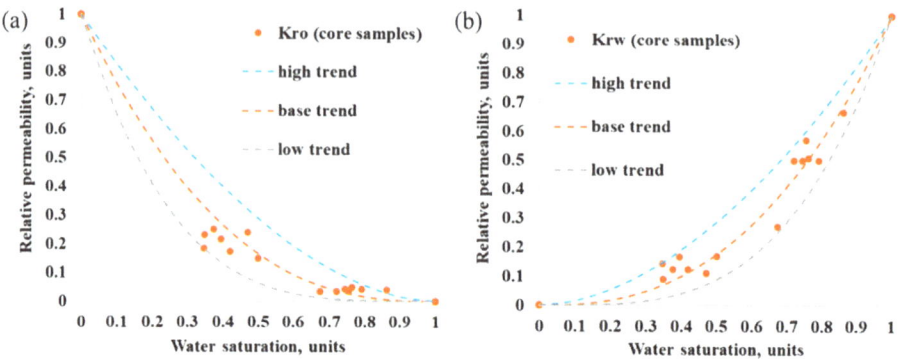

Figure 18. Relative permeability: (**a**)—for oil, (**b**)—for water.

The history of oilfield development is 10 years: total wells-58, producers-43, injectors-15. Four wells have a horizontal ending. The main completion intervals are D3el3, D3el1, and D3zd. Fractures and karst zones were included in the modelling of superreservoir distributions.

Further history-matching results were compared on all 4 permeability arrays by reproducing the development history with control by actual bottom hole pressure, without liquid flow rate limitations. To assess the history matching quality comparison of bottom hole pressure dynamics, it is shown in Figure 19.

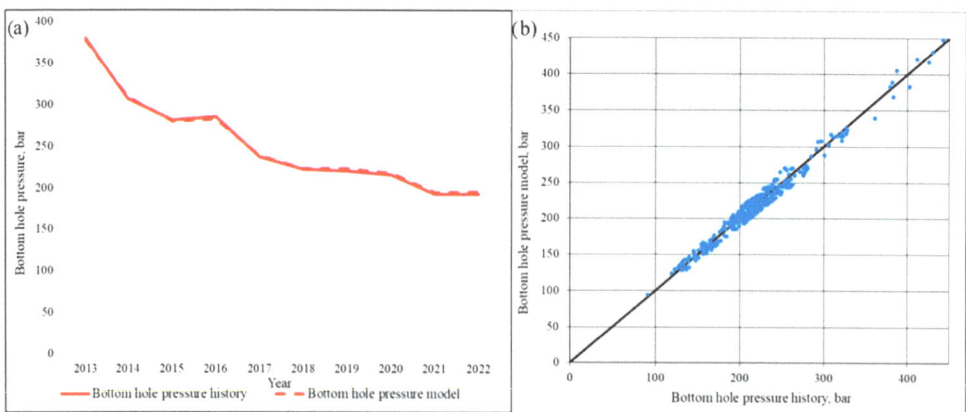

Figure 19. Comparison of simulated and historical bottomhole pressures: (**a**) comparison of the bottomhole pressure dynamics; (**b**) cross-plot between simulated and historical data.

Figure 19 shows a high convergence of calculated and historical bottomhole pressures, which allows us to assert that the initial conditions for carrying out calculations with bottomhole pressure control are correct.

The comparison was carried out in equal conditions after the first iteration of history matching.

When high permeability streaks were factored into the model, in just one iteration, it was possible to achieve better convergence with historical oil and liquid production than when using the permeability array derived from standard petrophysical dependence (Figure 20, Table 4).

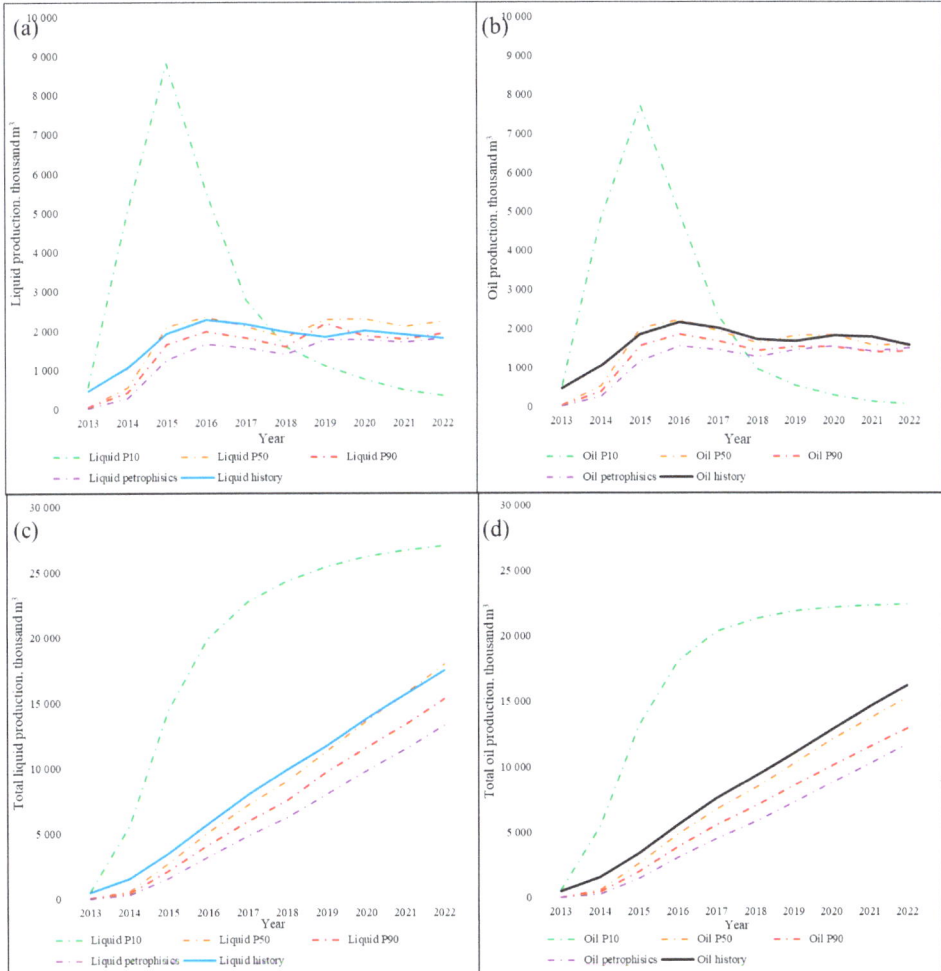

Figure 20. Comparison of actual and model oil and liquid production using all permeability arrays: (**a**) comparison of annual liquid production; (**b**) comparison of annual oil production; (**c**) comparison of cumulative liquid production; (**d**) comparison of cumulative oil production.

Table 4. Comparison of historical and model-calculated cumulative oil and liquid production for each permeability array.

	History	Petrophisics	P10	P50	P90
Total liquid production, thousand m^3	17,478	13,266	27,037	17,969	15,303
Total oil production, thousand m^3	16,191	11,760	22,428	15,304	12,972
Deviation of the liquid from history, %		24.1	−54.7	−2.8	12.4
Deviation of oil from history, %		27.4	−38.5	5.5	19.9

The best convergence with historical oil and liquid production was achieved using a permeability array with a super-reservoir distributed according to the P50 probability (deviation in cumulative liquid production—−2.8%, deviation in cumulative oil production—5.5%).

Figure 20 and Table 4 show that variants with permeability calculation according to standard petrophysical dependence and super-reservoir distribution with probability P90 underestimate formation potential; wells produce less liquid than according to history; hence, the productivity of wells with this permeability array is lower than real. The opposite is the situation with the distribution of super-reservoir with probability P10: this permeability array significantly overestimates the formation potential and well productivity at the peak production moment, thereby depleting the energy of the deposit. As a consequence, there are sharp rates of liquid and oil production decline, as well as faster water breakthroughs. The most favourable results are obtained in reservoir distribution with a probability of P50. Using this version of the permeability array, it was possible to reproduce the reservoir potential and, as a consequence, productivity in the most accurate way.

5. Conclusions

This study gives a detailed description of the geological structure of the Alpha field, which has a carbonate reef reservoir with a complex void structure. The core was analysed, and a detailed description was given to the nature of fractures and vuggy pores whose high density leads to the formation of super-reservoirs.

Super-reservoirs have a huge impact on reservoir management. They both ensure high well productivity and impose the risk of premature water breakthroughs. In order to reduce uncertainty in the distribution of highly productive intervals in the model volume, a methodology for probabilistic identification of super-reservoirs was proposed.

The methodology includes consideration of 10 studies of various scales. At the initial stage, possible super-reservoirs are identified for each study separately, after which the super-reservoirs are summed to form a general probability curve of super-reservoir distribution. The probability curve is then used to identify super-reservoir intervals with different degrees of uncertainty: P10, P50, and P90.

In the next step, all three cases of super-reservoir identification are distributed in the model volume, and permeability arrays are calculated for each variant; besides, a permeability array is calculated through a standard petrophysical dependence for comparison.

Then, the obtained arrays were loaded into the existing dynamic model, and the convergence of calculated and actual indicators of the Alpha field development after the first iteration of history matching was compared. Calculations with each permeability array were carried out in equal conditions, with control by drawdown pressure and without control by liquid rates. The best convergence with actual oil and liquid production was shown by the calculation using permeability array P50; deviations on accumulated liquid and oil were only −2.8% and 5%, respectively, which indicates the most accurate and correct distribution of permeability in the model volume.

Thus, the application of this technique reduces the uncertainty of the distribution of highly productive intervals in the volume of the reservoir, which makes it possible to take into account the risks of premature watering of wells as well as to predict the rate of oil production. Using this technique, it was possible to recreate reservoir potential and adjust well productivity with high accuracy in the model, which increased the overall adaptability of the model.

Author Contributions: Conceptualization, D.S. and S.K.; methodology, A.K.; software, D.S.; validation, A.K., A.B. and N.K.; formal analysis, N.K. and E.O.; investigation, A.B.; resources, D.S. and I.P.; data curation, N.K.; writing—original draft preparation, D.S.; writing—review and editing, E.O.; visualization, A.B. and E.O.; supervision, A.K. and I.P.; project administration, S.K.; funding acquisition, S.K. All authors have read and agreed to the published version of the manuscript.

Funding: This study was conducted under Russian Science Foundation grant No. 22-17-00111.

Data Availability Statement: Data are contained within the article.

Conflicts of Interest: The authors declare no conflict of interest.

References

1. Li, Q.; Han, Y.; Liu, X.; Ansari, U.; Cheng, Y.; Yan, C. Hydrate as a by-product in CO_2 leakage during the long-term sub-seabed sequestration and its role in preventing further leakage. *Environ. Sci. Pollut. Res.* **2022**, *29*, 77737–77754. [CrossRef] [PubMed]
2. Li, Q.; Wang, F.; Wang, Y.; Forson, K.; Cao, L.; Zhang, C.; Zhou, C.; Zhao, B.; Chen, J. Experimental investigation on the high-pressure sand suspension and adsorption capacity of guar gum fracturing fluid in low-permeability shale reservoirs: Factor analysis and mechanism disclosure. *Environ. Sci. Pollut. Res.* **2022**, *29*, 53050–53062. [CrossRef] [PubMed]
3. Tavakoli, V. *Carbonate Reservoir Heterogeneity: Overcoming the Challenges*; Springer International Publishing: Tehran, Iran, 2019; pp. 1–108.
4. Lucia, F.J.; Kerans, C.; Jennings, J.W. Carbonate reservoir characterization. *J. Pet. Technol.* **2003**, *55*, 70–72. [CrossRef]
5. Masalmeh, S.K.; Jing, X.D. Improved characterisation and modelling of carbonate reservoirs for predicting waterflood performance. In Proceedings of the International Petroleum Technology Conference, Dubai, United Arab Emirates, 4–6 December 2007. [CrossRef]
6. Martin, A.J.; Solomon, S.T.; Hartmann, D.J. Characterization of petrophysical flow units in carbonate reservoirs. *AAPG Bull.* **1997**, *81*, 734–759. [CrossRef]
7. Dominguez, G.C.; Fernando, S.V.; Chilingarian, G.V. Simulation of carbonate reservoirs. *Dev. Pet. Sci.* **1992**, *30*, 543–588. [CrossRef]
8. Correia, M.G.; Maschio, C.; Schiozer, D.J. Integration of multiscale carbonate reservoir heterogeneities in reservoir simulation. *J. Pet. Sci. Eng.* **2015**, *131*, 34–50. [CrossRef]
9. Massonnat, G.J.; Michel, J.; Gatel, P.; Ruiu, J.; Danquigny, C.; Lesueur, J.L.; Borgomano, J. Multi-Scale Sedimentary forward Reservoir Modelling: A Disruptive Solution for Simulating Heterogeneity in Carbonates. Application to the Kharaib-2 Reservoir Unit. In Proceedings of the Abu Dhabi International Petroleum Exhibition & Conference, Abu Dhabi, United Arab Emirates, 2–5 October 2023. [CrossRef]
10. Beltiukov, D.A.; Kochnev, A.A.; Galkin, S.V. The combining different-scale studies in a reservoir simulation model of a deposit with a fractured-cavernous type of carbonate reservoir. *IOP Conf. Ser. Earth Environ. Sci.* **2022**, *1021*, 012027. [CrossRef]
11. Zhang, W.; He, Z.; Duan, T.; Li, M.; Zhao, H. Hierarchical modeling of carbonate fault-controlled Paleokarst systems: A case study of the Ordovician reservoir in the Tahe Oilfield, Tarim Basin. *Front. Earth Sci.* **2022**, *10*, 840661. [CrossRef]
12. Naseer, M.T. Seismic attributes and reservoir simulation' application to image the shallow-marine reservoirs of Middle-Eocene carbonates, SW Pakistan. *J. Pet. Sci. Eng.* **2020**, *195*, 107711. [CrossRef]
13. Baker, H.A.; Al-Jawad, S.N.; Murtadha, Z.I. Permeability Prediction in Carbonate Reservoir Rock Using FZI. *Iraqi J. Chem. Pet. Eng.* **2013**, *14*, 49–54. [CrossRef]
14. Babadagli, T.; Al-Salmi, S. A review of permeability-prediction methods for carbonate reservoirs using well-log data. *SPE Reserv. Eval. Eng.* **2004**, *7*, 75–88. [CrossRef]
15. Corbett, P.; Potter, D. Petrotyping: A basemap and atlas for navigating through permeability and porosity data for reservoir comparison and permeability prediction. In Proceedings of the International Symposium of the Society of Core Analysts, Abu Dhabi, United Arab Emirates, 5–9 September 2004.
16. Guerriero, V.; Mazzoli, S.; Iannace, A.; Vitale, S.; Carravetta, A.; Strauss, C. A permeability model for naturally fractured carbonate reservoirs. *Mar. Pet. Geol.* **2013**, *40*, 115–134. [CrossRef]
17. Zambrano, M.; Volatili, T.; Mancini, L.; Pitts, A.; Giorgioni, M.; Tondi, E. Pore-scale dual-porosity and dual-permeability modeling in an exposed multi-facies porous carbonate reservoir. *Mar. Pet. Geol.* **2021**, *128*, 105004. [CrossRef]
18. Uba, H.M.; Chiffoleau, Y.; Pham, T.; Divry, V.; Kaabi, A.; Thuwaini, J. Application of a Hybrid Dual Porosity/Dual-Permeability Representation of Large-Scale Fractures to the Simulation of a Giant Carbonate Reservoir. In Proceedings of the SPE Middle East Oil and Gas Show and Conference, Manama, Bahrain, 11 March 2007. [CrossRef]
19. Takougang, E.M.T.; Bouzidi, Y.; Ali, M.Y. Characterization of small faults and fractures in a carbonate reservoir using waveform inversion, reverse time migration, and seismic attributes. *J. Appl. Geophys.* **2019**, *161*, 116–123. [CrossRef]

20. Subasi, A.; El-Amin, M.F.; Darwich, T.; Dossary, M. Permeability prediction of petroleum reservoirs using stochastic gradient boosting regression. *J. Ambient Intell. Humaniz. Comput.* **2020**, *13*, 3555–3564. [CrossRef]
21. Chen, G.; Meng, Y.; Huan, J.; Wang, Y.; Xiao, L.; Zhang, L.; Feng, D. A new predrilling reservoir permeability prediction model and its application. *J. Pet. Sci. Eng.* **2022**, *210*, 110086. [CrossRef]
22. Zolotukhin, A.; Gayubov, A. Machine learning in reservoir permeability prediction and modelling of fluid flow in porous media. *IOP Conf. Ser. Mater. Sci. Eng.* **2019**, *700*, 012023. [CrossRef]
23. Benetatos, C.; Giglio, G. Coping with uncertainties through an automated workflow for 3D reservoir modelling of carbonate reservoirs. *Geosci. Front.* **2021**, *12*, 100913. [CrossRef]
24. Arnold, D.; Demyanov, V.; Christie, M.; Bakay, A.; Gopa, K. Optimisation of decision making under uncertainty throughout field lifetime: A fractured reservoir example. *Comput. Geosci.* **2016**, *95*, 123–139. [CrossRef]
25. Arnold, D.; Demyanov, V.; Tatum, D.; Christie, M.; Rojas, T.; Geiger, S.; Corbett, P. Hierarchical benchmark case study for history matching, uncertainty quantification and reservoir characterisation. *Comput. Geosci.* **2013**, *50*, 4–15. [CrossRef]
26. Matveev, I.; Shishaev, G.; Eremyan, G.; Demyanov, V.; Popova, O.; Kaygorodov, S.; Belozerov, B.; Uzhegova, I.; Konoshonkin, D.; Korovin, M. Geology driven history matching. In Proceedings of the SPE Russian Petroleum Technology Conference, Moscow, Russia, 22–24 October 2019. [CrossRef]
27. Rashid, F.; Hussein, D.; Glover, P.W.J.; Lorinczi, P.; Lawrence, J.A. Quantitative diagenesis: Methods for studying the evolution of the physical properties of tight carbonate reservoir rocks. *Mar. Pet. Geol.* **2022**, *139*, 105603. [CrossRef]
28. Sajed, O.K.M.; Glover, P.W. Dolomitisation, cementation and reservoir quality in three Jurassic and Cretaceous carbonate reservoirs in north-western Iraq. *Mar. Pet. Geol.* **2020**, *115*, 104256. [CrossRef]
29. Shibayama, A.; Hamami, M.; Yamada, T.; Kohda, A.; Farhan, Z.; Bellah, S.; Shibasaki, T.; Jasmi, S. The Application of Geological Concepts for Various Types of High-Permeability Streaks to the Full-Field Simulation Model History Matching of Carbonate Reservoir, Offshore Abu Dhabi. In Proceedings of the Abu Dhabi International Petroleum Exhibition & Conference, Abu Dhabi, United Arab Emirates, 13–16 November 2017. [CrossRef]
30. Hu, D.; Rui, G.; Songhao, H.; Yuanbing, W.; Zhaowu, Z. Integrated management and Application of Horizontal Well Water Flooding Technology in a Large-scale Complicated Carbonate Oilfield Containing High permeability Streaks. In Proceedings of the SPE/IATMI Asia Pacific Oil & Gas Conference and Exhibition, Nusa Dua, Indonesia, 20–22 October 2015. [CrossRef]
31. Ding, S.; Jiang, H.; Wang, L.; Liu, G.; Li, N.; Liang, B. Identification and Characterization of High-permeability Zones in Waterflooding Reservoirs With an Ensemble of Methodologies. In Proceedings of the SPE/IATMI Asia Pacific Oil & Gas Conference and Exhibition, Nusa Dua, Indonesia, 20–22 October 2015. [CrossRef]
32. Correia, M.G.; Hohendorff, J.C.; Schiozer, D.J. Multiscale integration for Karst-reservoir flow-simulation models. *SPE Reserv. Eval. Eng.* **2020**, *23*, 518–533. [CrossRef]
33. Pantou, I. *Impact of Stratigraphic Heterogeneity on Hydrocarbon Recovery in Carbonate Reservoirs: Effect of Karst*; Imperial College London: London, UK, 2014; pp. 1–56.
34. Bigoni, F.; Pirrone, M.; Trombin, G.; Vinci, F.F.; Raimondi Cominesi, N.; Guglielmelli, A.; Ali Hassan, A.A.; Ibrahim Uatouf, K.S.; Bazzana, M.; Viviani, E. Middle East karst Carbonate: An Integrated Workflow for Prediction of Karst Enhancement Distribution. In Proceedings of the SPE Reservoir Characterisation and Simulation Conference and Exhibition, Abu Dhabi, United Arab Emirates, 17–19 September 2019. [CrossRef]
35. La Bruna, V.; Bezerra, F.; Souza, V.; Maia, R.; Auler, A.; Araújo, R.; Cazarin, C.; Rodrigues, M.; Vieira, L.; Sousa, M. High-permeability zones in folded and faulted silicified carbonate rocks—Implications for karstified carbonate reservoirs. *Mar. Pet. Geol.* **2021**, *128*, 105046. [CrossRef]
36. Zhang, L.; Zhang, W.; Li, Y.; Song, B.; Liu, D.; Deng, Y.; Xu, J.; Wang, Y. Sequence Stratigraphy, Sedimentology, and Reservoir Characteristics of the Middle Cretaceous Mishrif Formation, South Iraq. *J. Mar. Sci. Eng.* **2023**, *11*, 1255. [CrossRef]
37. Galkin, V.I.; Ponomareva, I.N.; Repina, V.A. Study of the process of oil recovery in reservoirs of various types of voids using multivariate statistical analysis. *Bull. Perm Natl. Res. Polytech. Univ. Geol. Oil Gas Eng. Min.* **2016**, *15*, 145–154. [CrossRef]
38. Putilov, I.; Kozyrev, N.; Demyanov, V.; Krivoshchekov, S.; Kochnev, A. Factoring in Scale Effect of Core Permeability at Reservoir Simulation Modeling. *SPE J.* **2022**, *27*, 1930–1942. [CrossRef]
39. Amaefule, J.O.; Altunbay, M.; Tiab, D.; Kersey, D.; Keelan, D. Enhanced Reservoir Description: Using Core and Log Data to identify Hydraulic (Flow) Units and Predict Permeability in Uncored Intervals/Wells. In Proceedings of the SPE Annual Technical Conference and Exhibition, Houston, TX, USA, 3–6 October 1993. [CrossRef]
40. Garrouch, A.A.; Al-Sultan, A.A. Exploring the link between the flow zone indicator and key open-hole log measurements: An application of dimensional analysis. *Pet. Geosci.* **2019**, *25*, 219–234. [CrossRef]

Disclaimer/Publisher's Note: The statements, opinions and data contained in all publications are solely those of the individual author(s) and contributor(s) and not of MDPI and/or the editor(s). MDPI and/or the editor(s) disclaim responsibility for any injury to people or property resulting from any ideas, methods, instructions or products referred to in the content.

Article

A Case Study on the CO_2 Sequestration in Shenhua Block Reservoir: The Impacts of Injection Rates and Modes

Ligen Tang [1,2], Guosheng Ding [1,2], Shijie Song [3], Huimin Wang [3], Wuqiang Xie [3] and Jiulong Wang [4,*]

1. Research Institute of Petroleum Exploration & Development, PetroChina, Beijing 100083, China
2. National Energy Underground Gas Storage R&D Center, Beijing 100083, China
3. Shaanxi Coal and Chemical Industry Group Co., Ltd., Xian 710100, China
4. Computer Network Information Center, Chinese Academy of Sciences, Beijing 100190, China
* Correspondence: jlwang@cnic.cn

Citation: Tang, L.; Ding, G.; Song, S.; Wang, H.; Xie, W.; Wang, J. A Case Study on the CO_2 Sequestration in Shenhua Block Reservoir: The Impacts of Injection Rates and Modes. *Energies* 2024, 17, 122. https://doi.org/10.3390/en17010122

Academic Editor: Nikolaos K. Koukouzas

Received: 10 November 2023
Revised: 19 December 2023
Accepted: 20 December 2023
Published: 25 December 2023

Copyright: © 2023 by the authors. Licensee MDPI, Basel, Switzerland. This article is an open access article distributed under the terms and conditions of the Creative Commons Attribution (CC BY) license (https://creativecommons.org/licenses/by/4.0/).

Abstract: Carbon capture and storage (CCS) is the most promising method of curbing atmospheric carbon dioxide levels from 2020 to 2050. Accurate predictions of geology and sealing capabilities play a key role in the safe execution of CCS projects. However, popular forecasting methods often oversimplify the process and fail to guide actual CCS projects in the right direction. This study takes a specific block in Shenhua, China as an example. The relative permeability of CO_2 and brine is measured experimentally, and a multi-field coupling CO_2 storage prediction model is constructed, focusing on analyzing the sealing ability of the block from the perspective of injection modes. The results show that when injected at a constant speed, the average formation pressure and wellbore pressure are positively correlated with the CO_2 injection rate and time; when the injection rate is 0.5 kg/s for 50 years, the average formation pressure increases by 38% and the wellbore pressure increases by 68%. For different injection modes, the average formation pressures of various injection methods are similar during injection. Among them, the pressure increases around the well in the decreasing injection mode is the smallest. The CO_2 concentration around the wellbore is the largest, and the CO_2 diffusion range continues to expand with injection time. In summary, formation pressure increases with the increase in injection rate and injection time, and the decreasing injection mode has the least impact on the increase in formation pressure. The CO_2 concentration is the largest around the well, and the CO_2 concentration gradually decreases. The conclusion helps determine the geological carrying capacity of injection volumes and provides insights into the selection of more appropriate injection modes. Accurate predictions of CO_2 storage capacity are critical to ensuring project safety and monitoring potentially hazardous sites based on reservoir characteristics.

Keywords: Shenhua block; saline aquifer; carbon capture and storage (CCS); multi-field coupling

1. Introduction

CCS aims to mitigate human-induced carbon dioxide emissions by injecting and storing carbon dioxide in specific geological structures [1,2]. In the pursuit of achieving carbon neutrality by the mid-21st century, CCS stands out as a pivotal carbon-negative technology, garnering significant attention and interest from countries globally [3,4]. Before CO_2 injection can proceed, a proper assessment of the risk of CO_2 leakage from injection wells and geological storage sites must be conducted [1,5]. The Shenhua Carbon Capture and Storage Demonstration Project in China's Ordos Basin stands as Asia's first and largest full-chain saline aquifer carbon dioxide storage project. There is a lot of engineering and research going on there. These studies include stress and deformation changes induced by injection, potential damage modes and safety factors, interactions between coal mining and carbon dioxide geology storage, and determination of injection pressure limits, and the upper limit of wellhead pressure is 18 MPa, which is reliable [6,7]. Prior to project implementation, a rigorous consideration of the impact of fluid flow in the formation and an accurate prediction of the formation's storage capacity are essential to ensure the safe development of the project [8,9]. CCS is a complex process that requires focus on its impact on formation pressure and CO_2 distribution. Different injection modes have important effects on the formation.

Geological formations such as basalts, coal seams, depleted oil reserves, soils, deep saline aquifers, and sedimentary basins exhibit vast potential for carbon dioxide storage [10]. There is potential for carbon dioxide (CO_2) recovery in ultra-deep water subsalt carbonate reservoirs for carbon capture and storage [11]. To facilitate the geological storage of CO_2, the pressure must exceed 7.38 MPa and the temperature must surpass 31.1 °C, indicating a theoretical storage depth exceeding 800 m. The Shenhua saline aquifer block in China satisfies these requirements. The profound geological structure's complexity, exploration extent, and limitations in indoor physical simulation experimental conditions pose challenges to reservoir characterization and geological modeling. Simultaneously, accurately predicting CO_2 migration patterns and ensuring reservoir safety are critical issues in current research. However, there is no very definite conclusion about the storage capacity of the reservoir. In particular, the research on the storage capacity under different injection modes is still blank.

Countries such as the United States, China, Russia, the United Kingdom, Croatia, and India are actively accelerating the global deployment of CCS, making significant contributions to the reduction in global greenhouse gas emissions [12–18]. Challenges such as carbon dioxide leakage, energy inefficiency, and high implementation costs pose significant obstacles to the development of CCS. The safety assessment of CCS represents one of the greatest challenges; accurate predictions of geological carrying capacity and storage capacity are essential prerequisites for formulating and implementing a viable plan [19–24].

Berrezueta conducted laboratory studies on carbon dioxide–brine–rock interactions and performed some sensitivity analyses [25]. Xie investigated the influence of geological and engineering parameters on CO_2 migration and flow characteristics through indoor injection experiments, supplemented by X-ray computed tomography (CT) and scanning electron microscopy (SEM) experiments and computational fluid dynamics (CFD) numerical simulations [26]. These experimental studies did not conduct real simulations under formation conditions and could not directly guide the project. Hu integrated CFD simulation technology into the experimental study of atmospheric CO_2 diffusion in full-scale blasting emission tests of high-pressure supercritical phase CO_2 pipelines, and quantitatively analyzed the relationship between supercritical CO_2 leakage diameter and dangerous distance [27]. However, this study did not directly analyze the formation CO_2 distribution. Tutolo used high-performance computing techniques to study the coupled effects of cold CO_2 injection and background hydraulic head gradients on reservoir-scale mineral volume changes. Research has found that the migration and flow characteristics of CO_2 in sandstone during the geological storage process have a significant impact on the physical and mechanical properties of the rock [28]. Therefore, research on CO_2 storage must consider the physical and mechanical properties of rocks and fluid flow characteristics. Yang utilized the VOF (Volume of Fluid) method, capable of tracking dynamic changes in the two-phase interface, to establish two-dimensional and three-dimensional models and numerically simulate a supercritical CO_2–brine two-phase flow [29]. Without considering the influence of reservoir mechanical properties on seepage characteristics, it is inaccurate to simply study two-phase flow. In water–mechanical–chemical-coupled simulations, simplified flow mechanisms can lead to significant deviations in predicted throughput and storage performance [30]. Ratnakar and Omosebi et al. developed a machine learning-based workflow to inject single-phase supercritical carbon dioxide into deep saline aquifers to assess leakage risks [31–36]. The shortcoming is that these studies did not conduct sufficient and effective analysis and research on formation pressure changes.

Given the limitations of the current body of research, this study addresses the relatively singular factors considered and explores other issues. It entails experimental measurements of CO_2 migration and brine flow characteristics under different driving pressures. Additionally, a comprehensive large-scale multi-field coupling model of the reservoir (encompassing seepage, chemical diffusion, and solid mechanics fields) was established. Experimental data were incorporated into the model, and the influence of formation pressure on rock permeability characteristics was thoroughly examined. The primary focus of this study is the innovative exploration of various CO_2 injection modes, with an evaluation of reservoir storage capacity and risk conducted through the analysis of changes in pressure around the well, average formation pressure alterations, and CO_2 distribution. Taking a specific CO_2 geological storage project in the Shenhua saline aquifer as the research subject, the study integrates experiments and simulations, aligning with the actual engineering background and conditions. This approach aims to elucidate the CO_2 geological storage mechanism in saline aquifers. The research methods and conclusions derived from this study provide valuable insights into the geological storage mechanism and seepage laws of CO_2 in saline aquifers, playing a pivotal role in informing the scientific and safe implementation of storage projects.

The framework of this article is structured as follows: Section 2 delves into core methods, including mathematical models, physical models, physical properties, seepage characteristic parameters, introduction to injection methods, and an overview of the block. Section 3 engages in a discussion of the results, covering model verification, reservoir pressure comparison, and CO_2 distribution. Finally, the article concludes with a summary.

2. Methodology

This study fully considered the physical properties of the reservoir and fluid. Multi-fields mainly include multiphase transfer field, Darcy seepage field, and solid mechanics field. The coupling method is introduced in detail in the mathematical model. Figure 1 is the flow of the process.

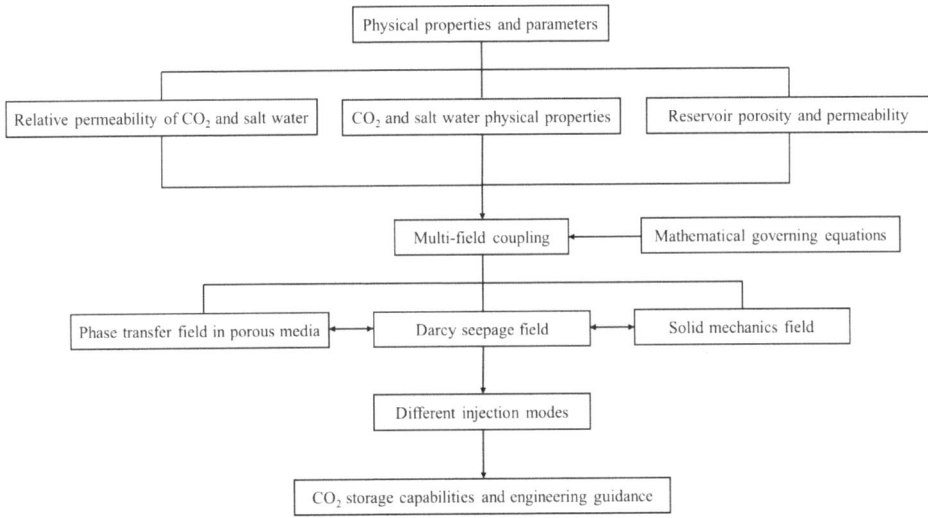

Figure 1. Flow of the process.

2.1. Mathematical Model

This study takes into account multi-field coupling (encompassing seepage, chemical diffusion, and solid mechanics fields), diffusion effects, and effective stress. The mathematical model comprises the multiphase fluid flow mass conservation theory, seepage mechanics momentum equation, solid mechanics stress balance differential equation, constitutive equation, geometric equation, Terzaghi effective stress principle, and diffusion equation. This section primarily introduces each equation and its physical meaning, elucidating how they are coupled to establish connections.

The continuity equation of multiphase seepage delineates the mass conservation of multiphase mixed fluids. This equation is articulated in terms of the volume fraction of each phase [37]:

$$\frac{\partial \epsilon_p \rho_{s_i} s_i}{\partial t} + \nabla \cdot N_i = 0 \quad (1)$$

Among them, ϵ_p is the porosity, ρ_{s_i} is the fluid density, and s_i is the volume fraction.

$$N_i = \rho_{s_i} u_i \quad (2)$$

u_i is the fluid velocity.

Darcy's seepage flux equation is a constitutive equation that characterizes the flow of liquid through porous media. This equation finds extensive application in petroleum engineering and groundwater engineering:

$$u_i = -\frac{\kappa_{rs_i}}{\mu_{s_i}} \kappa (\nabla p - \rho_i g) \quad (3)$$

$i = 1, 2, 3$. s_1 is brine saturation. s_2 is carbon dioxide saturation. s_3 is bound brine saturation.

$$s_1 + s_2 + s_3 = 1 \quad (4)$$

If Darcy's multiphase seepage equation considers diffusion effects, then:

$$u_i = -\frac{\kappa_{rs_i}}{\mu_{s_i}} \kappa (\nabla p - \rho_i g) - D_{cs_i} \nabla s_i \quad (5)$$

D_{cs_i} is the diffusion coefficient, and the value here is 6×10^{-9} m^2/s from the literature [38].

The stress balance equation of a solid elucidates the equilibrium of forces at each point within a stationary solid. In a three-dimensional space, for a solid within a volume element and considering three directions (x, y, and z), the stress balance equation can be expressed as [39]:

$$\left. \begin{array}{l} \frac{\partial \sigma_x}{\partial x} + \frac{\partial \tau_{yx}}{\partial y} + \frac{\partial \tau_{zx}}{\partial z} + f_x = 0 \\ \frac{\partial \sigma_y}{\partial y} + \frac{\partial \tau_{zy}}{\partial z} + \frac{\partial \tau_{xy}}{\partial x} + f_y = 0 \\ \frac{\partial \sigma_z}{\partial z} + \frac{\partial \tau_{xz}}{\partial x} + \frac{\partial \tau_{yz}}{\partial y} + f_z = 0 \end{array} \right\} \quad (6)$$

σ_x, σ_y, σ_z, τ_{xy}, τ_{yz}, τ_{zx} are the stress components, and f_x, f_y, f_z are the body force components. Here, we only consider gravity:

$$\left. \begin{array}{l} f_x = 0 \\ f_y = 0 \\ f_z = \rho g \end{array} \right\} \quad (7)$$

The relationship between the shear stress components and displacement components:

$$\left. \begin{array}{l} \tau_{yx} = \tau_{xy} = \frac{E}{2(1+\nu)} \left(\frac{\partial u_x}{\partial y} + \frac{\partial u_y}{\partial x} \right) \\ \tau_{xz} = \tau_{zx} = \frac{E}{2(1+\nu)} \left(\frac{\partial u_z}{\partial x} + \frac{\partial u_x}{\partial z} \right) \\ \tau_{zy} = \tau_{yz} = \frac{E}{2(1+\nu)} \left(\frac{\partial u_y}{\partial z} + \frac{\partial u_z}{\partial y} \right) \end{array} \right\} \quad (8)$$

E is the elastic modulus; ν is the Poisson's ratio, and u_x, u_y, u_z are the displacement components.

Terzaghi's effective stress principle asserts that while the stress in the soil is borne by both the soil skeleton and the water vapor in the soil, only the effective stress transmitted through the soil particles induces soil deformation. The pressure transmitted through water vapor in the pores does not contribute to the strength of the soil [39]:

$$\left.\begin{array}{l}\sigma_x = \sigma'_x + \alpha p \\ \sigma_y = \sigma'_y + \alpha p \\ \sigma_z = \sigma'_z + \alpha p\end{array}\right\} \quad (9)$$

$\sigma'_x, \sigma'_y, \sigma'_z$ are the effective stress components; α is the Biot coefficient, and p is the reservoir pressure. Equilibrium Equation (10) by bagging (7), (8), and (9) into (6):

$$\left.\begin{array}{l}\frac{\partial \sigma'_x}{\partial x} + \alpha \frac{\partial p}{\partial x} + \frac{E}{2(1+v)}\left(\frac{\partial^2 u_x}{\partial y^2} + \frac{\partial^2 u_y}{\partial x \partial y}\right) + \frac{E}{2(1+v)}\left(\frac{\partial^2 u_x}{\partial z^2} + \frac{\partial^2 u_z}{\partial x \partial z}\right) = 0 \\ \frac{\partial \sigma'_y}{\partial y} + \alpha \frac{\partial p}{\partial y} + \frac{E}{2(1+v)}\left(\frac{\partial^2 u_y}{\partial z^2} + \frac{\partial^2 u_z}{\partial y \partial z}\right) + \frac{E}{2(1+v)}\left(\frac{\partial^2 u_y}{\partial x^2} + \frac{\partial^2 u_x}{\partial y \partial x}\right) = 0 \\ \frac{\partial \sigma'_z}{\partial z} + \alpha \frac{\partial p}{\partial z} + \frac{E}{2(1+v)}\left(\frac{\partial^2 u_z}{\partial x^2} + \frac{\partial^2 u_x}{\partial z \partial x}\right) + \frac{E}{2(1+v)}\left(\frac{\partial^2 u_z}{\partial y^2} + \frac{\partial^2 u_y}{\partial z \partial y}\right) + \rho g = 0\end{array}\right\} \quad (10)$$

And three-dimensional partial differential equilibrium equation, denoted as Equation (11):

$$\left.\begin{array}{l}\frac{Ev}{(1+v)(1-2v)}\left(\frac{\partial^2 u_x}{\partial x^2} + \frac{\partial^2 u_y}{\partial x \partial y} + \frac{\partial^2 u_z}{\partial x \partial z}\right) + \frac{E}{(1+v)}\frac{\partial^2 u_x}{\partial x^2} + \alpha \frac{\partial p}{\partial x} + \frac{E}{2(1+v)}\left(\frac{\partial^2 u_x}{\partial y^2} + \frac{\partial^2 u_y}{\partial x \partial y}\right) + \frac{E}{2(1+v)}\left(\frac{\partial^2 u_x}{\partial z^2} + \frac{\partial^2 u_z}{\partial x \partial z}\right) = 0 \\ \frac{Ev}{(1+v)(1-2v)}\left(\frac{\partial^2 u_x}{\partial y \partial x} + \frac{\partial^2 u_y}{\partial y^2} + \frac{\partial^2 u_z}{\partial y \partial z}\right) + \frac{E}{(1+v)}\frac{\partial^2 u_y}{\partial y^2} + \alpha \frac{\partial p}{\partial y} + \frac{E}{2(1+v)}\left(\frac{\partial^2 u_y}{\partial z^2} + \frac{\partial^2 u_z}{\partial y \partial z}\right) + \frac{E}{2(1+v)}\left(\frac{\partial^2 u_y}{\partial x^2} + \frac{\partial^2 u_x}{\partial y \partial x}\right) = 0 \\ \frac{Ev}{(1+v)(1-2v)}\left(\frac{\partial^2 u_x}{\partial z \partial x} + \frac{\partial^2 u_y}{\partial z \partial y} + \frac{\partial^2 u_z}{\partial z^2}\right) + \frac{E}{(1+v)}\frac{\partial^2 u_z}{\partial z^2} + \alpha \frac{\partial p}{\partial z} + \frac{E}{2(1+v)}\left(\frac{\partial^2 u_z}{\partial x^2} + \frac{\partial^2 u_x}{\partial z \partial x}\right) + \frac{E}{2(1+v)}\left(\frac{\partial^2 u_z}{\partial y^2} + \frac{\partial^2 u_y}{\partial z \partial y}\right) + \rho g = 0\end{array}\right\} \quad (11)$$

We bring $\nabla^2 = \frac{\partial^2}{\partial x^2} + \frac{\partial^2}{\partial y^2} + \frac{\partial^2}{\partial z^2}$ into (11) and simplify the solid (stress field) Equation (12) for carbon dioxide reservoir calculation considering the effect of effective stress:

$$\left.\begin{array}{l}\frac{E}{2(1+v)(1-2v)}\frac{\partial \varepsilon_V}{\partial x} + \frac{E}{2(1+v)}\nabla^2 u_x + \alpha \frac{\partial p}{\partial x} = 0 \\ \frac{E}{2(1+v)(1-2v)}\frac{\partial \varepsilon_V}{\partial y} + \frac{E}{2(1+v)}\nabla^2 u_y + \alpha \frac{\partial p}{\partial y} = 0 \\ \frac{E}{2(1+v)(1-2v)}\frac{\partial \varepsilon_V}{\partial z} + \frac{E}{2(1+v)}\nabla^2 u_z + \alpha \frac{\partial p}{\partial z} + \rho g = 0\end{array}\right\} \quad (12)$$

$\varepsilon_x, \varepsilon_y$ and ε_z are the strain components, and ε_V is the volume component:

$$\left.\begin{array}{l}\varepsilon_x = \frac{\partial u_x}{\partial x} \\ \varepsilon_y = \frac{\partial u_y}{\partial y} \\ \varepsilon_z = \frac{\partial u_z}{\partial z} \\ \varepsilon_V = \varepsilon_x + \varepsilon_y + \varepsilon_z\end{array}\right\} \quad (13)$$

α takes 1, which is experience from the engineering site, then:

$$\left.\begin{array}{l}\frac{E}{2(1+v)(1-2v)}\frac{\partial \varepsilon_V}{\partial x} + \frac{E}{2(1+v)}\nabla^2 u_x + \frac{\partial p}{\partial x} = 0 \\ \frac{E}{2(1+v)(1-2v)}\frac{\partial \varepsilon_V}{\partial y} + \frac{E}{2(1+v)}\nabla^2 u_y + \frac{\partial p}{\partial y} = 0 \\ \frac{E}{2(1+v)(1-2v)}\frac{\partial \varepsilon_V}{\partial z} + \frac{E}{2(1+v)}\nabla^2 u_z + \frac{\partial p}{\partial z} + \rho g = 0\end{array}\right\} \quad (14)$$

Subsequently, the connection between the multiphase seepage field and the solid mechanical field can be established through Equations (5) and (14), considering both the effective stress principle and diffusion principle.

2.2. Physical Models and Numerical Methods

This multi-field coupling model employs the finite element method (FEM, COMSOL) to numerically solve the aforementioned equations. As illustrated in Figure 2, the geometric model dimensions are 2400 m × 2400 m × 300 m. A 1/4 symmetrical structure was utilized for the study. Our research hypothesis involves incorporating roller supports around and at the bottom of the reservoir to restrict normal movement. The upper surface of the reservoir is free, and the outer boundary serves for outflow. No flux is present at the boundary, the internal boundary of the reservoir, and the upper and lower boundaries. FEM calculations were executed based on the mathematical and physical models, with grid divisions as depicted in Figure 3. The total number of units is 242,895, and the grid around the well is dense. The maximum element size is 31.8 m; the minimum element size is 6 m; the maximum element growth rate is 1.13; the curvature factor is 0.5; the minimum element quality is 0.1842; the average element quality is 0.6632; and the overall quality of the grid is good.

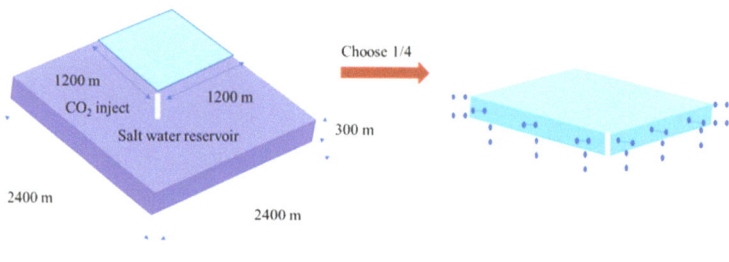

Figure 2. Model establishment.

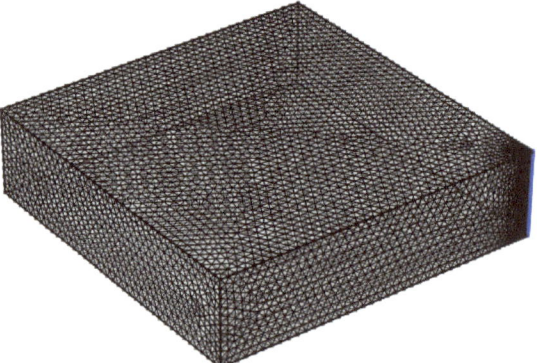

Figure 3. Mesh division.

According to engineering site data, the Young's modulus of the entire reservoir is 1.3 GPa, the Poisson's ratio is 0.23, and the density is 2560 kg/m^3. The salinity of brine is 23.5 g/L with a density of 0.984 g·cm^{-3}. The main ionic components are Na$^+$ and Cl$^-$. The relationship between porosity and permeability and reservoir pressure is as follows:

$$\left.\begin{array}{l} epsilon = 0.2984 \times (1 + (2 \times 10^{(-2.5)}) \times p/1\ [MPa]) \\ kappa = 1 \times 10^{(-12)} \times \left(\frac{epsilon}{0.2984}\right)^3 (m2) \end{array}\right\} \quad (15)$$

Epsilon is porosity, and *kappa* is permeability. And the empirical formula is given by the project.

2.3. Physical Properties and Seepage Characteristic Parameters

Figure 4 illustrates the experimental system designed for testing the relative permeability of supercritical CO_2–brine. The system comprises a CO_2 storage tank, booster pump, pressure gauge, incubator, core holder, confining pressure pump, back pressure valve, back pressure pump, gas–liquid separator, and gas–water metering device (Appendix A shows the experimental equipment and procedure). The experiment simulates a temperature of 70 °C (temperature of the reservoir), with inlet and outlet pressures set at 10 MPa and 8 MPa, respectively. Under these conditions, CO_2 attains a supercritical state. Figure 5 illustrates the result of relative permeability.

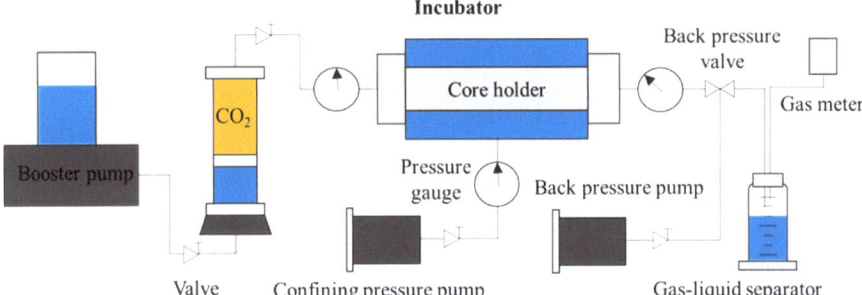

Figure 4. Diagram of the experimental system for supercritical CO_2–brine relative permeability testing.

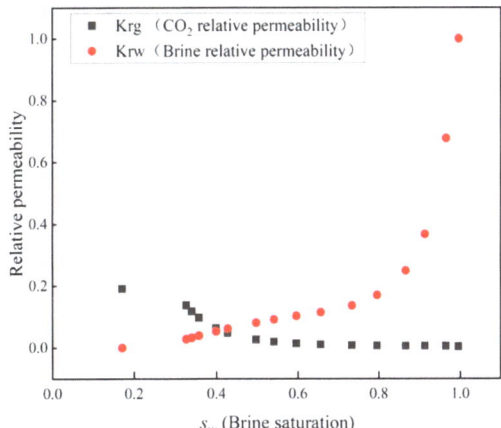

Figure 5. The experimental result of relative permeability.

The relative permeability of brine and CO_2 is obtained:

$$\left. \begin{array}{l} K_{rw} = \left(\frac{(s_w - 0.1706)}{0.8294}\right)^{7.661} \\ K_{rg} = 0.1909 \times \left(1 - \frac{(s_w - 0.1706)}{0.8294}\right)^{3.502} \end{array} \right\} \quad (16)$$

The density and viscosity change curves of supercritical carbon dioxide with pressure at 70 °C are plotted based on the thermophysical parameters from the National Institute of Standards and Technology (NIST) in Figures 6 and 7.

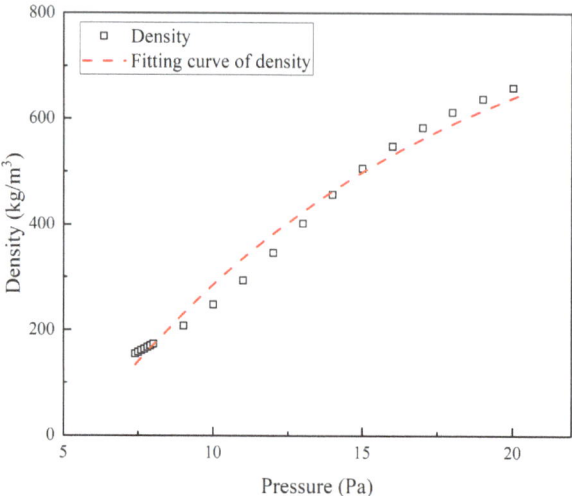

Figure 6. Carbon dioxide density changes with pressure.

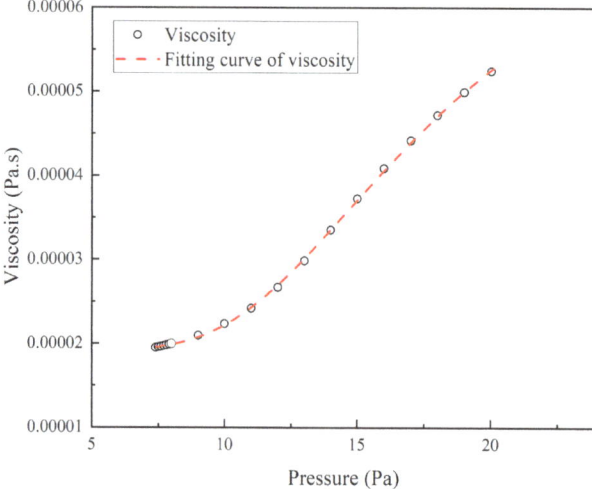

Figure 7. Carbon dioxide viscosity changes with pressure.

A magnetic levitation balance with precision 1 µg was employed for measuring the density of saltwater in a measuring cylinder. This electromagnetic levitation balance utilizes a combination of an electromagnet positioned outside the measuring container and a permanent magnet inside the measuring container to directly measure the absolute density of the fluid within the isolated and closed measuring container.

In the investigation of the impact of injection pressure on the density of saltwater under CO_2 storage conditions within the saltwater layer, the baseline temperature was set to 70 °C. Each set of experiments was conducted over a time period of 120 h, and the density under various pressures is illustrated in Figure 8.

Under identical conditions, the viscosity of the carbon dioxide aqueous solution, as measured with a viscometer, is presented in Figure 9 [40].

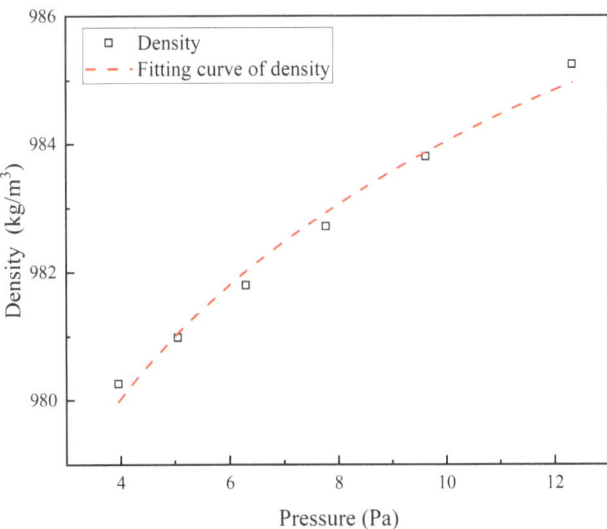

Figure 8. The density of carbon dioxide–brine solution changes with pressure.

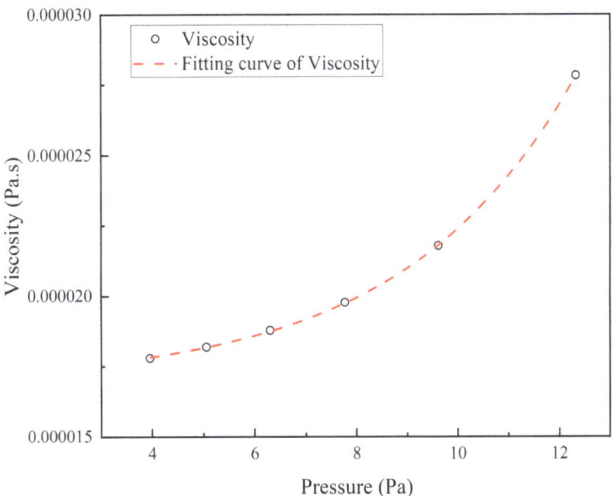

Figure 9. The viscosity of carbon dioxide–brine solution changes with pressure.

2.4. Introduction to Injection Methods

This study delves into the carbon sequestration capacity of the reservoir from the perspectives of injection amount and injection mode.

Table 1 shows all injection modes. Five injection rates modes: 0.1, 0.2, 0.3, 0.4, and 0.5 kg/s. Seven injection modes: 10 years at intervals of 0.5–0.1 kg/s; 10 years at intervals of 0.1–0.5 kg/s; 25 years at 0.5 kg/s, 25 years at 0.1 kg/s; 25 years at 0.4 kg/s and 25 years at 0.2 kg/s; 0.1 kg/s in 25 years, 0.5 kg/s in 25 years; 0.2 kg/s in 25 years and 0.4 kg/s in 25 years, and Mode 3. We compared the changes in pressure over time: pressure around the well, average formation pressure, and changes in CO_2 distribution over time.

Table 1. Injection modes.

	0–10 Years	10–20 Years	20–25 Years	25–30 Years	30–40 Years	40–50 Years
Mode 1	0.1	0.1	0.1	0.1	0.1	0.1
Mode 2	0.2	0.2	0.2	0.2	0.2	0.2
Mode 3	0.3	0.3	0.3	0.3	0.3	0.3
Mode 4	0.4	0.4	0.4	0.4	0.4	0.4
Mode 5	0.5	0.5	0.5	0.5	0.5	0.5
Mode 6	0.5	0.4	0.3	0.3	0.2	0.1
Mode 7	0.1	0.2	0.3	0.3	0.4	0.5
Mode 8	0.5	0.5	0.5	0.1	0.1	0.1
Mode 9	0.4	0.4	0.4	0.2	0.2	0.2
Mode 10	0.1	0.1	0.1	0.5	0.5	0.5
Mode 11	0.2	0.2	0.2	0.4	0.4	0.4

2.5. Block Introduction

The Shenhua Group is currently executing China's inaugural full-chain carbon dioxide capture and geological storage demonstration project, situated in the Ordos Basin in the eastern part of northwest China. As depicted in Figure 10, it spans five provinces (autonomous regions), including Shaanxi, Shanxi, and Inner Mongolia, covering a total area of more than 27.6×10^4 km^2. The Ordos Basin can be divided into six primary tectonic unit structures based on the history of geological structural changes. These include the Yishaan slope, the western margin thrust belt, the Shanxi burned skirt belt, the Tianhuan depression, the Yimeng uplift in the north, and the Yimeng uplift in the south. The Ordos Basin stands as one of the largest terrestrial sedimentary basins in China, characterized as a craton sedimentary basin. It lacks major fault zones traversing the entire basin, exhibits geological stability, even stress distribution, and boasts a thick sedimentary layer (with an average thickness of about 6000 m). Given these geological characteristics, it can conservatively be inferred that the Ordos formation possesses a significant geological storage capacity for CO_2 [39].

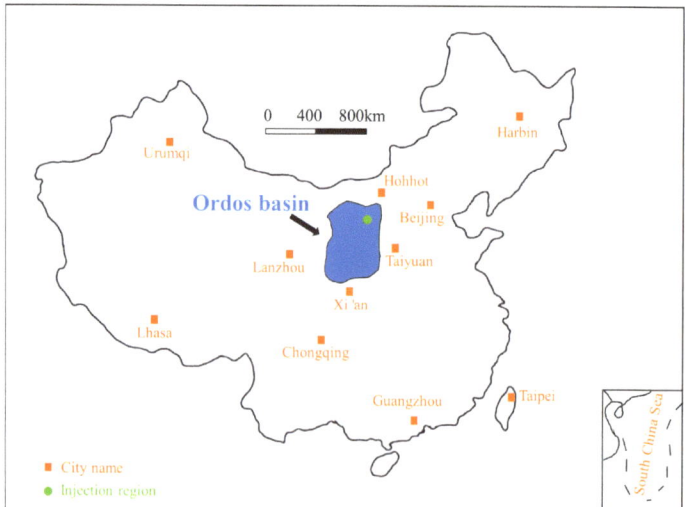

Figure 10. Location map of the CCS demonstration project in the Shenhua Ordos Basin.

In this project, carbon dioxide from coal tail gases is captured through liquid processing and stored in deep brine aquifers. The primary target layer for CO_2 injection is the saline aquifer beneath the mined coal seam. The formation receiving the carbon dioxide injection is characterized by low porosity, low permeability, and high heterogeneity. The project is currently operating successfully, with no reported CO_2 leaks or associated environmental hazards, and only minor pressure build-up has been observed.

3. Results and Discussion

3.1. Model Validation

To validate the accuracy of the multi-field coupling simulation, we conducted permeability experiments and simulations on cores, comparing them based on the relationship between permeability and reservoir pressure at the engineering site. Water injection experiments were carried out using on-site provided cores, with a core length of 6 cm and a diameter of 2.5 cm. The outlet pressure was set to 8 MPa, and the inlet pressure varied at 10 MPa, 12 MPa, 14 MPa, 16 MPa, 18 MPa, and 20 MPa.

Figure 11 illustrates the relationship between the average core flow velocity and pressure for three experiment cases: one without considering solid mechanics, one considering solid mechanics in simulation, and one with experiments. We observed that simulations considering solid mechanics align closely with experimental results. However, simulations neglecting solid mechanics introduce increasing errors as the pressure rises. Therefore, to ensure simulation accuracy, accounting for the influence of rock mass solid mechanics is essential. According to Equation (15), as the pressure increases, the permeability increases.

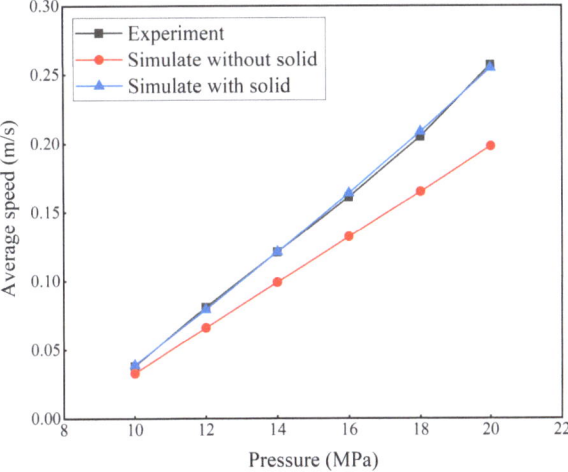

Figure 11. Comparison between core penetration experiments and simulations.

3.2. Reservoir Pressure Comparison

In the simulations, the initial formation pressure is 8 MPa, and the temperature is 70 °C, placing CO_2 in a supercritical state. Injection was conducted over 50 years at five rates (0.1–0.5 kg/s) with intervals of 0.1 kg/s. Figure 12a,b illustrate the changes in pressure around the well and in the formation at different injection rates. It is evident that both the pressure around the well and the average pressure in the formation increase with CO_2 injection. The higher the injection amount per unit time, the greater the pressure change.

Figure 12c,d depict pressure cloud diagrams with injection rates of 0.1 kg/s and 0.5 kg/s, respectively. The pressure in the reservoir rises annually with injection, and the pressure around the well is notably higher than in other locations. Due to the influence of gravity, the pressure value in the lower layer of the reservoir is higher. Key findings for different injection rates after 50 years include:

- 0.1 kg/s: Average max formation pressure is about 8.6 MPa (7% higher), and max wellbore pressure is about 9.1 MPa (14% higher, the average pressure value of the wellbore).
- 0.2 kg/s: Average max formation pressure is about 9.2 MPa (15% higher), and max wellbore pressure is about 10.4 MPa (30% higher).
- 0.3 kg/s: Average max formation pressure is about 9.8 MPa (23% higher), and max wellbore pressure is about 11.4 MPa (42% higher).

- 0.4 kg/s: Average max formation pressure is about 10.6 MPa (32% higher), and max wellbore pressure is about 12.6 MPa (57% higher).
- 0.5 kg/s: Average max formation pressure is about 11.0 MPa (38% higher), and max wellbore pressure is about 13.5 MPa (68% higher).

Therefore, in CCS projects, considering the specific working conditions of the reservoir is crucial to estimating the maximum injection rate. This consideration becomes particularly important for controlling the injection rate and determining the appropriate injection time.

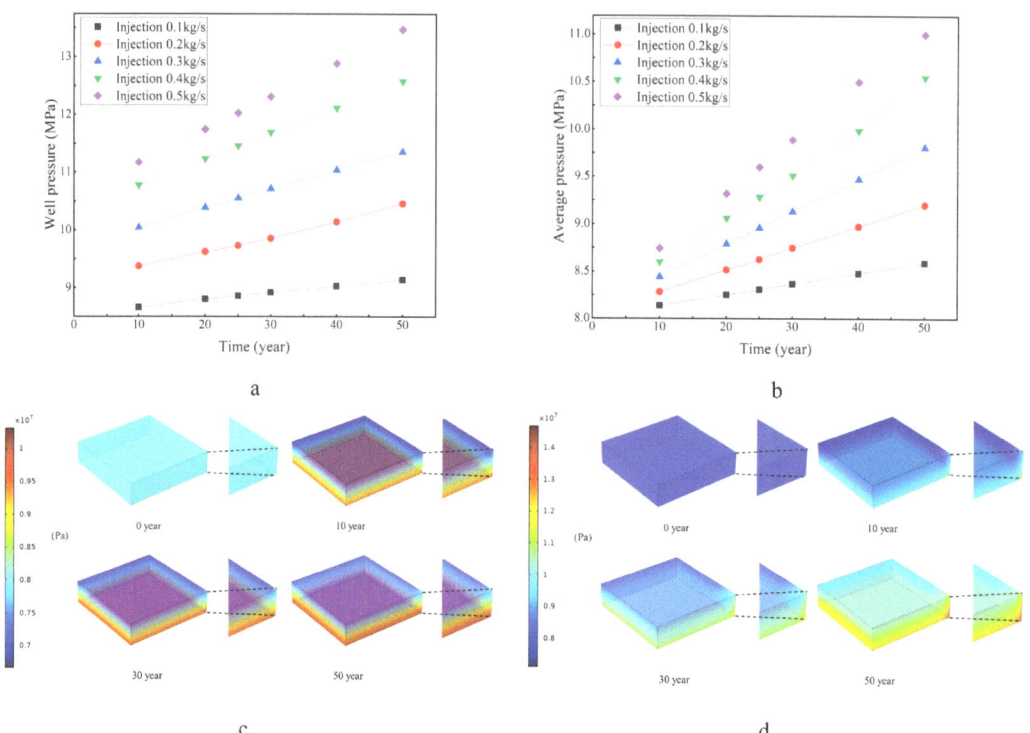

Figure 12. Changes in pressure around the well and in the formation with different injection volumes. (**a**) Pressure around the well; (**b**) Average pressure in the formation; (**c**) Pressure cloud chart at an injection rate of 0.1 kg/s; (**d**) Pressure cloud chart at an injection rate of 0.5 kg/s.

Similarly, for the other six injection modes mentioned in Section 2.4, the average injection rate is 0.3 kg/s: 10 years is an interval of 0.5–0.1 kg/s; 10 years is an interval of 0.1–0.5 kg/s; 25 years is 0.5 kg/s, 0.1 kg/s in 25 years; 0.4 kg/s in 25 years and 0.2 kg/s in 25 years; 0.1 kg/s in 25 years, 0.5 kg/s in 25 years; 0.2 kg/s in 25 years, 0.4 kg/s in 25 years. Analysis of the results. Examining Figure 13, which shows the average pressure changes around the well and in the formation with different injection modes, several observations can be made:

- Average Formation Pressure: It increases with injection time, peaking at 50 years for various injection modes. The maximum values at 50 years are 9.74, 9.89, 9.73, 9.77, 9.89, and 9.86 MPa. These values are relatively close to the case of a constant injection rate of 0.3 kg/s, which reaches 9.81 MPa after 50 years.
- Wellbore Pressure: The pressure around the well does not exhibit a simple monotonic change over time. The maximum value occurs at different times for various injection modes, and there is a considerable gap between these maximum values. It is worth noting that Mode 6, Mode 8, and Mode 9 each experienced a decrease in wellbore

pressure in different years, which was due to their reduced injection rates. The pressure around the well is affected by both the injection time and injection rate. The pressure around the well becomes higher as the injection time and injection rate increase.

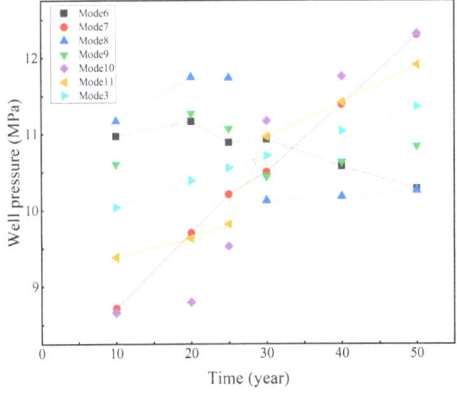
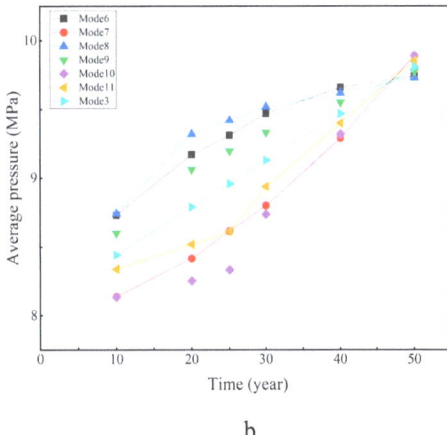

a b

Figure 13. Changes in pressure around the well and in the formation under injection Modes 6–11. (**a**) Pressure around the well; (**b**) Average pressure in the formation.

Considering these findings, the injection mode that decreases year by year seems to be the most suitable for this model. This mode results in a maximum wellbore pressure of only 10.93 MPa, making it more conducive to the safe development of the project. This information is valuable for optimizing injection strategies and ensuring the project's safety and efficacy.

3.3. CO_2 Distribution Analysis

In this section, we discuss the CO_2 distribution. Figure 14 shows the distribution of CO_2 injection volume and rate. Figure 14a shows the distribution of injection rate 0.1 kg/s CO_2 over time. Figure 14b shows the distribution of injection rate 0.5 kg/s CO_2 over time. Figure 14c shows the distribution cloud diagram of CO_2 with an injection volume of 0.1 kg/s; Figure 14d is the distribution cloud diagram of CO_2 with an injection volume of 0.5 kg/s. We can find that as the injection time increases, the diffusion range of CO_2 becomes larger and larger, and the concentration of CO_2 on the diffusion path becomes larger, and the maximum concentration is around the well. And, the greater the injection rate, the wider the diffusion range and the higher the concentration. As the diffusion range increases, the concentration of carbon dioxide becomes smaller and smaller. When the volume fraction of carbon dioxide is less than 1%, we consider it to be no longer diffusing. Due to the effect of gravity, the CO_2 concentration in the upper layer of the reservoir is greater than that in the lower layer and is distributed in a circular cone. For the five injection rates of 0.1–0.5 kg/s, the maximum diffusion ranges are 596 m, 608 m, 622 m, 621 m, and 640 m, respectively, in 50 years (The distance between the uppermost layer of the reservoir and the well is marked by an orange double arrow). For the other six injection modes, the CO_2 distribution is shown in Figure 15. As the injection time increases, the CO_2 diffusion range increases year by year. The maximum concentration is also around the well and occurs in the year of maximum injection volume. The diffusion ranges under these six working conditions are 683 m, 576 m, 696 m, 690 m, 564 m, and 557 m.

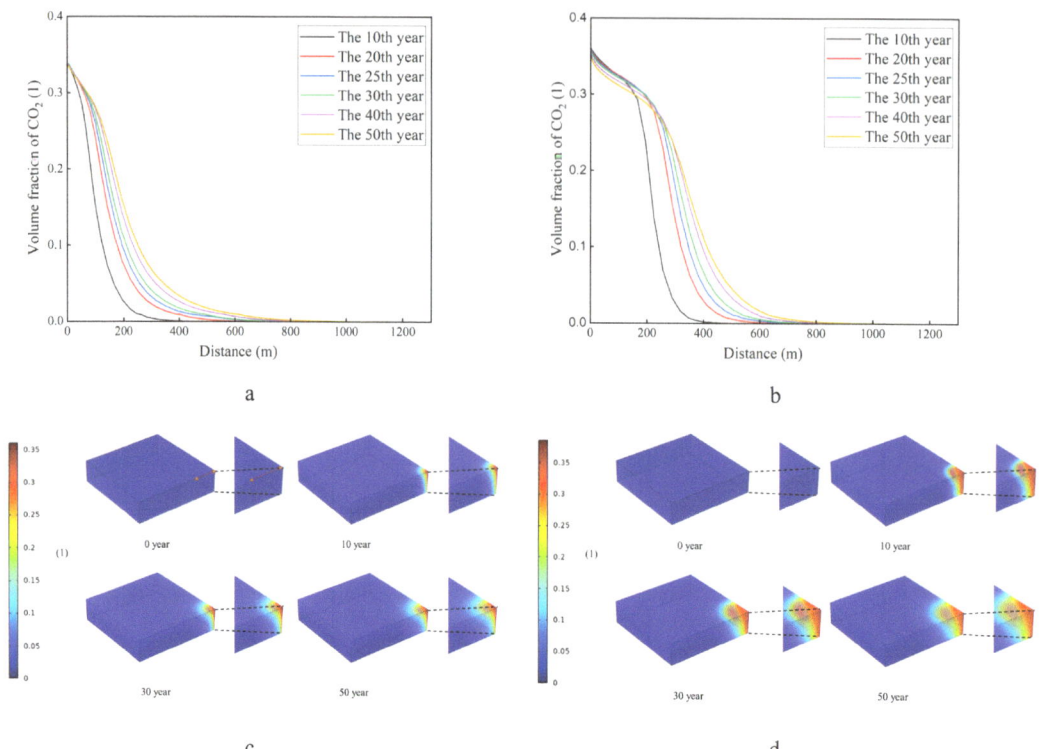

Figure 14. Distribution of CO_2 with different injection amounts. (**a**) Distribution of injection rate 0.1 kg/s CO_2 over time; (**b**) Injection rate 0.5 kg/s CO_2 distribution over time; (**c**) Injection rate 0.1 kg/s CO_2 distribution cloud chart; and (**d**) Distribution cloud chart of CO_2 injection volume 0.5 kg/s.

In summary, we should pay more attention to whether CO_2 leakage occurs around the well, and the greater the injection rate, the higher the frequency of attention. For the reservoir after 690 m, monitoring can be relatively reduced.

Although this study helps determine the geological carrying capacity of the injection volume, provides insights into selecting a more appropriate injection mode, and has a good guiding role for engineering, there are still some limitations: Failure to consider the impact of changes in reservoir temperature. The boundary conditions and parameters we used were all from the site and meet the engineering requirements to the greatest extent. The calculation results have good convergence and are in line with the site's basic understanding of the pressure and CO_2 distribution around the well. It has good engineering guidance analysis. Subsequent engineering development will also be closely integrated with simulation.

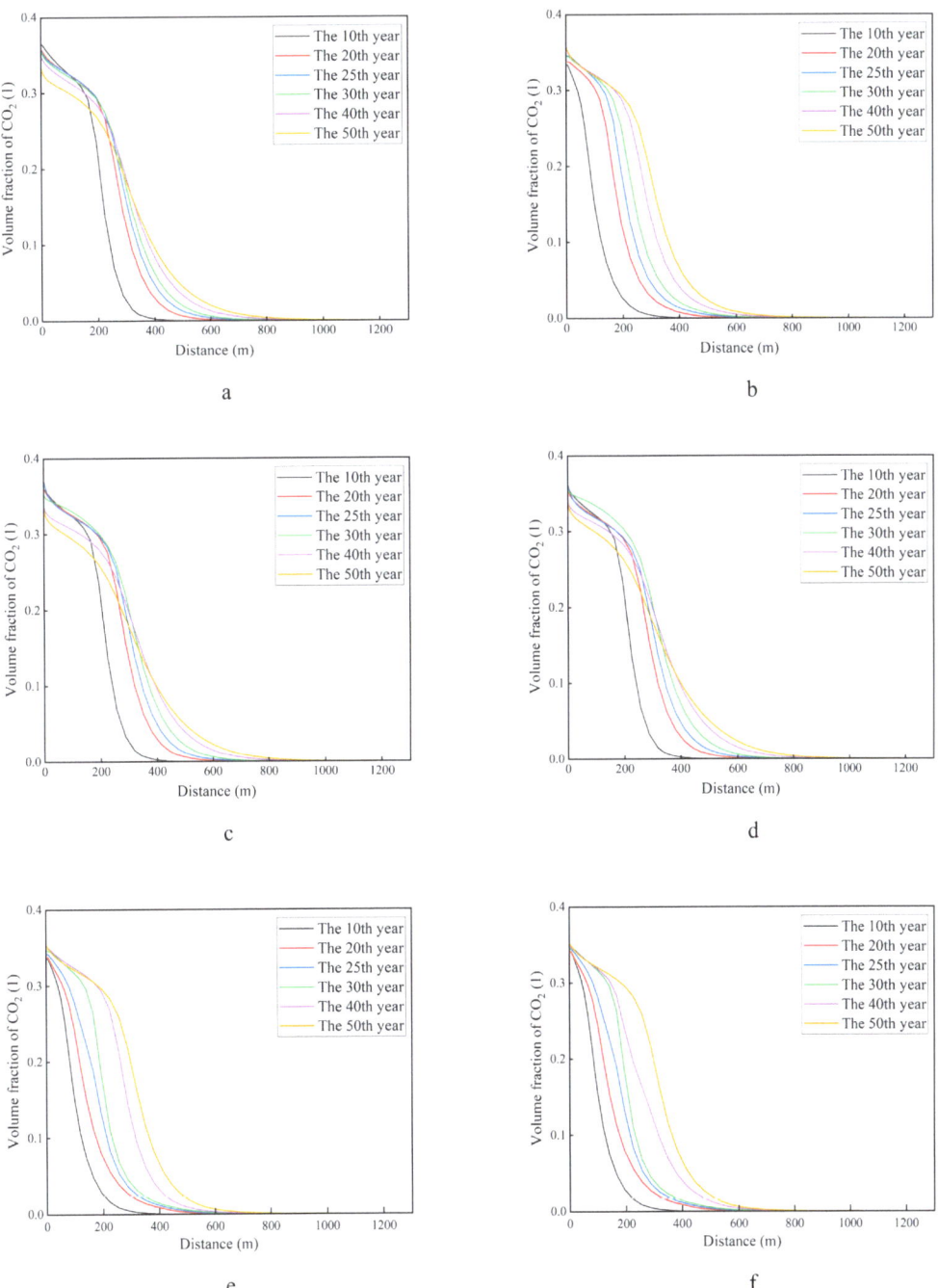

Figure 15. CO_2 distribution in different injection modes. (**a**) Injection Mode 6; (**b**) Injection Mode 7; (**c**) Injection Mode 8; (**d**) Injection Mode 9; (**e**) Injection Mode 10; and (**f**) Injection Mode 11.

4. Conclusions

To precisely anticipate the CO_2 migration pattern, assess the CO_2 storage capacity and formation safety, and ensure the project's seamless advancement, we amalgamated experiments and simulations to formulate a multi-field coupling CO_2 storage prediction model for a specific block in Shenhua. Our study focused on the carbon storage capacity of the reservoir concerning injection volume and mode. The key findings are as follows:

The average pressure in the formation and around the well rises proportionally to the total volume of injected CO_2, with the pressure around the well within the reservoir exhibiting the most significant increase. Additionally, higher injection rates correlate with elevated reservoir pressures. For instance, injecting at a rate of 0.1 kg/s for 50 years resulted in an approximately 7% increase in the average formation pressure compared to pre-injection levels, accompanied by a 14% increase in the maximum pressure around the well. In contrast, injecting at a rate of 0.5 kg/s for the same duration led to a roughly 38% surge in the average maximum formation pressure and a 68% increase in the maximum pressure around the well compared to pre-injection levels.

In the case of various injection modes, the average formation pressure rises with the total injection volume. After 50 years of injection, the maximum average pressure values in the formation become quite similar. Among the modes, the decreasing injection mode with a 10-year interval results in the smallest maximum pressure value around the well, measuring only 10.93 MPa.

The maximum concentration of CO_2 within the reservoir is concentrated around the well, and the extent of CO_2 diffusion expands with the cumulative injection volume. Larger injection rates per unit time led to higher maximum concentrations of CO_2 around the well, increased concentrations along the diffusion path, and broader diffusion ranges. The maximum diffusion range remains under 690 m. Enhanced CO_2 leakage monitoring is recommended around the well and within a 690 m radius from the well.

Author Contributions: L.T. is responsible for article conceptualization, methodology, and writing—review and editing; G.D. is responsible for coding of models. S.S. is responsible for data curation and writing—original draft; H.W. is responsible for investigation; W.X. is responsible for methodology and data curation; J.W. is responsible for conceptualization, writing—review and editing, and supervision. All authors have read and agreed to the published version of the manuscript.

Funding: This study was funded by the China Postdoctoral Science Foundation, No. 2022M713204.

Data Availability Statement: Data is contained within the article.

Conflicts of Interest: Authors Ligen Tang and Guosheng Ding were employed by the company Research Institute of Petroleum Explora-tion & Development, PetroChina. Authors Shijie Song, Huimin Wang and Wuqiang Xie were employed by the company Shaanxi Coal and Chemical Industry Group Co., Ltd. The remaining authors declare that the research was conducted in the absence of any commercial or financial relationships that could be construed as a potential conflict of interest.

Abbreviations

The following abbreviations are used in this manuscript:

CCS	Carbon capture and storage
CO_2	Carbon dioxide
CT	Computed tomography
SEM	Scanning electron microscopy
CFD	Computational fluid dynamics
VOF	Volume of Fluid
FEM	Finite element method
NIST	National Institute of Standards and Technology

Nomenclature

The following variables are used in this manuscript:

Variable	Meaning
ϵ_p	porosity

κ	permeability
ρ_{s_i}	fluid density
s_i	volume fraction
u_i	fluid velocity
D_{cs_i}	diffusion coefficient
σ_i	normal stress component
τ_{ij}	shear stress component
f_i	body force component
g	gravity
E	elastic modulus
v	Poisson's ratio
u_i	displacement component
σ'_i	effective stress component
p	reservoir pressure
ε_i	strain component
ε_V	volume strain
α	Biot coefficient
K_{rw}	water relative permeability
K_{rg}	gas relative permeability

Appendix A

Figure A1 shows the experimental equipment, and it mainly comprises a booster pump, pressure gauges, incubator, core holder, confining pressure pump, back pressure pump, intermediate containers, and gas–water metering device. The experimental procedure is as below:

Measuring brine phase permeability: Put the rock sample that has been saturated with simulated formation brine into the core holder, use a displacement pump to make the formation brine pass through the rock sample at a certain pressure or flow rate, and wait until the pressure difference between the inlet and outlet of the rock stabilizes. The brine phase permeability is measured three times in a row, and the relative error is less than 3%.

Establishing bound brine: Use humidified nitrogen or compressed air to drive brine, establish the irreducible brine saturation of the rock sample, and measure the effective permeability of the gas phase in the bound brine state.

Inject gas and brine into the rock sample at a certain ratio, and when the flow is stable, measure the inlet and outlet pressure difference, the gas and brine flow rates, and the quality of the brine rock sample.

The proportion of brine gradually increases. After the experiment reaches the gas phase relative permeability value less than 0.005, the brine phase permeability is measured and the experiment ends.

Figure A1. Experimental equipment. 1. Gas–water metering device; 2. Measuring cylinder; 3. Back pressure valve; 4. Back pressure pump; 5. Pressure gauge; 6. Intermediate container; 7. Core holder; 8. Confining pressure pump; and 9. Incubator.

References

1. Song, Y.; Jun, S.; Na, Y.; Kim, K.; Jang, Y.; Wang, J. Geomechanical challenges during geological CO_2 storage: A review. *Chem. Eng. J.* **2023**, *456*, 140968. [CrossRef]
2. Agaton, C.B. Application of real options in carbon capture and storage literature: Valuation techniques and research hotspots. *Sci. Total Environ.* **2021**, *795*, 148683. [CrossRef] [PubMed]
3. Zuch, M.; Ladenburg, J. Navigating the information pathway to carbon capture and storage acceptance: Patterns and insights from a literature review. *Energy Res. Soc. Sci.* **2023**, *105*, 103283. [CrossRef]
4. Li, M.; He, N.; Xu, L.; Peng, C.; Chen, H.; Yu, G. Eco-CCUS: A cost-effective pathway towards carbon neutrality in China. *Renew. Sustain. Energy Rev.* **2023**, *183*, 113512. [CrossRef]
5. Marbun, B.; Sinaga, S.Z.; Purbantanu, B.; Santoso, D.; Kadir, W.; Sule, R.; Prasetyo, D.E.; Prabowo, H.; Susilo, D.; Firmansyah, F.R. Lesson learned from the assessment of planned converted CO_2 injection well integrity in Indonesia-CCUS project. *Heliyon* **2023**, *9*, e18505. [CrossRef]
6. Li, X.; Li, Q.; Bai, B.; Wei, N.; Yuan, W. The geomechanics of Shenhua carbon dioxide capture and storage (CCS) demonstration project in Ordos Basin, China. *J. Rock. Mech. Geotech.* **2016**, *8*, 948–966. [CrossRef]
7. Tan, Y.; Nookuea, W.; Li, H.; Thorin, E.; Yan, J. Property impacts on Carbon Capture and Storage (CCS) processes: A review. *Energy Convers. Manag.* **2016**, *118*, 204–222. [CrossRef]
8. Vatalis, K.I.; Laaksonen, A.; Charalampides, G.; Benetis, N.P. Intermediate technologies towards low-carbon economy. The Greek zeolite CCS outlook into the EU commitments. *Renew. Sustain. Energy Rev.* **2012**, *16*, 3391–3400. [CrossRef]
9. Lai, N.; Yap, E.H.; Lee, C.W. Viability of CCS: A broad-based assessment for Malaysia. *Renew. Sustain. Energy Rev.* **2011**, *15*, 3608–3616. [CrossRef]
10. Shaw, R.; Mukherjee, S. The development of carbon capture and storage (CCS) in India: A critical review. *Carbon Capture Sci. Technol.* **2022**, *2*, 100036. [CrossRef]
11. Rodrigues, H.W.; Mackay, E.J.; Arnold, D.P. Multi-objective optimization of CO_2 recycling operations for CCUS in pre-salt carbonate reservoirs. *Int. J. Greenh. Gas Control* **2022**, *119*, 103719. [CrossRef]
12. Leonzio, G.; Bogle, D.; Foscolo, P.U.; Zondervan, E. Optimization of CCUS supply chains in the UK: A strategic role for emissions reduction. *Chem. Eng. Res. Des.* **2020**, *155*, 211–228. [CrossRef]
13. Vulin, D.; Mocilac, I.K.; Jukic, L.; Arnaut, M.; Vodopic, F.; Saftic, B.; Sedlar, D.K.; Cvetkovic, M. Development of CCUS clusters in Croatia. *Int. J. Greenh. Gas Control* **2023**, *124*, 103857. [CrossRef]
14. Bazhenov, S.; Chuboksarov, V.; Maximov, A.; Zhdaneev, O. Technical and economic prospects of CCUS projects in Russia. *Sustain. Mater. Technol.* **2022**, *33*, e00452. [CrossRef]
15. Yuan, J.; Lyon, T.P. Promoting global CCS RDD&D by stronger U.S.-China collaboration. *Renew. Sustain. Energy Rev.* **2012**, *16*, 6746–6769.
16. Maheen, R.; Cai, L.; Zhang, Y.S.; Zhao, M. Quantitative analysis of carbon dioxide emission reduction pathways: Towards carbon neutrality in China's power sector. *Carbon Capture Sci. Technol.* **2023**, *7*, 100112. [CrossRef]
17. Liu, Z.; Gao, M.; Zhang, X.; Liang, Y.; Guo, Y.; Liu, W.; Bao, J. CCUS and CO_2 injection field application in abroad and China: Status and progress. *Geoenergy Sci. Eng.* **2023**, *229*, 212011. [CrossRef]
18. Jiang, K.; Ashworth, P.; Zhang, S.; Liang, X.; Sun, Y.; Angus, D. China's carbon capture, utilization and storage (CCUS) policy: A critical review. *Renew. Sustain. Energy Rev.* **2020**, *119*, 109601. [CrossRef]
19. Terwel, B.W.; Harinck, F.; Ellemers, N.; Daamen, D.D. Going beyond the properties of CO_2 capture and storage (CCS) technology: How trust in stakeholders affects public acceptance of CCS. *Int. J. Greenh. Gas Control* **2011**, *5*, 181–188. [CrossRef]

20. Tcvetkov, P.; Cherepovitsyn, A.; Fedoseev, S. Public perception of carbon capture and storage: A state-of-the-art overview. *Heliyon* **2019**, *5*, e02845. [CrossRef]
21. van Os, H.W.; Herber, R.; Scholtens, B. Not Under Our Back Yards? A case study of social acceptance of the Northern Netherlands CCS initiative. *Renew. Sustain. Energy Rev.* **2014**, *30*, 923–942. [CrossRef]
22. Jiang, K.; Ashworth, P.; Zhang, S.; Hu, G. Print media representations of carbon capture utilization and storage (CCUS) technology in China. *Renew. Sustain. Energy Rev.* **2022**, *155*, 111938. [CrossRef]
23. Xu, Y.; Liu, B.; Chen, Y.; Lu, S. Public perceived risks and benefits of carbon capture, utilization, and storage (CCUS): Scale development and validation. *J. Environ. Manag.* **2023**, *347*, 119109. [CrossRef] [PubMed]
24. Storrs, K.; Lyhne, I.; Drustrup, R. A comprehensive framework for feasibility of CCUS deployment: A meta-review of literature on factors impacting CCUS deployment. *Int. J. Greenh. Gas Control* **2023**, *125*, 103878. [CrossRef]
25. Berrezueta, E.; Kovacs, T.; Herrera-Franco, G.; Mora-Frank, C.; Caicedo-Potosi, J.; Carrion-Mero, P.; Carneiro, J. Laboratory studies on CO_2-brine-rock interaction: An analysis of research trends and current knowledge. *Int. J. Greenh. Gas Control* **2023**, *123*, 103842. [CrossRef]
26. Xie, J.; Yang, X.; Qiao, W.; Peng, S.; Yue, Y.; Chen, Q.; Cai, J.; Jiang, G.; Liu, Y. Investigations on CO_2 migration and flow characteristics in sandstone during geological storage based on laboratory injection experiment and CFD simulation. *Gas Sci. Eng.* **2023**, *117*, 205058. [CrossRef]
27. Hu, Y.; Yan, X.; Chen, L.; Yu, S.; Liu, C.; Yu, J. Leakage hazard distance of supercritical CO_2 pipelines through experimental and numerical studies. *Int. J. Greenh. Gas Control* **2022**, *119*, 103730. [CrossRef]
28. Tutolo, B.M.; Kong, X.; Seyfried, W.E., Jr.; Saar, M.O. High performance reactive transport simulations examining the effects of thermal, hydraulic, and chemical (THC) gradients on fluid injectivity at carbonate CCUS reservoir scales. *Int. J. Greenh. Gas Control* **2015**, *39*, 285–301. [CrossRef]
29. Yang, Y.; Wang, J.; Wang, J.; Zhang, Q.; Yao, J. Pore-scale numerical simulation of supercritical CO_2-brine two-phase flow based on VOF method. *Nat. Gas Ind. B* **2023**, *10*, 466–475. [CrossRef]
30. Cai, M.; Li, X.; Zhang, K.; Su, Y.; Wang, D.; Yang, S.; Liu, S. Coupled hydro-mechanical-chemical simulation of CCUS-EOR with static and dynamic microscale effects in tight reservoirs. *Fuel* **2024**, *357*, 129888. [CrossRef]
31. Ratnakar, R.R.; Chaubey, V.; Dindoruk, B. A novel computational strategy to estimate CO_2 solubility in brine solutions for CCUS applications. *Appl. Energy* **2023**, *342*, 121134. [CrossRef]
32. Omosebi, O.A.; Oldenburg, C.M.; Reagan, M. Development of lean, efficient, and fast physics-framed deep-learning-based proxy models for subsurface carbon storage. *Int. J. Greenh. Gas Control* **2022**, *114*, 103562. [CrossRef]
33. Du, S.; Wang, J.; Wang, M.; Yang, J.; Zhang, C.; Zhao, Y.; Song, H. A systematic data-driven approach for production forecasting of coalbed methane incorporating deep learning and ensemble learning adapted to complex production patterns. *Energy* **2023**, *263*, 126121. [CrossRef]
34. Du, S.; Wang, M.; Yang, J.; Zhao, Y.; Wang, J.; Yue, M.; Xie, C.; Song, H. An enhanced prediction framework for coalbed methane production incorporating deep learning and transfer learning. *Energy* **2023**, *282*, 128877. [CrossRef]
35. Xie, C.; Du, S.; Wang, J.; Lao, J.; Song, H. Intelligent modeling with physics-informed machine learning for petroleum engineering problems. *Adv. Geo-Energy Res.* **2023**, *8*, 71–75. [CrossRef]
36. Xie, C.; Zhu, J.; Yang, H.; Wang, J.; Liu, L.; Song, H. Relative permeability curve prediction from digital rocks with variable sizes using deep learning. *Phys. Fluids* **2023**, *35*, 096605. [CrossRef]
37. Nassan, T.H.; Amro, M. Finite Element Simulation of Multiphase Flow in Oil Reservoirs-Comsol Multiphysics as Fast Prototyping Tool in Reservoir Simulation. *Gorn. Nauk. I Tekhnologii Min. Sci. Technol.* **2020**, *4*, 220–226. [CrossRef]
38. Sharafi, M.S.; Ghasemi, M.; Ahmadi, M.; Kazemi, A. An experimental approach for measuring carbon dioxide diffusion coefficient in water and oil under supercritical conditions. *Chin. J. Chem. Eng.* **2021**, *34*, 160–170. [CrossRef]
39. Mase, G.T.; Smelser, R.E.; Mase, G.E. *Continuum Mechanics for Engineers*; CRC Press: Boca Raton, FL, USA, 2009.
40. Ebrahiminejadhasanabadi, M.; Nelson, W.M.; Naidoo, P.; Mohammadi, A.H.; Ramjugernath, D. Experimental measurements of CO_2 solubility, viscosity, density, sound velocity and evaporation rate for 2-(2-aminoethoxy) ethanol (DGA)+ 1-methylpyrrolidin-2-one (NMP)/water+ ionic liquid systems. *Fluid Phase Equilibria* **2022**, *559*, 113475. [CrossRef]

Disclaimer/Publisher's Note: The statements, opinions and data contained in all publications are solely those of the individual author(s) and contributor(s) and not of MDPI and/or the editor(s). MDPI and/or the editor(s) disclaim responsibility for any injury to people or property resulting from any ideas, methods, instructions or products referred to in the content.

Article

Fracture Spacing Optimization Method for Multi-Stage Fractured Horizontal Wells in Shale Oil Reservoir Based on Dynamic Production Data Analysis

Wenchao Liu [1,*], Chen Liu [1,*], Yaoyao Duan [2], Xuemei Yan [2], Yuping Sun [3] and Hedong Sun [2,*]

1 School of Civil and Resource Engineering, University of Science and Technology Beijing, Beijing 100083, China
2 PetroChina Research Institute of Petroleum Exploration and Development, Langfang 065007, China; duanyy69@petrochina.com.cn (Y.D.); yanxuemei69@petrochina.com.cn (X.Y.)
3 PetroChina Research Institute of Petroleum Exploration and Development, Beijing 100083, China; sunyuping01@petrochina.com.cn
* Correspondence: wcliu_2008@126.com (W.L.); m202110025@xs.ustb.edu.cn (C.L.); sunhed@petrochina.com.cn (H.S.)

Citation: Liu, W.; Liu, C.; Duan, Y.; Yan, X.; Sun, Y.; Sun, H. Fracture Spacing Optimization Method for Multi-Stage Fractured Horizontal Wells in Shale Oil Reservoir Based on Dynamic Production Data Analysis. *Energies* **2023**, *16*, 7922. https://doi.org/10.3390/en16247922

Academic Editor: Roland W Lewis

Received: 5 November 2023
Revised: 30 November 2023
Accepted: 3 December 2023
Published: 5 December 2023

Copyright: © 2023 by the authors. Licensee MDPI, Basel, Switzerland. This article is an open access article distributed under the terms and conditions of the Creative Commons Attribution (CC BY) license (https://creativecommons.org/licenses/by/4.0/).

Abstract: In order to improve the shale oil production rate and save fracturing costs, based on dynamic production data, a production-oriented optimization method for fracture spacing of multi-stage fractured horizontal wells is proposed in this study. First, M. Brown et al.'s trilinear seepage flow models and their pressure and flow rate solutions are applied. Second, deconvolution theory is introduced to normalize the production data. The data of variable pressure and variable flow rate are, respectively, transformed into the pressure data under unit flow rate and the flow rate data under unit production pressure drop; and the influence of data error is eliminated. Two kinds of typical curve of the normalized data are analyzed using the pressure and flow rate solutions of M. Brown et al.'s models. The two fitting methods constrain each other. Thus, reservoir and fracture parameters are interpreted. A practical model has been established to more accurately describe the seepage flow behavior in shale oil reservoirs. Third, using Duhamel's principle and the rate solution, the daily and cumulative production rate under any variable production pressure can be obtained. The productivity can be more accurately predicted. Finally, the analysis method is applied to analyze the actual dynamic production data. The fracture spacing of a shale oil producing well in an actual block is optimized from the aspects of production life, cumulative production, economic benefits and other influencing factors, and some significant conclusions are obtained. The research results show that with the goal of maximum cumulative production, the optimal fracture spacing is 5.5 m for 5 years and 11.4 m for 10 years. All in all, the fracture spacing optimization and design theory of multi-stage fractured horizontal wells is enriched.

Keywords: shale oil; multi-stage fractured horizontal well; fracture spacing optimization; deconvolution; dynamic production data analysis

1. Introduction

The global shale reservoir resources are rich and valuable to exploit [1]. However, shale reservoirs have a low permeability, which makes crude oil flow and reservoir production difficult [2]. Therefore, shale reservoirs are often exploited by the multi-stage fracturing of horizontal wells [3]. Through hydraulic fracturing, the reservoir can be stimulated, several main fractures are formed perpendicular to the horizontal wells and the fracture network is formed around the main fractures. Therefore, the stimulated reservoir volume area is formed. Reservoir permeability and porosity are increased, oil flows more easily and oil recovery efficiency is improved. The higher the fracture numbers, the higher the production rate [4], but the higher the production cost [5,6]. The fracture spacing not only needs to meet the requirements of a high production rate, but also needs to make the production cost not too high. Therefore, it is very necessary to optimize the fracture spacing [7]. The schematic diagram of fracture spacing optimization is shown in Figure 1. By optimizing the fracture spacing, efficient reservoir development can be achieved.

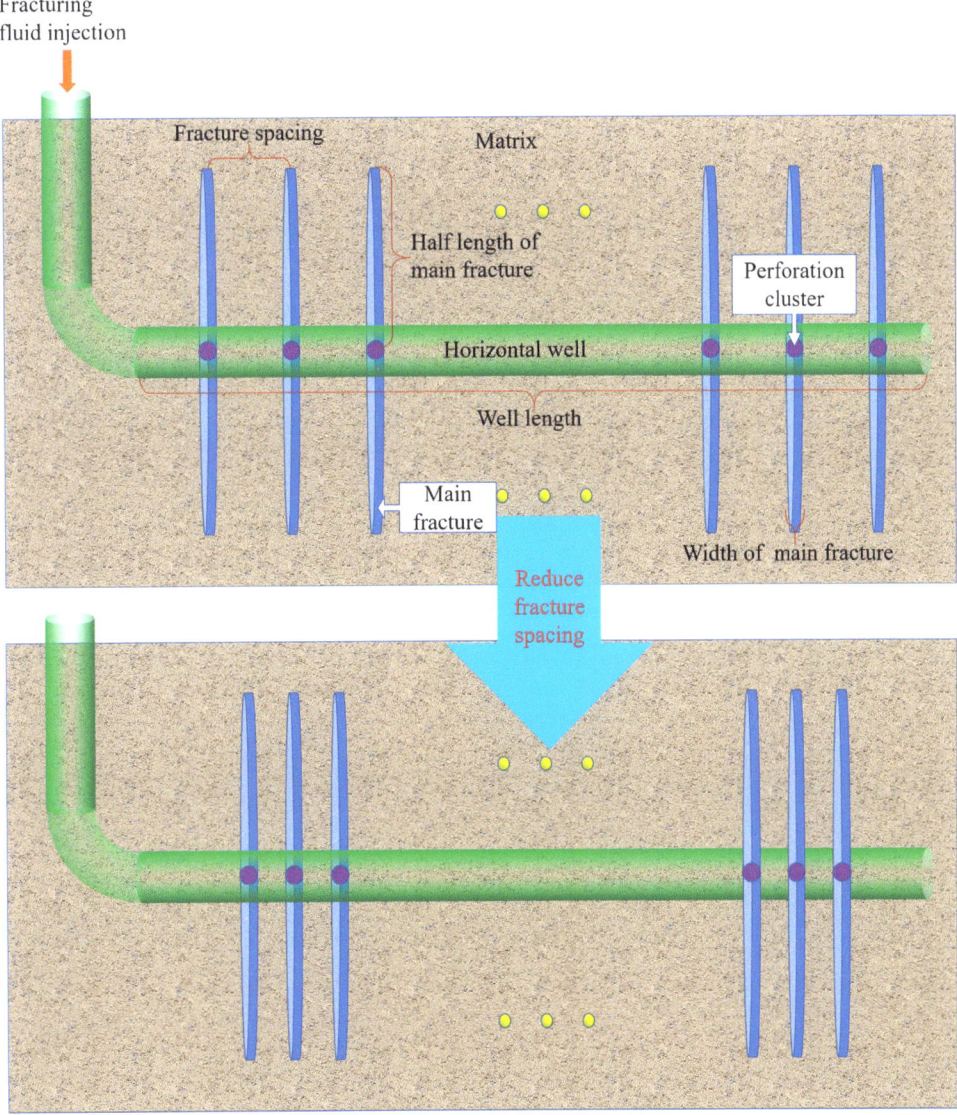

Figure 1. Schematic diagram of fracture spacing optimization.

In recent years, scholars have carried out a series of studies for optimizing fracture spacing [8–15]. From 2017 to 2022, some scholars took fracture propagation in the process of hydraulic fracturing as the research object. They presented the numerical model considering elastic fluid mechanics and stress disturbances and different fracture flow distributions [8], the computational model of embedded discrete fractures [9], the mathematical models of the coupling effect of rock and fluid dynamics [10], the computational optimization model based on intelligent variable-fidelity radial basis function [11], the mathematical model of fully coupled deformation and seepage flow in porous media [12], the 3D solid mechanics model of hydraulic fracturing [13], the numerical model considering fracture geometry and proppant flow dynamics [14] and the numerical model considering stress variation with depth [15] in order to optimize the fracture spacing in shale reservoirs from different research angles. However, these models involve fluid viscosity, reservoir permeability, reservoir porosity, reservoir thickness, fracture width, fracture length, fracture number, total stress, effective stress, rock storage coefficient and other parameters, which are difficult to obtain in the actual production process.

With the development of field measurement technology, a large amount of on-site pressure and production rate data can be obtained. The reservoir and fracture parameters can be interpreted by inversion method using the dynamic production data. Then the production rate can be calculated through the forward computation of the seepage model by using the interpreted values of reservoir and fracture parameters. Then the fracture spacing can be optimized. In 2006, through modeling the flow in each streamline independently in real time, the Wang–Kovscek [16] streamline method for production data inversion had been improved by Vegard R. Stenerud and Nut-Andreas Lie [17]. The results showed that this method had better matching and faster convergence rate. In 2016, aiming at the problem that nonlinearity and variable production rate should be considered when interpreting production data of shale gas reservoirs, a classical trilinear flow model was modified and a method for comprehensively analyzing variable production rate data was proposed by Wu et al. [18]. In this method, considering the desorption and the nonlinearity of compressibility, modified material balance equation and material balance time were used to process the production data. It was proved through a field case that this method could more accurately interpret the production data. In 2017, aiming at the problem of fracture inversion, a fracture inversion method was proposed by using production data based on Griffith failure criterion and ground stress correlation by Zhang et al. [19]. Theoretical examples showed that this method was effective for the accurate inversion of fractures, but as the fracture numbers are more, the inversion results become worse. In 2018, aiming at the problem of the significant discontinuities in production data caused by frequent shut-ins, a new production data analysis method for solving the discontinuous problem based on pseudo time was proposed by Li et al. [20]. Duhamel's principle, Laplace transform and inversion and the Newman method were used to solve the model used for production data analysis, and the analytical and numerical solutions were verified. The results showed that this method had great potential in estimating formation parameters and predicting the well production dynamics more effectively. In 2021, an improved spatial inversion method of data was proposed by Liu et al. [21]. The reservoir state fields can be quickly predicted by observing the production data. The method was also tested in the field. The results showed that this method had high computational efficiency and accuracy. The above studies [16–21] proposed some new inversion methods of dynamic production data or improved the existing methods for interpretating reservoir parameters or fracture parameters after fracturing, and a certain theoretical basis for the dynamic production data inversion technology of multi-stage fractured horizontal wells was provided. However, due to the dramatic changes in flow pressure and the production rate in unconventional oil and gas production data with large errors [22], the normalized typical data points in the above dynamic production data inversion method were scattered, smooth typical curves were difficult to obtain and the data fitting effect was also poor, which resulted in great uncertainty in the fitting results. In addition, the interpreted post-hydraulic fracturing models of seepage flow during production in the aforementioned studies was rarely further applied to the optimization of productivity enhancement in the oil field.

In 2020, Mohammed and Joseph combined data analysis with theoretical models to establish a hybrid hydraulic fracturing model that combines data and theory. The results showed that the hybrid model has higher accuracy. It is feasible to combine data analysis with theoretical models [23]. Therefore, based on the dynamic production data inversion, a new production-oriented optimization method for the fracture spacing of multi-stage fractured horizontal wells in shale oil is proposed. In particular, deconvolution algorithm [24–31] is introduced to normalize the pressure data. Not only can the data of variable pressure and variable flow rate be directly converted into pressure data under unit flow, but also the regularization of deconvolution calculation can be performed, which can eliminate the influence of data error and expand the investigation distance of production data analysis. As a result, more information for production data analysis can be provided, and thus the fitting effect is improved and the uncertainty of parameter interpretation is reduced. The main research contents of this study include: first, a three-linear seepage mathematical model for multi-stage fractured horizontal wells in shale reservoir, the pressure solution under constant flow rate and the flow rate solution under constant production pressure in Laplace space are introduced [32]. Second, the abundant on-site production data are fully utilized for dynamic production data inversion, and the deconvolution theory to normalize the production data is also introduced. By referring to the specific algorithm of pressure deconvolution for data normalization, the data of variable pressure and variable flow rate are converted into the pressure data under unit flow rate, and the influence of data errors is also eliminated. According to the pressure under unit flow rate, the typical curve analysis of the pressure data under unit flow rate is carried out. The reservoir parameters and fracture parameters after hydraulic fracturing are interpreted, so that the interpreted seepage model is more in line with the reality and the seepage flow behavior can be represented more accurately. Then, the Duhamel's principle and the analytical solution of the interpretation model are used to calculate the flow rate per unit production pressure drop. The daily and cumulative production rate of horizontal wells under any production pressure system can be obtained, which can predict the productivity more accurately and efficiently. According to the productivity obtained, the fracture spacing can be optimized, and an optimization method for the fracture spacing of multi-stage fractured horizontal wells is proposed. Finally, the proposed fracture spacing optimization method was used to analyze the dynamic production data of a shale oil production well in the actual block. The fracture spacing was optimized from the aspects of production life, cumulative production, total economic benefit [33], balance of payments, fracturing cost, oil price and other influential factors [34]. The optimization method of fracture spacing proposed in this paper has its own advantages compared with the optimization method of fracture spacing based on fracture propagation in solid mechanics [35], and they can complement each other. By comprehensively utilizing these two methods, better fracture spacing can be obtained. Significant reference for the design of adjacent well fracture spacing in the same block in the future is provided. Some technical guidance is provided for later production and secondary fracturing of reservoirs.

2. Evaluation of Optimal Fracture Spacing for Multi-Stage Fractured Horizontal Wells in Shale Oil

2.1. Mathematical Model of Shale Oil Seepage Flow in Multi-Stage Fractured Horizontal Wells

The physical model of trilinear seepage flow in multi-stage fractured horizontal wells in shale reservoirs [32] is shown in Figure 2. The wellbore direction of the horizontal well is parallel to the Y-axis and the radius of the wellbore is r_w. A number of equally spaced primary fractures have been hydraulically fractured perpendicular to the wellbore (i.e., along the x direction). All fractures penetrate the reservoir completely. The top and bottom of the reservoir are closed and the ambient temperature is constant. Fluid flow in the reservoir is divided into reservoir flow area, inter-fracture flow area and main fracture flow area. The fluid first flows from the reservoir flow area into the inter-fracture flow area, and then flows from the inter-fracture flow area into the main fracture and finally flows through the main fracture into the wellbore of the horizontal well. The horizontal well is in production with constant flow rate and variable pressure. Due to geometric symmetry, the inter-fracture interference is not considered.

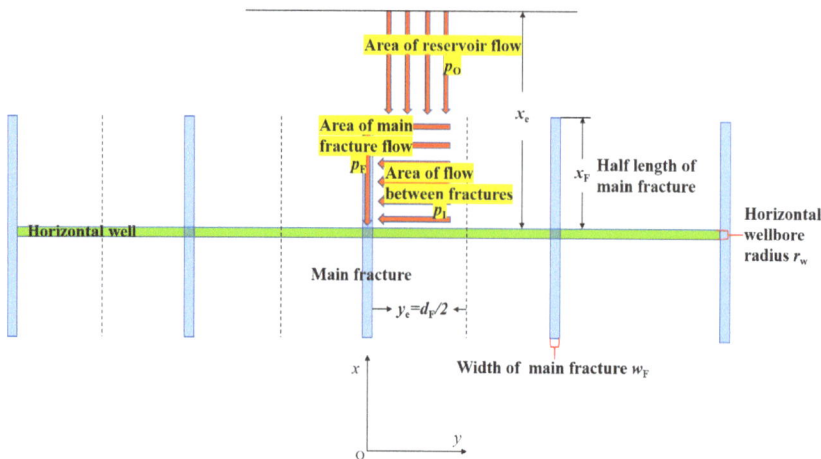

Figure 2. Physical model of trilinear seepage flow in shale oil reservoir developed by multi-stage fractured horizontal wells.

In 2011, a mathematical model of trilinear seepage flow in multi-stage fractured horizontal wells in shale reservoir based on the above physical model was established by M. Brown et al. [32]. The dimensionless variables in M. Brown et al.'s model is defined as follows:

$$C_{FD} = \frac{k_F w_F}{k_I x_F};\ y_e = \frac{d_F}{2};\ \eta_{FD} = \frac{\eta_F}{\eta_I};\ \eta_{OD} = \frac{\eta_O}{\eta_I};\ x_D = \frac{x}{x_F};\ y_D = \frac{y}{x_F};\ x_{eD} = \frac{x_e}{x_F};$$

$$w_D = \frac{w_F}{x_F};\ q_F = \frac{q}{n_F};\ \eta_F = \frac{k_F}{\phi_F c_{tF} \mu};\ \eta_I = \frac{k_I}{\phi_I c_{tI} \mu};\ \eta_O = \frac{k_O}{\phi_O c_{tO} \mu};\ y_{eD} = \frac{y_e}{x_F};$$

$$t_D = \frac{\eta_I}{x_F^2} t;\ p_{OD} = \frac{2\pi k_I h}{q_F B \mu}(p_{ini} - p_O);\ p_{ID} = \frac{2\pi k_I h}{q_F B \mu}(p_{ini} - p_I);$$

$$p_{FD} = \frac{2\pi k_I h}{q_F B \mu}(p_{ini} - p_F);\ C_D = \frac{C}{2\pi(\phi c_t)_I h x_F^2}.$$

where C_{tO} is the comprehensive compression coefficient of seepage flow in the reservoir flow area, atm^{-1}; C_{tI} is the comprehensive compression coefficient of seepage flow in the inter-fracture flow area, atm^{-1}; C_{tF} is the comprehensive compression coefficient of seepage flow in the main fracture flow area, atm^{-1}; ϕ_O is the porosity in the reservoir flow area; ϕ_I is the porosity in the inter-fracture flow area; ϕ_F is the porosity in the main fracture flow area; k_O is the permeability in the reservoir flow area, D; k_I is the permeability in the inter-fracture flow area, D; k_F is the permeability in the main fracture flow area, D; n_F is the number of primary fractures; q is the horizontal well production, cm^3/s; B is the volume coefficient; p_w is the bottom hole pressure, atm; w_F is the fracture width, cm; t_D is dimensionless time; x_D is the dimensionless distance in the x direction; x_{eD} is the dimensionless outer boundary distance; η_{OD} is a defined dimensionless variable; y_D is the dimensionless distance in the y direction; w_D is the dimensionless fracture width; y_{eD} is the dimensionless distance at 1/2 of the fracture spacing; η_{FD} is a defined dimensionless parameter; C_{FD} is a defined dimensionless quantity; and C_D is the dimensionless wellbore storage factor. The unit system of the formulas in the manuscript is the Darcy unit system.

The Laplace transformation solution of dimensionless pressure in reservoir flow area is as follows [32]:

$$\tilde{p}_{OD} = \tilde{p}_{ID}\big|_{x_D=1} \cdot \frac{\cosh\left[\sqrt{s/\eta_{OD}}(x_{eD} - x_D)\right]}{\cosh\left[\sqrt{s/\eta_{OD}}(x_{eD} - 1)\right]}, \qquad (1)$$

where $\tilde{p}_{OD}$ is Laplace transformation of dimensionless pressure p_{OD} in the reservoir flow area; $\tilde{p}_{ID}$ is the Laplace transformation of dimensionless pressure p_{ID} in the inter-fracture flow area; s is the complex variable of Laplace transformation.

The Laplace transform solution of dimensionless pressure in the inter-fracture flow area is [32]:

$$\tilde{p}_{ID} = \tilde{p}_{FD}\bigg|_{y_D = w_D/2} \cdot \frac{\cosh\left[\sqrt{\alpha_O}(y_{eD} - y_D)\right]}{\cosh\left[\sqrt{\alpha_O}\left(y_{eD} - \frac{w_D}{2}\right)\right]}, \tag{2}$$

where $\tilde{p}_{FD}$ is Laplace transformation of dimensionless pressure p_{FD} in main fracture flow area; α_O is a defined dimensionless parameter.

The Laplace transform solution of dimensionless pressure in main fracture flow area is as follows [32]:

$$\tilde{p}_{FD} = \frac{\pi}{C_{FD}s\sqrt{\alpha_F}} \cdot \frac{\cosh\left[\sqrt{\alpha_F}(1 - x_D)\right]}{\sinh(\sqrt{\alpha_F})}, \tag{3}$$

where α_F is a defined dimensionless parameter.

According to Equation (3), the dimensionless bottom hole pressure $\tilde{p}_{wD}$ at a constant flow rate can be obtained as follows [32]:

$$\tilde{p}_{wD} = \tilde{p}_{FD}(x_D = 0) = \frac{\pi}{C_{FD}s\sqrt{\alpha_F}\tanh(\sqrt{\alpha_F})}, \tag{4}$$

where $\tilde{p}_{wD}$ is the dimensionless bottom hole pressure.

The flow in the fracture of the trilinear seepage flow model is one-dimensional linear flow. However, the fluid flow along the fracture surface into the wellbore of the horizontal well will produce the radial flow. For solving the contradiction, a choking skin factor for approximating the choking resistance generated by the radial flow was proposed by Mukherjee and Economides [36]. The choking skin factor s_c can be calculated as follows [36]:

$$s_c = \frac{k_I h}{k_F w_F}\left[\ln\left(\frac{h}{2r_w}\right) - \frac{\pi}{2}\right], \tag{5}$$

According to Equations (4) and (5), the dimensionless bottom hole pressure with constant flow rate considering the radial flow at the fracture surface can be obtained as follows [32]:

$$\tilde{p}_{wD} = \frac{\pi}{C_{FD}s\sqrt{\alpha_F}\tanh(\sqrt{\alpha_F})} + \frac{s_c}{s}, \tag{6}$$

In order to make the solution more practical, the wellbore storage effect should be considered. $\tilde{p}_{wD}$ in Equation (6) can be substituted into the following convolution expression in the Laplacian domain [32]:

$$\tilde{p}_{wD,\text{storage}} = \frac{\tilde{p}_{wD}}{1 + C_D s^2 \tilde{p}_{wD}}, \tag{7}$$

Applying Duhamel's principle, Equation (7) can be used to obtain the dimensionless flow rate solution per unit production pressure drop as follows [32]:

$$\tilde{q}_D = \frac{1}{\tilde{p}_{wD}} \cdot \frac{1}{s^2} \cdot p_{wD,\text{const}}, \tag{8}$$

where $\tilde{q}_D$ is the Laplace transformation of dimensionless flow q_D corresponding to fixed bottom hole pressure; $P_{wD,\text{const}}$ is a fixed dimensionless bottom hole pressure.

The dimensionless variables in the above model solution are defined as follows:

$$p_{wD,\text{const}} = \frac{2\pi k_I h}{q_F B \mu}; \quad q_D = \frac{q_u}{q_F}$$

where q_u is the flow rate corresponding to the fixed bottom hole pressure, cm^3/s; q_D is the dimensionless flow rate corresponding to the fixed bottom hole pressure.

Equations (7) and (8) are, respectively, the pressure under constant flow and the flow rate under constant pressure in Laplace space, but not the solution in real space. Therefore, the Stehfest algorithm [37,38] is introduced to invert the solution of Laplace space to obtain the solution of the real space [39]. The calculation formula of Stehfest algorithm is as follows [37,38]:

$$f(T) = \frac{\ln 2}{T} \sum_{i=1}^{N} V_i \bar{f}\left(\frac{\ln 2}{T} i\right), \tag{9}$$

where:

$$V_i = (-1)^{N/2+i} \sum_{k=[\frac{i+1}{2}]}^{Min(i,N/2)} \frac{k^{N/2+1}(2k)!}{(N/2-k)!k!(k-1)!(i-k)!(2k-i)!}, \tag{10}$$

where the larger value of N is, the more accurate the calculation will be; generally, N is an even integer between 6 and 18 [39].

2.2. Deconvolution

In the actual production process of shale oil, due to the low permeability of shale reservoir and the change in flow dynamics in the production process, it is difficult to keep the pressure and flow rate of dynamic production data constant in reality. However, the mathematical model used in this article is of constant flow rate or constant pressure. The deconvolution algorithm is introduced to normalize the bottom hole flow pressure data, and then the data of variable pressure and variable flow rate are transformed into the pressure data under unit flow rate; the influence of data errors can also be eliminated by the data normalization process, so more information for production data analysis can be provided and ultimately the fitting effect can be improved. The deconvolution principle for well test interpretation or production dynamics data analysis is as follows:

According to Duhamel's principle, the pressure-flow rate convolution relation is obtained as follows [25]:

$$p_{ini} - p_{wD}(t) = \int_0^t q(t-\tau) \frac{\partial p_u(\tau)}{\partial t} d\tau, \tag{11}$$

where p is the bottom hole pressure under variable flow rate, atm; t is the production time, s; p_u is the flow response per unit flow rate, atm.

According to Duhamel's principle, the flow rate function under variable bottom hole pressure is obtained as follows [40]:

$$q(t) = \int_0^t \Delta p_{wD}(t-\tau) \cdot q'_u(\tau) d\tau \tag{12}$$

where Δp_{wD} is the production pressure drop, atm.

In the case of known variable flow rate q and bottom hole pressure p under the variable flow q rate, Equation (11) can be used to obtain the transient pressure response p_u in the oil reservoir for the whole production time [24]). It is worth noting that when using deconvolution calculations in practical applications, it is necessary to exclude the influence of stimulation measures during production, and the changes in reservoir physical property, fluid property and variable wellbore storage effect [25,26]. A series of studies on the deconvolution algorithm for inversion of reservoir production data [25–31] were carried out by many scholars. In 2004, based on Tikhonov regularization objective function [27], a new deconvolution algorithm was proposed by Schroeter et al. [28]. In 2006, the practical application of a B-spline-based deconvolution algorithm in well test analysis was investigated by Ilk et al. [30,31]. From 2017 to 2018, an improved B-spline deconvolution algorithm was proposed by Liu et al. [25,26]. Adding a nonlinear regularization based on curvature minimization, the stability of B-spline deconvolution algorithm was improved by Liu et al. [25,26]. Some theoretical bases for the practical application of deconvolution algorithm were provided by the above researchers. In this study, the improved B-spline deconvolution algorithm [25,26] is used, which has fast calculation speed and good stability. Through this deconvolution algorithm, the variable flow rate and variable pressure data can be transformed into the pressure data under unit flow rate and flow rate data under unit production pressure drop, and the error can be eliminated [26].

The specific pressure deconvolution algorithm [25,26] is as follows:

(1) The pressure derivative per unit flow rate is reconstructed by employing the IIK second-order B-spline function weight sum.
(2) The convolution integral property is adopted so that the sensitivity matrix for the deconvolution calculation can be solved quickly and analytically.
(3) The idea of curvature minimization is introduced to increase nonlinear constraints, which reduces errors and improves stability.

The specific algorithm can be referred to in the literature [25,26].

The deconvolution algorithm applied to production decline [26] is as follows:

(1) The flow rate derivative per unit production pressure drop is reconstructed by employing the IIK second-order B-spline function weight sum.
(2) The convolution integral property is adopted so that the sensitivity matrix for the deconvolution calculation can be solved quickly and analytically by piecewise integration according to the pressure drop section.

The specific algorithm can be referred to in the literature [26].

2.3. Optimization Method of Fracture Spacing in Multi-Stage Fractured Horizontal Well of Shale Oil

In the actual production process of the reservoir, the pressure and flow rate in the formation are not constant. However, the flow model introduced in this study has a constant flow, so a method based on deconvolution is proposed to optimize the fracture spacing in shale oil. The operational and constant parameters are the reservoir parameters and fracture parameters explained through production data, fracture pacing and number of fractures, fracturing cost and oil price. The optimal fracture spacing can be determined with production and economic benefits as the objective functions. The operational and constant parameters can be obtained through seismic data, production data analysis and experimental testing. Firstly, by using the actual production data in the field and Duhamel's principle, the deconvolution algorithm is used to normalize these actual production data by Equations (11) and (12), so that the actual production data with variable flow rate and variable pressure can be transformed into pressure data under unit flow rate and flow rate data under unit production pressure drop, and the influence of data error can also be eliminated. Then, based on the theoretical model of seepage flow and the pressure solution under unit flow rate introduced above (i.e., Equation (7)), the pressure data (pressure drop and pressure derivative) under unit flow rate calculated by deconvolution is analyzed by double logarithmic typical curve fitting method, so as to interpret the reservoir parameters and fracture parameters. At the same time, based on the theoretical model of seepage flow and the flow rate data under unit production pressure drop introduced above (i.e., Equation (8)), the flow rate data under unit production pressure drop calculated by deconvolution are analyzed by Blasingame production decline typical curve fitting method, so as to interpret the reservoir parameters and fracture parameters. The two fitting methods can constrain each other and significantly reduce the uncertainty of model interpretation results. In addition, during the fitting process of the feature curve, the B-spline cardinality and smoothing factor are used as constraints, so that the normalization parameter tuning and the theoretical model calculation parameter tuning are mutually constrained. At the same time, seismic data and on-site data are used as conditional constraints, and through the combined action of multiple constraints, the multiplicity of interpretation results is greatly reduced, and the fitting degree of the double logarithmic typical curve is improved. Therefore, the interpretation results have high accuracy. Finally, using Duhamel's principle (i.e., Equation (12)) and the model analytical solution (i.e., Equation (8)) to calculate the flow rate under the unit production pressure drop, the daily production rate q and cumulative production rate of horizontal well under any production pressure control can be calculated. Then productivity can be predicted more accurately and efficiently. Based on the model productivity calculation, under different production life, fracturing cost and oil price, the fracture spacing is optimized with the goal of maximum cumulative production and break-even, respectively. The flow chart of fracture spacing optimization method for multi-stage fractured horizontal wells in shale oil is shown in Figure 3.

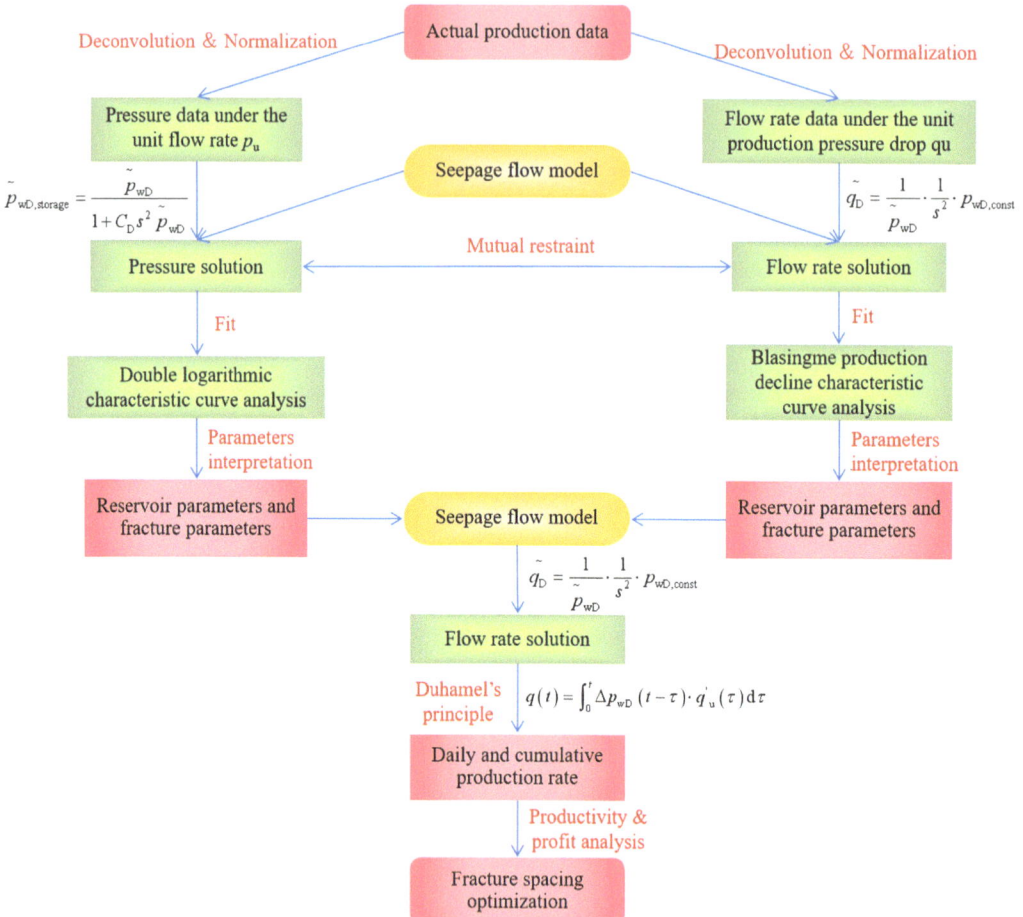

Figure 3. Flow chart of fracture spacing optimization method for multi-stage fractured horizontal wells in shale oil.

There are some precautions to apply this calculation method to practical engineering. The optimization method of fracture spacing is only applicable to single-phase flow and cannot occur in oil gas two-phase flow. The pressure of the reservoir cannot be lower than the bubble point pressure. The fracture spacing has little effect on the fracture propagation ability. In addition, obtaining a certain amount of data is necessary.

3. Practical Application

In this section, the fracture spacing of a multi-stage fractured horizontal well in a shale oil block at a China oilfield is optimized. Due to the low water production rate during the long-time production period of this well, fluid flow is considered as single-phase oil flow. And due to the absence of other wells around the well, interference between wells is not considered. Therefore, this well is suitable for the mathematical model in this article. It is known that the initial pressure of reservoir is 12.5 MPa, the length of the horizontal wellbore is 1740 m, the wellbore radius is 0.076 m, the number of main fractures is 60, the width of the main fracture is 0.001 m, the effective reservoir thickness is 9.9 m, the porosity of the shale matrix is 10%, the fluid viscosity of oil is 0.5 mPa·s, the average water cut of the production well is 0.35, the shale oil density is 850 kg/m^3, the fracture cost of hydraulic fracturing is 160,000 Yuan per cluster and the current shale oil price is 3800 Yuan/ton. The matrix permeability is less than 1.0 mD and the volume coefficient is 1.3. The information can be used as constraint for dynamic production data analysis.

3.1. Dynamic Production Data Analysis

The multi-stage fractured horizontal well in Ordos Basin was tested for long-time flowing pressure without well shut-in. The dynamic data of bottom hole pressure and daily production rate are shown in Figure 4.

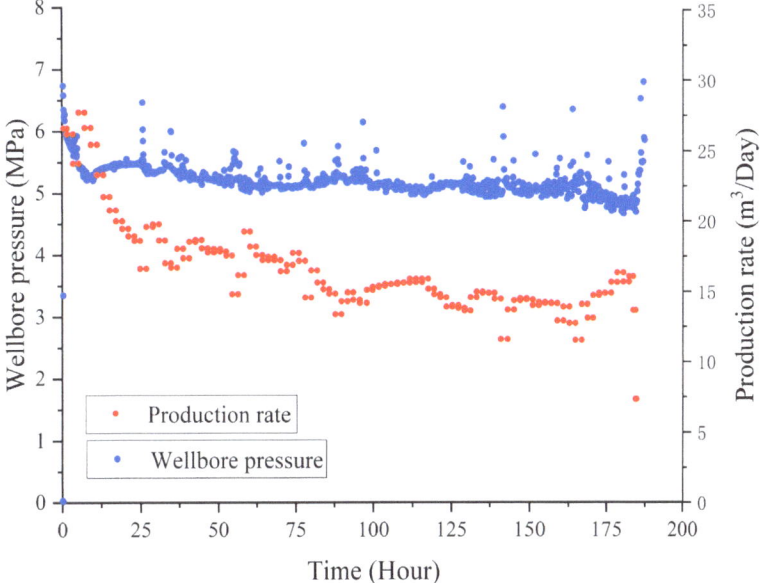

Figure 4. Dynamic data of bottom hole pressure and daily production rate.

No other stimulation measures are implemented during the well's production; furthermore, the seepage flow process can be approximated as a single-phase oil flow. Thus, the production data can meet the requirements for deconvolution application. The production data were analyzed according to the aforementioned analysis method based on deconvolution.

First of all, the dynamic data of variable pressure and variable production rate in Figure 4 can be normalized by the application of deconvolution algorithm and Equation (11). As a result, the deconvolved bottom hole pressure data per unit flow rate are obtained, which is shown in Figure 5. It can be seen from Figure 5 that the application of deconvolution eliminates the impact of data errors and a smooth pressure drop curve is obtained.

The typical curve analysis (pressure drop and pressure drop derivative) can be performed using the unit-rate bottom hole pressure solution (i.e., Equation (7)) of the model obtained. The analysis result is shown in Figure 6. It can be seen from the Figures 5 and 6 that the model obtained fits the normalized production data very well; the application of deconvolution eliminates the impact of data error, and data divergence is effectively prevented. Smooth typical curves are obtained, and the bottom hole pressure drop behavior in the reservoir development can be clearly reflected.

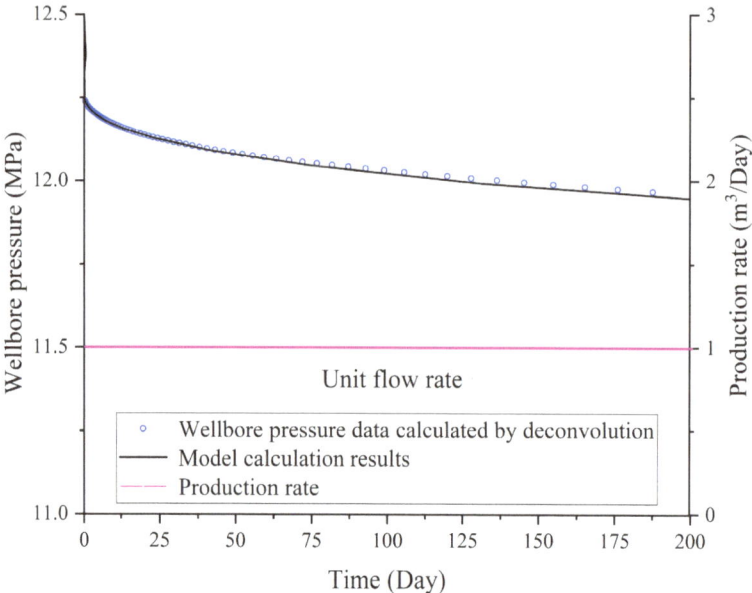

Figure 5. Deconvolved bottom hole pressure data per unit flow rate.

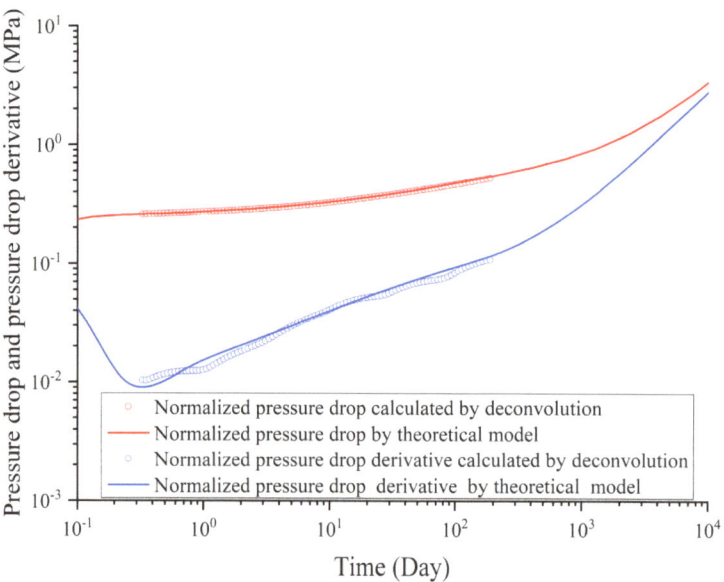

Figure 6. Data of pressure drop and pressure drop derivative at unit flow rate.

Then, normalizing the dynamic data of variable pressure and variable production rate in Figure 4 by the application of deconvolution algorithm and Equation (12), the deconvolved production rate data per unit production pressure drop is obtained, which is shown in Figure 7. It can be seen from Figure 7 that the impact of data errors is eliminated and a smooth production decline curve is obtained.

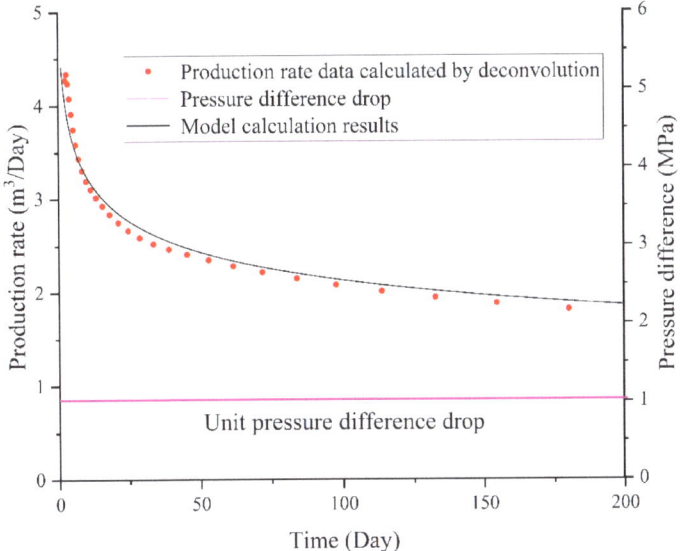

Figure 7. Deconvolved flow rate data per unit production pressure drop.

The typical curve analysis of production decline under unit production pressure drop can be performed using the flow rate solution (i.e., Equation (8)) under unit production pressure drop of the model obtained. The analysis result is shown in Figure 8. It can be seen from Figures 7 and 8 that the obtained model fits well with the normalized production data; the application of deconvolution eliminates the impact of data errors and effectively prevents data divergence. A smooth typical curve has been obtained, which can clearly reflect the production rate behavior during reservoir development.

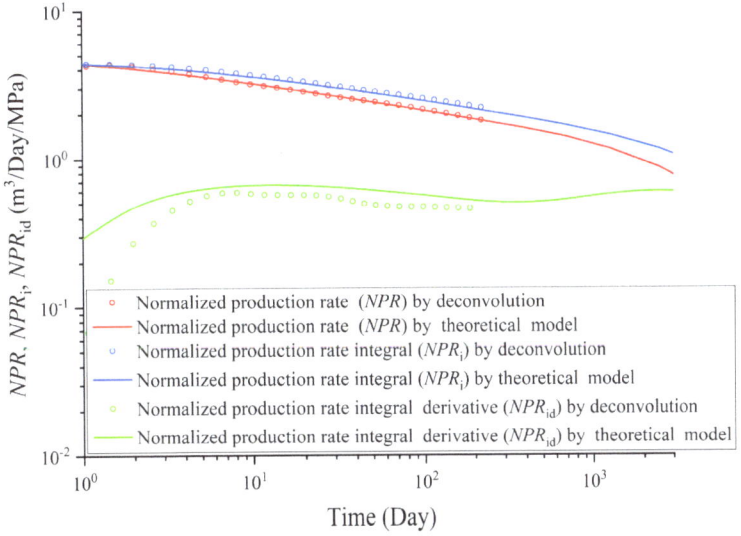

Figure 8. Analysis curve of production decline under unit production pressure drop.

The fitting method in Figure 6 is performed from the perspective of pressure drop and pressure drop derivative. The fitting method in Figure 8 is performed from the perspective of production rate. The two fitting methods can constrain each other and significantly reduce the uncertainty of model interpretation results.

Based on Duhamel's principle and Equation (12), the historical fitting data of production rate are obtained, which are shown in Figure 9. It can be seen from Figure 9 that the fitting effect of productivity history data is very good. The dynamic geological reserve is evaluated as 3.17×10^6 m^3.

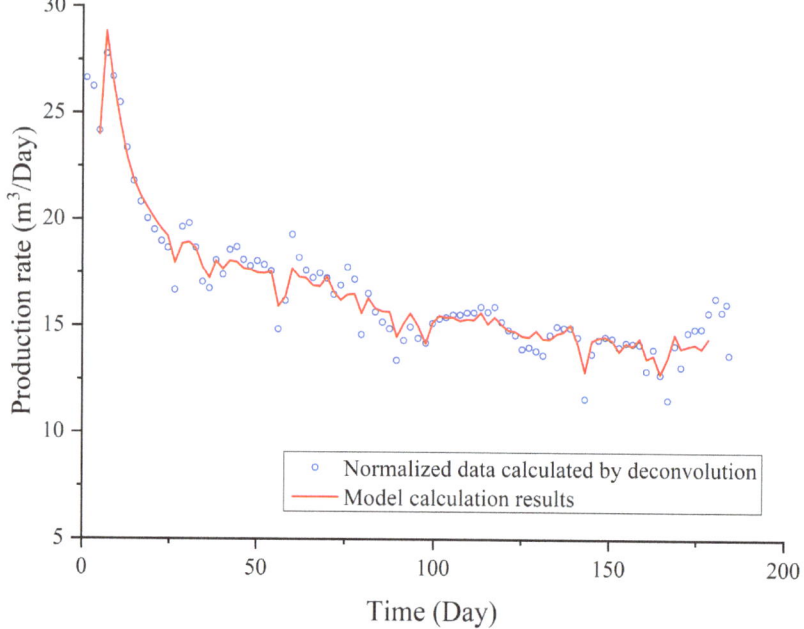

Figure 9. Historical matching of production rate data.

The seepage flow behavior can be characterized more accurately. The reservoir parameters and fracture parameters are shown in Table 1.

Table 1. Interpretation results of reservoir parameters and fracturing parameters for a shale oil reservoir.

Parameters	Values Obtained by Typical Curve Analysis
Initial reservoir pressure	12.5
Reservoir thickness	10
Half-length of fracture	50
Fracture width	0.001
Fracture conductivity	30
Shale matrix permeability	0.75
Fluid viscosity	0.5
Dynamic geological reserves	3.17×10^6

3.2. Optimization of Fracture Spacing for Multi-Stage Fractured Horizontal Wells in Shale Oil

The production pressure drop is set as 7.5 Mpa in the future for the well. Then according to Duhamel's principle (i.e., Equation (12)) and the flow rate solution under unit production pressure drop calculated by the interpretation model (i.e., Equation (8)), the daily production rate of horizontal well under any production pressure control can be obtained quickly. The cumulative production rate of horizontal well can be obtained through an integral calculation. Then the fracture spacing can be optimized according to the well productivity.

In the following, the fracture spacing are optimized from the aspects of production life, cumulative production, total economic benefits (it is equal to the production multiplied by oil price), balance of payments (the total economic benefit is equal to the fracturing cost.), fracturing cost, oil price and other influencing factors. Research on the variation in daily production rate and cumulative production rate with different fracture spacing (or different fractures number under the same horizontal well length) with production time is conducted. The effect of fracture spacing (or fracture numbers) on cumulative production under different production life is studied. The optimal fracture spacing (or fractures number) is determined with the goal of the maximum cumulative production and the balance of payments. By changing the oil price, the effect of oil price on the optimal fracture spacing is studied, with the goal of maximum total cumulative production and the balance of payments. By changing the fracturing cost, the effect of fracturing cost on the optimal fracture spacing is studied, with the goal of maximum total cumulative production and the balance of payments. Finally, the results of fracture spacing optimization are compared and analyzed.

3.2.1. Production Rate Change with Production Time under Different Fracture Spacing

The variation in daily oil production rate with production time under different fractures number is shown in Figure 10. It can be seen from Figure 10 that daily shale oil production rate declines continuously as production time goes on, with a rapid decline rate in the first year and then a relatively slow decline rate. The reason is that with the continuous production of the reservoir, the reservoir pressure gradient gradually decreases; then, the flow rate of shale oil and the recovery rate of the reservoir decreases. In the early stage of reservoir exploitation, the more the fracture number, the smaller the fracture spacing, the greater the daily shale oil production rate; however, in the later stage of exploitation, the more the fracture number and the smaller daily shale oil production rate. The reason is that in the early stage of production, the more the fracture numbers, the greater the stimulated reservoir volume area that can introduce a higher rate of oil recovery. Therefore, the daily production rate is higher. However, with the increase in production time, the more the hydraulic fractures, the smaller the fracture spacing, the easier the reservoir is to be mined out and the faster the production decline.

The variation in cumulative shale oil production with production time under different fracture numbers is shown in Figure 11. It can be seen from Figure 11 that as production time goes by, the cumulative production of shale oil gradually increases, but the growth rate gradually decreases and the cumulative production growth rate is faster in the early stage of exploitation. This is because the daily production rate of shale oil is large in the early stage of exploitation, and the daily production rate of shale oil gradually decreases with the growth of time. At the same time, the more the fracture number, the higher the cumulative production of shale oil; however, with the increase in the fracture number, the increase in the cumulative production becomes less. This indicates that with the increase in the fracture number (that is, the fracture spacing decreases), the effect of the increase in the fracture number on the increase in cumulative shale oil production will become lower and lower, and the economic benefits brought by the increase in the fracture number will become less and less.

3.2.2. Relationship between Cumulative Production and Fracture Spacing under Different Production Life

The curves of cumulative production and fracture number under two different production lives are shown in Figure 12. The reservoir settings, fracture conductivities and economic parameters under different production life are same. It can be seen from Figure 12 that the cumulative production gradually increases with the increase in the fracture number (the fracture spacing decreases), and the cumulative production basically remains unchanged when the fracture number increases to a certain extent. In the case of 5 years of production, when the fracture number is more than 320, the cumulative shale oil production remains basically unchanged, which can be regarded to be very close to the well-controlled reserve, and the increase in the fracture number cannot bring more economic benefits. In the case of 10 years of production, when the fracture number is more than 160, the cumulative shale oil production remains basically the same, and the increase in the fracture number does not bring any more economic benefits. And the fracture number corresponding to the maximum cumulative production for 5 years of production is greater than that for 10 years of production.

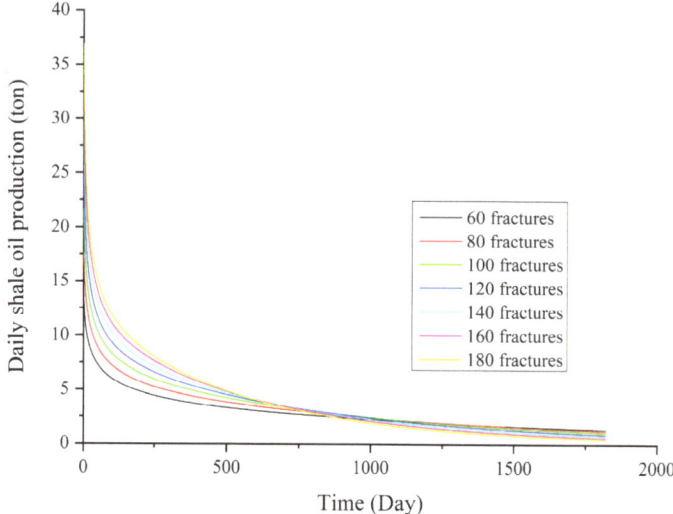

Figure 10. Variation in daily shale oil production with production time under different fracture numbers.

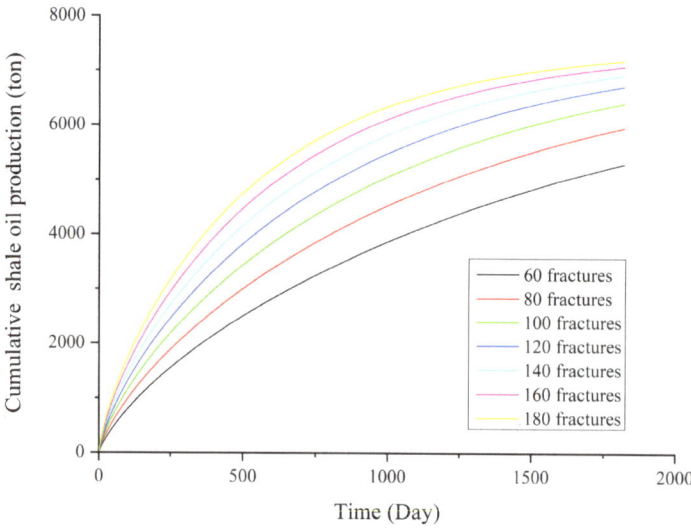

Figure 11. Variation in cumulative shale oil production with production time under different fracture numbers.

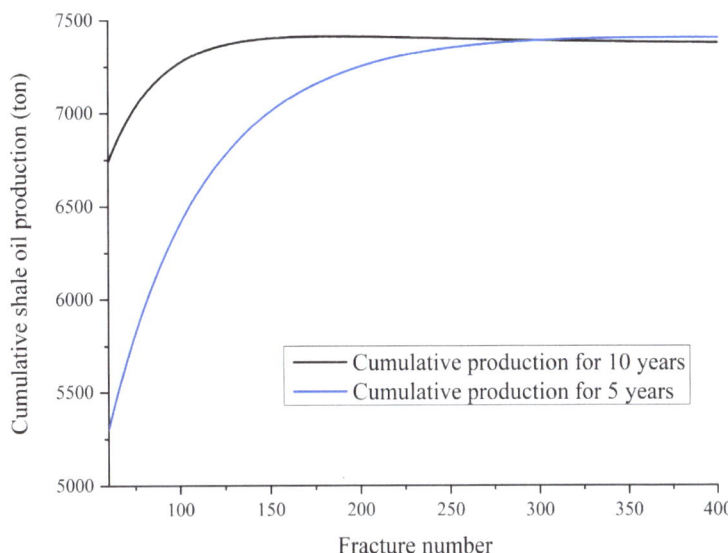

Figure 12. Curves of cumulative production and fractures number under two different production life.

The cumulative production difference between 5 years and 10 years is equal to the vertical distance between the two curves in Figure 12. Therefore, it can be concluded that as the fracture number increases (the fracture spacing decreases), the cumulative production difference decreases gradually. And as the fracture number reaches 320, almost no difference exists. This indicates that when the fracture number is greater than or equal to 320, almost no oil can be produced in the second 5 years of production, which means that the reservoir has been fully exploited in the first 5 years. As the fracture number increases, the 5-year cumulative production increases faster than the 10-year cumulative production. If more oil is needed in the short term, the fracture number can be increased.

3.2.3. Fracture Spacing Optimization under Different Production Life

(1) Cumulative Production as the Optimization Objective

The curve of the relationship between cumulative production and fracture spacing under different production life is shown in Figure 13. It can be seen that with the increase in fracture spacing, the cumulative production gradually decreases. When the fracture spacing is less than 5.5 m, the 5-year cumulative production stays at the maximum, which means the reservoir is almost fully developed based on 5-year production life. When the fracture spacing is less than 11.4 m, the reservoir is almost fully developed based on 10-year production life. Since the cost of fracturing increases with the fracture number, the fracture number should not be too high, that is, the fracture spacing should not be too small. Therefore, the minimum spacing of fractures should be 5.5 m, which corresponds to a 5-year cumulative production of 7394 tons; and the minimum spacing of fractures should be 11.4 m, which corresponds to a 10-year cumulative production of 7430 tons. The minimum fracture spacing for 5 years is smaller than that for 10 years. If the production period is shorter, the maximum fracture spacing should be smaller.

(2) Balance of Payments as the Optimization Objective

The curves of relationship between total economic benefit under different production life and fracturing cost and fracture spacing is shown in Figure 14. The total economic benefit is obtained by multiplying the cumulative production with the shale oil price of 3800 Yuan/ton. The higher the cumulative production, the higher the total economic benefit. With the increase in fracture spacing, the total economic benefit gradually decreases. This is because as the fracture spacing increases, the fracture number decreases, the mining rate decreases, the cumulative production decreases and the total economic benefits also decrease. The larger the fracture spacing, the smaller the fracture number and the smaller the total fracturing cost. The fracture cost is selected as 160,000 Yuan per cluster. When the fracture spacing is too small, the fracture number is too high, which will lead to higher production cost than the total economic benefit, that is, a deficit emerges. When the fracture spacing is 10.3 m, the 5-year payments are exactly balanced, and the total economic benefit for 5 years and fracturing cost are both 27.08 million Yuan. When the fracture spacing is 9.8 m, the 10-year payments are exactly balanced, and the total economic benefit for 10 years and fracturing cost are both 28.16 million Yuan. The longer the production life, the smaller the minimum fracture spacing and the higher the total economic benefit and fracturing cost.

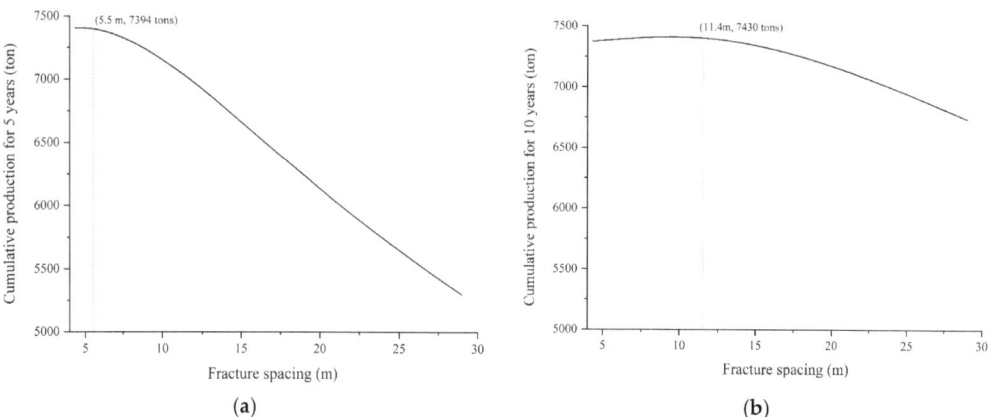

Figure 13. (**a**) The curve of the relationship between cumulative production and fracture spacing for 5 years; (**b**) the curve of the relationship between cumulative production and fracture spacing for 10 years.

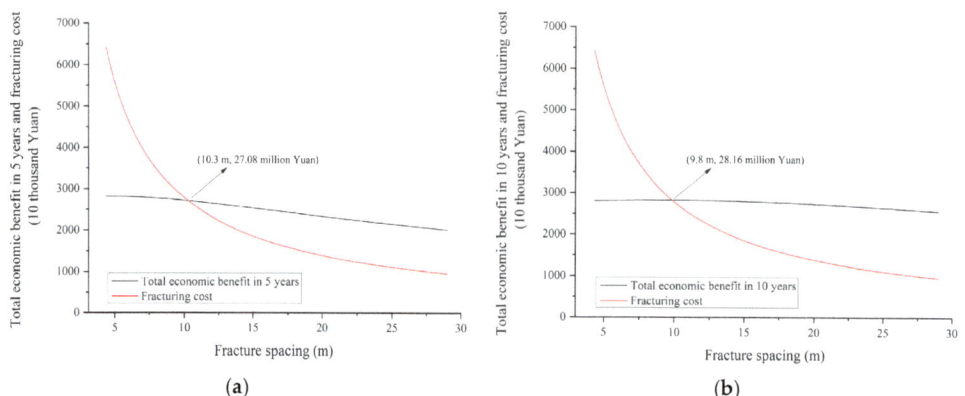

Figure 14. (**a**) Curves of relationship between total economic benefit for 5 years and fracturing cost and fracture spacing; (**b**) curves of relationship between total economic benefit for 10 years and fracturing cost and fracture spacing.

The curves of relationship between total economic benefit under different production lives and fracturing cost and fracture spacing at different oil price are shown in Figure 15;

the oil price is set as 3300 Yuan/ton, 3800 Yuan/ton and 4300 Yuan/ton. It can be seen that as the oil price rises, economic benefit will also increase. In the equilibrium state of payments, the higher the oil price, the smaller the fracture spacing and the more the produced oil. The fracture spacing under the zero profit constraint for 10 years is smaller than that for 5 years.

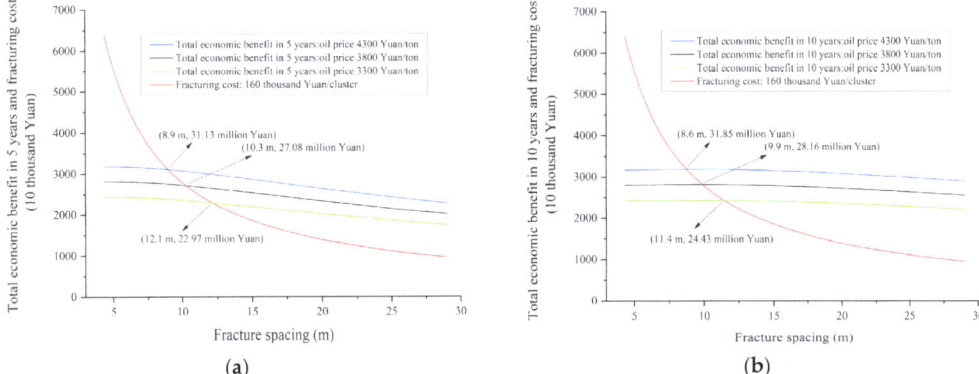

Figure 15. (**a**) Curves of relationship between total economic benefit for 5 years and fracturing cost and fracture spacing at different oil price; (**b**) curves of relationship between total economic benefit for 10 years and fracturing cost and fracture spacing at different oil price.

The curves of relationship between total economic benefit under different production life and total fracturing cost and fracture spacing at different fracturing cost per cluster are shown in Figure 16. The fracturing cost is set as 140,000 Yuan per cluster, 160,000 Yuan per cluster and 180,000 Yuan per cluster, respectively. The oil price keeps constant. It can be seen that as the fracturing cost increases, the fracture spacing corresponding to the equilibrium state of payments is larger and the produced cumulative oil is less. With the advancement of technology, the fracturing cost will reduce largely and then the fracture number corresponding to the equilibrium state of payments will increase, the fracture spacing will reduce and, thus, more produced cumulative oil will be obtained. The fracture spacing under the zero profit constraint for 10 years is smaller than that for 5 years, but their difference is not big.

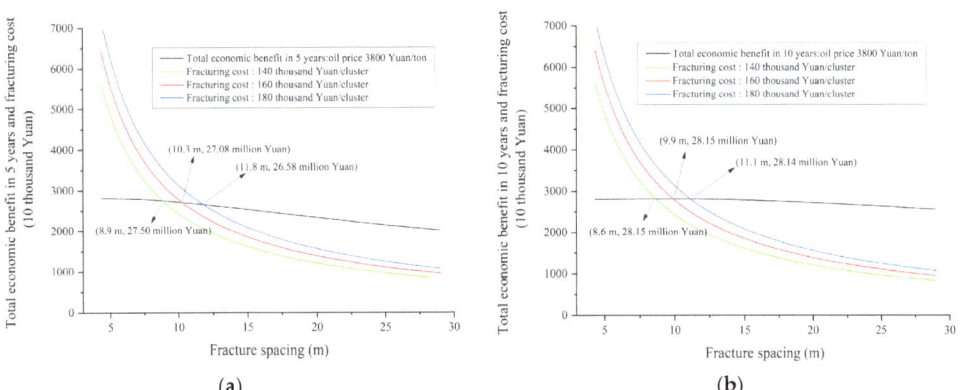

Figure 16. (**a**) Curves of relationship between total economic benefit for 5 years and total fracturing cost and fracture spacing at different fracturing cost per cluster; (**b**) curves of relationship between total economic benefit for 10 years and total fracturing cost and fracture spacing at different fracturing cost per cluster.

(3) The Optimization Result Analysis

The statistical results of fracture spacing optimization are shown in Table 2. According to Table 2, in comparison with 5-year production life, the optimal fracture spacing aiming at maximum cumulative production corresponding to 10-year production life is larger. There is little difference between the maximum cumulative production for 5 years and 10 years. The optimal fracture spacing aiming at the highest profit is equal for 5 years and 10 years. The fracture spacing under the zero profit constraint for 10 years is smaller than that for 5 years, but their difference is not big. The cumulative production under the zero profit constraint for 10 years is larger than that for 5 years. When the oil price increases by 500 Yuan per ton, in comparison with the case of 5 years of production, fracturing spacing under the zero profit constraint for the case of 10 years of production needs to be reduced by a smaller value and the total production increase is also smaller. And when the average fracturing cost of each cluster is reduced by 20,000 Yuan, the reduction in 10-year fracture spacing under the zero profit constraint is small. The smaller the production life, the greater the impact of increasing the same oil price or fracturing cost on the optimal fracture spacing, and the higher the sensitivity.

Table 2. Statistical results of fracture spacing optimization.

Production Lifetime	Optimal Fracture Spacing Aiming at Maximum Cumulative Production	Maximum Cumulative Production	Fracture Spacing under the Zero Profit Constraint	Cumulative Production under the Zero Profit Constraint	Effect of Oil Price on Fracture Spacing under Zero Profit Constraint	Effect of Fracturing Cost on Fracture Spacing under the Zero Profit Constraint
5 years	5.5 m	7394 tons	10.3 m	7126 tons	For a 500 Yuan increase in oil price per ton, the fracture spacing should be reduced by 1.6 m, and the total oil production will be increased by 140 tons.	For a 20,000 Yuan increase in average fracturing cost per cluster, the fracture spacing should be reduced by 1.5 m, and the total oil production will be increased by 121 tons.
10 years	11.4 m	7430 tons	9.8 m	7410 tons	For a 500 Yuan increase in oil price per ton, the fracture spacing should be reduced by 1.4 m, and the total oil production will be increased by 3 tons.	For a 20,000 Yuan increase in average fracturing cost per cluster, the fracture spacing should be reduced by 1.3 m and the total oil production will be increased by 1.3 tons.

4. Conclusions

In this study, an optimization method for fracture spacing of multi-stage fractured horizontal well based on dynamic production data inversion is proposed by making full use of the abundant production data in shale oil field. A deconvolution algorithm is applied to normalize the production data in order to transform the data of variable pressure and variable flow into the pressure data under the unit flow rate and flow rate data under unit production pressure drop. And the influence of data error can be eliminated.

First, a trilinear seepage flow mathematical model of multi-stage fractured horizontal well in shale oil reservoirs and its pressure solution under the unit rate in Laplace domain are introduced. Then, the pressure solution under the unit flow rate is used to fit the normalized pressure data under the unit flow rate by double logarithmic typical curve fitting method. The flow rate solution under unit production pressure drop is used to fit the normalized data under the unit production pressure drop by a Blasingame production decline typical curve fitting method. And some parameters of reservoir and fracture are interpreted. The two fitting methods can constrain each other and significantly reduce the uncertainty of model interpretation results. The interpreted mathematical model is more in line with the reality and the seepage flow behavior can be depicted more accurately. Furthermore, using the Duhamel's principle and the rate solution under unit production pressure drop of the interpreted model, the daily production rate and cumulative production of horizontal well under any production pressure regime can be obtained quickly, and, thus, the productivity can be predicted more accurately and efficiently. Based on the productivity calculation of the model, the fracture spacing is optimized from the perspective of productivity.

Finally, based on the above optimization method for fracture spacing, the fracture spacing of a production well in an actual shale oil block is optimized from the aspects of production life, cumulative oil production, total economic benefit, net profit, fracturing cost, oil price and other influencing factors. The research results show that if the maximum cumulative production is taken as the goal, the optimal fracture spacing is 5.5 m for 5 years and 11.4 m for 10 years. The maximum cumulative production for 5 years is 7394 tons. The maximum cumulative production for 10 years is 7430 tons. The optimal fracture spacing for 5 years is less than that for 10 years. If the production period is shorter, the maximum fracture spacing should be smaller. And if the balance of income and expenditure is taken as the constraint, the optimal fracture spacing is 10.3 m for 5 years and 9.8 m for 10 years. The cumulative production under the zero profit constraint for 5 years is 7126 tons. The cumulative production under the zero profit constraint for 10 years is 7410 tons. The shorter the production life, the more sensitive the effect of oil price and fracturing cost towards the selection of the optimal fracture spacing. All in all, a significant reference value for hydraulic fracturing and the optimization of fracture spacing of adjacent wells in the same shale oil block is provided in this study, and some technical guidance for the later production and secondary fracturing of reservoir is also provided.

Author Contributions: Conceptualization, W.L.; methodology, W.L. and H.S.; software, W.L.; validation, W.L. and C.L.; formal analysis, C.L. and W.L.; investigation, W.L. and C.L.; resources, Y.D.; data curation, X.Y. and Y.S; writing—original draft preparation, C.L.; writing—review and editing, W.L.; visualization, W.L. and C.L.; supervision, W.L.; project administration, Y.D., X.Y., Y.S. and H.S.; funding acquisition, W.L. All authors have read and agreed to the published version of the manuscript.

Funding: This research was funded by China National Petroleum Corporation Innovation Foundation, grant number 2021DQ02-0901.

Data Availability Statement: The data presented in this study are available on request from the corresponding author.

Acknowledgments: We thank the reviewers who have taken the time to provide valuable suggestions, and thank the editors of Energies for their hard work and excellent support.

Conflicts of Interest: The authors declare no conflict of interest.

References

1. Wu, T.; Pan, Z.J.; Liu, B.; Connell, L.D.; Sander, R.; Fu, X.F. Laboratory Characterization of Shale Oil Storage Behavior: A Comprehensive Review. *Energy Fuels* **2021**, *35*, 7305–7318. [CrossRef]
2. Shams, K.; Clement, A.; Jaber, A.J.; Osama, M.S.; Zeeshan, T.; Mohamed, M.; Abdulazeez, A. A review on non-aqueous fracturing techniques in unconventional reservoirs. *J. Nat. Gas Sci. Eng.* **2021**, *95*, 104223.
3. Wang, X.Z.; Peng, X.L.; Zhang, S.J.; Liu, Y.; Peng, F.; Zeng, F.H. Guidelines for Economic Design of Multistage Hydraulic Fracturing, Yanchang Tight Formation, Ordos Basin. *Nat. Resour. Res.* **2020**, *29*, 1413–1426. [CrossRef]
4. Liang, Y.B.; Cheng, Y.F.; Han, Z.Y.; Pidho, J.J.; Yan, C.L. Study on Multiscale Fluid–Solid Coupling Theoretical Model and Productivity Analysis of Horizontal Well in Shale Gas Reservoirs. *Energy Fuels* **2023**, *37*, 5059–5077. [CrossRef]
5. Luo, S.G.; Zhao, Y.L.; Zhang, L.H.; Chen, Z.X.; Zhang, X.Y. Integrated Simulation for Hydraulic Fracturing, Productivity Prediction, and Optimization in Tight Conglomerate Reservoirs. *Energy Fuels* **2021**, *35*, 14658–14670. [CrossRef]

6. Al-Nakhli, A.; Zeeshan, T.; Mahmoud, M.; Abdulaziz, A. Reducing Breakdown Pressure of Tight Reservoirs Via in-Situ Pulses: Impact of Mineralogy. In Proceedings of the SPE/IATMI Asia Pacific Oil & Gas Conference and Exhibition, Bali, Indonesia, 29–31 October 2019.
7. Xu, J.X.; Yang, L.F.; Liu, Z.; Ding, Y.H.; Gao, R.; Wang, Z.; Mo, S.Y. A New Approach to Embed Complex Fracture Network in Tight Oil Reservoir and Well Productivity Analysis. *Nat. Resour. Res.* **2021**, *30*, 2575–2586. [CrossRef]
8. Zhao, J.Z.; Chen, X.Y.; Li, Y.M.; Fu, B.; Xu, W.J. Numerical simulation of multi-stage fracturing and optimization of perforation in a horizontal well. *Pet. Explor. Dev.* **2017**, *44*, 119–126. [CrossRef]
9. Xu, S.Q.; Feng, Q.H.; Wang, S.; Javadpour, F.; Li, Y.Y. Optimization of multistage fractured horizontal well in tight oil based on embedded discrete fracture model. *Comput. Chem. Eng.* **2018**, *117*, 291–308. [CrossRef]
10. Zhang, H.; Sheng, J. Optimization of horizontal well fracturing in shale gas reservoir based on stimulated reservoir volume. *J. Pet. Sci. Eng.* **2020**, *190*, 107059. [CrossRef]
11. Zhao, G.X.; Yao, Y.D.; Wang, L.; Adenutsi, C.D.; Feng, D.; Wu, W.W. Optimization design of horizontal well fracture stage placement in shale gas reservoirs based on an efficient variable-fidelity surrogate model and intelligent algorithm. *Energy Rep.* **2022**, *8*, 3589–3599. [CrossRef]
12. Zhang, D.X.; Zhang, L.H.; Tang, H.Y.; Zhao, Y.L. Fully coupled fluid-solid productivity numerical simulation of multistage fractured horizontal well in tight oil reservoirs. *Pet. Explor. Dev.* **2022**, *49*, 382–393. [CrossRef]
13. Wu, X.T.; Li, Y.C.; Tang, C.A. Fracture spacing in horizontal well multi-perforation fracturing optimized by heat extraction. *Geothermics* **2022**, *101*, 102376. [CrossRef]
14. Bochkarev, A.V.; Budennyy, S.A.; Nikitin, R.N.; Mitrushkin, D.A.; Erofeev, A.A.; Zhukov, V.V. Optimization of multi-stage hydraulic fracturing design in conditions of Bazhenov formation (Russian). *Oil Ind. J.* **2017**, *2017*, 50–53.
15. Singh, A.; Zoback, M.; Mark, M.C. Optimization of Multi-Stage Hydraulic Fracturing in Unconventional Reservoirs in the Context of Stress Variations with Depth. In Proceedings of the SPE Annual Technical Conference and Exhibition, Online, 30 March–5 May 2020.
16. Caers, J.; Gross, H.; Kovscek, A.R. A direct sequential simulation approach to streamline based history matching. In *Geostatistics Banff 2004: Proceedings to the Seventh International Geostatistics Congress*; Leuangthong, O., Deutsch, C.V., Eds.; Springer: Berlin/Heidelberg, Germany, 2004.
17. Stenerud, V.R.; Lie, K.A. A multiscale streamline method for inversion of production data. *J. Pet. Sci. Eng.* **2006**, *54*, 79–92. [CrossRef]
18. Wu, Y.H.; Cheng, L.S.; Huang, S.J.; Jia, P.; Zhang, J.; Lan, X.; Huang, H.L. A practical method for production data analysis from multistage fractured horizontal wells in shale gas reservoirs. *Fuel* **2016**, *186*, 821–829. [CrossRef]
19. Zhang, L.M.; Zhang, X.M.; Zhang, K.; Zhang, H.; Yao, J. Inversion of fractures with combination of production performance and in-situ stress analysis data. *J. Nat. Gas Sci. Eng.* **2017**, *42*, 232–242. [CrossRef]
20. Li, Q.Y.; Li, P.C.; Pang, W.; Li, D.L.; Liang, H.B.; Lu, D.T. A new method for production data analysis in shale gas reservoirs. *J. Nat. Gas Sci. Eng.* **2018**, *56*, 368–383. [CrossRef]
21. Liu, D.; Rao, X.; Zhao, H.; Xu, Y.F.; Gong, R.X. An improved data space inversion method to predict reservoir state fields via observed production data. *Pet. Sci.* **2021**, *18*, 1127–1142. [CrossRef]
22. Clarkson, C.R. Chapter Five—Type-curve analysis methods. In *Unconventional Reservoir Rate-Transient Analysis, Gulf Professional Publishing*; Clarkson, C.R., Ed.; Gulf Professional Publishing: Houston, TX, USA, 2021; pp. 373–484.
23. Mohammed, S.F.B.; Joseph, S.K. Deep hybrid modeling of chemical process: Application to hydraulic fracturing. *Comput. Chem. Eng.* **2020**, *134*, 106696.
24. Wang, F.; Pan, Z.Q. Deconvolution-based well test model for the fractured horizontal wells in tight gas reservoirs. *Acta Petrol. Sin.* **2016**, *37*, 898–902+938.
25. Liu, W.C.; Liu, Y.W.; Han, G.F.; Zhang, J.Y.; Wan, Y.Z. An improved deconvolution algorithm using B-splines for well-test data analysis in petroleum engineering. *J. Pet. Sci. Eng.* **2017**, *149*, 306–314. [CrossRef]
26. Liu, W.C.; Liu, Y.W.; Zhu, W.Y.; Sun, H.D. A stability-improved efficient deconvolution algorithm based on B-splines by appending a nonlinear regularization. *J. Pet. Sci. Eng.* **2018**, *164*, 400–416. [CrossRef]
27. Tikhonov, A.N. Regularization of incorrectly posed problems and the regularization method. *Dokl. Akad. Nauk SSSR* **1963**, *151*, 501–504.
28. Schroeter, T.V.; Hollaender, F.; Gringarten, A.C. Deconvolution of well test data as a nonlinear total least squares problem. *SPE J.* **2004**, *9*, 375–390. [CrossRef]
29. Levitan, M.M.; Crawford, G.E.; Hardwick, A. Practical considerations for pressure-rate deconvolution of well-test data. *SPE J.* **2006**, *11*, 35–47. [CrossRef]
30. Ilk, D.; Anderson, D.M.; Valko, P.P.; Blasingame, T.A. Analysis of Gas-Well Reservoir Performance Data Using B-Spline Deconvolution. In Proceedings of the SPE Gas Technology Symposium, Calgary, AB, Canada, 15–17 May 2006.
31. Ilk, D.; Valko, P.P.; Blasingame, T.A. A Deconvolution Method Based on Cumulative Production for Continuously Measured Flowrate and Pressure Data. In Proceedings of the Eastern Regional Meeting, Lexington, KY, USA, 17–19 October 2007.
32. Brown, M.; Ozkan, E.; Raghavan, R.; Kazemi, H. Comparison of fractured horizontal-well performance in tight sand and shale reservoirs. *SPE Reserv. Eval. Eng.* **2011**, *14*, 248–259.

33. Wachtmeister, H.; Lund, L.; Aleklett, K.; Mikael, H. Production Decline Curves of Tight Oil Wells in Eagle Ford Shale. *Nat. Resour. Res.* **2017**, *26*, 365–377. [CrossRef]
34. Li, W.G.; Yue, H.; Sun, Y.P.; Guo, Y.; Wu, T.P.; Zhang, N.Q.; Chen, Y. Development Evaluation and Optimization of Deep Shale Gas Reservoir with Horizontal Wells Based on Production Data. *Geofluids* **2021**, *2021*, 4815559. [CrossRef]
35. Han, X.; Feng, F.P.; Zhang, X.C.; Cao, J.; Zhang, J.; Suo, Y.; Yan, Y.; Yan, M.S. An unequal fracturing stage spacing optimization model for hydraulic fracturing that considers cementing interface integrity. *Pet. Sci.* **2023**, *20*, 1995–8226. [CrossRef]
36. Mukherjee, H.; Economides, M.J. A Parametric Comparison of Horizontal and Vertical Well Performance. *SPE Form. Eval.* **1991**, *6*, 209–216. [CrossRef]
37. Harald, S. Algorithm 368: Numerical inversion of Laplace transforms. *Commun. ACM* **1970**, *13*, 47–49.
38. Harald, S. Remark on algorithm 368: Numerical inversion of Laplace transforms. *Commun. ACM* **1970**, *13*, 624.
39. Tong, D.K.; Chen, Q.L. Remark on Stehfest numerical inversion method of Laplace transforms. *Acta Petrol. Sin.* **2001**, *22*, 91–92.
40. Liu, W.C.; Liu, Y.W.; Zhu, W.Y.; Sun, H.D. Improvement and application of ILK flow-rate deconvolution algorithm based on the second-order B-splines. *Acta Pet. Sin.* **2018**, *39*, 327–334.

Disclaimer/Publisher's Note: The statements, opinions and data contained in all publications are solely those of the individual author(s) and contributor(s) and not of MDPI and/or the editor(s). MDPI and/or the editor(s) disclaim responsibility for any injury to people or property resulting from any ideas, methods, instructions or products referred to in the content.

Article

Dynamic Productivity Prediction Method of Shale Condensate Gas Reservoir Based on Convolution Equation

Ping Wang [1,*], Wenchao Liu [2,*], Wensong Huang [1], Chengcheng Qiao [2], Yuepeng Jia [1] and Chen Liu [2]

[1] PetroChina Research Institute of Petroleum Exploration and Development, Beijing 100083, China
[2] School of Civil and Resource Engineering, University of Science and Technology Beijing, Beijing 100083, China
* Correspondence: wp2011@petrochina.com.cn (P.W.); wcliu_2008@126.com (W.L.)

Abstract: The dynamic productivity prediction of shale condensate gas reservoirs is of great significance to the optimization of stimulation measures. Therefore, in this study, a dynamic productivity prediction method for shale condensate gas reservoirs based on a convolution equation is proposed. The method has been used to predict the dynamic production of 10 multi-stage fractured horizontal wells in the Duvernay shale condensate gas reservoir. The results show that flow-rate deconvolution algorithms can greatly improve the fitting effect of the Blasingame production decline curve when applied to the analysis of unstable production of shale gas condensate reservoirs. Compared with the production decline analysis method in commercial software HIS Harmony RTA, the productivity prediction method based on a convolution equation of shale condensate gas reservoirs has better fitting affect and higher accuracy of recoverable reserves prediction. Compared with the actual production, the error of production predicted by the convolution equation is generally within 10%. This means it is a fast and accurate method. This study enriches the productivity prediction methods of shale condensate gas reservoirs and has important practical significance for the productivity prediction and stimulation optimization of shale condensate gas reservoirs.

Keywords: shale condensate gas reservoirs; multi-stage fractured horizontal wells; flow-rate deconvolution; Blasingame production decline typical curve; productivity prediction

Citation: Wang, P.; Liu, W.; Huang, W.; Qiao, C.; Jia, Y.; Liu, C. Dynamic Productivity Prediction Method of Shale Condensate Gas Reservoir Based on Convolution Equation. *Energies* **2023**, *16*, 1479. https://doi.org/10.3390/en16031479

Academic Editor: Reza Rezaee

Received: 4 January 2023
Revised: 19 January 2023
Accepted: 26 January 2023
Published: 2 February 2023

Copyright: © 2023 by the authors. Licensee MDPI, Basel, Switzerland. This article is an open access article distributed under the terms and conditions of the Creative Commons Attribution (CC BY) license (https://creativecommons.org/licenses/by/4.0/).

1. Introduction

Despite the strong trend of zero carbon, the world still needs a lot of fossil fuels [1]. Compared with traditional fossil energy, natural gas produces less carbon dioxide when releasing the same calorific value. The shale gas reservoir is an unconventional reservoir, which has self-generation, self-storage, and a large area of continuous accumulation [2]. Global shale gas resources are abundant, and with continuous exploration and development, the amount of resources is increasing. At present, the global resource of shale gas is 4.57×10^{14} m^3 which is equivalent to conventional natural gas resources, and is mainly distributed in North America, Central Asia, China, Latin America, the Middle East, North Africa, and the former Soviet Union [3]. With the progress of exploration technology, the exploration and development of shale condensate gas reservoirs have gradually become a hot research topic. As an important growth point of condensate oil reserves, the global technology's recoverable resources of shale oil are 738×10^8 t, and the recoverable resources of associated condensate oil is predicted to be $(140 \sim 192) \times 10^8$ t. It can be seen that the condensate oil formed by condensation of shale condensate gas reservoirs contributes greatly to global oil and gas resources [4]. Increasing the proportion of shale condensate gas in energy consumption can effectively reduce carbon emissions. Although shale condensate gas reservoirs are abundant, the overall pore structure scale of shale reservoirs is extremely low, and effective shale gas seepage channels within micro-scale pore throats are difficult to be formed. Then, ultra-low porosity and permeability are exhibited [2]. Therefore, the reservoir is stimulated in order to achieve economic exploitation of shale gas through horizontal wells and volume fracturing technology [5]. With research in recent years, the multi-stage fractured horizontal well technology has been greatly improved and has been widely applied. The production of shale gas is effectively improved. Production of shale gas reservoirs is the core index to evaluate the development effect of shale gas reservoirs. How to predict it accurately and efficiently is a hot and key issue in shale gas development.

At present, there are four main methods for predicting the production of shale gas reservoirs. The first is the empirical method, including the Arps production decline analysis method [6], power exponential decline model [7], extended exponential decline (SEPD) model [8], Duong decline model [9], and universal exponential decline model [10]. The empirical method has the advantages of simple and convenient calculation. However, the empirical method lacks theoretical basis and is greatly affected by some factors such as geographical location, basic geological conditions of gas reservoirs, mining schemes, mining tools, and on-site operators. Therefore, the empirical method has obvious limitations. Not only that, but the empirical model also has the defect of limited applicable conditions. It must be ensured that the production data meet the conditions of the model before using the empirical method to predict the productivity of shale gas reservoirs. The second is a method for predicting the productivity of shale gas reservoirs based on artificial intelligence. In 2017, Zhu et al. established a neural network that predicts shale gas production with better accuracy and stability than traditional BP neural networks [11]. In 2021, according to hydraulic fracturing data, production data, and evaluation of final recoverable reserves (EUR) of 282 wells in WY shale gas reservoirs, an EUR evaluation algorithm for shale gas wells based on deep learning has been obtained by Liu et al. [12]. In 2022, Song et al. has established a model for the productivity prediction of shale gas reservoirs, which combines the BP neural network and genetic algorithm [13]. The productivity after horizontal wells fracturing has been predicted by forward and reverse training and parameter correction by error identification. In 2022, a hyper-parameters optimized long short-term memory (LSTM) network to effectively forecast daily gas production has been proposed by Qiu et al. [14]. The production data analysis method based on artificial intelligence is adopted to improve the reliability of prediction results. However, its interpretability is poor, and it cannot explain which factor was the main influencing factor. In addition, some learning samples are needed. Numerical modeling is also often used to explain engineering problems [15]. Considering the influence of shale gas adsorption and desorption, Wei et al. has obtained the Blasingame typical curve of shale gas reservoirs by the numerical method [16]. The study expands the understanding of shale gas flow characteristics and can be used for reserves prediction. Considering the multi-scale flow characteristics of shale gas, a 3D numerical model for the seepage of shale gas has been established discretely using the finite volume method, and the production performance of multi-stage fractured horizontal wells in shale gas reservoirs have been simulated by Chen et al. [17]. Zhang et al. have established a continuum-discrete fracture network coupling model considering the multi-scale seepage mechanism of shale, and have used the controlled volume finite element method and unstructured triangular prism grid for numerical solution. The numerical simulation model has successfully been applied to comprehensive and field examples [18]. Although numerical simulation can directly represent shale gas seepage characteristics, it also has high accuracy for productivity prediction. However, the mathematical modeling, grid division, and numerical solution of shale reservoirs after hydraulic fracturing are relatively difficult. The fourth is the typical curve analysis method. In 2020, considering the adsorption/desorption, diffusion, threshold pressure gradient, and stress sensitivity of gas flow in shale, Wang et al. have established a five-linear seepage mathematical model for multi-stage fractured horizontal wells in shale gas reservoirs. Then, the flow regime is divided by using the typical curve, which provides a scientific basis for single well productivity prediction [19]. Jiang et al. have established a matrix fracture fluid flow model considering the stress-sensitive effect, and have obtained the Blasingame typical curve to predict the productivity of multi-stage fractured horizontal wells in tight reservoirs [20]. The typical curve analysis method is widely used in engineering, but the use of the model is closely related to the mining stage, as the choice of the model has higher requirements. In summary, the methods for the productivity analysis of shale gas reservoirs have some defects. Not only that, but most methods for productivity prediction are aimed at shale dry gas, but now shale condensate gas also has great development value. At present, the productivity prediction methods of shale condensate gas reservoirs based on characteristic curve analysis are relatively few.

The special seepage mechanism and complex development methods for shale condensate gas reservoirs lead to serious fluctuations in production data. In addition, the special seepage law and production mechanism of shale condensate gas reservoirs lead to the pressure data and production data measured on site, which are constantly changing with time. However, the internal boundary conditions of the mathematical model used for production data analysis are generally constant pressure (or constant flow-rate) conditions. Shut-in is usually used to solve the problem that production data do not match the theoretical mathematical model in conventional oil and gas reservoirs. However, for shale gas development, the production of shale gas wells will be greatly affected by shut-in and economic losses will be caused. Therefore, when analyzing the production data of shale gas reservoirs, the normalization method is often used to process the production data. The traditional normalization method extracts equivalent discrete flow-rate data corresponding to constant pressure from variable pressure production by introducing material balance time or converting the flow rate corresponding to variable pressure into the flow-rate corresponding to constant pressure by the pressure normalized rate (PNR) and superposition time combination method [21]. However, for the production data of shale gas with rapid changes, the production data processed by the traditional normalization method still has the defects of scattered distribution of characteristic data points and poor smoothness, which leads to great uncertainty in data fitting. The smooth characteristic data curve generated after processing production data by deconvolution algorithm can improve the fitting effect of data, reduce the uncertainty of interpretation results, and eliminate the influence of data errors. A series of progress on deconvolution has been made. In 2007, using the cumulative flow-rate data and the second-order B-spline function weight sum, the flow rate corresponding to unit pressure difference has been reconstructed by ILK et al. [22]. However, the accuracy of the algorithm is not high, and the error in the initial stage and the final stage is large. In 2016, the von Schroeter deconvolution model [23] has been improved by Wang et al. [24]. The formation flow rate is converted with the surface flow rate and the wellbore flow rate so that the wellbore storage effect can be removed by the transient pressure response after deconvolution. However, there are also defects of low accuracy. In 2018, inspired by the ILK deconvolution algorithm with high calculation accuracy, the flow rate data are used to replace the cumulative flow rate data of the original algorithm for the deconvolution calculation by Liu et al. [25]. A fast solution method has been presented, which greatly improves the computational efficiency and accuracy, and significantly improves the fitting effect of data.

In order to evaluate the shale gas and condensate oil productivity of multi-stage fractured horizontal wells in shale condensate gas reservoirs, this study proposes a set of typical curve analysis processes for dynamic production data. Firstly, a trilinear seepage mathematical model for multi-stage fractured horizontal wells is introduced in this work. The linearization of mathematical model and Laplace transformation have been used to obtain the flow rate solution of the model corresponding to per unit pseudo bottom hole flowing pressure drop. Secondly, a calculation method is used to convert condensate oil into equivalent gas. The condensate oil and gas two-phase production data of shale condensate gas reservoir are converted into equivalent gas production data. The pressure data is calculated as pseudo pressure data to realize the linearization of production data. On this basis, the improved flow-rate deconvolution algorithm [25] is introduced to normalize the production data. Then, the normalized dynamic production data and the Blasingame production decline typical curve are fitted with the flow rate solution corresponding to per unit pseudo bottom hole flowing pressure drop, and some of the reservoir and fracture parameters are explained. Finally, according to the explained parameters and the flow-rate solution corresponding to variable pseudo bottom hole flowing pressure calculated by the seepage mathematical model, the shale gas and condensate oil predicted recoverable reserves results of multi-stage fractured horizontal wells in shale condensate gas reservoirs are obtained. It has important reference value for studying the main controlling factors of the productivity of shale condensate gas reservoirs and optimizing production systems.

2. Dynamic Productivity Prediction Method of Shale Condensate Gas Reservoir Based on Convolution Equation

2.1. Seepage Model of Multi-Stage Fractured Horizontal Well in Shale Condensate Gas Reservoir

After the hydraulic fracturing of shale condensate gas wells, a very complex fracture network will be formed in shale reservoirs. According to Brown et al. [26], it is a very effective way of simplification to deal with the complex fracture network of the stimulated reservoir volume (SRV) of the fractured zone. After staged fracturing of shale condensate gas reservoirs, the reservoir will show obvious linear flow, which can last for one year or even several years. Therefore, the mining process of shale condensate gas reservoirs is simplified as a trilinear flow from the main fracture flow zone to the wellbore, from the inter-fracture flow zone to the main fracture flow zone, and from the reservoir flow zone to the inter-fracture flow zone. The established trilinear seepage physical model is shown in Figure 1. It is considered that all the main fractures are of equal length and equal spacing distribution. Without considering the flow in the horizontal well, the horizontal well is considered to be infinitely conductive; the confinement effect of micro-nano pores in shale reservoirs is not considered [27].

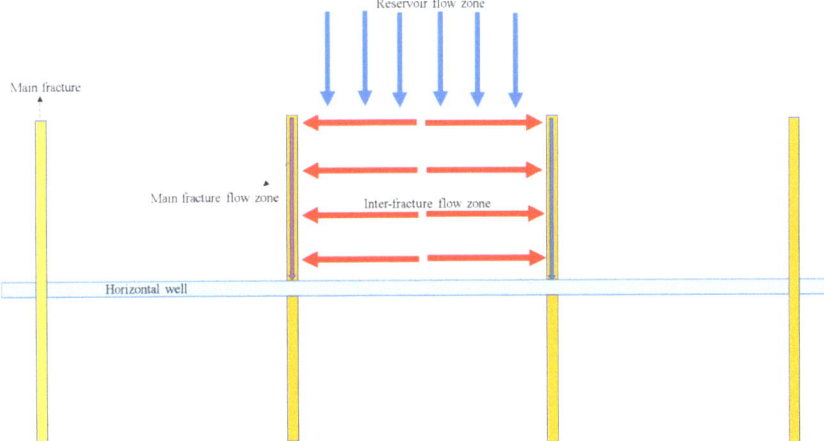

Figure 1. Trilinear seepage physical model of staged fracturing horizontal wells in shale condensate gas reservoirs.

Based on the trilinear seepage physical model of multi-stage fractured horizontal wells in shale condensate gas reservoirs, the unsteady seepage mathematical models of reservoir flow zone, inter-fracture flow zone, and main fracture flow zone are established as follows.

For the linear flow area of the reservoir [26], the linearized seepage mathematical model is as follows:

$$\frac{\partial^2 m_O}{\partial x^2} = \frac{1}{\eta_O} \frac{\partial m_O}{\partial t} \quad (1)$$

$$m_O|_{t=0} = m_i \quad (2)$$

$$\left.\frac{\partial m_O}{\partial x}\right|_{x=x_e} = 0 \tag{3}$$

$$m_O|_{x=x_F} = m_I|_{x=x_F} \tag{4}$$

where, the subscript O represents the reservoir flow area; subscript I represents the flow area between main fractures; the subscript i represents in the initial state; η_O is the diffusion coefficient of the matrix area, cm^2/s, defined as Equation (5); x is the distance from the horizontal wellbore, cm; x_e is the size of the reservoir in the x direction, cm; x_F is the fracture half length, cm; t is time, s; m represents the pseudo pressure, atm^2/cp, defined as Equation (6):

$$\eta_O = \frac{K_O}{\phi_O \left(\frac{1}{P_O} - \frac{1}{Z}\frac{\partial Z}{\partial P_O}\right)\mu} \tag{5}$$

$$m = 2\int_0^P \frac{P}{\mu Z} dP \tag{6}$$

where, K is the permeability, D; ϕ is the porosity; P represents the pressure, atm; p represents the real production pressure, atm; μ is the viscosity of shale gas, cp; Z is the compression factor of shale gas, dimensionless.

It can be seen from Equation (5) that the diffusion coefficient is a variable related to pressure. It has strong nonlinearity. In order to make the model suitable for the deconvolution algorithm, the diffusion coefficient of this study is averaged and considered to be approximately constant.

For the linear flow area between main fractures [26], the linearized seepage mathematical model is as follows:

$$\frac{\partial^2 m_I}{\partial y^2} + \frac{K_O}{K_I x_F}\left.\frac{\partial m_O}{\partial x}\right|_{x=x_F} = \frac{1}{\eta_I}\frac{\partial m_I}{\partial t} \tag{7}$$

$$m_I|_{t=0} = m_i \tag{8}$$

$$\left.\frac{\partial m_I}{\partial y}\right|_{y=\frac{d_F}{2}} = 0 \tag{9}$$

$$m_I|_{y=\frac{w_F}{2}} = m_F|_{y=\frac{w_F}{2}} \tag{10}$$

where, subscript F represents the flow area of main fracture; y represents the distance in the y direction, cm; η_I is the diffusion coefficient of the flow area between main fractures, cm^2/s, defined as Equation (11); w_F is the width of hydraulic fracture, cm; d_F is the distance between two adjacent main fractures, cm.

$$\eta_I = \frac{K_I}{\phi_I \left(\frac{1}{P_I} - \frac{1}{Z}\frac{\partial Z}{\partial P_I}\right)\mu} \tag{11}$$

For the linear flow area of main fracture [26], the linearized seepage mathematical model is as follows:

$$\frac{\partial^2 m_F}{\partial x^2} + \frac{2K_I}{K_F w_F}\left.\frac{\partial m_I}{\partial y}\right|_{y=\frac{w_F}{2}} = \frac{1}{\eta_F}\frac{\partial m_F}{\partial t} \tag{12}$$

$$m_F|_{t=0} = m_i \tag{13}$$

$$\left.\frac{\partial m_F}{\partial x}\right|_{x=x_F} = 0 \tag{14}$$

$$\left.\frac{\partial m_F}{\partial x}\right|_{x=0} = \frac{q_F P_{sc} T}{k_F w_F h T_{sc}} \tag{15}$$

where, η_F is the diffusion coefficient of hydraulic fracture, cm^2/s, defined as Equation (16); h represents the reservoir thickness, cm; T_{sc} is the temperature under standard conditions, K; P_{sc} is the pressure under standard conditions, atm; q_F is the flow in the fracture, cm^3/s.

$$\eta_F = \frac{K_F}{\phi_F \left(\frac{1}{P_F} - \frac{1}{Z}\frac{\partial Z}{\partial P_F}\right)\mu} \tag{16}$$

The linearized mathematical model of the three linear flow area can be solved by Laplace space [26], and then the flow-rate solution corresponding to the production condition of constant pressure in Laplace space can be obtained as follows:

$$\tilde{q} = \frac{1 + Cs^2 \tilde{m}_w}{\tilde{m}_w} \cdot \frac{1}{s^2} \cdot m_{w,const} \tag{17}$$

where, ~ represents the corresponding variable in Laplace space; q is the flow-rate corresponding to constant pressure production condition, cm^3/s; m_w is the pseudo bottom hole pressure solution corresponding to the constant production rate, atm^2/cp which is equal to m_F at x = 0; C is wellbore storage coefficient, cm^3/atm; s is the Laplace transform parameter; $m_{w,const}$ is the pseudo bottom hole flowing pressure corresponding to constant production pressure, atm^2/cp.

2.2. Consider the Production of Condensate Oil in Model Solution

The mathematical model of multi-stage fractured horizontal wells in shale condensate gas reservoir established in Section 2.1 is based on single-phase flow with only shale gas flow. However, in actual production, with the formation pressure of shale condensate gas reservoirs gradually lower than the dew point pressure, condensate will be precipitated near the wellbore [28]. When the formation pressure is lower than the critical flow pressure, the precipitated condensate oil forms a continuous phase, and then forms a seepage state of oil and gas two-phase flow. Therefore, shale gas and condensate oil will be produced simultaneously in real conditions. This paper uses a gas equivalent conversion method [29]. The production of shale condensate is equivalently converted to the production of shale gas so that only single-phase gas production data analysis is required. The conversion method of gas equivalent is as follows.

It is assumed that the relative density of the condensate oil is γ_o and the molecular weight is M_o; assuming that the volume of 1 cm^3 condensate oil in the standard state (p = 0.101 MPa, T = 293 K) before liquefaction is V (cm^3); since the weight of 1 cm^3 condensate oil is γ_o g, then:

$$n = \frac{\gamma_o}{M_o} (g \cdot mol) \tag{18}$$

$$V = \frac{nZRT}{p} = \frac{\gamma_o}{M_o} \times \frac{1 \times 8.314 \times 10^{-6} \times 293}{0.101} = 0.0241 \frac{\gamma_o}{M_o} (cm^3) \tag{19}$$

where, n is the mole number of gas, g·mol. That is to say: before the liquefaction of 1 cm^3 condensate oil with a relative density of γ_o and molecular weight of M_o, the volume of condensate oil in standard state is shown in Equation (20), which is the conversion coefficient of converting condensate oil volume into shale gas volume in standard state. Therefore, if the condensate oil production of the gas well is q_o cm^3/s, the converted gas production is as follows:

$$q_{GE} = 0.0241 \frac{\gamma_o}{M_o} \cdot q_o (cm^3/s) \tag{20}$$

If only the relative density of condensate oil is known, and the molecular weight of condensate oil is unknown, the Gragoe equation [29] can be used to replace M_o:

$$M_o = \frac{44.29 \gamma_o}{1.03 - \gamma_o} \tag{21}$$

Substituting Equation (21) into Equation (20), we can obtain that:

$$q_{GE} = 5.367 \times 10^{-4}(1.03 - \gamma_o) \cdot q_o \tag{22}$$

The total gas production equivalent is as follows:

$$q_{total} = 5.367 \times 10^{-4}(1.03 - \gamma_o) \cdot q_o + q_g \tag{23}$$

where, q_{total} is the total gas production equivalent, cm³/s; q_g is the production of shale gas, cm³/s.

2.3. Flow-Rate Deconvolution

In order to solve the problem that the dynamic production data of shale condensate gas reservoirs do not match the internal boundary conditions of the mathematical model of the productivity prediction theory, and to solve the defect that the traditional normalization method of production data is not suitable for the production data with sharp changes in shale gas reservoirs, an improved ILK flow rate deconvolution algorithm [25] is introduced in this study. Based on Duhamel's principle, the algorithm converts the flow rate data corresponding to variable bottom hole pressure drop into the flow rate corresponding to per unit bottom hole pressure drop by Equation (24):

$$q(t) = \int_0^t \Delta p(t-\tau) \cdot q'_u(\tau) d\tau \tag{24}$$

The implementation process of the improved ILK flow rate deconvolution algorithm is as follows: first, based on Equation (24), the second-order B-spline function weight and the flow-rate derivative corresponding to unit pressure drop are used to reconstruct. Then, using the mathematical properties of the convolution integral, the sensitivity matrix in the deconvolution calculation process is solved analytically and quickly by using the technique of piecewise integral according to the pressure drop segment. Then the dichotomy method is used to quickly find the pressure drop segment of each group of flow rate data points to further improve the computational efficiency.

The integral derivative of regularized production in unsteady seepage segments will be affected by the error of dynamic production data of shale condensate gas reservoirs, and the data fluctuating is caused. The direct use of original flow rate data will result in the poor effect of typical curve fitting, and the error of the parameters of the reservoir and fracture obtained by inversion is large. The flow rate data corresponding to variable bottom hole pressure drop can be transformed into the flow rate data corresponding to unit bottom hole pressure drop by using the improved ILK flow rate deconvolution algorithm, so as to match the inner boundary conditions of the theoretical seepage model, and the processed flow-rate data is smoother, which conforms to the law of flow rate decline corresponding to unit bottom hole pressure drop. Therefore, the effect of the Blasingame production decline typical curve [30] fitting is better; the inverted reservoir and fracture parameters are more accurate and the accuracy of productivity prediction is improved. In addition, when using the improved ILK flow-rate deconvolution algorithm to normalize the dynamic production data, the long-time shut-in operation of the traditional production data normalization method can be avoided; the impact of the shut-in test on shale gas production is avoided and the economic losses are reduced. Moreover, the improved ILK flow rate deconvolution algorithm has fast calculation speed and high stability, and a large number of field data can be processed quickly and accurately.

Although the improved flow rate deconvolution algorithm has the above advantages, there are some limitations in the calculation of flow rate deconvolution: first, the flow rate deconvolution algorithm can only be used in linear problems or linearization problems (such as single-phase flow, Darcy flow, material balance, etc.) and must be applied to the Duhamel's principle. Second, it must be ensured that the interpretation model does not change from beginning to end in the process of interpreting the production history data.

2.4. Dynamic Production Data Analysis Method for Shale Condensate Gas Reservoir Based on Flow-Rate Deconvolution

First, the trilinear seepage physical model for shale condensate gas reservoirs and the corresponding mathematical model are established. The unsteady seepage mathematical model is linearized and the flow rate solution corresponding to constant pseudo bottom hole pressure drop is obtained. Second, the production data of shale gas and condensate oil are transformed into the total equivalent gas production data by using Equation (23). The data of bottom hole flowing pressure is transformed into the data of bottom hole pressure by using Equation (6) of the pseudo pressure in the theoretical mathematical model. The method of Lee et al. [31] is used to calculate the viscosity of shale gas. The method of Yarborough and Hall [32] is used to calculate the compressibility factor of shale gas. The linearization of the nonlinear dynamic production data measured in the field is realized. Then, based on Equation (24), based on Duhamel's principle, the improved ILK flow rate deconvolution algorithm is used to convert the flow rate corresponding to the variable pseudo bottom hole pressure drop into the flow rate data corresponding to the constant pseudo bottom hole pressure drop, and the normalization of production data is realized. Finally, the typical curve fitting of Blasingame production decline is performed on the flow rate solution corresponding to the pseudo bottom hole pressure drop and the normalized dynamic production data. Based on the basic data of reservoir and the construction data of fracturing in the developed analysis program of flow instability, by adjusting the main fracture conductivity, fracture half length, outer boundary distance, matrix permeability, matrix porosity, the matrix comprehensive compression coefficient. and other parameters, the real data points of normalized production, normalized production integral, and normalized production integral derivative are fitted with the data points of the same type of the theoretical model flow rate solution, and some parameters of the fracture and reservoir are explained. The flow chart of the dynamic production data analysis method for shale condensate gas reservoirs based on flow deconvolution is shown in Figure 2:

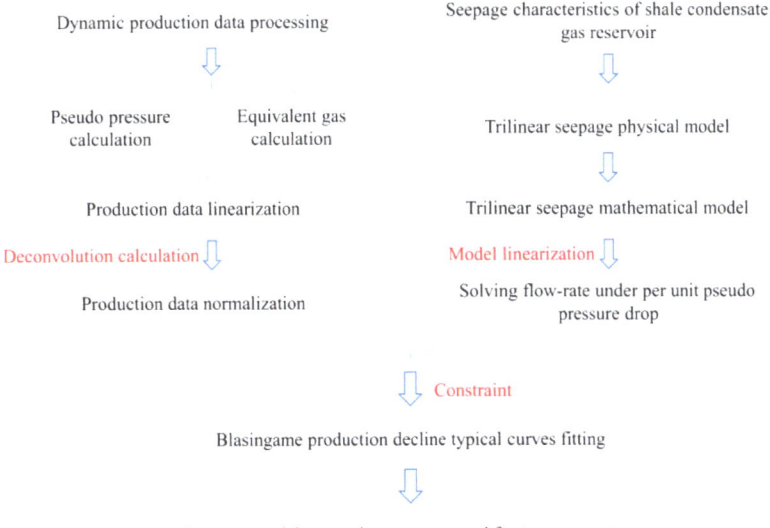

Figure 2. The flow chart of the production dynamic data analysis method for shale condensate gas reservoirs based on flow rate deconvolution.

The typical curve fitting of Blasingame production decline is realized by adjusting the main fracture conductivity, fracture half length, outer boundary distance, matrix permeability, matrix porosity, the matrix comprehensive compression coefficient, and other parameters. In the process, the known data of fracturing construction and the basic data of the reservoir are used as condition constraints, and the normalized parameter debugging of flow rate deconvolution calculation and the model parameter debugging of seepage theory model calculation are mutually restricted in the process of typical curve fitting. Therefore, more reliable parameter results can be analyzed. Not only that, but the analytical method used in the flow rate deconvolution algorithm is used in this dynamic production data analysis method; the calculation speed is faster and a large number of dynamic production data can be processed quickly. In addition, the flow rate deconvolution algorithm is used for normalization to make the typical curve fitting better, and the parameter results analyzed by the dynamic production data are more accurate.

2.5. Proposal of Dynamic Productivity Prediction Method for Shale Condensate Gas Reservoir Based on Convolution Equation

The dynamic productivity prediction method of shale condensate gas reservoirs is of great significance to the evaluation of the fracturing effect of staged fracturing horizontal wells and the analysis of the main control factors of productivity. The basis for the formulation of reasonable development technology policy of shale condensate gas reservoirs and the optimization suggestion of staged fractured horizontal wells can be provided. Therefore, a dynamic productivity prediction method for shale condensate gas reservoirs based on the convolution equation is proposed, and the recoverable reserves of shale gas and condensate oil can be predicted.

The dynamic productivity prediction process of shale condensate gas reservoirs is as follows: first, based on the Duhamel principle, the convolution Equation (24) is used to calculate the flow rate solution corresponding to the pseudo bottom hole flowing pressure obtained in Section 2.1, and the flow rate solution corresponding to variable pseudo bottom hole flowing pressure is obtained. Then, the parameter results of the analysis and inversion of the dynamic production data in Section 2.4 are brought into the flow-rate solution corresponding to the variable pseudo bottom hole flowing pressure, and the total gas equivalent flow rate can be predicted. For the later production stage of horizontal wells, it is set to continue production corresponding to constant pressure, and CGR is approximately a fixed value [33,34]. Therefore, after the predicted total gas production equivalent flow rate is obtained, the recoverable reserves of shale gas and condensate oil of staged fractured horizontal wells in shale condensate gas reservoirs are predicted by using Equations (25) and (26), according to the method of condensate oil converted gas equivalent and condensate oil-gas ratio CGR.

$$q_g = \frac{q_{tot}}{1 + 5.367 \times 10^{-4} \times (1.03 - \gamma_o) \times \text{CGR}} \quad (25)$$

$$q_o = \text{CGR} \times q_g \quad (26)$$

The flow chart of the dynamic productivity prediction method for shale condensate gas reservoirs based on the convolution equation is shown in Figure 3.

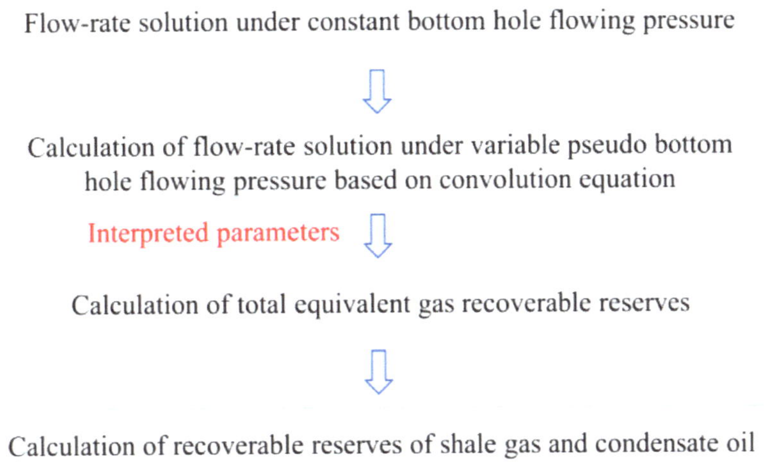

Flow-rate solution under constant bottom hole flowing pressure

Calculation of flow-rate solution under variable pseudo bottom hole flowing pressure based on convolution equation

Interpreted parameters

Calculation of total equivalent gas recoverable reserves

Calculation of recoverable reserves of shale gas and condensate oil

Figure 3. The flow chart of the dynamic productivity prediction method for shale condensate gas reservoirs based on the convolution equation.

3. Practical Application

The horizontal well length of a multi-stage fractured horizontal well in the Duvernay shale reservoir is 1962 m, the wellbore radius is 0.107 m, the number of hydraulically fractured segments is 21, and the distance from the nearest well is about 3700 m. The reservoir where the well is located has a thickness of 41 m and a reservoir temperature of 115 °C. The initial gas reservoir pressure is 62 MPa. The multi-stage fractured horizontal well has been put into production for about 2500 days with a cumulative shale gas production of 36.39×10^6 m³ and a daily average shale gas production of 13.07×10^3 m³/d. The cumulative production of condensate oil is 26.91×10^3 m³, and the average daily production of condensate oil is 9.66 m³/d. The dynamic monitoring data of daily production of condensate oil and shale gas are shown in Figure 4. The pseudo bottom hole flowing pressure vs. bottom hole flowing pressure curve is shown in Figure 5. The bottom hole flowing pressure and CGR of the well are shown in Figure 6. It can be seen that the relationship between bottom hole flowing pressure and CGR is consistent with that described in Section 2.5. CGR is closely related to the bottom hole flowing pressure. When the bottom hole flowing pressure remains constant, CGR is also approximately a constant value. Therefore, after predicting the recoverable reserves of gas equivalent, the recoverable reserves of condensate oil corresponding to constant pressure can be predicted by using constant CGR. According to Equation (23), the condensate oil production is converted into equivalent gas production, and the shale gas production is added. Then the total cumulative equivalent gas production is calculated to be 40.36×10^6 m³. The converted daily equivalent gas production data and bottom hole flowing pressure data are shown in Figure 7.

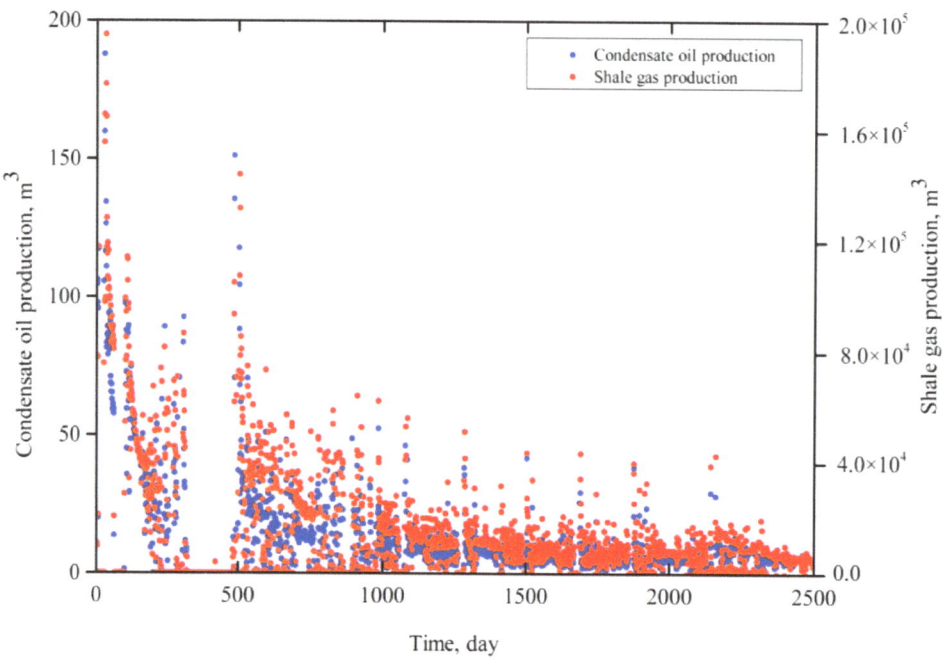

Figure 4. Daily output of condensate oil and shale gas in a multi-stage fractured horizontal well.

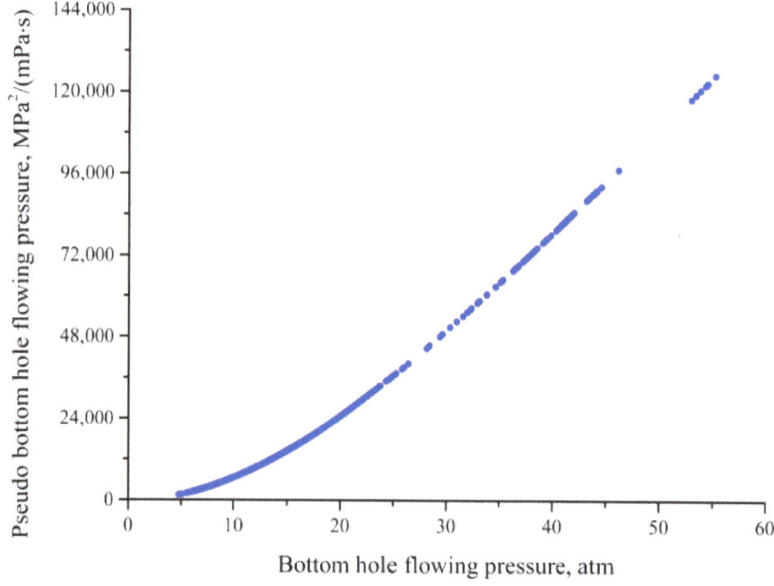

Figure 5. Pseudo bottom hole flowing pressure vs. bottom hole flowing pressure curve.

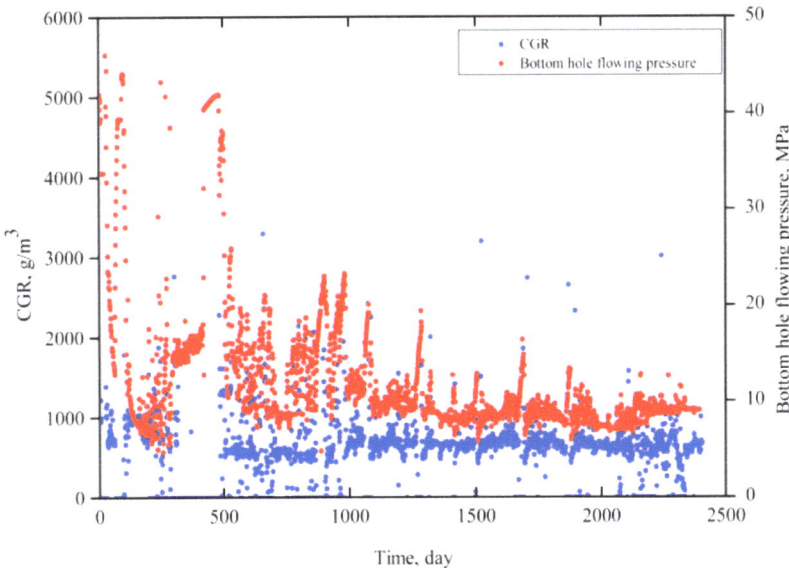

Figure 6. Bottom hole flowing pressure and CGR of the multi-stage fractured horizontal well.

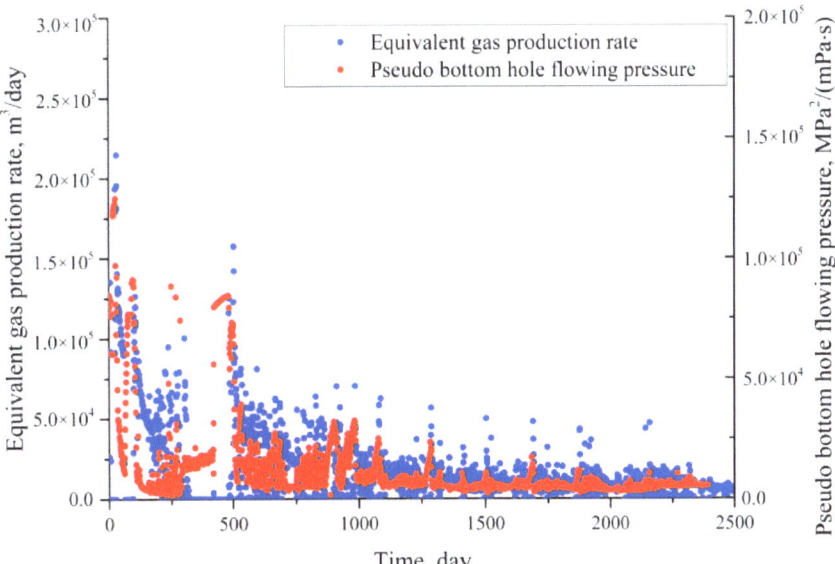

Figure 7. Equivalent gas production rate and pseudo bottom hole flowing pressure of the multi-stage fractured horizontal well.

It can be seen from Figure 7 that the daily production rate data and the bottom hole flowing pressure data are seriously fluctuated, and the data error is large. In addition, the shut-in operation has carried out during the period, and the production data are intermittent. The traditional pressure normalized rate and material balance time combination method is used to normalize the production data, and the obtained Blasingame production decline typical curve is shown in Figure 8. The improved ILK flow-rate deconvolution algorithm [25] is used to normalize the production rate data. Figure 9 shows the production rate data corresponding to unit pseudo bottom hole flowing pressure drop calculated by the deconvolution algorithm. The Blasingame production decline typical curve of the deconvolution algorithm output is shown in Figure 10. It can be seen that the flow rate deconvolution algorithm eliminates the noise of the dynamic production data and the calculated output Blasingame production decline typical curve is smoother. Then, based on the basic reservoir data and fracturing construction data, the normalized dynamic production data is fitted to the Blasingame production decline typical curve of the seepage theoretical model by adjusting the fracture half-length, fracture conductivity, matrix permeability, and other parameters. It is interpreted that the fracture half-length is 45 m, the main fracture conductivity is 16.5 mD·cm, the reservoir matrix permeability is 0.0006 mD, and the reservoir porosity is 0.04. Figure 11a shows the fitting results of the Blasingame production decline typical curve calculated by traditional PNR-MBT method and the ones calculated by the seepage theoretical model. Figure 11b shows the fitting results of the Blasingame production decline typical curve calculated by the flowrate deconvolution algorithm and the ones calculated by the seepage theoretical model. Obviously, the dynamic production data processed by the flowrate deconvolution algorithm is smoother, and the Blasingame production decline typical curve fitting effect is better. The equivalent gas production corresponding to the setting pressure can be obtained by putting the parameters interpreted from the Blasingame production decline typical curve fitting into the flow rate solution corresponding to the variable bottom hole flowing pressure, calculated according to the seepage theoretical model. The pressure setting diagram of the whole production stage is shown in Figure 12. In the later stage of the setting, the constant bottom hole flowing pressure of 9 MPa is used for 30 years of production, and the CGR is 0.6 L/m^3. Figure 13 shows the equivalent gas production calculated by the mathematical model and the real equivalent gas production. It can be seen that the equivalent gas production calculated by the seepage theoretical model and the real equivalent gas production are well fitted. Even in the shut-in stage, the fitting effect is also good, which proves the accuracy of later production prediction. At this time, the output of the late constant pressure production is calculated by the mathematical model, while the output of the early stage is measured by the field data. This further improves the accuracy of predicting equivalent gas recoverable reserves over 30 years of production. After obtaining the predicted equivalent gas recoverable reserves, the predicted recoverable reserves of shale gas and condensate oil can be calculated by using the contents in Section 2.5. The recoverable reserves of equivalent gas are 5.5050×10^7 m^3, the recoverable reserves of shale gas are 4.9679×10^7 m^3, and the recoverable reserves of condensate oil are 3.4794×10^4 m^3.

In addition to the dynamic productivity prediction of single well, the dynamic production prediction method of shale condensate gas reservoirs based on the convolution equation is used to predict the production of 10 multi-stage fractured horizontal wells in the Duvernay area. Moreover, the business software HIS Harmony RTA is used to analyze the production data of these wells by the multi-stage Arps production decline curve, and then the predicted shale gas production and condensate oil production are obtained. The results and errors of the two methods are shown in Table 1.

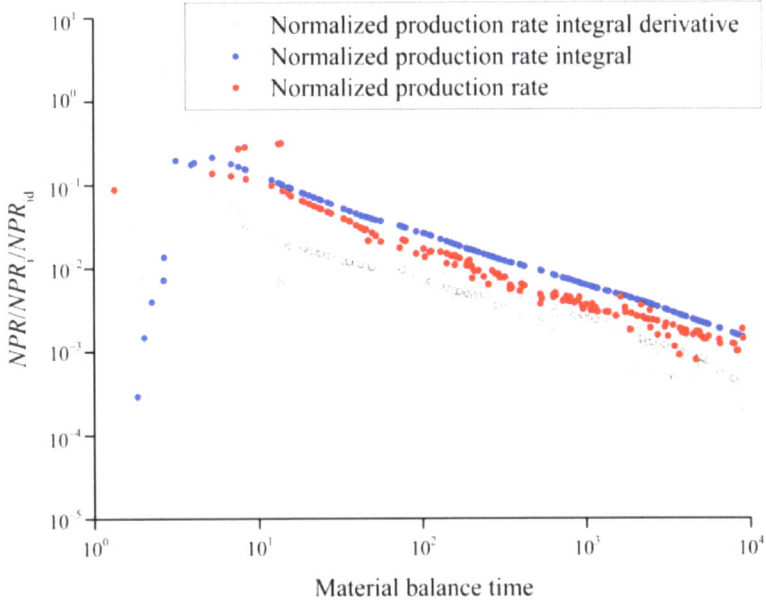

Figure 8. Blasingame production decline typical curve of the single well corresponding to PNR-MBT method.

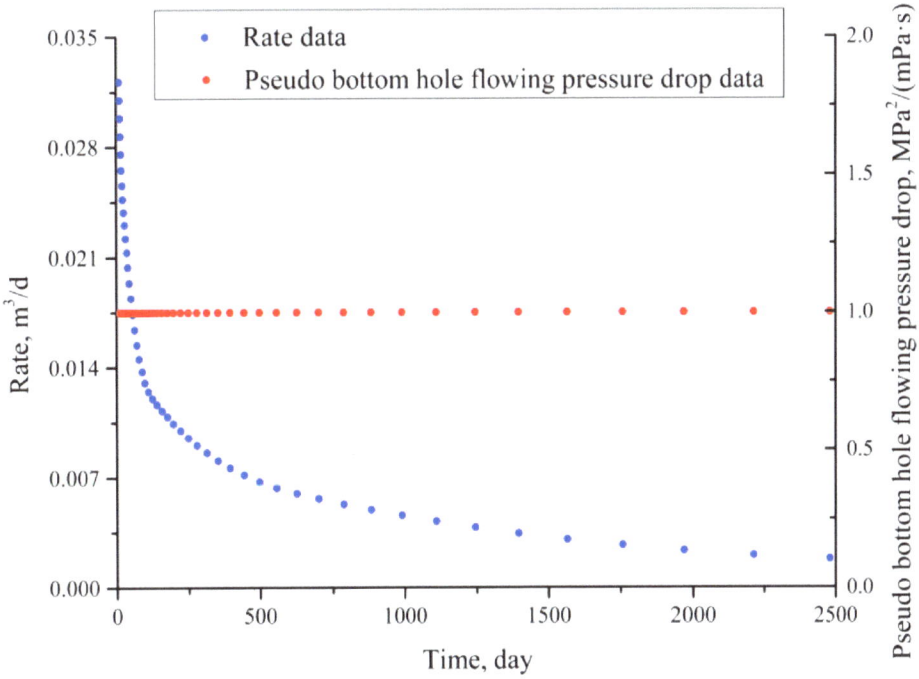

Figure 9. Flow rate data per unit pseudo bottom hole flowing pressure drop.

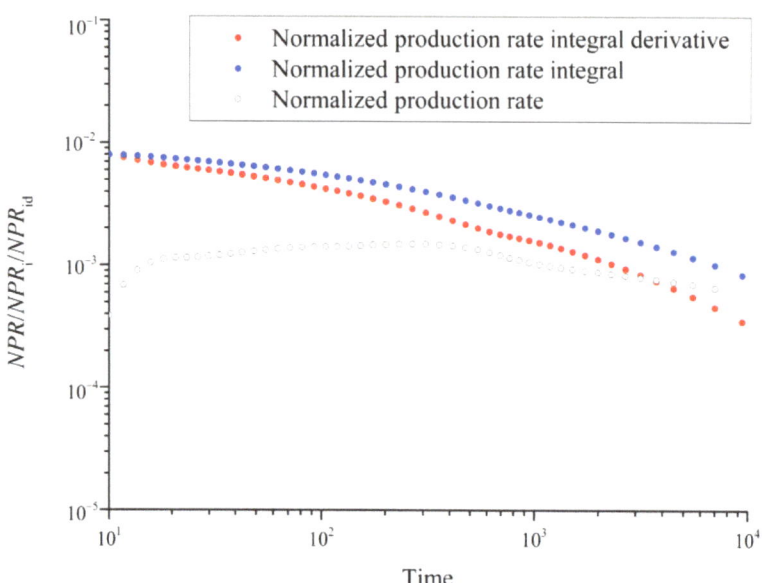

Figure 10. Blasingame production decline typical curve of the single well corresponding to flow rate deconvolution algorithm.

Table 1. Comparison of prediction results between the two methods.

Well Number	Actual Production ($10^4 \times m^3$)			Decline Curve Analysis Method ($10^4 \times m^3$) & Error (%)			Method Based on Convolution Equation ($10^4 \times m^3$) & Error (%)		
	Equivalent Gas	Shale Gas	Condensate Oil	Equivalent Gas	Shale Gas	Condensate Oil	Equivalent Gas	Shale Gas	Condensate Oil
Well 1	4036.24	3564.95	2.6369	6267.51 (55.3%)	5573.6 (38.1%)	3.88252 (47.2%)	3901.14 (3.4%)	3635.23 (2.0%)	2.25646 (14.4%)
Well 2	3663.72	3253.88	2.4906	4238.35 (15.7%)	3768.03 (15.8%)	2.85809 (14.8%)	3653.61 (0.3%)	3356.13 (3.1%)	2.12563 (14.6%)
Well 3	3242.82	2893.69	2.1216	3718.84 (14.7%)	3358.36 (16.1%)	2.19074 (3.3%)	3538.81 (9.1%)	3224.65 (11.4%)	2.09555 (1.2%)
Well 4	3631.22	3272.39	2.1805	4621.96 (27.3%)	4215.65 (28.8%)	2.46904 (13.2%)	3354.19 (7.6%)	2995.52 (8.5%)	1.97988 (9.2%)
Well 5	4480.10	4026.30	2.7577	4987.70 (11.3%)	4477.17 (11.2%)	3.10243 (12.5%)	4653.17 (3.9%)	4225.09 (4.9%)	2.64729 (4.0%)
Well 6	6128.55	5633.07	3.0109	6224.16 (1.6%)	5727.29 (1.7%)	3.01939 (0.3%)	5820.20 (5.0%)	5320.48 (5.5%)	2.82470 (6.2%)
Well 7	4307.33	3898.02	2.2694	4981.07 (15.6%)	4127.80 (5.9%)	2.99860 (32.1%)	3924.89 (8.9%)	3594.27 (7.8%)	1.83308 (19.2%)
Well 8	2633.71	2390.25	1.3622	3277.84 (24.4%)	3039.89 (27.2%)	1.33133 (2.3%)	2839.40 (7.8%)	2593.71 (8.5%)	1.37467 (0.9%)
Well 9	4162.75	3823.97	1.8955	4741.44 (13.9%)	4368.07 (14.2%)	2.08903 (10.2%)	4415.87 (6.1%)	4053.64 (6.0%)	2.02681 (6.9%)
Well 10	3850.55	3481.67	2.0639	4281.17 (11.2%)	3869.36 (11.1%)	2.30411 (11.6%)	3917.84 (1.7%)	3567.19 (2.5%)	1.96195 (4.9%)

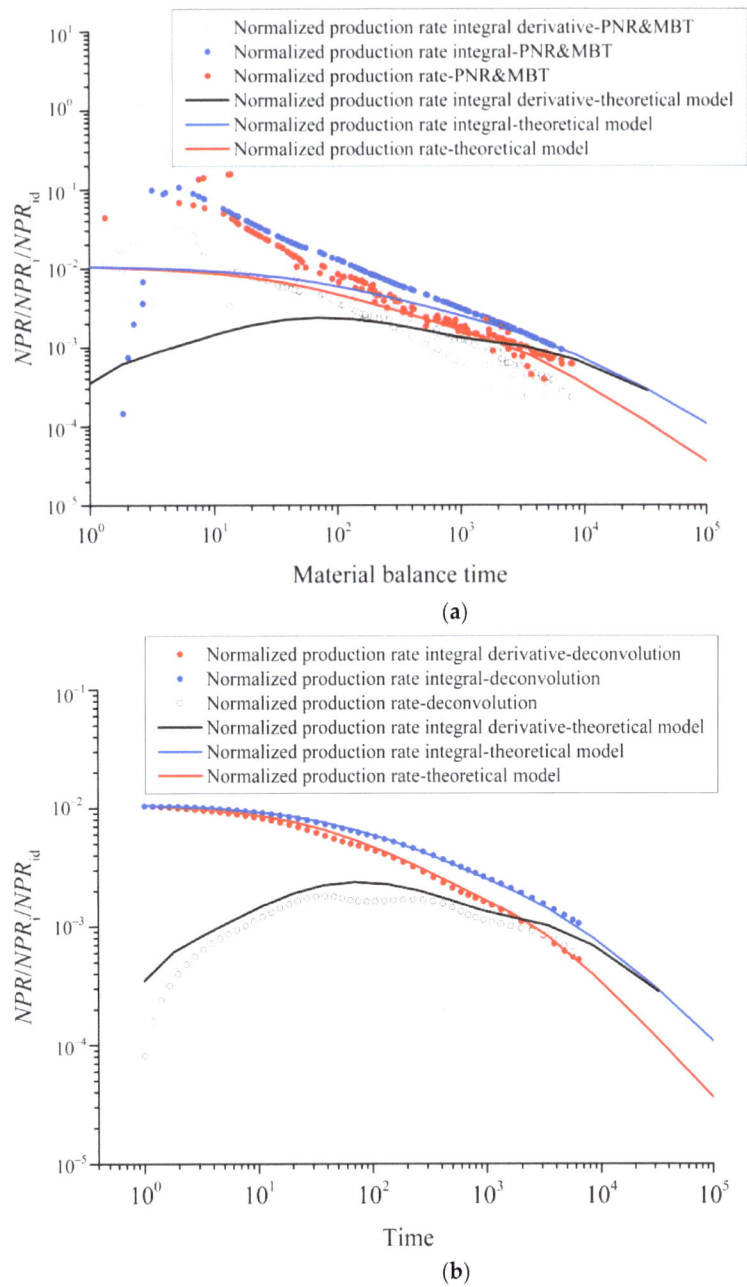

Figure 11. Blasingame production decline typical curve fitting results of the production data processed by two different normalization methods and the seepage theoretical model: (**a**) PNR-MBT method; (**b**) flow rate deconvolution algorithm.

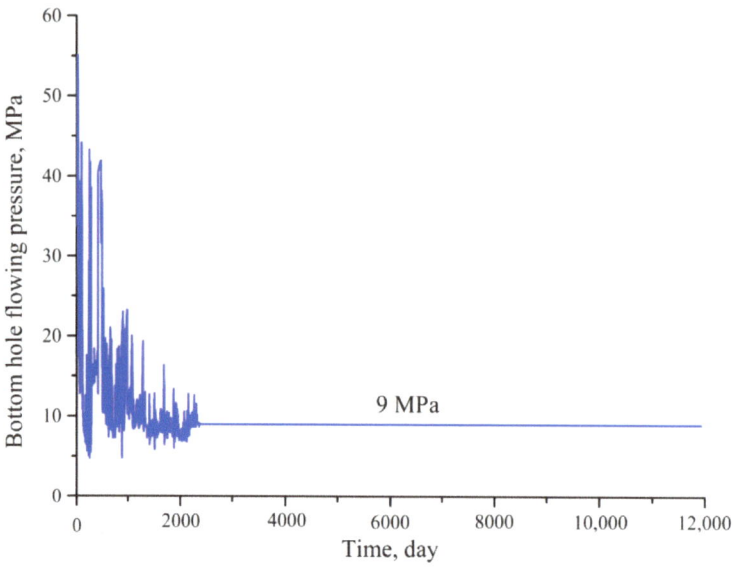

Figure 12. Schematic diagram of production pressure settings.

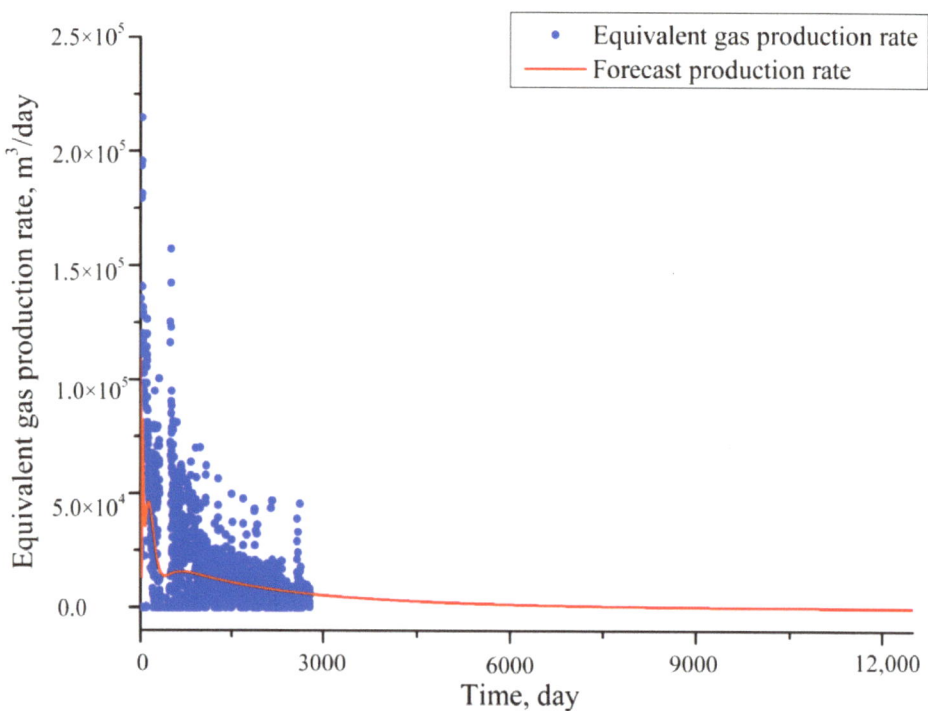

Figure 13. The fitting effect between the equivalent gas production calculated by the mathematical model and the real equivalent gas production.

For equivalent gas, the error of production predicted by the convolution equation is within 10%. In particular for well 2 and well 10, the errors are only 0.3% and 1.7%, respectively. For shale gas and condensate, the error of production predicted by the convolution equation is generally within 10%, too. Especially for well 1, the error of shale gas production is only 2%. For well 8, the error of condensate oil production is only 0.9%. The calculation results of two production prediction methods show that the production decline curve fitting effect of the method based on the convolution equation is better. Compared with the traditional production decline curve method, the predicted cumulative recoverable reserves of this method are more accurate. Moreover, the production prediction method based on the convolution equation uses the flow rate deconvolution algorithm to normalize the production data. Therefore, especially for the wells with serious production data oscillation, the Blasingame production decline typical curve fitting effect is much better than the traditional normalization method. The accuracy of the predicted cumulative recoverable reserves of shale gas and condensate oil is much higher than that predicted by the traditional production decline analysis method. By comparing the production predicted by the convolution equation with the actual production, the accuracy of the gas equivalent conversion method in Section 2.2 is also partially confirmed. This also confirms that the assumption that the relative density of condensate is constant has little influence on the prediction results of recoverable reserves.

4. Conclusions

In this paper, a method for predicting dynamic productivity of shale condensate gas reservoir based on convolution equation is proposed. It aims to obtain some reservoir parameters and hydraulic fracturing parameters through the analysis of production rate data and bottom hole flowing pressure, so as to predict the accumulative recoverable reserves of shale gas and condensate gas. It provides an accurate and efficient method for the production prediction of shale gas condensate reservoirs. Firstly, a trilinear seepage mathematical model for multi-stage fractured horizontal wells has been introduced in this paper. Linearization of mathematical model and Laplace transformation have been used to obtain the flow rate solution of the model corresponding to unit pseudo bottom hole flowing pressure drop. Secondly, a calculation method has been used to convert condensate oil into equivalent gas. The condensate oil and gas two-phase production data of shale condensate gas reservoirs were converted into equivalent gas production data. The pressure data was calculated as pseudo pressure data to realize the linearization of production data. On this basis, the flow rate deconvolution algorithm has been introduced to convert the production rate data corresponding to variable pseudo bottom hole flowing pressure into the production rate data corresponding to per unit pseudo bottom hole flowing pressure drop, which realizes the normalization of dynamic production data. Then, in the process of typical curve fitting, some reservoir parameters and fracture parameters have been adjusted with the basic reservoir parameters and fracturing construction data as constraints. The normalized dynamic production data and the Blasingame production decline typical curve have been fitted with the flow rate solution corresponding to per unit pseudo bottom hole flowing pressure drop, and some of the reservoir and fracture parameters have been explained. Finally, according to the explained parameters and the flow rate solution corresponding to variable pseudo bottom hole flowing pressure calculated based on the Duhamel's principle, the shale gas and condensate oil predicted recoverable reserves results of multi-stage fractured horizontal wells in shale condensate gas reservoirs are obtained.

This method has been used to predict the dynamic productivity of 10 multi-stage fractured horizontal wells in the Duvernay shale condensate gas reservoir. The results show that compared with the production decline analysis method in commercial software HIS Harmony RTA, the dynamic productivity prediction method based on the convolution equation of shale condensate gas reservoirs has better fitting effect and higher accuracy of recoverable reserves prediction. For equivalent gas, the error of production predicted by convolution equation is within 10%. In particular for well 2 and well 10, the errors are only 0.3% and 1.7%, respectively. For shale gas and condensate, the error of production predicted by convolution equation is generally within 10%, too. Especially for well 1, the error of shale gas production is only 2%. For well 8, the error of condensate oil production is only 0.9%. It proves the accuracy of this method.

The convolution equation-based dynamic productivity prediction method of shale gas condensate reservoirs can effectively evaluate the hydraulic fracturing effect. It can accurately and efficiently predict the recoverable reserves of multi-stage fractured horizontal wells in shale condensate gas reservoirs. Compared to empirical methods, it is much more accurate. Compared to previous methods of typical curve analysis, due to the introduction of the deconvolution algorithm in this method, the data processing speed is faster and the fitting effect is better. Compared to artificial intelligence analysis methods, it does not require a large sample of production data. In addition, some reservoir and fracture parameters can be explained to facilitate the analysis of productivity factors. In addition, compared with the numerical simulation method, this method is simpler and more efficient. It enriches the production decline analysis method of shale condensate gas reservoirs and has important practical significance for the production prediction and stimulation optimization of shale condensate gas reservoirs.

Author Contributions: Conceptualization, P.W. and W.L.; methodology, W.L.; software, W.L.; validation, C.Q.; formal analysis, C.Q. and W.L.; investigation, P.W., W.L., C.Q. and C.L.; resources, P.W.; data curation, P.W. and W.H; writing—original draft preparation, C.Q. and C.L.; writing—review and editing, W.L.; visualization, C.Q.; supervision, W.L.; project administration, P.W., W.H. and Y.J.; funding acquisition, W.L. All authors have read and agreed to the published version of the manuscript.

Funding: This research was funded by [CNPC Innovation Found] grant number [2021DQ02-0901] And The APC was funded by [PetroChina Research Institute of Petroleum Exploration and Development].

Data Availability Statement: The data presented in this study are available on request from the corresponding author.

Acknowledgments: The authors acknowledge the financial support from CNPC Innovation Found (Grant No. 2021DQ02-0901).

Conflicts of Interest: The authors declare no conflict of interest.

References

1. Jia, B.; Xian, C.; Tsau, J.-S.; Zuo, X.; Jia, W. Status and Outlook of Oil Field Chemistry-Assisted Analysis during the Energy Transition Period. *Energy Fuels* **2022**, *36*, 12917–12945. [CrossRef]
2. Liu, Y.; Gao, D.; Li, Q.; Wan, Y.; Duan, W.; Zeng, X.; Li, M.; Su, Y.; Fan, Y.; Li, S.; et al. Mechanical frontiers in shale-gas development. *Adv. Mech.* **2019**, *49*, 1–236.
3. Zhou, D. Continuous deepening of world shale gas exploration and development. *Sino-Glob. Energy* **2019**, *24*, 64.
4. Wang, Z.; Wen, Z.; He, Z.; Song, C.; Liu, X.; Chen, R.; Liu, Z.; Bian, H.; Shi, H. Global condensate oil resource potential and exploration fields. *Acta Pet. Sin.* **2021**, *42*, 1556–1565.
5. Lei, Q.; Xu, Y.; Cai, B.; Guan, B.; Wang, X.; Bi, G.; Li, H.; Li, S.; Ding, B.; Fu, H.; et al. Progress and prospects of horizontal well fracturing technology for shale oil and gas reservoirs. *Pet. Explor. Dev.* **2022**, *49*, 191–199. [CrossRef]
6. Arps, J.J. Analysis of decline curves. *Pet. Trans.* **1945**, *160*, 228–247. [CrossRef]
7. Ilk, D.; Rushing, J.A.; Perego, A.D.; Blasingame, T.A. Exponential vs. hyperbolic decline in tight gas sands: Understanding the origin and implications for reserve estimates using Arps' decline curves. In Proceedings of the SPE Annual Technical Conference and Exhibition, Denver, CO, USA, 21–24 September 2008.
8. Valko, P.P. Assigning value to stimulation in the Barnett Shale: A simultaneous analysis of 7000 plus production hystories and well completion records. In Proceedings of the SPE Hydraulic Fracturing Technology Conference, The Woodlands, TX, USA, 19–21 January 2009.
9. Duong, A.N. An unconventional rate decline approach for tight and fracture-dominated gas wells. In Proceedings of the SPE Canadian Unconventional Resources and International Petroleum Conference, Calgary, AB, Canada, 19–21 October 2010.
10. Chen, Y.; Xu, L.; Wang, L. Applications of the generic exponential production decline model on estimating well-controlled recoverable reserves of shale gas fields in the United States. *Pet. Reserv. Eval. Dev.* **2021**, *11*, 469–475.

11. Zhu, H.; Kong, D.; Qian, X. Shale Gas Production Prediction Method Based on Adaptive Threshold Denoising BP Neural Network. *Sci. Technol. Eng.* **2017**, *17*, 128–132.
12. Liu, Y.Y.; Ma, X.H.; Zhang, X.W.; Guo, W.; Kang, L.; Yu, R.; Sun, Y. A deep-learning-based prediction method of the estimated ultimate recovery (EUR) of shale gas wells. *Pet. Sci.* **2021**, *18*, 1450–1464. [CrossRef]
13. Song, L.; Wang, J.; Liu, C. BP-GA algorithm assisted intelligent horizontal well fracturing design. *Fault-Block Oil Gas Field* **2022**, *29*, 417–421, +426.
14. Qiu, K.; Li, J.; Chen, D. Optimized long short-term memory (LSTM) network for performance prediction in unconventional reservoirs. *Energy Rep.* **2022**, *8*, 15436–15445. [CrossRef]
15. Alberti, L.; Angelotti, A.; Antelmi, M.; La Licata, I. Borehole Heat Exchangers in aquifers: Simulation of the grout material impact. *Rend. Online Soc. Geol. Ital.* **2016**, *41*, 268–271. [CrossRef]
16. Wei, M.; Duan, Y.; Dong, M.; Fang, Q. Blasingame decline type curves with material balance pseudo-time modified for multi-fractured horizontal wells in shale gas reservoirs. *J. Nat. Gas Sci. Eng.* **2016**, *31*, 340–350. [CrossRef]
17. Chen, X.; Tang, C.; Du, Z.; Tang, L.; Wei, J.; Ma, X. Numerical simulation on multi-stage fractured horizontal wells in shale gas reservoirs based on the finite volume method. *Nat. Gas Ind. B* **2019**, *6*, 347–356. [CrossRef]
18. Zhang, R.-H.; Chen, M.; Tang, H.-Y.; Xiao, H.-S.; Zhang, D.-L. Production performance simulation of a horizontal well in a shale gas reservoir considering the propagation of hydraulic fractures. *Geoenergy Sci. Eng.* **2023**, *221*, 111272. [CrossRef]
19. Wang, Q.; Ye, Y.; Dong, J.; Wan, J. Unsteady flow model of fractured horizontal well in shale gas reservoir. *Sci. Technol. Eng.* **2020**, *20*, 10225–10234.
20. Jiang, L.; Liu, J.; Liu, T.; Yang, D. Semi-analytical modeling of transient rate behaviour of a horizontal well with multistage fractures in tight formations considering stress-sensitive effect. *J. Nat. Gas Sci. Eng.* **2020**, *82*, 103461. [CrossRef]
21. Clarkson, C.R.; Yuan, B.; Zhang, Z. A new straight-line analysis method for estimating fracture/reservoir properties using dynamic fluid-in-place calculations. *SPE Reserv. Eval. Eng.* **2020**, *23*, 606–626. [CrossRef]
22. Ilk, D.; Valko, P.P.; Blasingame, T.A. A deconvolution method based on cumulative production for continuously measured flowrate and pressure data. In Proceedings of the SPE Eastern Regional Meeting, Lexington, KY, USA, 17–18 October 2007.
23. Von Schroeter, T.; Hollaender, F.; Gringarten, A.C. Deconvolution of well test data as a nonlinear total least squares problem. *SPE J.* **2004**, *9*, 375–390. [CrossRef]
24. Wang, F.; Pan, Z. Deconvolution-based well test model for the fractured horizontal wells in tight gas reservoirs. *Acta Pet. Sin.* **2016**, *37*, 898–902+938.
25. Liu, W.; Liu, Y.; Zhu, W.; Sun, H. Improvement and application of ILK flow-rate deconvolution algorithm based on the second-order B-splines. *Acta Pet. Sin.* **2018**, *39*, 327–334.
26. Brown, M.; Ozkan, E.; Raghavan, R.; Kazemi, H. Practical solutions for pressure-transient responses of fractured horizontal wells in unconventional shale reservoirs. *SPE Reserv. Eval. Eng.* **2011**, *14*, 663–676. [CrossRef]
27. Du, F.; Nojabaei, B. Estimating diffusion coefficients of shale oil, gas, and condensate with nano-confinement effect. *J. Pet. Sci. Eng.* **2020**, *193*, 107362. [CrossRef]
28. Mirzaie, M.; Esfandyari, H.; Tatar, A. Dew point pressure of gas condensates, modeling and a comprehensive review on literature data. *J. Pet. Sci. Eng.* **2021**, *211*, 110072. [CrossRef]
29. Liu, N. Condensate oil conversion processing in condensate gas well test interpretation. *Well Test.* **1994**, *3*, 19–21.
30. Lu, T.; Li, Z.; Lai, F.; Meng, Y.; Ma, W.; Sun, Y.; Wei, M. Blasingame decline analysis for variable rate/variable pressure drop: A multiple fractured horizontal well case in shale gas reservoirs. *J. Pet. Sci. Eng.* **2019**, *178*, 193–204. [CrossRef]
31. Lee, A.L.; Gonzalez, M.H.; Eakin, B.E. The Viscosity of Natural Gases. *JPT* **1966**, *18*, 997–1000. [CrossRef]
32. Yarborough, L.; Hall, K.R. How to Solve EOS for Z-factors. *Oil Gas J.* **1974**, *72*, 86.
33. Whitson, C.H.; Sunjerga, S. PVT in liquid-rich shale reservoirs. In Proceedings of the SPE Annual Technical Conference and Exhibition, San Antonio, TX, USA, 8–10 October 2012.
34. Barzin, Y.; Walker, G.J. Choke management—The driver of liquid yield trends in shale reservoirs. In Proceedings of the Unconventional Resources Technology Conference, Unconventional Resources Technology Conference (URTeC), Houston, TX, USA, 20–22 June 2022; pp. 960–976.

Disclaimer/Publisher's Note: The statements, opinions and data contained in all publications are solely those of the individual author(s) and contributor(s) and not of MDPI and/or the editor(s). MDPI and/or the editor(s) disclaim responsibility for any injury to people or property resulting from any ideas, methods, instructions or products referred to in the content.

Article

Numerical Simulation Study on Temporary Well Shut-In Methods in the Development of Shale Oil Reservoirs

Qitao Zhang [1,2], Wenchao Liu [1,*], Jiaxin Wei [3], Arash Dahi Taleghani [2], Hai Sun [4] and Daobing Wang [5]

1. School of Civil and Resource Engineering, University of Science and Technology Beijing, Beijing 100083, China
2. John and Willie Leone Family Department of Energy and Mineral Engineering, The Pennsylvania State University, State College, PA 16801, USA
3. Geology Institute, No. 2 Oil Production Plant, Changqing Oilfield Company, Qingcheng, Qingyang 745100, China
4. School of Petroleum Engineering, China University of Petroleum (East China), Qingdao 266580, China
5. School of Mechanical Engineering, Beijing Institute of Petrochemical Technology, Beijing 102617, China
* Correspondence: wcliu_2008@126.com

Citation: Zhang, Q.; Liu, W.; Wei, J.; Taleghani, A.D.; Sun, H.; Wang, D. Numerical Simulation Study on Temporary Well Shut-In Methods in the Development of Shale Oil Reservoirs. *Energies* **2022**, *15*, 9161. https://doi.org/10.3390/en15239161

Academic Editor: Reza Rezaee

Received: 18 October 2022
Accepted: 17 November 2022
Published: 2 December 2022

Publisher's Note: MDPI stays neutral with regard to jurisdictional claims in published maps and institutional affiliations.

Copyright: © 2022 by the authors. Licensee MDPI, Basel, Switzerland. This article is an open access article distributed under the terms and conditions of the Creative Commons Attribution (CC BY) license (https://creativecommons.org/licenses/by/4.0/).

Abstract: Field tests indicate that temporary well shut-ins may enhance oil recovery from a shale reservoir; however, there is currently no systematic research to specifically guide such detailed operations in the field, especially for the design of the shut-in scheme and multiple rounds of shut-ins. In this study, the applicability of well shut-in operations for shale oil reservoirs is studied, and a numerical model is built using the finite element method. In order to simulate the production in a shale oil reservoir, two separate modules (i.e., Darcy's law and phase transport) were two-way coupled together. The established model was validated by comparing its results with the analytical Buckley–Leverett equation. In this paper, the geological background and parameters of a shale oil reservoir in Chang-7 Member (Chenghao, China) were used for the analyses. The simulation results show that temporary well shut-in during production can significantly affect well performance. Implementing well shut-in could decrease the initial oil rate while decreasing the oil decline rate, which is conducive to long-term production. After continuous production for 1000 days, the oil rate with 120 days shut-in was 9.85% larger than the case with no shut-in. Besides, an optimal shut-in time has been identified as 60 days under our modeling conditions. In addition, the potential of several rounds of well shut-in operations was also tested in this study; it is recommended that one or two rounds of shut-ins be performed during development. When two rounds of shut-ins are implemented, it is recommended that the second round shut-in be performed after 300 days of production. In summary, this study reveals the feasibility of temporary well shut-in operations in the development of a shale oil reservoir and provides quantitative guidance to optimize these development scenarios.

Keywords: temporary shut-in; well performance; shale oil reservoir; oil–water displacement; optimized shut-in scheme

1. Introduction

Shale oil is considered a critical unconventional source to meet future energy demands [1]. For the development of a shale oil reservoir, multistage fracturing in the horizontal wells (MFHW) has been broadly utilized to produce the trapped hydrocarbon resources [2]. Through fracturing treatment, fractures with high conductivity are created in the formation rock, which is crucial for the development of unconventional reservoirs [3,4]. Field evidence shows that, after hydraulic fracturing, temporary well shut-ins can improve well performance and enhance ultimate recovery in low-permeability unconventional reservoirs. Due to recent fluctuations in the oil price due to the COVID pandemic, a considerable number of these wells around the world were temporarily shut-in over the past few years [5,6], and this validates our conclusion. Therefore, the well shut-in has become a hot topic in the development of shale and tight oil reservoirs, especially in China. In the

Changqing oil field in the northwest of China, well shut-in has been utilized extensively during the development of its shale oil reservoirs. Since 2017, there have been hundreds of new horizontal wells fractured within the Chang-7 Member in the upper Triassic Yanchang formation. All those production wells underwent a shut-in before flowback and formal production were resumed. Even though temporary well shut-in has been widely used in this region, the mechanism behind this process is still unclear to the operators.

Currently, imbibition, including dynamic and spontaneous imbibition, is considered one of the major causes and can explain the mechanism behind the well shut-in [7]. Therefore, understanding how to take an advantage of the imbibition effect to enhance well performance and achieve better long-term productivity could be substantial to the development of shale oil reservoirs [8]. In terms of the studies concerning imbibition, nuclear magnetic resonance (NMR) imaging techniques [9] are widely used in the experiments. Generally speaking, in those experiments, researchers saturated core samples with oil and then immersed the core samples into water to see the imbibition. During the experiment process, NMR testing is used to image the oil–water displacement [10–15]. Karimi et al. [16] leveraged centrifugation and NMR techniques at the same time to study the oil–water displacement pattern, focusing on the effect of capillary force. They found that, due to the imbibition effect, water enters into the tiny pores much more easily than oil in a water-wet environment. Tu and Sheng [17] studied the effect of pressure on imbibition in a shale oil reservoir, utilizing experimental and numerical methods. Based on the NMR imaging results, Cheng et al. [18] claimed that submicropores contribute more to the ultimate recovery of spontaneous and dynamic imbibition, despite the nanopores having stronger capillary forces. In addition to experimental studies on imbibition, there has been some progress made in the theoretical and numerical study of this phenomenon [19]. Schmid et al. [20,21] presented a semianalytical solution to describe spontaneous imbibition behavior. In their study, they also provided a numerical solution for this problem. Their simulation efforts considered two wettability cases and three viscosity ratio cases [21]. Besides, Khan et al. [22] simulated the oil–water displacement based on a fully implicit black oil simulator, considering strongly water-wet, weakly-wet, and mixed wet cases, respectively.

In terms of the effect of shut-in on well performance in unconventional reservoirs, studies are still relatively limited. Wang [23] studied the characteristic of reservoir energy balance and energy storage after a well shut-in. It can be concluded that a temporary well shut-in after fracturing could accelerate the diffusion and pressure propagation and slow down the reservoir energy depletion. Zhang et al. [24] evaluated the potential for oil recovery enhancement when considering imbibition and the corresponding time delay in a shale oil reservoir for a shut-in. In their study, the effects of the pore geometry and clay content were focused on. Liu [25] used NMR to study the effect of well shut-in on the imbibition rate and recovery of tuffite, shale, tight sandstone, and clastic volcanic rock. Many studies looked at the effect of well shut-in on aqueous phase trapping (APT) caused by formation damage due to drilling, completion, etc. Based on the experimental results, they found that APT could be "auto-removed" after temporary well shut-in [26–28] and the optimal postfrac shut-in time could be determined [29]. Wang et al. [30] presented a pressure drop model for postfracturing shut-in simulation, considering multiple effects, including fluid imbibition and oil replacements. Their research revealed that the pressure drop during shut-in can be divided into eight sequential stages, and their results can be used for the interpretation of fracture and reservoir parameters. Eltahan et al. [31] studied the impact of well shut-in after hydraulic-fracture treatments on the productivity and recovery of tight oil reservoirs. In their study, the embedded discrete fracture model (EDFM) [32] was used for fracture representation, and their results indicated that the well shut-in could improve the recovery by as much as 5%. Jia et al. [33] investigated the shut-in effect on production performance, considering stress sensitivity. They found that reservoir permeability and capillary force have the most obvious effects on shut-in performance.

Regarding the gaps in the current research on well shut-in methods, although a great amount of research has been presented, there still exist some problems to be solved in this area. First, most of the experiments mentioned before were carried out in a core scale, for which the conclusion may not be convincing when repeated at the field scale. Second, it is known that well shut-in operation can be effective for the long-term production of shale oil reservoirs. However, several key points still need to be addressed for optimizing future production scenarios, including determining the optimum shut-in interval, what is the difference between optimum shut-in time in different reservoirs, and the potential for multiple rounds of shut-ins. Finally, for The Chang-7 Member, the shale oil reservoir in the Changqing oil reservoir, extensive field tests of well shut-in have been carried out while there lacks specific guidance for field production, especially for the potential of multiple rounds of well shut-ins. This issue demands we present a specific investigation [34]. In summary, the studies mentioned before still cannot fully explain the mechanism behind the well shut-in and provides guidance on the shut-in operation in the field. Further studies are still necessary in this area.

The objective of this research is to study the feasibility of the temporary shut-in method in the development of shale oil reservoirs. Besides, we also want to design reasonable production schemes and explore the potential for multiple rounds of well shut-ins. This paper is organized as follows: First, the geological background of the research area will be described. Then, a coupled mathematical model for the simulation of a well shut-in and its numerical implementation is explained in Section 3. The numerical model is verified by comparing the results with the Buckley–Leverett equation in Section 4.1. After that, the detailed results and discussion section are presented in Section 5, including oil–water displacement during shut-in, determining the optimal well shut-in interval, and the potential for multiple rounds of well shut-ins. This study provides new insight into the optimal well shut-in time and the potential benefits of multiple rounds of well shut-ins. The results and conclusions from this paper can be expected to provide quantitative guidance to optimize the operation scenario in the development of shale oil reservoirs. Some suggestions derived from this study can be directly applied in the Chenghao area and for the Changqing oil field.

2. Geological Background

This study focuses on the shale oil developments in the Chenghao area, which is located in the Ordos Basin, northwest of China. In this area, production wells are mainly drilled to exploit shale oil in the Chang-7 Member in the upper Triassic Yanchang formation [35]. The Chang-7 Member consists primarily of profundal laminated shale, occasional tuffaceous mudstone, and muddy siltstone interbeds. It is one of the most organic-rich intervals in the central and southern parts of the Ordos Basin [36]. The Chang-7 Member in the Ordos Basin has porosities ranging from 6% to 11% and an average porosity of 8.8%. The permeability of the reservoir matrix ranges from 0.08 to 0.3 mD, and the average permeability value is 0.13 mD [35]. According to the high-pressure mercury intrusion tests carried out for this field, the pore throat radius mainly distributes from 0.02 to 1 μm. Besides, the organic matter carbon content in the matrix ranges from 0.45% to 35.85%, with an average of 9.02%. Hydrocarbon generation potential ranges from 0.19 to 116.17 mg/g, with an average value of 34.55 mg/g [37]. Since 2017, there are now hundreds of new horizontal wells fractured in the Chang-7 Member in the upper Triassic Yanchang formation. It is worth mentioning that all the production wells underwent shut-ins before flowback and production. Those newly developed wells are (on average) drilled with 1709 m of horizontal length and fracked with an injection of 28612 m^3 of slickwater and 3268 m^3 of proppants placed into 23 clusters with a total of 111 induced fractures [38–40]. Geological information of the Chang-7 Member can be found in the reference [41].

3. Mathematical Modeling

In order to describe the two-phase fluid flow in a shale oil reservoir, some idealizations and assumptions are made as follows: (1) the model is simplified into two-phase isothermal water–oil flows in a shale oil reservoir. (2) It is assumed that the permeability of the matrix is isotropic. (3) The formation is fully stimulated (penetrated) by the hydraulic fractures with finite conductivity. The fracture width is assumed to be uniform. (4) The fluid flow is presumed to be into the horizontal plane, and gravity drainage is ignored; the fluid flow in the horizontal wellbore is not taken into account. (5) There is no mass transfer/exchange happening between the water phase and the oil phase.

3.1. Pressure Formulation of Oil–Water Mixture

In this study, two independent numerical modules were utilized to calculate the pressure distribution and saturation distribution in a shale oil reservoir, respectively. For the phase transport in the shale oil reservoir, a finite element method (FEM) based model, called "Phase Transport in Porous Media" (PTPM), was utilized to calculate the saturation distribution. For the pressure distribution in the shale oil reservoir, Darcy's law was adopted to solve the pressure distribution of the water–oil mixture. Then, the calculated pressure and saturation field are coupled together to constitute a complete numerical model for further analysis. Compared to the classical two-phase Darcy's model, this method has some advantages, including less computational demand, lower manipulation difficulty, and good convergent performance [42].

In order to calculate the pressure distribution of the oil–water mixture, the properties of the mixture need to be generated as follows [43].

$$\bar{\rho} = \sum_\alpha \rho_\alpha S_\alpha, \ \alpha = o, w, \tag{1}$$

$$\bar{\mu} = \frac{\bar{\rho}}{\sum_\alpha \frac{k_{r\alpha} \rho_\alpha}{\mu_\alpha}}, \ \alpha = o, w, \tag{2}$$

when $\alpha = o$, this represents the oil phase, and when $\alpha = w$, this indicates the water phase. $\bar{\rho}$ denotes the density of the oil–water mixture. ρ_α represents the density of the corresponding phase. S_α represents the saturation of the corresponding phase, and $\bar{\mu}$ denotes the viscosity of the oil–water mixture. $k_{r\alpha}$ represents the relative permeability for the oil and water phases. μ_α represents the viscosity of the corresponding phase.

Darcy's law for the oil–water mixture can be expressed as follows.

$$\bar{u} = -\frac{k}{\bar{\mu}} \nabla \bar{p}, \tag{3}$$

where $\bar{u}$ is Darcy's velocity vector for the oil–water mixture. k is the permeability of porous media. $\bar{p}$ is the pressure of the oil–water mixture.

The continuity equation for the fluid flow can be expressed as

$$\nabla \cdot (\overline{\rho u}) + \frac{\partial(\phi \bar{\rho})}{\partial t} = 0, \tag{4}$$

where ϕ is the reservoir porosity; t is the time.

Through Equation (4), the pressure of the oil–water mixture can be calculated. However, what we really need is the pressure of the oil and water phases rather than the average pressure of the oil–water mixture. Therefore, additional steps should be taken. The pressure of the water phase p_w and the oil phase p_o can be calculated through the equation of weighted fluid pressure [43] and the capillary force equation [44].

$$\bar{p} = S_o p_o + S_w p_w, \tag{5}$$

$$p_o - p_w = p_c = p_{entry} \left(\frac{S_w - S_{wi}}{1 - S_{wi} - S_{or}} \right)^l, \tag{6}$$

where p_c is the capillary pressure. p_{entry} is the entry capillary pressure. S_{wi} is the irreducible water saturation. S_{or} is the residual oil saturation. l is the capillary pressure exponent. Since the saturation information would be transferred from the calculation of PTPM module, the only unknown values in Equations (5) and (6) are p_w and p_o, and thus, the pressure of oil and water phases can be solved.

3.2. Phase Transport in Porous Media

Based on the oil pressure p_w and water pressure p_o calculated in the previous section, the flow velocity for the oil phase u_o and the water phase u_w in the PTPM module can be calculated as follows.

$$u_\alpha = -\frac{kk_{r\alpha}}{\mu_\alpha}\nabla p_\alpha, \ \alpha = o, w, \tag{7}$$

Then, the continuity equation for each phase can be expressed as

$$\frac{\partial}{\partial t}(\phi\rho_\alpha S_\alpha) + \nabla\cdot(\rho_\alpha u_\alpha) = \rho_\alpha q_\alpha, \ \alpha = o, w, \tag{8}$$

where q_α is the sink or source term in the reservoir.

Besides, an additional constraint equation needs to be added to the model, i.e.,

$$S_o + S_w = 1, \tag{9}$$

Here, the only unknown variable in Equation (8) is water saturation, and thus, the saturation profile can be obtained. By coupling Darcy's flow module and PTPM module, the pressure and saturation distributions of oil and water phases can be acquired.

3.3. Fluid Flow in Hydraulic Fractures

For the development of a shale oil reservoir with multistage hydraulic fracturing, fractures play a crucial role in the production. In terms of the simulation of fluid flow inside the hydraulic fractures, here a discrete fracture model [45] is adopted.

The kinematic equation in the discrete fracture model can be written as follows in order to replace Equation (3).

$$\overline{u}_f = -\frac{k_f}{\mu}\nabla_T \overline{p}_f, \tag{10}$$

where $\overline{u}_f$ is the fluid velocity for the oil–water mixture inside the fractures. $\nabla_T \overline{p}_f$ is the pressure gradient tangent to the fracture surface [46]. k_f is the permeability of the fracture.

The continuity equation for fluid flow along the hydraulic fractures is used to replace Equation (4), which is

$$\nabla_T\cdot(d_f\overline{\rho u}_f) + d_f\frac{\partial}{\partial t}(\phi_f\overline{\rho}) = 0, \tag{11}$$

where d_f is the fracture width. ϕ_f is the porosity of the propped fracture.

3.4. Auxiliary Boundary Conditions

The outer boundaries of the model, $\partial\Omega_1$, are assumed to form a closed Neumann boundary condition as follows [47]

$$-n\cdot\rho\overline{u}|\partial\Omega_1 = 0, \tag{12}$$

where n is the normal vector to the outer boundary $\partial\Omega_1$.

The simulation task can be divided into several steps, including (1) water injection, (2) well shut-in, and (3) production. For each step, different well treatments are implemented. In this study, we are assuming the horizontal well is infinitely conductive, while the fractures have a finite fracture conductivity.

(1) For the water injection step, the boundary condition $\partial\Omega_2$, i.e., the perforation holes, for a horizontal well is defined as an inhomogeneous Neumann condition with a flow rate.

$$-n\cdot\rho\overline{u}|\partial\Omega_2 = \overline{\rho}u_0, \tag{13}$$

$$S_w|\partial\Omega_2 = 1, \quad (14)$$

where u_0 is an input parameter to represent the injected water influx. Here we distribute the total injection rate average to each fracture.

(2) For the shut-in step, the boundary condition $\partial\Omega_2$ for the horizontal well is defined as the homogeneous Neumann condition or closed boundary.

$$-n \cdot \rho\bar{u}|\partial\Omega_2 = 0, \quad (15)$$

(3) For the production step, the boundary condition $\partial\Omega_2$ again for a horizontal well is defined as Dirichlet boundary condition with a bottomhole pressure.

$$\bar{p}|\partial\Omega_2 = p_w, \quad (16)$$

where p_w is an input parameter to represent the bottomhole pressure at the horizontal well. The initial conditions for the reservoir would be

$$\bar{p}|_{t=0} = p_i, \quad (17)$$

$$S_w|_{t=0} = S_i, \quad (18)$$

where p_i is the initial reservoir pressure; S_i is the initial reservoir water saturation.

In order to achieve numerical stability and avoid any numerical diffusion/oscillation, detailed numerical settings need to be designed in advance. In this study, the coupling of fluid flow and phase transfer in the reservoir should be especially highlighted. In terms of the coupling of PTPM for the saturation field and Darcy's law for the pressure field, the Segregated algorithm is recommended in this model rather than using Fully Coupled algorithms, considering the numerical performance [48]. The Fully Coupled approach forms a single large system of equations that solve for all of the unknowns (the fields) and includes all of the couplings between the unknowns (the multiphysics effects) at once, within a single iteration. On the other hand, the Segregated approach will not solve all of the unknowns at one time. Instead, it subdivides the problem into two or more segregated steps. Each step will usually represent a single physics, but sometimes even a single physics can be subdivided into steps, and sometimes one step can contain multiple physics. These individual segregated steps are smaller than the full system of equations that are formed with the Fully Coupled approach. The segregated steps are solved sequentially within a single iteration, and thus, less memory is required. Since the Fully Coupled approach includes all coupling terms between the unknowns, it often converges more robustly and in fewer iterations as compared to the Segregated approach. However, each iteration will require relatively more memory and time to solve, so the Segregated approach can be faster overall. In the Segregated algorithm, the constraint for the saturation range, i.e., 0–1, can be easily prescribed mathematically using its value-limit function. In addition, an Anderson acceleration method is also used to improve numerical stability [49].

The calculation process is divided into three steps: the water injection process, well shut-in process, and the production process, respectively. Therefore, it is necessary to consider the delivery of the simulation results. When the calculation of the water injection study is terminated, the simulation results in the last time step should be transported into the well shut-in study as the initial value by using the "study reference" interface in COMSOL Multiphysics. The only change between these two steps is the boundary condition. Then, the same process needs to be set for the other studies. By doing so, different physical processes with different boundary conditions and numerical settings could be connected together, as shown in Figure 1.

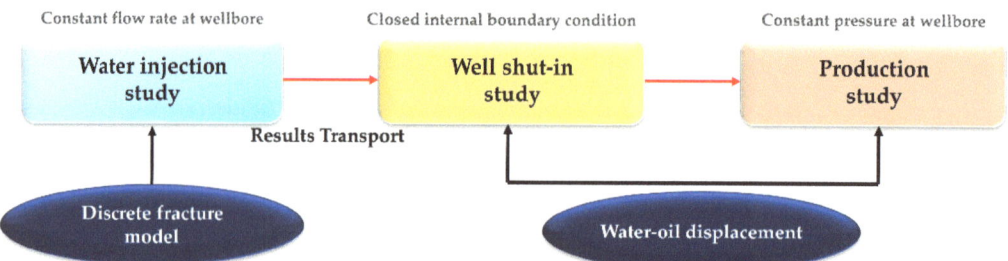

Figure 1. Flow-process diagram for the transport of the simulation results in the model.

4. Numerical Implementation

4.1. Model Validation

In order to validate the numerical model, a simulation case is built to compare its calculation results with the analytical solutions. For two-phase flow, the most widely used analytical model is the Buckley–Leverett model [50]. This model estimates the advance of a fluid displacement front in an immiscible displacement process. According to the Buckley–Leverett equation, the distance x traveled by a particular S1 contour can be expressed as the following.

$$x - x_0 = \frac{\left(\frac{df_1}{dS_1}\right)}{A \cdot \phi} \cdot \int_0^t q(t)dt, \tag{19}$$

where x_0 is the radius of the wellbore. f_1 is the fractional flow of phase 1. A is the cross-section in the flow direction. q is the injection rate. ϕ is the porosity of the reservoir matrix. t represents time.

In this validation section, we assume that the relative permeability of phase 1 is expressed as $k_{r1} = S_1^2$, and the relative permeability of phase 2 is expressed as $k_{r2} = (1 - S_1)^2$. Then, the fraction flow of phase 1 can be derived as follows.

$$f_1 = \frac{1}{1 + \frac{S_1^2 \mu_1}{(1-S_1)^2 \mu_2}}, \tag{20}$$

Then, the derivative of f_1 can be expressed as follows.

$$\frac{df_1}{dS_1} = \frac{\frac{2S_1 \mu_1}{(1-S_1)^2 \mu_2} + \frac{2S_1^2 \mu_1}{(1-S_1)^3 \mu_2}}{\left(1 + \frac{S_1^2 \mu_1}{(1-S_1)^3 \mu_2}\right)^2}, \tag{21}$$

Therefore, the Buckley–Leverett equation utilized in this section can be derived as follows.

$$x - x_0 = \frac{\left(\frac{2S_1 \mu_1}{(1-S_1)^2 \mu_2} + \frac{2S_1^2 \mu_1}{(1-S_1)^3 \mu_2}\right) \cdot q \cdot t}{\left(1 + \frac{S_1^2 \mu_1}{(1-S_1)^3 \mu_2}\right)^2 \cdot A \cdot \phi}, \tag{22}$$

The parameters for the numerical simulation and analytical calculation are as shown in Table 1.

Table 1. Parameters for the validation of the numerical model [51].

Parameter	Value	Unit
Injection rate of phase 1	1×10^{-6}	m/s
Area of cross section	1	m^2
Viscosity of phase 1	1	mPa·s
Viscosity of phase 2	1	mPa·s
Porosity of reservoir matrix	15	%
Total calculation time	300	days
Permeability of reservoir matrix	100	10^{-3} μm^2
Initial saturation of phase 2 in reservoir	100	%

The numerical model is validated by comparing its results with the analytical Buckley–Leverett equation. Figure 2 shows the comparison of the flood front curve between the numerical model and the analytical model. It can be seen from Figure 2 that the numerical model runs strictly according to the setting regimes. In general, the numerical model has relatively high accuracy.

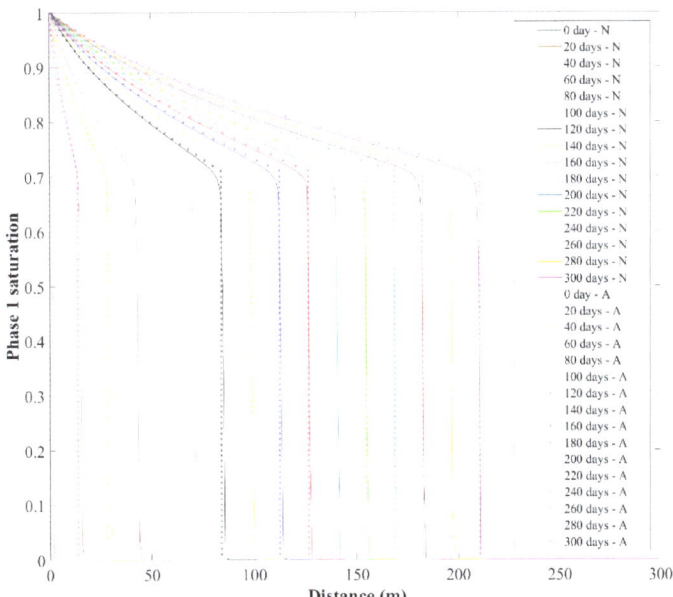

Figure 2. Comparison of flood front curve between the numerical model and analytical model over 300 days: N represents numerical calculation results; A represents analytical calculation results.

4.2. Description of the Simulation Model

In the development of a shale oil reservoir, multistage hydraulic fracturing technology is widely adopted. After the hydraulic fracturing, the reservoir matrix area can be mainly divided into two separate flow regions, which are the stimulated reservoir area (SRV) and the unstimulated reservoir matrix area, as shown in Figure 3. Based on the mathematical model presented in Section 3, a numerical model for the shale oil reservoir was established for the simulations.

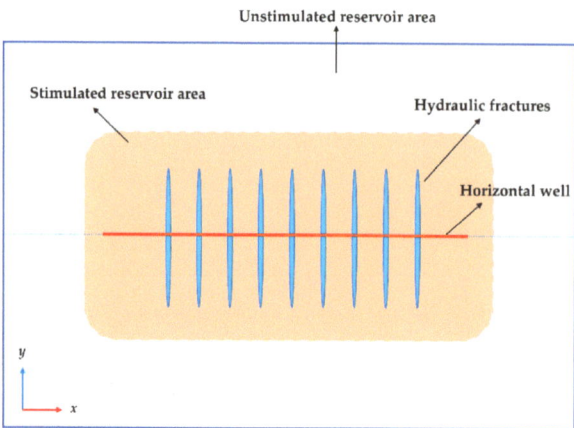

Figure 3. The diagram of shale oil reservoir with a multistage fracturing horizontal well.

Due to the symmetry of the physical model, the model geometry can be simplified to reduce the computational quantity by only calculating half of the reservoir area, as shown in Figure 4a. The horizontal well is presumed to be located at the center of the reservoir, which is the bottom of the geometry model after the simplification and is represented using blue lines, as shown in Figure 4a. The main hydraulic fractures are distributed alongside the horizontal well. Unstructured grids were generated for model discretization, as shown in Figure 4b. The mesh grids near the wellbore and hydraulic fractures are refined significantly, as shown in the figure.

The parameters for the simulation of the multistage fracturing horizontal well in a shale oil reservoir are derived from a typical production well (H-1 in The Chang-7 Member, Chenghao area, northwest of China). The detailed parameter settings are as listed in Table 2.

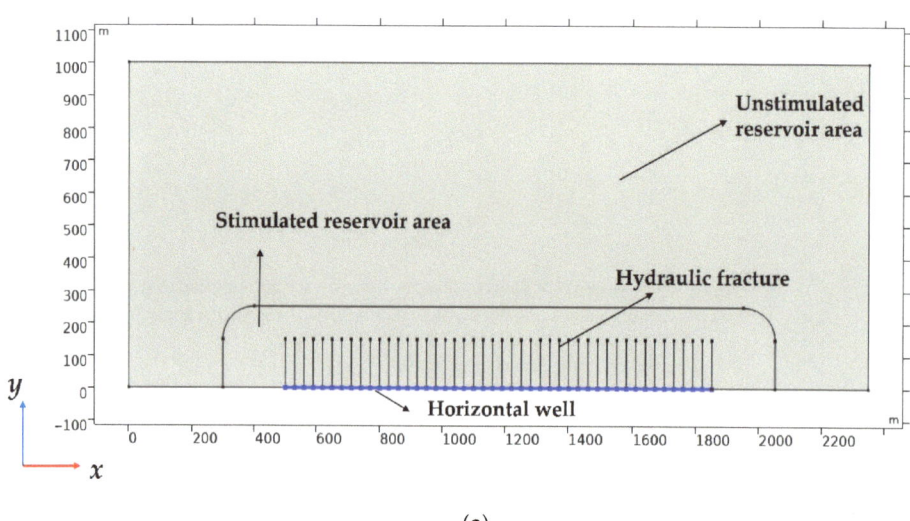

(a)

Figure 4. *Cont.*

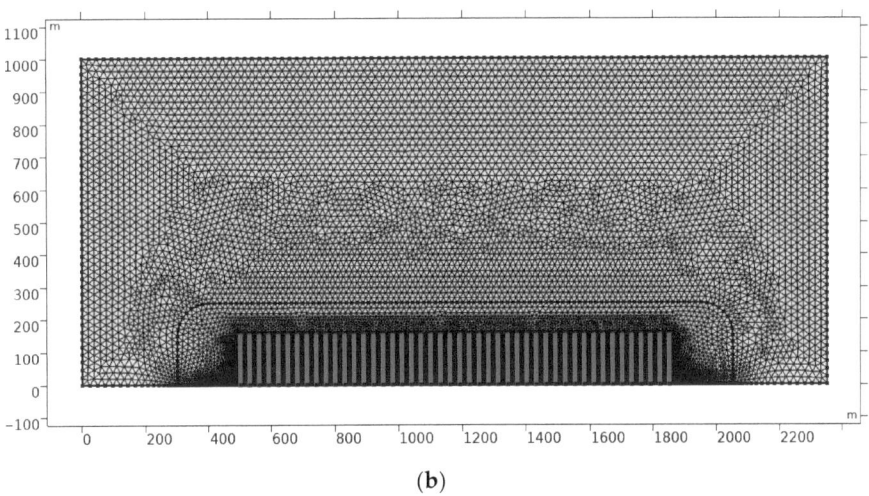

(b)

Figure 4. The illustration of the geometry and mesh generation for the simulation model. (**a**) The diagram of the half-geometry for the simulation model (2-D plan view). (**b**)The diagram of the half-geometry for the simulation model (2-D plan view).

Table 2. Parameters for the simulation model of a multistage fracturing horizontal well in a shale oil reservoir.

Parameter	Value	Unit
Reservoir area after symmetrical processing	2350 × 1000	m^2
Main hydraulic fracture width	0.5	cm
Permeability of main hydraulic fractures	1800	10^{-3} μm^2
Permeability of unstimulated reservoir area	0.2	10^{-3} μm^2
Permeability of stimulated reservoir area	50	10^{-3} μm^2
Initial porosity of the unstimulated reservoir area	7.62	%
Initial porosity of the stimulated reservoir area	10.4	%
Initial porosity of the main hydraulic fractures	15.3	%
Initial reservoir pressure	2.5×10^7	Pa
Capillary force at $S_w = 0.5$	1.87×10^6	Pa
Production pressure of the horizontal well	1.5×10^7	Pa
Half-length of the main hydraulic fractures	150	m
Initial oil density	850	kg/m^3
Oil viscosity	1	mPa·s
Water density	1000	kg/m^3
Water viscosity	0.2	mPa·s
Comprehensive compressibility coefficient for matrix area	5×10^{-9}	1/Pa
Comprehensive compressibility coefficient for fractures	1.0×10^{-8}	1/Pa
Oil saturation range	20–80	%
Calculation time of well shut-in	60	days
Calculation time of production	1000	days

The model is discretized using finite element methods. The governing equations for the reservoir domain and the fracture domain are formulated separately and were coupled at COMSOL's interface for the subsurface flow module and the discrete fracture model. Detailed numerical setting details can be found in Table 3. By implementing the presented numerical settings and mesh generation scenarios, the numerical simulations of well shut-in operation in a shale oil reservoir are conducted. The convergence plot is shown in Figure 5. One may see from the figure that the model solution converges fast and smoothly for the given parameters.

Table 3. Parameters for the numerical setting.

Parameter	Value	Unit
Time-dependent solver	MUMPS [52]	#
Time stepping method	Backward Differentiation Formulas (BDF)	#
Maximum BDF order	1	#
Minimum BDF order	5	#
Event tolerance	0.01	#
Consistence initialization	Backward Euler	#
Fraction of initial step for Backward Euler	0.01	#
Initial time step	0.001	day
Maximum time step	10	day

The symbol # means null.

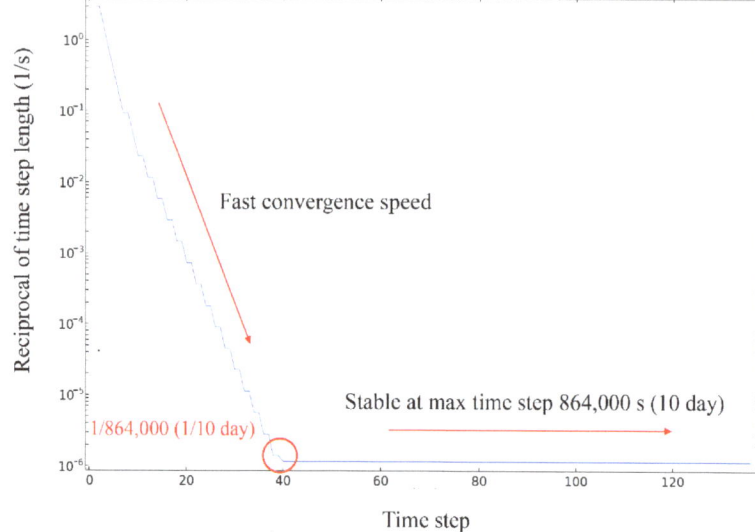

Figure 5. Convergence plot of the numerical model.

5. Results and Discussion

5.1. Analysis of Oil–Water Displacement during the Well Shut-In Process

After the water injection, a well shut-in interval of 60 days was considered in the numerical model. The oil saturation distribution at the beginning time of the shut-in is shown in Figure 6a. The distribution of oil saturation after a continuous shut-in for 60 days is shown in Figure 6b.

In Figure 6a, we see that there is a clear boundary between the reservoir oil and the front edge of the injected water. By contrast, it can be seen in Figure 6b that the boundary between the oil in the reservoir and the injected water becomes less distinctive or even vanishes after shut-in. In this process, the injected water moves deep into the reservoir matrix due to capillary forces and creates higher water pressure near the (horizontal) well.

A magnified comparison between Figure 6a,b is presented in Figure 7 to reveal the effect of the shut-in. Figure 7 indicates the oil saturation distribution near the wellbore and fractures for the 1st day and 60th day of the shut-in. According to the comparison, we can see that an oil–water displacement is clear. After a continuous shut-in for 60 days, the isosaturation line, $S_o = 0.75$, moves from $y = 203$ m to $y = 251$ m, which means the injected water is imbibed into the reservoir matrix. At the same time, the isosaturation line, $S_o = 0.57$, moves back slightly, from 181 m to 169 m, which means some oil in the reservoir is displaced to areas near the wellbore and fractures. These characteristic movements of the isosaturation lines indicate oil–water displacement during the shut-in due to a dynamic imbibition mechanism.

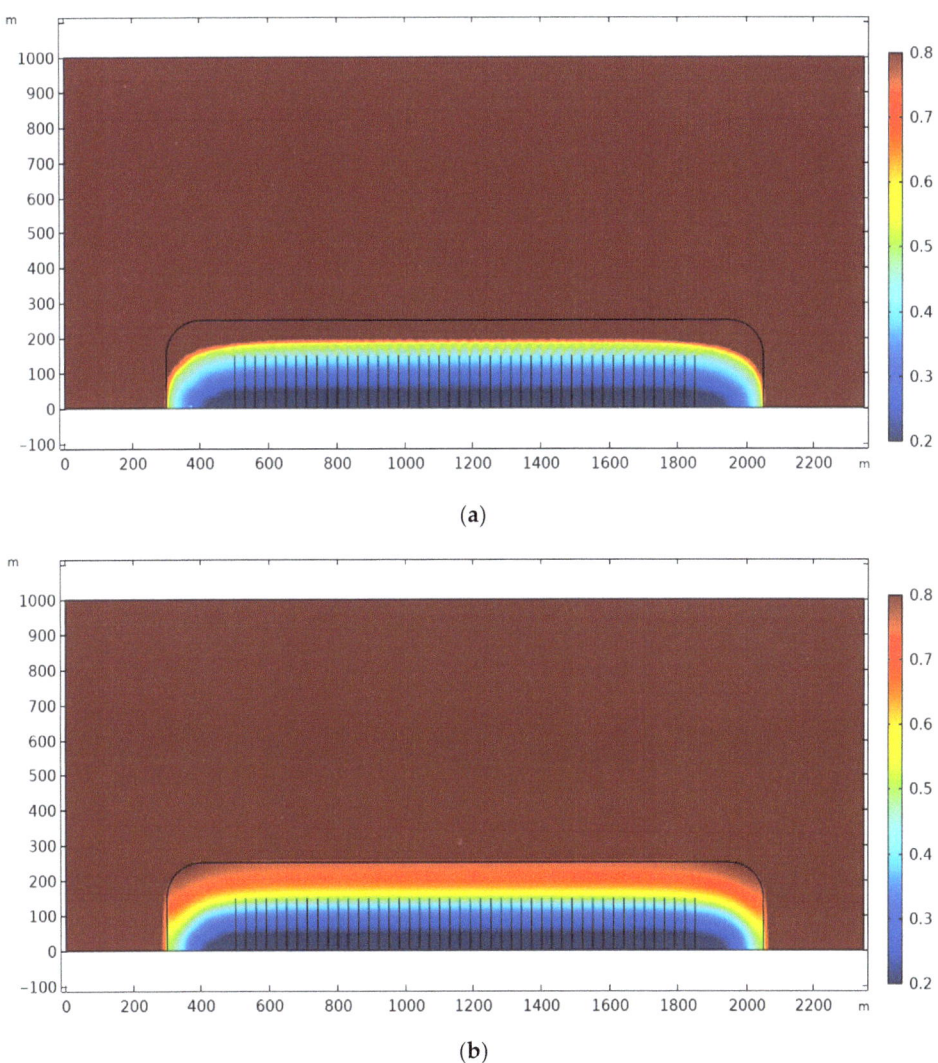

Figure 6. The comparison of oil saturation distribution before and after well shut-in operation. (**a**) Oil saturation distribution on the first day of the well shut-in operation. (**b**) Oil saturation distribution after 60th days of well shut-in operation.

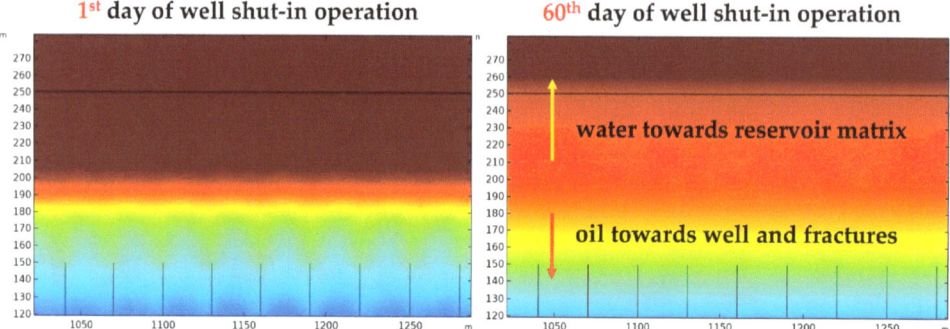

Figure 7. Magnified diagrams to compare oil saturation distribution. (**Left**): oil saturation at the beginning of shut-in; (**Right**): oil saturation after 60 days of shut-in.

5.2. Effect of Shut-In Time on Well Performance

In the last section, we compared the distribution pattern of oil saturation during shut-ins. In this section, we are going to study the effect of the duration of shut-in on well performance. Five simulation cases with different shut-in intervals are implemented for comparison. The shut-in time in those cases varies from no shut-in to 120 days.

The daily water rate for the first 1000 days after shut-in is shown in Figure 8. It can be seen that as the shut-in time increases, the decline rate of the daily water rate decreases. At the same time, as the shut-in time increases, the change in the decline rate becomes insignificant, and it is difficult to distinguish the difference between the water rate after 90 days of shut-in versus 120 days of shut-in.

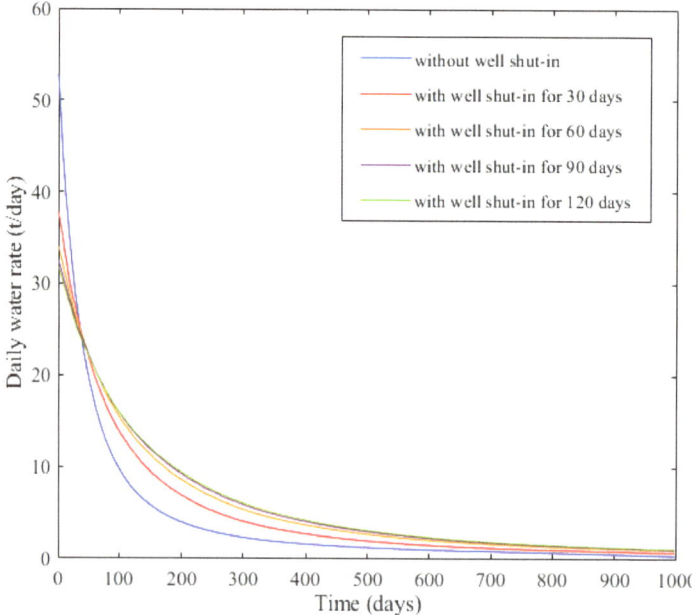

Figure 8. Water production rate for the first 1000 days after different shut-in times.

The daily oil production rate for the first 1000 days after shut-in (for different shut-in times) is plotted in Figure 9. It can be seen that, in the case of no shut-in, the oil production rate not only rises very fast at the beginning of the production process but also decreases severely after reaching the peak. In this simulation scenario, the reservoir energy is depleting quickly, which is detrimental to long-term oil production. In contrast, the oil rate curves with well shut-in vary with time at a relatively lower rate. The longer the shut-in time, the lower the rate for the curve. Besides, as the shut-in time increases, the decline rate of the oil production rate decreases. After continuous production for 1000 days, the oil production rate with 120 days shut-in was 9.85 % greater than that with no shut-in operation, which is considerable and should not be neglected when considering the long-term production of shale oil reservoirs. From Figure 9, We can also conclude that the temporary well shut-in method might lead to productivity loss at the early stage of production. However, the implementation of wells shut-in will decrease the production decline rate (or slow down the energy depletion) and increase the oil production rate after long-term production. As a result, we also need to consider the economic efficiency of the well shut-in method, given the estimated well production lifetime.

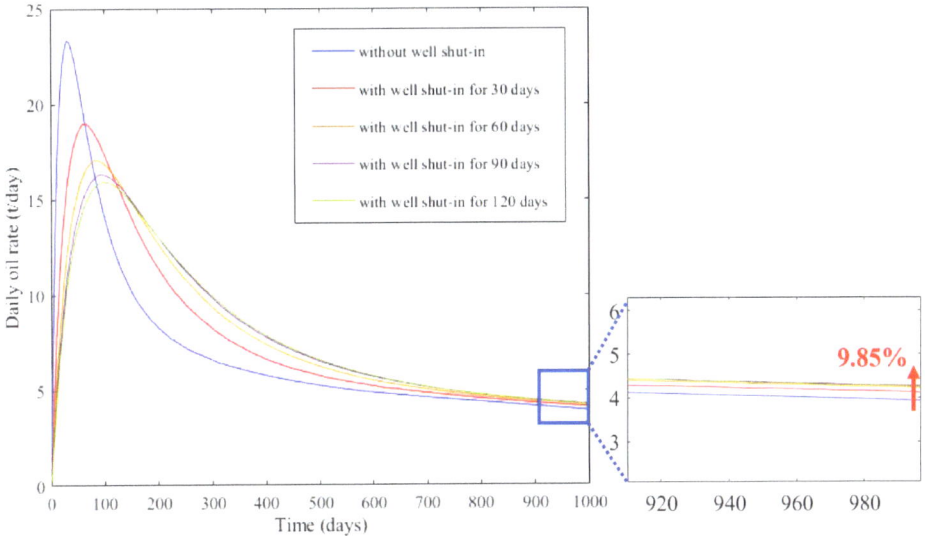

Figure 9. Oil production rate for the first 1000 days after different shut-in times.

Besides, Table 4 summarizes the information on oil production peak for different scenarios described in Figure 9. From the table, we can see that the peak of oil production and its timing for different scenarios vary with the shut-in time. As shut-in time increases from 0–30 days, the value of the oil rate peak decreases dramatically from 23.33 t/day to 18.96 t/day. Then, this change gradually sees a lower rate. As the shut-in time increases from 30 days to 60 days, the well productivity peak decreases from 18.96 t/day to 17.09 t/day. Then, as this time changes from 60 days to 120 days, the peak value drops slightly from 17.09 t/day to 15.94 t/day.

We can also look at the cumulative oil production rate for this period to decipher the importance of shut-in time as shown in Figure 10. As shut-in time increases, the cumulative oil production in the first 1000 days increases dramatically. Then, beyond 60 days shut-in time, cumulative oil production only sees a miniscule improvement. According to Figure 10, a positive relationship between the cumulative oil rate and well shut-in time can be identified up to a specific shut-in time.

Table 4. Peak production rate and their timings according to five simulation scenarios.

Shut-In Time (Day)	Oil Rate Peak (t/Day)	Timing (Day)
0	23.33	32
30	18.96	63
60	17.09	84
90	16.32	93
120	15.94	97

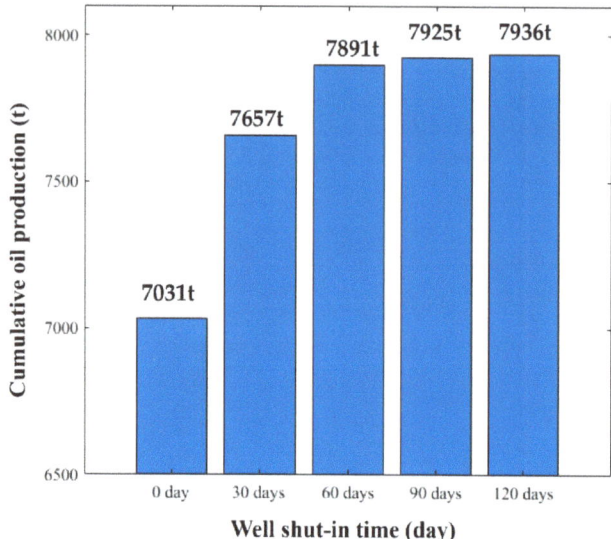

Figure 10. Comparing cumulative oil production rates after 1000 days for different shut-in times.

However, it is worth noting that the positive correlation between cumulative oil production and shut-in time does not mean that a longer shut-in time leads to better development efficiency. This is because different shut-in scenarios imply different operation times. Basically, if there is no shut-in, the total operation time is 1000 days. However, if the well shut-in time is 30 days, the total time would be 1030 days. When we calculate the average daily oil rate, we need to consider the time cost and calculate cumulative oil production by dividing it by the total days of operation. This approach may draw different conclusions from the analysis. The operation time for different simulation scenarios is shown in Table 5. The average daily oil rate according to different well shut-in times is shown in Figure 11.

Table 5. Average production rates for different simulation cases.

Well Shut-In Time (Days)	0	30	60	90	120
Cumulative oil production (t)	7031	7657	7891	7925	7936
Total operation time (days)	1000	1030	1060	1090	1120
Average daily oil rate (t/day)	7.031	7.434	7.445	7.271	7.086

It could be seen from Figure 11 that there exists an optimal well shut-in time, which is 60 days for the given parameters. In other words, for this specific production well, a well shut-in operation of 60 days could achieve the best development efficiency for the exploitation of the shale oil reservoir for the first 1000 days of production. To sum up, well shut-in operation could generate better development efficiency for the production well. A reasonable shut-in operation might be significantly beneficial to the long-term production in shale oil reservoirs.

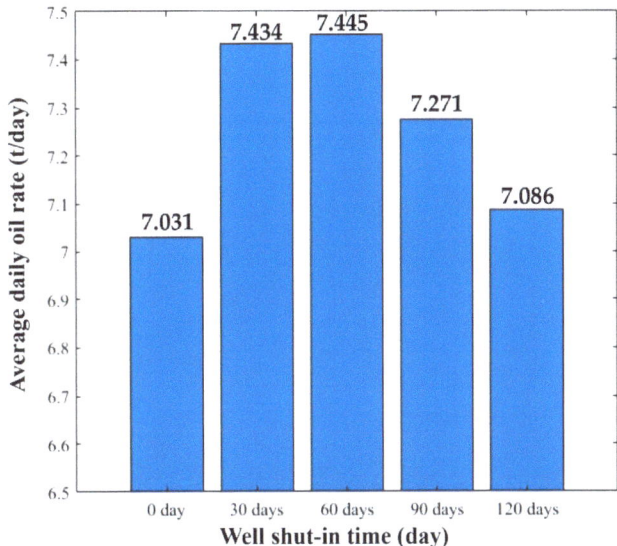

Figure 11. The comparison of average daily oil rate between different well shut-in time cases.

5.3. Potential for Multiple Rounds of Shut-In

According to the field tests and calculation results from Section 5.2, well shut-in has proved to be an effective method for enhancing well performance in shale oil reservoirs. Currently, most of the field tests in China are only carried out with a single round of shut-in to enhance well performance. Multiple rounds of shut-ins have never been tested yet. As a result, in this section, we are going to explore the potential of multiple rounds of shut-ins in the development of shale oil reservoirs. Simulation scenarios with one, two, and three rounds of well shut-in operations are shown in Table 6. Other simulation parameters are shown in Table 2. For these three simulation scenarios, a total operation time is set as 1000 days which is consistent with the real conditions considered by the operators of unconventional wells. It should be noted that the conclusion might be slightly different if a different total operation time is incorporated.

Table 6. Operation design for simulation scenarios with multiple rounds of shut-ins.

Rounds	Scenario Description
1	Shut-in 60 days → Produce 940 days (Total 1000 days)
2	Shut-in 60 days → Produce 300 days → Shut-in 60 days → Produce 580 days (Total 1000 days)
3	Shut-in 60 days → Produce 300 days → Shut-in 60 days → Produce 300 days → Shut-in 60 days → Produce 220 days (Total 1000 days)

The daily oil rate with one, two, and three rounds of well shut-ins are shown in Figures 12–14.

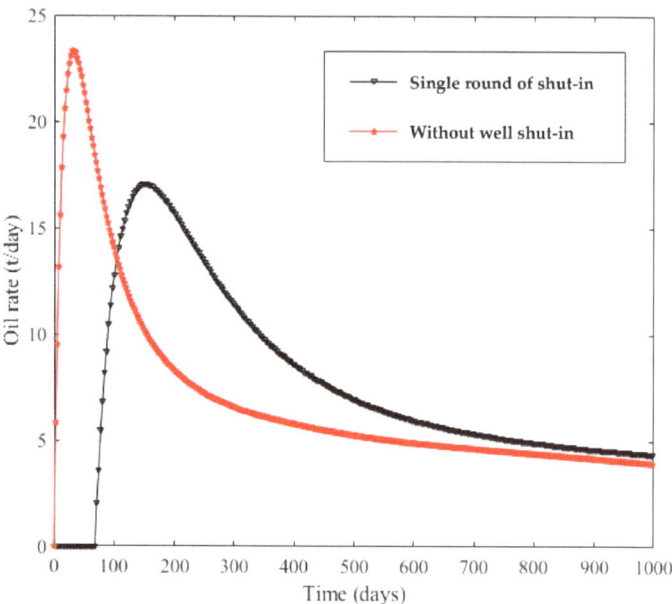

Figure 12. Daily oil rate with a single round of well shut-in operations.

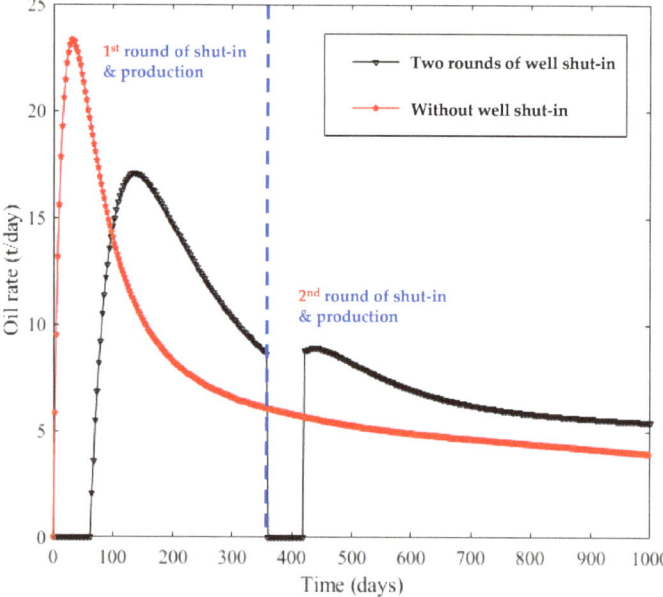

Figure 13. Daily oil rate with two rounds of well shut-in operations.

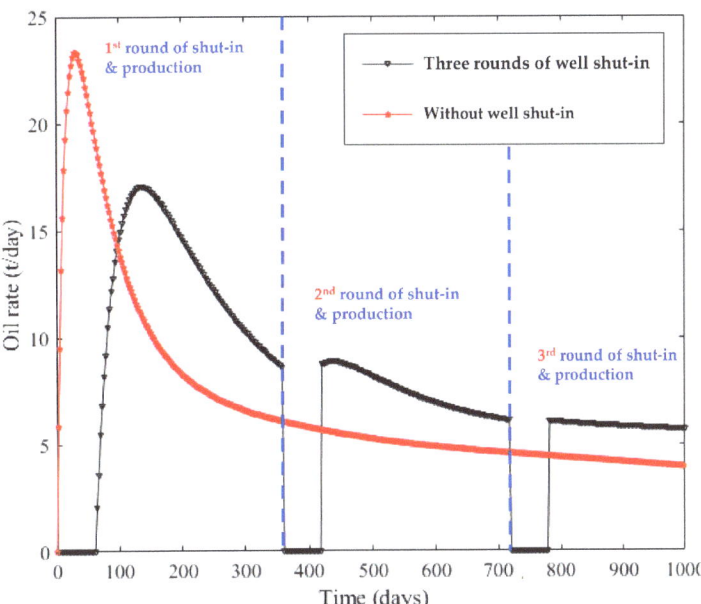

Figure 14. Daily oil rate with three rounds of well shut-in operations.

The red lines in the above figures represent the oil production rate without shut-in, which are provided for comparative purposes. Based on the comparison between Figures 12 and 13, it could be known that the second round of shut-ins could temporally enhance the oil production rate and reduce the decline rate. This change can be credited to the dynamic imbibition effect during the shut-in. By contrast, based on the comparison between Figures 13 and 14, it can be seen that the effect of the three rounds of shut-ins is limited. This is because after 700 days, a great amount of fracturing water around the wellbore has flowed back and the effect of the dynamic imbibition is limited. The comparison of average daily oil rates after 1000 days, including shut-in and production, is shown in Figure 15. According to the figure, when there is no shut-in, the average oil rate after 1000 days is 7.031 t/day. When two rounds of shut-ins are implemented, the average oil rate reaches its peak, 7.434 t/day, which is 5.75 % greater than the case with no shut-in. To sum up, under the simulation conditions of this paper, the oil production of one and two rounds of well shut-in operations are close to each other. Therefore, one or two rounds of shut-ins are equally beneficial and are recommended here.

Based on the calculated results in this section, we have already shown that multiple rounds of well shut-in might significantly benefit reservoir development. However, assuming two rounds of shut-ins are implemented in the operation process, the timing for the second round of shut-in would be of vital importance for production. Since there is no related discussion in previous studies, in this section, a simulation case with different timings for the second round shut-in, i.e., the time length of the first round of production is designed for further comparison and discussion. Different production times for the first round, i.e., 200 days, 300 days, 400 days, and 500 days, are selected for the simulations.

Figures 16–19 show the daily oil rate with different times (length) for first round of production, i.e., the timing of the second round of shut-in, respectively. It can be seen from the figures that, after the second round of shut-in, the daily oil rate would slightly rise again. However, as the time length of the first round of production increases, this production rise would diminish accordingly. When there are 500 days of production before the second round of shut-in, this rise in production rate could hardly be identified and under this condition, the second round of shut-in would have few positive effects on the oil production. Therefore, the timing of the shut-in is of vital importance for the reservoir development when multiple rounds of shut-ins are implemented.

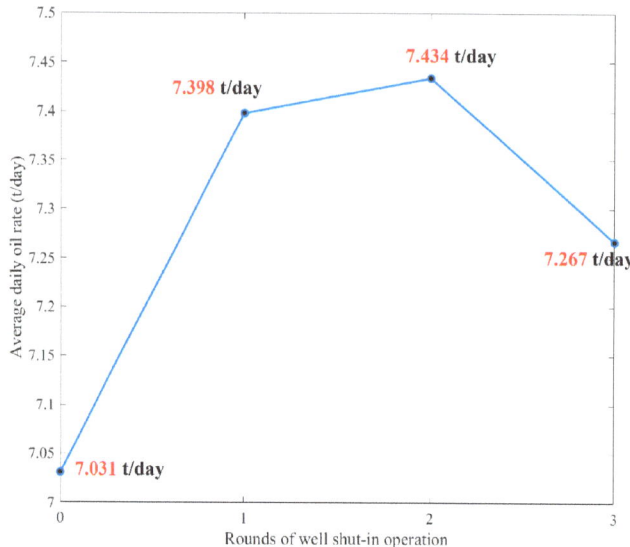

Figure 15. Average oil production rate between different rounds of shut-in scenarios.

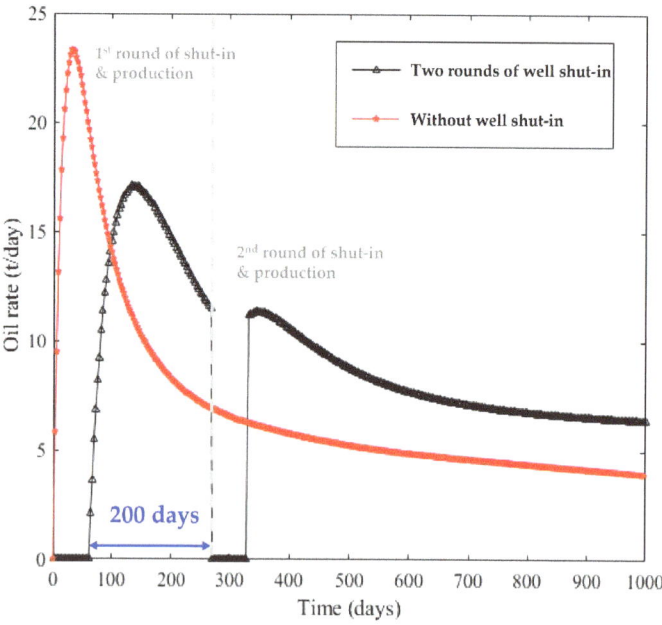

Figure 16. Daily oil rate with 200 days of first round of production.

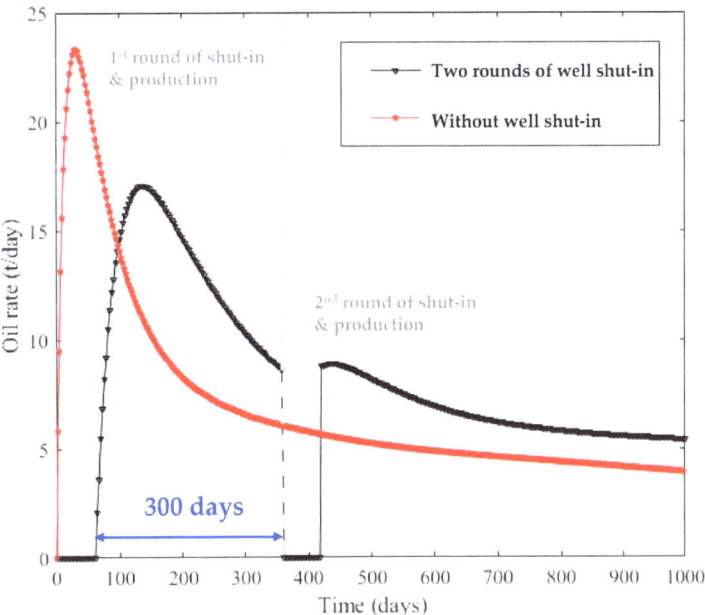

Figure 17. Daily oil rate with 300 days of first round of production.

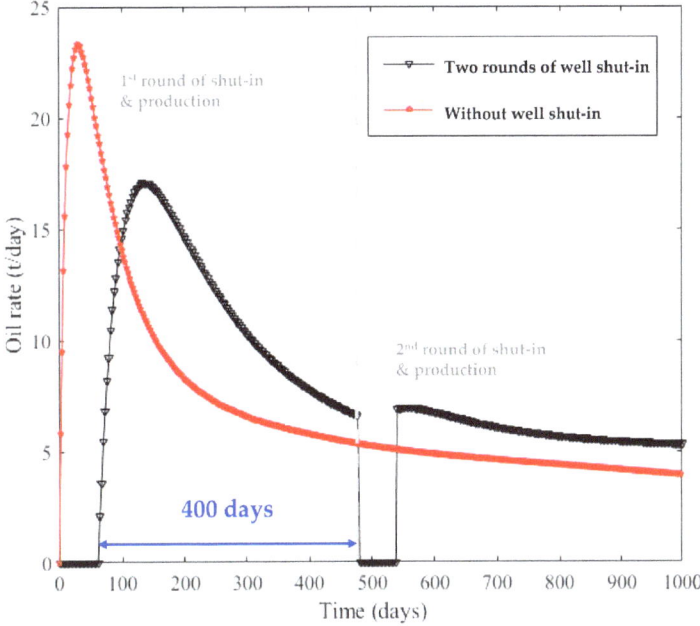

Figure 18. Daily oil rate with 400 days of first round of production.

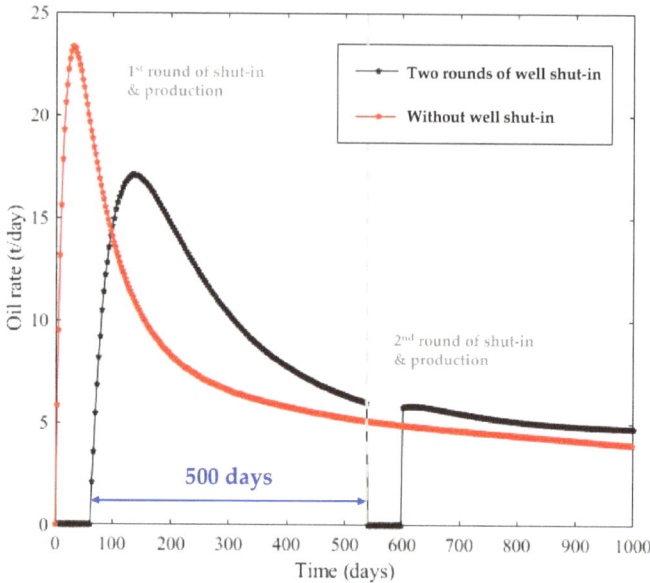

Figure 19. Daily oil rate with 500 days of first round of production.

Figure 20 shows a comparison between the average daily oil rate and different lengths of production for the first round. According to the figure, when the time length of the first round of production is 200 days, the average oil rate in 1000 days is 7.377 t/day; for 300 days, the average oil rate reaches its peak, i.e., 7.434 t/day. Then, as the time length keeps increasing, the average oil rate starts to decrease rapidly. To sum up, under the simulation conditions of this paper, it is recommended to perform 200 to 300 days of production before the second round of shut-ins.

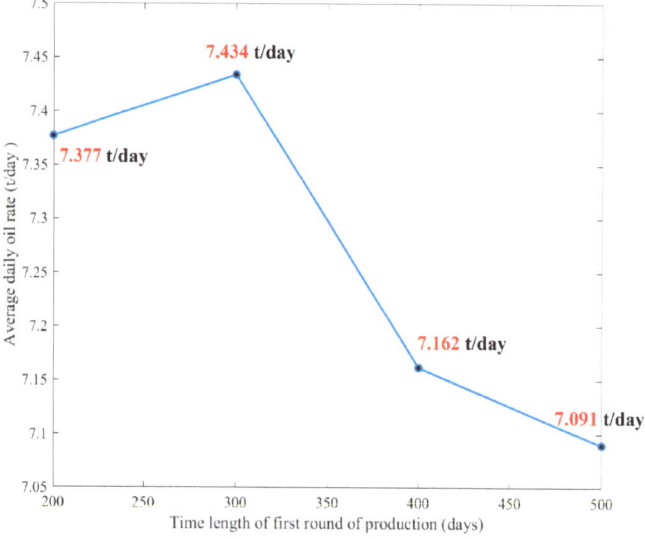

Figure 20. Average oil production rate between simulation cases with different timings of the second round of shut-ins.

6. Conclusions

In this study, the feasibility of the temporary well shut-in in a shale oil reservoir was studied using numerical modeling. A mathematical model for the simulation of immiscible two-phase flow in a shale oil reservoir was presented based on the coupling of the phase transport in the porous media module (PTPM) and Darcy's law. The model was validated by comparing its results with an available analytical solution. The geological background of the Chang-7 Member (Chenghao, China) is presented, and the corresponding parameters were built into the establishment of the model simulation. Based on the simulation results, a quantitative analysis was carried out, and several conclusions can be drawn.

According to the calculated saturation distribution, an oil–water displacement can be easily identified during well shut-ins, which reveals the effect of the dynamic imbibition. Based on the calculated production rates, it can be concluded that the implementation of a shut-in can decrease the initial oil rate; however, this decreases the oil decline rate as well, which is beneficial to long-term production. After 1000 days of production, the oil production rate, with 120 days of shut-ins, was 9.85 % larger than that with no shut-in operation. By judging the average daily production rates, the optimal shut-in time was determined as 60 days for our given field conditions. Besides, the potential of several rounds of shut-ins was also explored. Single and double rounds of shut-ins are equally beneficial to long-term production under the presented reservoir condition. When two rounds of shut-ins are implemented, it is recommended to perform the second shut-in round after 300 days of production. To sum up, this study reveals a workflow for feasibly studying temporary well shut-in operations in any shale oil reservoir and provides guidance for optimizing overall development scenarios.

Author Contributions: Conceptualization, W.L. and Q.Z.; methodology, J.W.; software, Q.Z.; validation, Q.Z. and W.L.; formal analysis, H.S.; investigation, Q.Z. and W.L.; resources, D.W.; writing—original draft preparation, Q.Z.; writing—review and editing, Q.Z. and A.D.T.; supervision, W.L.; project administration, W.L. All authors have read and agreed to the published version of the manuscript.

Funding: This research was funded by Fundamental Research Funds for Central Universities, grant number FRF-TP-17-023A1 and CNPC Innovation Found: 2021DQ02-0901.

Conflicts of Interest: The authors declare no conflict of interest.

References

1. Kilian, L. The impact of the shale oil revolution on US oil and gasoline prices. *Rev. Environ. Econ. Policy* **2016**, *10*, 185–205. [CrossRef]
2. Taleghani, A.D.; Gonzalez-Chavez, M.; Yu, H.; Asala, H. Numerical simulation of hydraulic fracture propagation in naturally fractured formations using the cohesive zone model. *J. Petrol. Sci. Eng.* **2018**, *165*, 42–57. [CrossRef]
3. Zhang, Q.; Zhu, W.; Liu, W.; Yue, M.; Song, H. Numerical simulation of fractured vertical well in low-permeable oil reservoir with proppant distribution in hydraulic fracture. *J. Petrol. Sci. Eng.* **2020**, *195*, 107587. [CrossRef]
4. Yue, M.; Zhang, Q.; Zhu, W.; Zhang, L.; Song, H.; Li, J. Effects of proppant distribution in fracture networks on horizontal well performance. *J. Petrol. Sci. Eng.* **2020**, *187*, 106816. [CrossRef]
5. Jefferson, M. A crude future? COVID-19s challenges for oil demand, supply and prices. *Energy Res. Soc. Sci.* **2020**, *68*, 101669. [CrossRef]
6. Kleit, A.; Taleghani, A.D. COVID Shut-In Choices Across Unconventional Reservoirs: Evidence from the Bakken and the Marcellus. *J. Energ. Resour. ASME* **2022**, *144*, 073009. [CrossRef]
7. Yang, L.; Ge, H.; Shi, X.; Cheng, Y.; Zhang, K.; Chen, H.; Shen, Y.; Zhang, J.; Qu, X. The effect of microstructure and rock mineralogy on water imbibition characteristics in tight reservoirs. *J. Nat. Gas. Sci. Eng.* **2016**, *34*, 1461–1471. [CrossRef]
8. Wang, J.; Liu, H.; Qian, G.; Peng, Y. Mechanisms and capacity of high-pressure soaking after hydraulic fracturing in tight/shale oil reservoirs. *Petrol. Sci.* **2020**, *18*, 546–564. [CrossRef]
9. Brown, R.; Fatt, I. Measurements of fractional wettability of oilfield rocks by the nuclear magnetic relaxation method. *Trans. AIME* **1956**, *207*, 262–264.
10. Sun, Y.; Li, Q.; Chang, C.; Wang, X.; Yang, X. NMR-Based Shale Core Imbibition Performance Study. *Energies* **2022**, *15*, 6319. [CrossRef]
11. Dai, C.; Cheng, R.; Sun, X.; Liu, Y.; Zhou, H.; Wu, Y.; You, Q.; Zhang, Y.; Sun, Y. Oil migration in nanometer to micrometer sized pores of tight oil sandstone during dynamic surfactant imbibition with online NMR. *Fuel* **2019**, *245*, 544–553. [CrossRef]

12. Yang, Z.; Liu, X.; Li, H.; Lei, Q.; Luo, Y.; Wang, X. Analysis on the influencing factors of imbibition and the effect evaluation of imbibition in tight reservoirs. *Petrol. Explor. Develop.* **2019**, *46*, 779–785. [CrossRef]
13. Guo, J.; Li, M.; Chen, C.; Tao, L.; Liu, Z.; Zhou, D. Experimental investigation of spontaneous imbibition in tight sandstone reservoirs. *J. Petrol. Sci. Eng.* **2020**, *193*, 107395. [CrossRef]
14. Wang, F.; Yang, K.; Zai, Y. Multifractal characteristics of shale and tight sandstone pore structures with nitrogen adsorption and nuclear magnetic resonance. *Petrol. Sci.* **2020**, *17*, 1209–1220. [CrossRef]
15. Peng, X.; Wang, X.; Zhou, X.; Lin, Z.; Zeng, F.; Huang, X. Lab-on-a-chip systems in imbibition processes: A review and applications/issues for studying tight formations. *Fuel* **2021**, *306*, 121603. [CrossRef]
16. Karimi, S.; Kazemi, H.; Simpson, G. Capillary Pressure, Fluid Distribution, and Oil Recovery in Preserved Middle Bakken Cores. In Proceedings of the SPE Oklahoma City Oil and Gas Symposium, Oklahoma City, OK, USA, 27–31 March 2017. SPE 185095.
17. Tu, J.; Sheng, J. Effect of Pressure on Imbibition in Shale Oil Reservoirs with Wettability Considered. *Energy Fuel* **2020**, *34*, 4260–4272. [CrossRef]
18. Cheng, Z.; Ning, Z.; Yu, X.; Wang, Q.; Zhang, W. New insights into spontaneous imbibition in tight oil sandstones with NMR. *J. Petrol. Sci. Eng.* **2019**, *179*, 455–464. [CrossRef]
19. Abd, A.; Elhafyan, E.; Siddiqui, A.; Alnoush, W.; Blunt, M.; Alyafei, N. A review of the phenomenon of counter-current spontaneous imbibition: Analysis and data interpretation. *J. Petrol. Sci. Eng.* **2019**, *180*, 456–470. [CrossRef]
20. Schmid, K.; Geiger, S.; Sorbie, K. Semianalytical solutions for co-current and countercurrent imbibition and dispersion of solutes in immiscible two-phase flow. *Water Resour. Res.* **2011**, *47*, W02550. [CrossRef]
21. Schmid, K.; Alyafei, N.; Geiger, S.; Blunt, M. Analytical solutions for spontaneous imbibition: Fractional-flow theory and experimental analysis. *SPE J.* **2016**, *21*, 2308–2316. [CrossRef]
22. Khan, A.; Siddiqui, A.; Abd, A.; Alyafei, N. Guidelines for numerically modeling Co-and counter-current spontaneous imbibition. *Transp. Porous Med.* **2018**, *124*, 743–766. [CrossRef]
23. Wang, R. *Numerical Simulation Study on Mechanism and Law of Energy Storage in Shut-in Schedule after Fracturing of Tight Oil*; China University of Petroleum (East China): Qingdao, China, 2016.
24. Zhang, Y.; Ge, H.; Shen, Y.; Jia, L.; Wang, J. Evaluating the potential for oil recovery by imbibition and time-delay effect in tight reservoirs during shut-in. *J. Petrol. Sci. Eng.* **2020**, *184*, 106557. [CrossRef]
25. Liu, D. *Research on Microcosmic Laws of Fracture Fluid Imbibition and Mechanisms of Productivity Enhancement by "Shut-in" in Unconventional Hydrocarbon Reservoir*; China University of Petroleum (Beijing): Beijing, China, 2017.
26. Cheng, Y. Impact of water dynamics in fractures on the performance of hydraulically fractured wells in gas-shale reservoirs. *J. Can. Pet. Technol.* **2012**, *51*, 143–151. [CrossRef]
27. Makhanov, K.; Habibi, A.; Dehghanpour, H.; Kuru, E. Liquid uptake of gas shales: A workflow to estimate water loss during shut-in periods after fracturing operations. *J. Unconv. Oil Gas Resour.* **2014**, *7*, 22–32. [CrossRef]
28. Yan, Q.; Lemanski, C.; Karpyn, Z.; Ayala, T. Experimental investigation of shale gas production impairment due to fracturing fluid migration during shut-in time. *J. Nat. Gas. Sci. Eng.* **2015**, *24*, 99–105. [CrossRef]
29. Zhou, Z.; Wei, S.; Lu, R.; Li, X. Numerical Study on the Effects of Imbibition on Gas Production and Shut-In Time Optimization in Woodford Shale Formation. *Energies* **2020**, *13*, 3222. [CrossRef]
30. Wang, F.; Ruan, Y.; Chen, Q.; Zhang, S. A pressure drop model of post-fracturing shut-in considering the effect of fracturing-fluid imbibition and oil replacement. *Petro. Explor. Dev.* **2021**, *48*, 1440–1449. [CrossRef]
31. Eltahan, E.; Rego, F.; Yu, W.; Sepehrnoori, K. Impact of well shut-in after hydraulic-fracture treatments on productivity and recovery of tight oil reservoirs. *J. Petrol. Sci. Eng.* **2021**, *203*, 108592. [CrossRef]
32. Zhao, Y.; Liu, L.; Zhang, L.; Zhang, X.; Li, B. Simulation of a multistage fractured horizontal well in a tight oil reservoir using an embedded discrete fracture model. *Energy Sci. Eng.* **2019**, *7*, 1485–1503. [CrossRef]
33. Jia, P.; Ke, X.; Niu, L.; Li, Y.; Cheng, L. Investigation of Shut-In Effect on Production Performance in Shale Oil Reservoirs With Key Mechanisms. *Adv. Phase Behav. Fluid* **2022**, *9*, 782279. [CrossRef]
34. Abdelsalam, S.; Zaher, A. On behavioral response of ciliated cervical canal on the development of electroosmotic forces in spermatic fluid. *Math. Model. Nat. Pheno.* **2022**, *17*, 27. [CrossRef]
35. Wang, F.; Chen, R.; Yu, W.; Tian, J.; Liang, X.; Tan, X.; Gong, L. Characteristics of lacustrine deepwater fine-grained lithofacies and source-reservoir combination of tight oil in the triassic chang 7 member in Ordos Basin, China. *J. Petrol. Sci. Eng.* **2021**, *202*, 108429. [CrossRef]
36. Pan, S.; Horsfield, B.; Zou, C.; Yang, Z. Upper Permian Junggar and Upper Triassic Ordos lacustrine source rocks in Northwest and Central China: Organic geochemistry, petroleum potential and predicted organofacies. *Int. J. Coal Geol.* **2016**, *158*, 90–106. [CrossRef]
37. Yang, H.; Fu, J.; He, H.; Liu, X.; Zhang, Z.; Deng, X. Formation and distribution of large low-permeability lithologic oil regions in Huaqing, Ordos Basin. *Petrol. Explor. Develop.* **2012**, *39*, 683–691. [CrossRef]
38. Wang, X.; Mazumder, R.; Salarieh, B.; Salman, A.; Shafieezadeh, A.; Li, Y. Machine Learning for Risk and Resilience Assessment in Structural Engineering: Progress and Future Trends. *J. Struct. Eng.* **2022**, *148*, 03122003. [CrossRef]
39. Mazumder, R.; Salman, A.; Li, Y. Reliability Assessment of Oil and Gas Pipeline Systems at Burst Limit State Under Active Corrosion. In *International Probabilistic Workshop*; Springer: Cham, Switzerland, 2021; pp. 653–660.

40. Mazumder, R.; Salman, A.; Li, Y. Failure risk analysis of pipelines using data-driven machine learning algorithms. *Struct. Saf.* **2021**, *89*, 102047. [CrossRef]
41. Fu, S.; Yao, J.; Li, S.; Zhou, X.; Li, M. Enrichment characteristics and resource potential of continental shale oil in Mesozoic Yanchang Formation, Ordos Basin. *Petrol. Geol. Exp.* **2020**, *42*, 698–710. (In Chinese)
42. Chen, Z. *Finite Element Methods and Their Applications*; Springer Science & Business Media: Berlin, Germany, 2005.
43. Chen, Z.; Huan, G.; Ma, Y. *Computational Methods for Multiphase Flows in Porous Media*; Society for Industrial and Applied Mathematics: Philadelphia, PA, USA, 2006.
44. Blunt, M. *Multiphase Flow in Permeable Media: A Pore-Scale Perspective*; Cambridge University Press: London, UK, 2017.
45. Karimi-Fard, M.; Durlofsky, L.; Aziz, K. An Efficient Discrete-Fracture Model Applicable for General-Purpose Reservoir Simulators. *SPE J.* **2004**, *9*, 227–236. [CrossRef]
46. COMSOL Multiphysics. Discrete Fracture in Rocks. 2013. Available online: https://www.comsol.com/blogs/discrete-fracture-in-rocks (accessed on 5 April 2013).
47. Ertekin, T.; Abou-Kassem, J.; King, G. *Basic Applied Reservoir Simulation*; Society of Petroleum Engineers: Richardson, TX, USA, 2001; Volume 7.
48. Frei, W. Improving Convergence of Multiphysics Problem. COMSOL Official Documentation. 2013. Available online: https://uk.comsol.com/blogs/improving-convergence-multiphysics-problems?setlang=1 (accessed on 23 December 2013).
49. Toth, A.; Kelley, C. Convergence analysis for Anderson acceleration. *SIAM J. Numer. Anal.* **2015**, *53*, 805–819. [CrossRef]
50. Buckley, S.; Leverett, M. Mechanism of fluid displacement in sands. *Trans. AIME.* **1942**, *146*, 107–116. [CrossRef]
51. Peters, E. *Advanced Petrophysics: Dispersion, Interfacial Phenomena*; Greenleaf Book Group: Austin, TX, USA, 2012; Volume 2.
52. Amestoy, P.; Duff, I.; L'Excellent, J. Multifrontal parallel distributed symmetric and unsymmetric solvers. *Comput. Methods Appl. Mech. Eng.* **2000**, *184*, 501–520. [CrossRef]

Article

Well Testing Methodology for Multiple Vertical Wells with Well Interference and Radially Composite Structure during Underground Gas Storage †

Hongyang Chu [1,2,3,*], Tianbi Ma [4,5], Zhen Chen [1], Wenchao Liu [1] and Yubao Gao [1]

[1] School of Civil and Resources Engineering, University of Science and Technology Beijing, Beijing 100083, China
[2] Harold Vance Department of Petroleum Engineering, Texas A&M University, College Station, TX 77843, USA
[3] State Key Laboratory of Petroleum Resources and Prospecting, China University of Petroleum, Beijing 102249, China
[4] Petroleum Exploration and Production Research Institute, SINOPEC, Beijing 100083, China
[5] Department of Geosciences, The University of Tulsa, Tulsa, OK 74104, USA
* Correspondence: hongyangchu@126.com or hongyangchu@ustb.edu.cn; Tel.: +86-15600277900
† This paper is an extended version of our paper published in the SPE Annual Technical Conference and Exhibition, Houston, TX, USA, 3–5 October 2022. Paper Number: SPE-210187-MS.

Abstract: To achieve the goal of decarbonized energy and greenhouse gas reduction, underground gas storage (UGS) has proven to be an important source for energy storage and regulation of natural gas supply. The special working conditions in UGS cause offset vertical wells to easily interfere with target vertical wells. The current well testing methodology assumes that there is only one well, and the interference from offset wells is ignored. This paper proposes a solution and analysis method for the interference from adjacent vertical wells to target vertical wells by analytical theory. The model solution is obtained by the solution with a constant rate and the Laplace transform method. The pressure superposition is used to deal with the interference from adjacent vertical wells. The model reliability in the gas injection and production stages is verified by commercial software. Pressure analysis shows that the heterogeneity and interference in the gas storage are caused by long-term gas injection and production. As both the adjacent well and the target well are in the gas production stage, the pressure derivative value in radial flow is related to production rate, mobility ratio, and 0.5. Gas injection from offset wells will cause the pressure derivative to drop later. Multiple vertical wells from the Hutubi UGS are used to illustrate the properties of vertical wells and the formation.

Keywords: underground gas storage; multiple vertical wells; flow behavior; well interference; case study; analytical model

Citation: Chu, H.; Ma, T.; Chen, Z.; Liu, W.; Gao, Y. Well Testing Methodology for Multiple Vertical Wells with Well Interference and Radially Composite Structure during Underground Gas Storage. *Energies* 2022, 15, 8403. https://doi.org/10.3390/en15228403

Academic Editor: Manoj Khandelwal

Received: 6 October 2022
Accepted: 4 November 2022
Published: 10 November 2022

Publisher's Note: MDPI stays neutral with regard to jurisdictional claims in published maps and institutional affiliations.

Copyright: © 2022 by the authors. Licensee MDPI, Basel, Switzerland. This article is an open access article distributed under the terms and conditions of the Creative Commons Attribution (CC BY) license (https:// creativecommons.org/licenses/by/ 4.0/).

1. Introduction

To pursue carbon-peaking and carbon-neutral goals, as well as meet the cycling energy demand on the electricity power grids, the major countries around the world typically employ large-scale energy storage systems [1]. These energy storage systems include pumped hydropower, compressed air, and UGS. In this context, UGS has been proven to be the most commercially mature large-scale energy storage technology and it has been implemented in many countries including China [2]. Natural gas is a fossil fuel with more potential for energy conversion and clean emissions than liquid petroleum [3,4]. The molecular formula of natural gas makes its combustion products virtually free of sulfur, dust, and other harmful substances, and it produces significantly less CO_2 than other fossil fuels. In addition, due to the recovery, transportation, and processing costs of natural gas, it is an attractive alternative to petroleum energy in the oil and gas industry [5]. Thus, from 2008 to 2018, natural gas consumption increased by 28.35% [6,7].

Although China has abundant natural gas resources, the supply of and demand for natural gas have been affected in the long term by technology and equipment limitations. In 2018, 42.9% of China's natural gas consumption was imported from overseas [8]. The limitations of the natural gas market also include seasonal and geographical factors. Most of China's natural gas reserves are located in the western region, while the principal areas of natural gas consumption are mostly developed cities along the eastern coast. Additionally, heating is one of the important purposes of natural gas, resulting in a much larger natural gas demand in winter than that in summer. To resolve these incongruities in the natural gas market, UGS is an important part of the natural gas industry [9].

Compared with the salt cavern type of UGS, the porous reservoir type of UGS has the advantages of a short UGS construction period and low operating cost. The limitation is that the stored gas cannot be recovered completely and the recovered gas in the surface requires further processing process. The injection and withdrawal rates in salt cavern type of UGS are fast, and most of the stored gas can be recovered. However, the total gas storage capacity in the salt cavern type of UGS is lower, and this UGS type has a higher probability of leakage risk. The treatment of brine in salt caverns requires additional technology. Therefore, porous reservoirs are more suitable large-scale storage sites. As shown in Table 1, the screen criteria include caprock lithology, tectonic activity, reservoir type, depth, and pore volume of the reservoir. The number of UGS facilities, working gas capacity, and maximal withdrawal rates of the porous reservoir (depleted oil and gas reservoir) type of UGS all largely dominate the total number of UGSs in North America, Europe, Commonwealth of Independent States (CIS), Middle East, and Asia–Oceania, as given in Table 2. In contrast to the recovery processes in these oil and gas reservoirs, the injection/production process in the UGS has the features of high rate, continuous injection/production in a short period, and collective well shut-in in a period [10,11]. The natural gas composition in UGS is related to the porous media type. As the UGS type is the acid gas reservoir, the acid gas content of the produced gas will gradually decrease. In an injection–withdrawal cycle, the acid gas content in the gas composition will increase with the increase in produced gas volume [12]. For the UGSs of the condensate gas reservoir type, the injected gas can evaporate and extract the condensate oil in the formation. This effect becomes significant with the increase in gas injection pressure. With the increase in the injection–withdrawal cycle, the contents of C2, C3 and C7+ components in the produced gas show a trend of first increasing and then decreasing [13]. These unconventional operating and composition conditions result in the unique flow behavior of UGS.

Table 1. The screen criteria for the depleted oil and gas reservoir type of UGS (modified from Lewandowska-Śmierzchalska et al. [14]).

	Caprock Lithology	Tectonic Activity	Reservoir Type	Depth	Pore Volume of Reservoir
Change from the most suitable type to the unsuitable type	Clay rock, clay shale, calcium sulfate rock, and salt rock	No faults	Gas reservoir	Large	Large
	Mudstone and mud shale	Single independent fault	Gas and oil reservoir	Medium	Medium
	Sandstones and siltstones	Plenty of faults in basement			
	Clay sandstone, limestone, and dolomite	The fault terminates in caprock			
	Sandstones	The fault extends to caprock	Oil reservoir	Small	Small

Table 2. The percentage of the depleted oil and gas reservoir type of UGS in the total number of UGSs (modified from Cedigaz [15]).

Regions	Number of UGS Facilities		Working Gas Capacity		Max. Withdrawal Rates	
	Salt Caverns	Porous Reservoir	Salt Caverns	Porous Reservoir	Salt Caverns	Porous Reservoir
North America	10%	90%	9%	91%	26%	74%
Europe	33%	67%	17%	83%	36%	64%
CIS	6%	94%	1%	99%	3%	97%
Middle East	0%	100%	0%	100%	0%	100%
Asia–Oceania	4%	96%	2%	98%	1%	99%
World Total	14%	86%	8%	92%	24%	76%

Conventional well testing studies are limited to single wells [16]. Multi-well testing mostly is within the scope of interference or pulse tests [17–20]. Interference testing typically requires one well to be active and the other well to be shut in to measure the pressure signal produced by the active well [21]. It is usually used to determine the degree of connectivity between wells or directional permeability. As adjacent wells are producing or injecting, Warren and Hartsock [22] first used an asymptotic approximation solution to describe interference between two production wells in an infinite reservoir. Onur et al. [23] proposed an analytical model for pressure buildup tests in multi-well systems with interference. A limitation of their model is that the multi-well system must achieve quasi-steady-state flow before well shut-in. Fokker and Verga [24] proposed a semi-analytical productivity test model that can consider vertical wells and horizontal wells. This method is not only suitable for oil and gas reservoirs but also "automatically" considers well interference. Aiming at the well interference caused by the adjacent well's water injection, Lin and Yang [25] established a well test model with an adjacent well's water injection by applying the material balance equation and superposition. Izadi and Yildiz [26] used a semi-analytical method to establish a transient model that could consider the multi-well system and natural fractures. For tight carbonate gas reservoirs, Wei et al. [27] proposed a multi-well model for hydraulically fractured wells. Chu et al. [28,29] proposed a semi-analytical model for multiple fractured horizontal wells with well interference. Their target domains include hydraulic fractures, natural fractures, and matrices in unconventional reservoirs. The well testing data of Hutubi UGS show that the well interference is serious [30]. Abnormal rising or falling characteristics appear in the late well testing data. Conventional single-well models cannot match field data from Hutubi UGS. The reason is that it violates the physical assumptions in single-well model (the study domain in single-well models contains only one well). The typical flow behaviors including the effects of well interference in UGS are still unclear. The unique flow behavior in UGS leads to limitations in storage volume calculation and energy storage capacity evaluation.

To fill these gaps, this paper uses an analytical method to establish transient models for the multi-well system with interference in the UGS. First, the governing equation for the multi-well system in a dimensionless domain was constructed. Laplace transforms were used to obtain the basic pressure solution for each well in the multi-well system. A model for the target well was further extended from a homogeneous model to a radially composite model to account for the continuing injection/production process in UGS. Adjacent gas injection/production interference in UGS is "automatically" taken into account by pressure superposition. We used commercial numerical simulators to verify the reliability of the proposed model under different operating conditions. We used flow regime and sensitivity analysis to describe typical flow behavior in UGS. We present a case study from the Hutubi UGS to further illustrate the model practicability. This work provides useful guidelines for storage volume calculation in UGS, energy storage capacity evaluation, and well location optimization.

The innovations of this work include the following: (1) a new analytical model of a multi-well system with well interference and radially composite structure is proposed;

(2) the unique flow regions in UGS are elucidated by sensitivity analysis; (3) a field case from the largest Hutubi UGS in China shows the method practicality.

2. Methodology

2.1. Physical Model

Figure 1a shows the technological process of a depleted oil and gas reservoir used for UGS. The storage process of UGS usually includes long-distance pipelines, filters, metering equipment, a compressor station, a cooler, and porous media. For gas injection process, it includes a heating device, dehydration equipment, metering equipment, and pipelines. Our research objectives mainly focus on the gas flow and storage behavior in porous media. Vertically, the anticline traps can ensure the integrity of UGS. The region of the entire UGS formation is bounded, and the properties of the boundary can be regarded as irregular. As the UGS is a depleted oil and gas reservoir, a large number of vertical wells are distributed in it. These wells can be used as production wells during gas storage, injection wells during gas injection, and monitoring wells during shut-in. The reservoir medium of the UGS can be considered a continuous homogeneous model or a dual-medium model proposed by Warren and Root (1963), as given in Figure 1b. Other assumptions are as follows:

(1) The initial pressure and the properties of the UGS are uniformly distributed.

(2) The UGS has closed top–bottom boundaries, and the outer boundary is irregular and closed.

(3) Gas flow in the UGS is single-phase and compressible and obeys Darcy's law.

(4) The thickness of the UGS is constant and the vertical well penetrates the formation completely.

(5) The effects of temperature and gravity on gas flow in UGS are negligible.

(6) Injection into or production from wells in the multi-well system is at a constant rate.

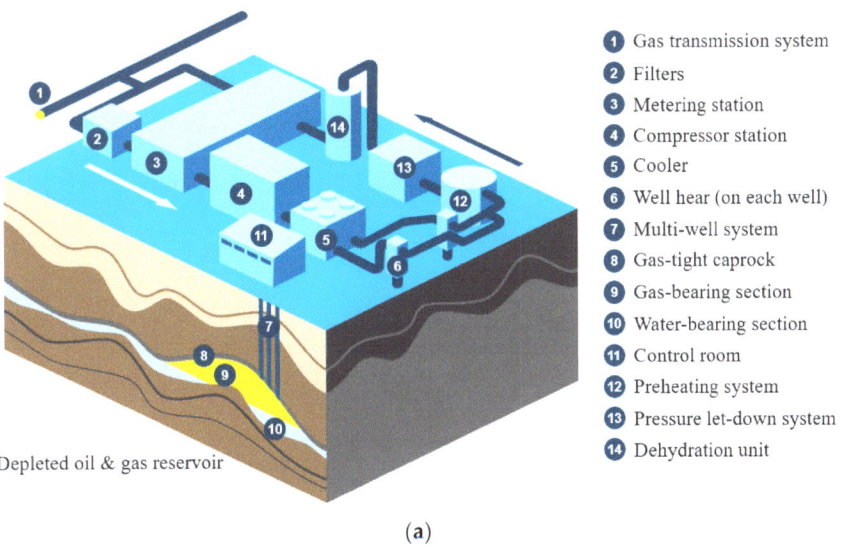

(a)

Figure 1. *Cont.*

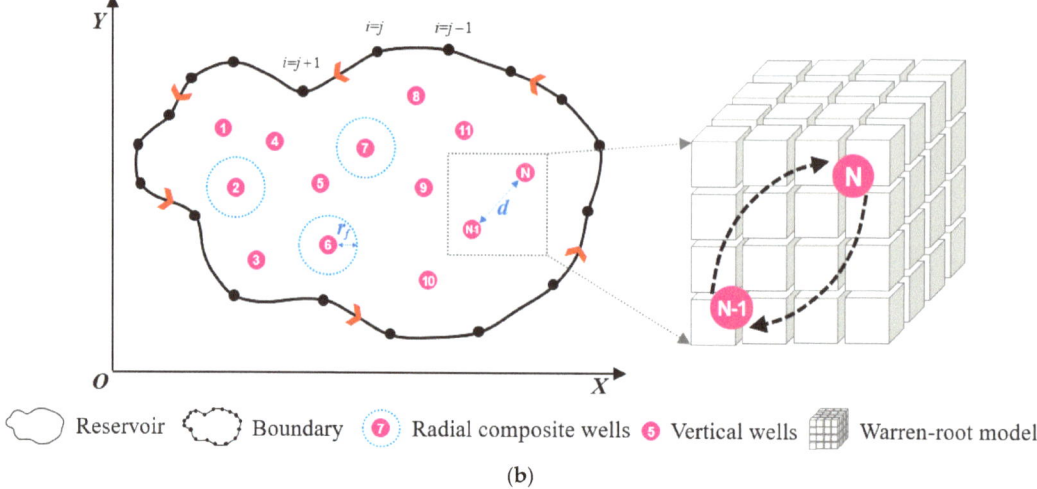

Reservoir　Boundary　⑦ Radial composite wells　⑤ Vertical wells　Warren-root model

(b)

Figure 1. Physical model of the multi-well system with well interference in UGS. (**a**) Schematic diagram (modified from Uniper [31]); (**b**) the multi-well system and reservoir media.

2.2. Mathematical Model

Using the definitions of dimensionless variables in the Appendix A, the continuity equation for a multi-well system in UGS can be written as

$$\frac{d^2\overline{m}_D}{dr_D^2} + \frac{1}{r_D}\frac{d\overline{m}_D}{dr_D} = u\frac{d\overline{m}_D}{dt_D} \tag{1}$$

where $\overline{m}_D$ is the dimensionless pseudo-pressure in the Laplace domain, t_D is the dimensionless time, r_D is the dimensionless distance, and u refers to the Laplace variable. The initial condition for the multi-well system can be written as

$$m_D(r_D, t_D)_{t_D \to 0} = 0 \tag{2}$$

The inner boundary condition including skin factor and wellbore storage effect is

$$C_D u \overline{m}_{wD} - r_D \frac{d\overline{m}_D}{dr_D}\bigg|_{r_D=1} = \frac{1}{u} \tag{3}$$

$$\overline{m}_{wD} = \left[\overline{m}_D - S\frac{d\overline{m}_D}{dr_D}\right]_{r_D=1} \tag{4}$$

where C_D is the wellbore storage coefficient, S is the skin factor, and $\overline{m}_{wD}$ is the dimensionless bottom-hole pseudo-pressure. The condition for the irregular closed outer boundary is

$$m_D(r_D, t_D)_{r_D \to \infty} = 0 \tag{5}$$

The solution for the homogeneous model with wellbore storage and skin factor is

$$\overline{m}_{wD} = \frac{K_0\left(\sqrt{f(u)}\right) + S\sqrt{f(u)}K_1\left(\sqrt{f(u)}\right)}{f(u)\left\{\sqrt{f(u)}K_1\left(\sqrt{f(u)}\right) + C_D f(u)\left[K_0\left(\sqrt{f(u)}\right) + S\sqrt{f(u)}K_1\left(\sqrt{f(u)}\right)\right]\right\}} \tag{6}$$

where K_0 refers to the zero-order first-class Bessel function and K_1 is the first-order first-class Bessel function. For a heterogeneous reservoir, Equation (1) for the inner and outer regions can be written as

$$\frac{d^2\overline{m}_{1D}}{dr_D^2} + \frac{1}{r_D}\frac{d\overline{m}_{1D}}{dr_D} = u\frac{d\overline{m}_{1D}}{dt_D} \tag{7}$$

$$\frac{d^2\overline{m}_{2D}}{dr_D^2} + \frac{1}{r_D}\frac{d\overline{m}_{2D}}{dr_D} = \frac{\omega_{12}}{M_{12}}u\frac{d\overline{m}_{2D}}{dt_D} \tag{8}$$

The inner boundary condition is

$$C_D u \overline{m}_{wD} - r_D \frac{d\overline{m}_{1D}}{dr_D}\bigg|_{r_D=1} = \frac{1}{u} \tag{9}$$

$$\overline{m}_{wD} = \left[\overline{m}_{1D} - S\frac{d\overline{m}_{1D}}{dr_D}\right]_{r_D=1} \tag{10}$$

The boundary conditions for the inner and outer regions are

$$\overline{m}_{1D}|_{r_D=r_{fD}} = \overline{m}_{2D}|_{r_D=r_{fD}} \tag{11}$$

$$\frac{d\overline{m}_{1D}}{dr_D}\bigg|_{r_D=r_{fD}} = \frac{1}{M_{12}}\frac{d\overline{m}_{2D}}{dr_D}\bigg|_{r_D=r_{fD}} \tag{12}$$

where the subscripts 1 and 2 represent the inner and outer regions, r_{fD} is the composite radius, and M_{12} is the mobility ratio. The outer boundary condition for the outer region is

$$m_{2D}(r_D, t_D)_{r_D \to \infty} = 0 \tag{13}$$

Thus, the radial composite model solution with wellbore storage and skin is

$$\overline{m}_{wD} = \frac{1}{u}\frac{1 + S \times R_D}{R_D + C_D u[1 + S \times R_D]} \tag{14}$$

where

$$R_D = I_{12} \times \sqrt{u}\frac{-KI \times KI_1 - 1}{KI \times KI_0 + 1} \tag{15}$$

$$I_{12} = \frac{I_1(\sqrt{u})}{I_0(\sqrt{u})} \tag{16}$$

$$KI_0 = \frac{K_0(\sqrt{u})}{I_0(\sqrt{u})} \tag{17}$$

$$KI_1 = \frac{K_1(\sqrt{u})}{I_0(\sqrt{u})} \tag{18}$$

$$KI = \frac{I_1(a)K_0(b) + \frac{\sqrt{x_{21}}}{M_{12}}I_0(a)K_1(b)}{K_1(a)K_0(b) - \frac{\sqrt{x_{21}}}{M_{12}}K_0(a)K_1(b)} \tag{19}$$

$$a = r_{fD}\sqrt{u} \tag{20}$$

$$b = r_{fD}\sqrt{x_{21}u} \tag{21}$$

$$x_{21} = \frac{\omega_{12}}{M_{12}} \tag{22}$$

where I_0 is the zero-order second-class Bessel function, I_1 is the first-order second-class Bessel function, and ω_{12} is the dispersion ratio. According to pressure superposition, the linear point source function method is chosen to consider the effect of adjacent wells (Ozkan and Raghavan [32]). The point source solution for the i-th offset well is

$$\overline{m}_{D,i} = \overline{q}_{D,i} K_0\left(\sqrt{f(u)} r_{D,i}\right) \tag{23}$$

where $-q_D$ refers to injection. For the Warren–Root model, $f(u)$ is

$$f(u) = \frac{wu(1-w) + \lambda}{u(1-w) + \lambda} \tag{24}$$

where λ is the interporosity flow coefficient and w is the storativity ratio. The tested well location in the multi-well system is considered to be the origin, and the dimensionless distance in Equation (23) is

$$r_{D,i} = \sqrt{(x_{D,i} - x_{wD})^2 + (y_{D,i} - y_{wD})^2} \tag{25}$$

Well interference can be treated with pressure superposition:

$$\overline{m}_{D,j} = \sum_{i=1}^{N} \overline{q}_{D,i} K_0\left(\sqrt{f(u)} r_{D,i,j}\right), j = 2, 3, 4 \ldots N \tag{26}$$

where N is the number of wells in the multi-well system, and the subscripts i and j represent the serial numbers. The bottom-hole pressure solution with the interference of offset wells is

$$\overline{m}_{wD,Final} = \overline{m}_{wD} + \overline{m}_{D,j} \tag{27}$$

We obtain the bottom-hole pressure in the time domain using the Stehfest numerical inversion algorithm (Stehfest [33]).

3. Method Verification

3.1. Homogeneous Model

Figures 2 and 3 compare pressure and pressure derivative results from a numerical simulator and the proposed model. In case 1 of the homogeneous model verification, the multi-well system includes two production wells. The tested well is producing at a constant production rate of 1 m^3/day. An adjacent well, 50 m from the tested well, is producing at the same constant production rate. In contrast, the adjacent well in case 2 is completing the gas injection process with the same injection rate. The reservoir boundary is set to infinity in two cases. The remaining input parameters of the two cases are shown in Table 3. As the offset well interferes with the testing process of the tested well, Figure 2 shows that the pressure derivative rises at later times and eventually stabilizes at a higher level. When the adjacent well is an injector, the derivative for the tested well in Figure 3 shows a continuous decrease, similar to the effect of a constant pressure boundary. The good agreement between the proposed model and the numerical model in Figures 2 and 3 shows that the proposed model is reliable.

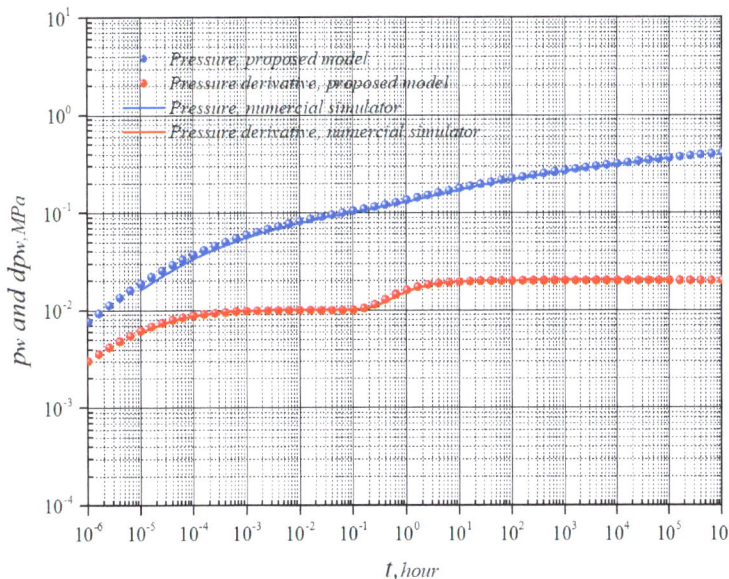

Figure 2. Comparison of the proposed model and numerical simulator for the homogeneous model case 1.

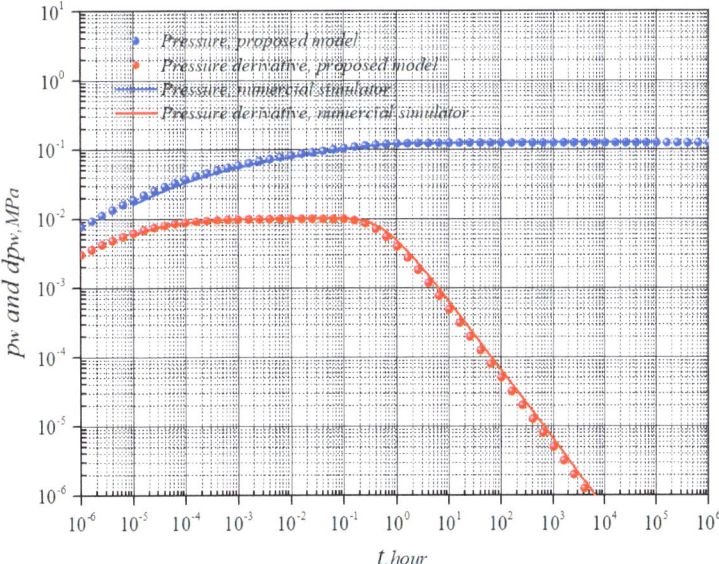

Figure 3. Comparison of the proposed model and numerical simulator for homogeneous model case 2.

Table 3. The reservoir, well, and gas properties in the method verification.

Group	Item	Case 1	Case 2	Case 3	Unit
Reservoir	Initial pressure	34.47	34.47	34.47	MPa
	Reservoir permeability	10	10	10	md
	Reservoir thickness	9.144	9.144	9.144	m
	Porosity	0.1	0.1	0.1	/
	Total compressibility	4.35×10^{-4}	4.35×10^{-4}	4.35×10^{-4}	MPa^{-1}
	Well spacing	50	50	500	m
	Mobility ratio	/	/	4	/
	Dispersion ratio	/	/	1	/
	Composite distance	/	/	100	m
Well	Tested well rate	1	1	1	m^3/day
	Adjacent well rate	1	−1	1	m^3/day
	Well radius	0.09	0.09	0.09	m
	Skin factor	0	0	0	/
	Wellbore storage coefficient	0	0	0	m^3/MPa
Gas	Viscosity	0.02	0.02	0.02	mpa.s
	Z-factor	0.0192	0.0192	0.0192	/

3.2. Radially Composite Model

Pressure transient testing results show that the continuous gas injection process in the UGS usually leads us to conclude that the formation is heterogeneous. Therefore, the radially composite model we used to describe the heterogeneous features is also appropriate for verification. We modeled the tested wells and adjacent wells in a multi-well system considered as a radially composite system and as a homogeneous system. The mobility ratio and dispersion ratio between the inner and outer regions were chosen as 4 and 1, as given in Table 4. The composite radius was initially chosen to be 100 m. The well spacing between the offset well and the tested well was selected as 500 m, and both wells produced at a constant rate of 1 m^3/day. The pressure derivative curve in Figure 4 shows that the derivative will eventually stabilize at a higher level, which may be related to the adjacent well interference and formation heterogeneity.

Table 4. The sequence, name, and feature of typical flow regimes in UGS.

Sequence	Name	Feature
(a)	Wellbore storage effect	Straight line with unit slope
(b)	Skin factor effect	Hump
(c)	Radial flow within the inner zone	Constant value of 0.5
(d)	Transitional flow	/
(e)	Radial flow in the outer region	Constant value of $0.5 \times M$
(f)	Transitional flow	/
(g)	Radial flow with well interference	Constant value of $0.5 \times M \times Q_{test,D} + Q_{adj,D} \times 0.5$

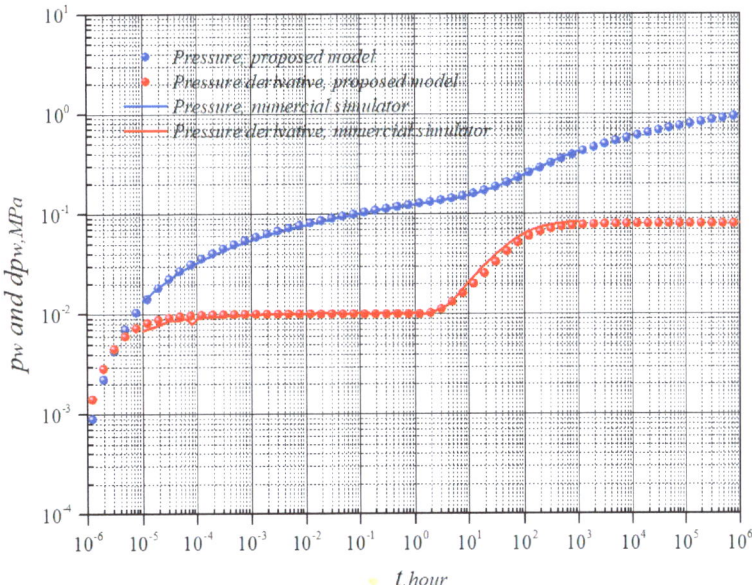

Figure 4. Comparison of the proposed model and numerical simulator for a radially composite model.

4. Results and Discussion

4.1. Flow Regimes

Figure 5 and Table 4 show that wellbore storage dominates the first flow regime. Due to the difference between the production rate at the wellhead and the well bottom, the pressure and its derivative both fall on a straight line with unit slope. The pressure derivative that follows exhibits the effect of a skin factor and includes a hump. The height of the hump is related to the wellbore storage coefficient C_D and the skin factor S. The third stage indicates radial flow within the inner zone, as the pressure derivative curve appears as a horizontal line with a value of 0.5. As the pressure response continues to propagate, the transition between the inner and outer regions begins gradually. When the pressure response propagating to the outer zone gradually becomes circular, radial flow in the outer region begins to appear. At this stage, the pressure derivative appears as a straight line, and its value is related to the ratio between 0.5 and the mobility ratio. After the radial flow in the outer region, the features of offset well interference gradually appear. The pressure derivative begins to rise continuously, indicating transitional flow. As well interference gradually increases, the characteristics of the entire multi-well system gradually appear. Finally, radial flow appears in the multi-well system. At this time, the horizontal derivative curve during radial flow of the multi-well system can be divided into two parts: the radial composite feature of the tested well and the adjacent well interference. The radially composite feature results in the horizontal line values related to 0.5, mobility ratio, and dimensionless production rate of the tested well. Well interference also causes the horizontal line value to relate to the dimensionless production rate of the offset well and 0.5. If offset wells are in the process of gas injection, the pressure derivative shows the characteristic of drop-off. The comparison of the newly added single-well model in Figure 5 shows that the pressure derivative of the single-well model finally stabilizes at the horizontal line with the value of 0.5 M.

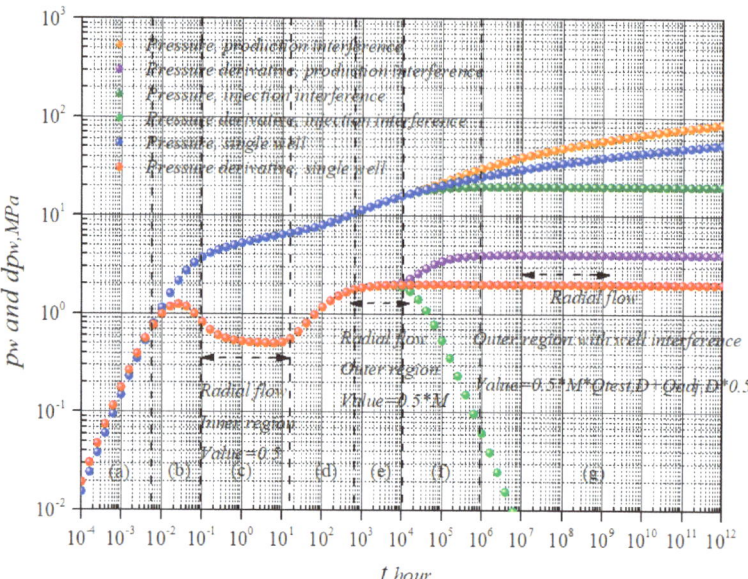

Figure 5. Flow regimes in the multi-well system with a radially composite reservoir and well interference during UGS. The pressure behaviors of the single well model are used for comparison.

4.2. Sensitivity Analysis

4.2.1. Effect of Well Spacing

Figure 6 shows the effect of well spacing on the pressure transient behavior of tested wells in the multi-well system as well spacing is gradually increased from 100 m to 5000 m. We can draw the conclusion that the variation of well spacing affects the radial flow regime in the inner region, the transitional flow between the inner and the outer regions, the radial flow in the outer region, and the pseudo-radial flow in the multi-well system. As well spacing increases, the duration of radial flow in the inner region increases gradually. The starting time of the radial flow of the multi-well system is proportional to the well spacing. When the well spacing is 100 m, the tested well and adjacent wells can easily communicate with each other to form a whole. In this case, the flow regimes are simple. Radial flow in the inner region is followed by a long transitional flow regime. This transitional flow is controlled by a combination of radial composite features and well interference. Since the well spacing is the same as the composite radius, the characteristics of well interference mask the radial flow feature of the outer region. The pressure derivative rises. With the gradual increase in well spacing, the transitional flow regime between the inner and outer regions, the radial flow regime in the outer region, and the transitional flow during well interference gradually emerge. As the well spacing increases to 1000 m, the transitional flow between the inner and outer regions begins to appear. When the well spacing increases to 5000 m, the pressure derivative exhibits both radial flow in the outer region and transitional flow that characterizes interference.

4.2.2. Effect of Neighboring Well's Production Rate

Figure 7 shows the effect of adjacent well production changes on the pressure derivative in the multi-well system. The dimensionless production rate of adjacent wells varies from 1 to 5, 10, and 20. A comparison of pressure derivatives shows that interference from adjacent well production affects the transition flow region of the test well and the pseudo-radial flow region of the multi-well system. As the production rate of adjacent wells increases, the slope of the pressure derivative curve increases gradually during the well interference transition flow regime. The pressure derivative curve also displays a horizontal line in the pseudo-radial flow regime of the multi-well system, and its value is proportional to the production rate of the adjacent well. Flow regime analysis shows that rate variation in adjacent wells affects the pseudo-radial flow during interference within the multi-well system.

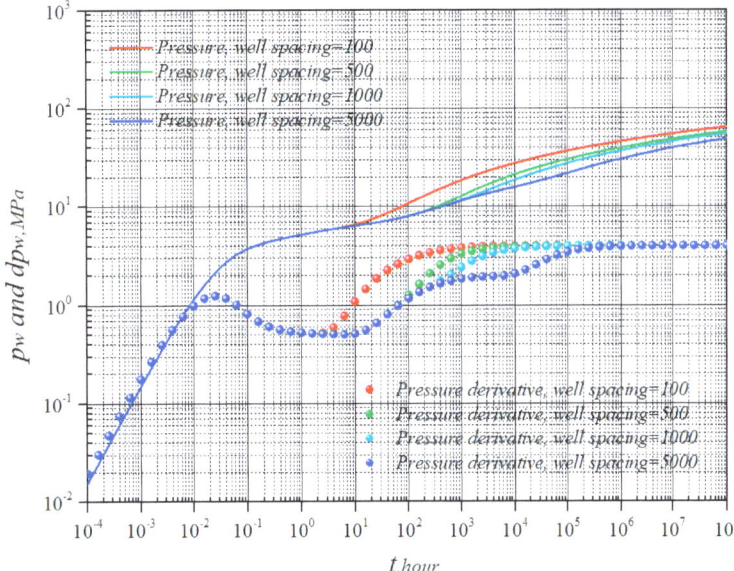

Figure 6. Effect of well spacing on the tested well's pressure behaviors in the multi-well system.

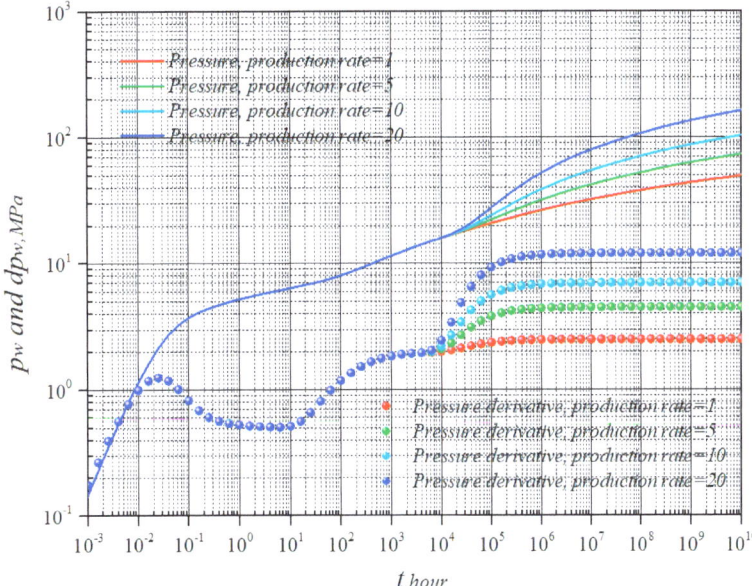

Figure 7. Effect of neighboring well's production rate on the tested well's pressure behaviors in the multi-well system.

4.2.3. Effect of Neighboring Well's Injection Rate

The effect of the neighboring well's injection rate on the pressure of the tested wells in a multi-well system is shown in Figure 8. The dimensionless neighbor well's injection rate increased from 1 to 6. An example of a neighbor well producing at a constant production rate is also shown for comparison. A comparison of pressure derivatives shows that the neighbor well affects the well interference transition flow regime and the pseudo-radial flow of the multi-well system. With the increase in the injection rate of adjacent wells, the rise of the pressure curve slows and the pressure derivative curve gradually rises to a lower horizontal level during pseudo-radial flow in the multi-well system. The tested well's dimensionless production rate is 1, and the mobility ratio between various regions is 4. As the dimensionless production rate of neighboring wells increases to 4, the pressure curve does not rise, and the derivative curve begins to trend downward. This downward trend becomes more pronounced as the injection rate into adjacent wells increases.

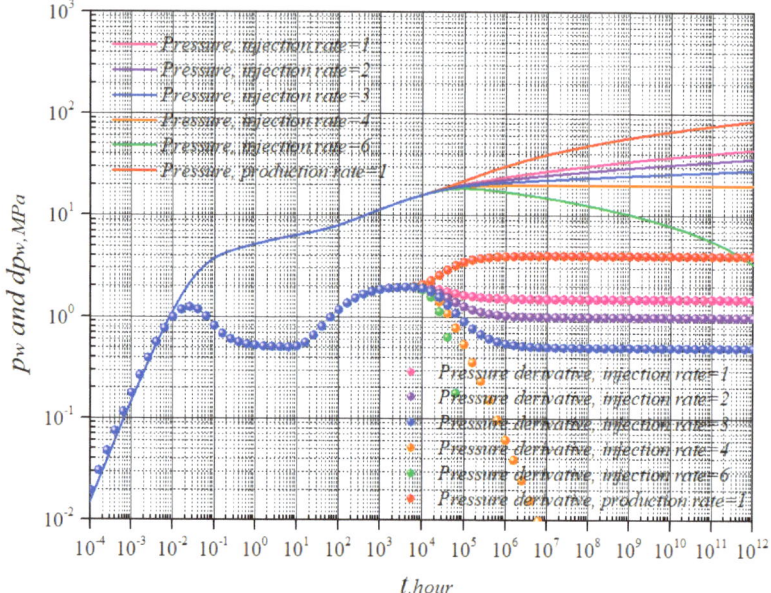

Figure 8. Effect of neighbor well's injection rate on the tested well's pressure behaviors in the multi-well system.

4.2.4. Effect of Composite Radius

The effect of composite radius changes on tested wells in a multi-well system is shown in Figure 9. The composite radius of the test well is gradually increased from 100 m to 500 m. The comparison of the pressure derivative curves in Figure 9 shows that the change of composite radius affects the radial flow in the inner region, the transitional flow between various regions, and the radial flow in the outer region. As the composite radius increases, the radial flow duration in the inner region gradually increases, and the starting and ending times of the transitional flow in the inner and outer regions are gradually delayed. As a result, radial flow in the outer region is masked. Both transitional flow in the inner and outer regions and radial flow in the outer region have the characteristics of the transitional flow.

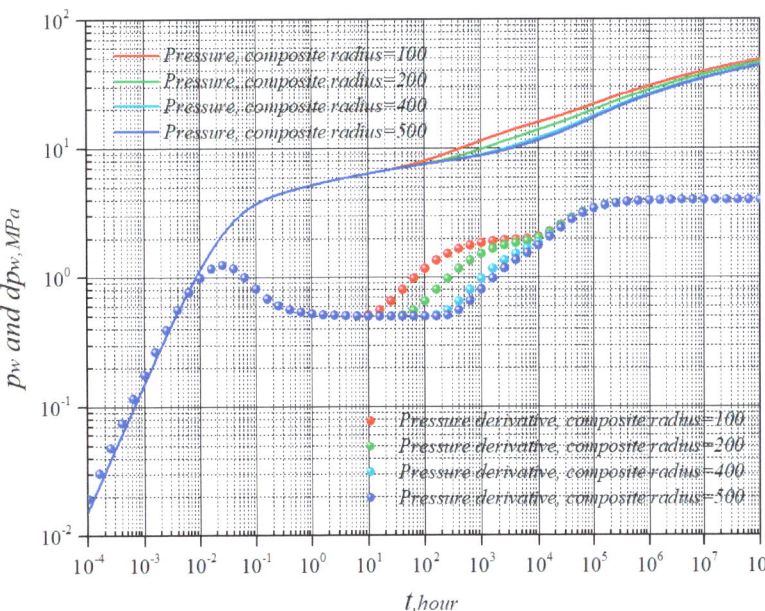

Figure 9. Effect of composite radius on the tested well's pressure behaviors in the multi-well system.

5. Field Application

5.1. Geological Background

Hutubi UGS is located in the southern margin of the Junggar Basin (piedmont depression of north Tianshan Mountain), as shown in Figure 10. It is a long axis anticline with an east–west direction of about 20 km in length and 3.5 km in width from north to south. It was cut along the long axis by two nearly parallel overthrust faults (Hutubi fault and Hutubi North fault). The gas layer of the Ziniquanzi Formation is a set of regressive delta deposits, about 355 m thick. The lithology is mainly brown, grayish-brown fine sandstone, unequal-grained sandstone, siltstone, pebbled unequal-grained sandstone, and pebbled argillaceous sandstone, and it has good sealing and trapping conditions. Before the UGS formed, it was a sandstone condensate reservoir with edge and bottom water. Geological data show that the average porosity is 20.9%, the permeability is 62.49 mD, and the original formation pressure is 33.96 MPa. The designed capacity of Hutubi gas storage is 107×10^8 m^3, and the working gas volume is 45.1×10^8 m^3.

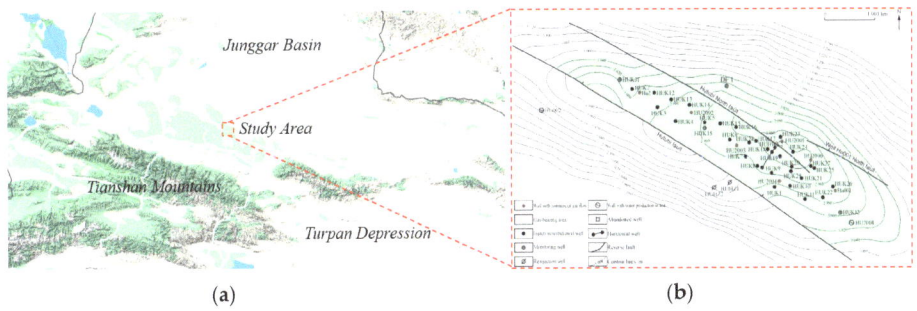

Figure 10. (**a**) Regional location of the study area. (**b**) Structure and well location map of the study area (modified from Zheng et al. [34]).

5.2. Pressure Transient Testing

As shown in Figure 10b, two wells in the UGS were selected for pressure transient testing. The target well (a) is in the fourth cycle of cumulative gas injection of 4970.46×10^4 m^3 before the pressure transient testing, and the daily gas injection rate is stable at 21.67×10^4 m^3/day. The pressure gauge was lowered to the well bottom hole of 3463 m on 31 August 2016, and the duration of the pressure transient test was 20 days. During testing, gas injection into the offset well began with varying rates. The cumulative gas injection into the target well (b) in the fourth cycle before the test was $10{,}580 \times 10^4$ m^3, and the daily gas injection rate was stable at 68.68×10^4 m^3/day. On 24 September 2016, the pressure gauge was lowered to the bottom hole at 3472 m, and the cumulative test duration was 19 days. During the test period, gas was injected into the adjacent well (a) at an average rate of 24.4341×10^4 m^3/day from 23 September to 1 October. Gas was injected into the adjacent well (b) from 24 September to 1 October at an average gas injection rate of 27.24×10^4 m^3/day. The pressure testing data in Figures 11 and 12 show that the derivatives decrease, and the difference between target wells (a) and (b) is that the test period is different. For target well (a), the impact time of falling down is the ending period of the pressure derivative. This test time for target well (b) is in the middle and late stages of the derivative curve. Combining the test period of the target well with the production period of the adjacent wells, we can draw the conclusion that there are well interference and formation heterogeneity. The single-well models in Figures 11 and 12 could not match the later stage of actual data. Further, we used the proposed model to match the field data and compared pressures and derivatives in Figures 11 and 12 which illustrate the applicability of the model. The reservoir, well, and gas properties are provided in Tables 5 and 6.

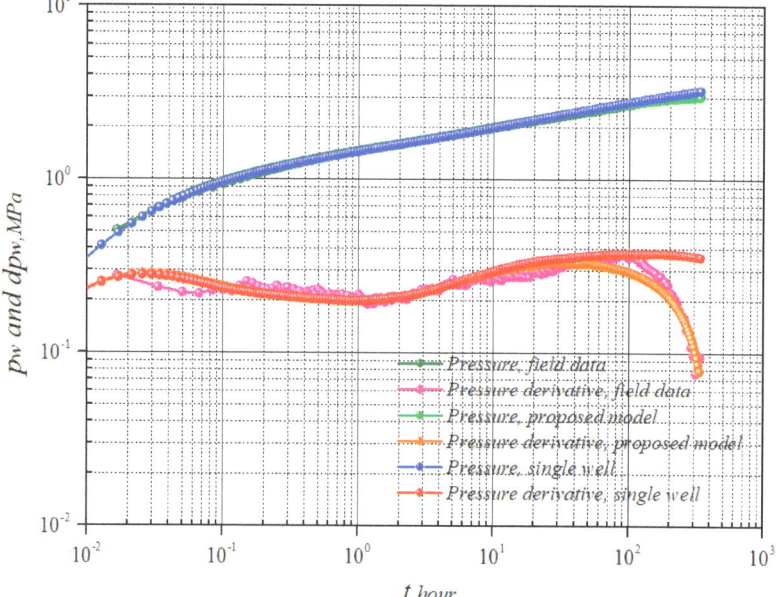

Figure 11. Pressure and pressure derivative match results in the fourth cycle of cumulative gas injection for case 1.

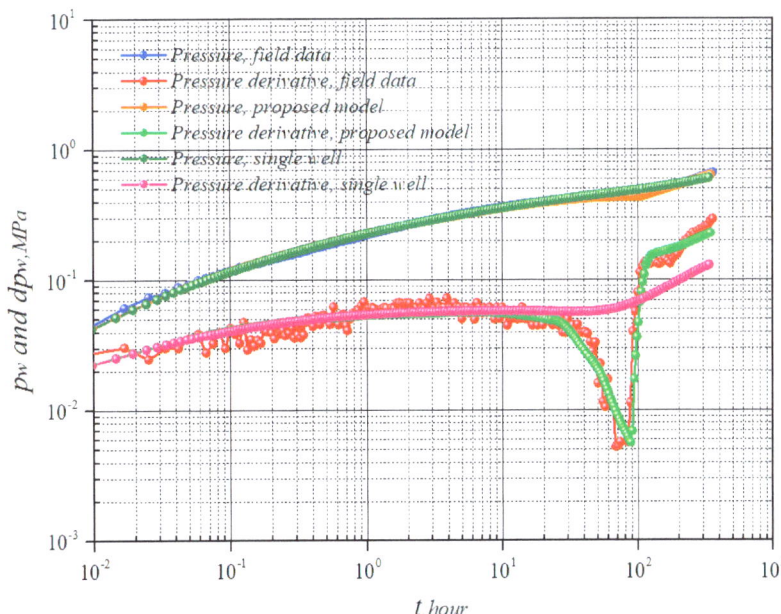

Figure 12. Pressure and pressure derivative match results in the fourth cycle of cumulative gas injection for case 2.

Table 5. The reservoir, well, and gas properties in pressure transient analysis of case 1.

Group	Item	Value	Unit
Reservoir	Initial pressure	29.47	MPa
	Reservoir permeability	4.31	md
	Reservoir thickness	35	m
	Porosity	0.2	
	Total compressibility	4.35×10^{-4}	MPa^{-1}
	Mobility ratio	2.1	
	Dispersion ratio	2.2	
	Composite distance	28	m
	Well spacing	300	m
	Adjacent well rate	1×10^6	m^3/day
Well	Well radius	0.1	m
	Skin factor	−1.37	
	Wellbore storage coefficient	0.63	m^3/MPa
Gas	Viscosity	0.02	mpa.s
	Z-factor	0.0192	

Table 6. The reservoir, well, and gas properties in pressure transient analysis of case 2.

Group	Item	Value	Unit
Reservoir	Initial pressure	32.45	MPa
	Reservoir permeability	70	md
	Reservoir thickness	22	m
	Porosity	0.2	
	Total compressibility	4.35×10^{-4}	MPa^{-1}
	Mobility ratio	10	
	Dispersion ratio	10	
	Composite distance	830	m
	Well spacing	470	m
Well	Adjacent well rate	5×10^5	m^3/day
	Well radius	0.1	m
	Skin factor	−4.75	
	Wellbore storage coefficient	0.15	m^3/MPa
Gas	Viscosity	0.02	mpa.s
	Z-factor	0.019	

6. Summary and Conclusions

This work uses an analytical approach to analyze the pressure transient behavior of a multi-well system in UGS. The model reliability was validated with a commercial numerical simulator. Typical flow regimes in UGS were diagnosed using Bourdet pressure derivatives. Sensitivity analysis and the field case from Hutubi UGS demonstrate the practical applicability of the method. Some key conclusions are as follows:

1. The typical flow regimes for a vertical well in the UGS include wellbore storage, skin effect, radial flow in an inner region, radial flow in an outer region, and effects of well interference.

2. Long-duration gas injection and production periods in the UGS amplify the influence of heterogeneities in a formation. With an increase in gas injection and production, heterogeneities exhibited by the reservoir increase, and the radially composite signature in the pressure transient test response becomes more apparent. The pressure derivative increases in the middle and later periods.

3. The typical signature of the flow regime during which well interference occurs depends on the operation of adjacent wells. As adjacent wells are producing, the pressure derivative finally exhibits the pseudo-radial flow of the multi-well system under the influence of well interference. The horizontal derivative value is related to the dimensionless production rate of the target well, adjacent wells, mobility ratio, and 0.5. When injection into adjacent wells is occurring, the final portion of the pressure derivative curve decreases.

4. Field application shows that well interference and formation heterogeneity are commonly observed in UGS, and the pressure derivative curve tends to have rising and falling features. The proposed model can be used to effectively analyze the transient pressure data with well interference and heterogeneity in the UGS.

This is our primary work on UGS. The study focuses on the pressure behavior of multiple vertical wells. Our future research work will be extended to complex situations of various well types, pressures, and rate behaviors. In addition, more UGS field data will be collected to form a multidisciplinary approach.

Author Contributions: Conceptualization, H.C. and T.M.; methodology, H.C.; software, Y.G.; validation, Y.G.; formal analysis, Z.C.; investigation, Z.C.; resources, W.L.; data curation, H.C.; writing—original draft preparation, H.C.; writing—review and editing, H.C.; visualization, H.C.; supervision, W.L.; project administration, T.M.; funding acquisition, H.C. All authors have read and agreed to the published version of the manuscript.

Funding: This research was funded by National Science Foundation, China (12202042); China Postdoctoral Science Foundations (2021M700391); and Fundamental Research Funds for the Central Universities (QNXM20220011). And The APC was funded by Fundamental Research Funds for the Central Universities (QNXM20220011).

Conflicts of Interest: The authors declare no conflict of interest.

Nomenclature

k	permeability, D
m_D	dimensionless pseudo-pressure pressure
t_D	dimensionless time
r_D	dimensionless distance
u	Laplace variable
C_D	wellbore storage coefficient
S	skin factor
m_{wD}	dimensionless bottom-hole pseudo-pressure
K_0	zero-order first-class Bessel function
K_1	first-order first-class Bessel function
r_{fD}	dimensionless composite radius
M_{12}	mobility ratio
ω_{12}	dispersion ratio
I_0	zero-order second-class Bessel function
I_1	first-order second-class Bessel function
λ	interporosity flow coefficient
ω	storativity ratio
N	well number
μ	viscosity, mpa.s
B	formation volume factor
h	thickness, m
C_t	total compressibility, MPa^{-1}
L	reference length, m
φ	porosity
d	well spacing, m
q	production rate, m^3/day
T	temperature, K
T_{sc}	temperature at standard conditions, K
p_{sc}	pressure at standard conditions, MPa

Appendix A. Dimensionless Variables

Dimensionless pseudo-pressure in UGS,

$$m_D = \frac{kT_{sc}h(m_i - m)}{3.684 \times 10^{-3} p_{sc} q_{t,sc} T} \tag{A1}$$

where

$$m = \int_{p_{sc}}^{p} \frac{p}{\mu Z} dp \tag{A2}$$

Dimensionless time,

$$t_D = \frac{3.6kt}{\mu \Lambda h^2} \tag{A3}$$

where

$$\Lambda = \varphi C_g + \frac{kh}{1.842 \times 10^{-3} q_{sc} \mu} \tag{A4}$$

Dimensionless wellbore storage coefficient,

$$C_D = \frac{0.159C}{\varphi C_t h L^2} \tag{A5}$$

Dimensionless well spacing,

$$d_D = \frac{d}{L} \tag{A6}$$

Dimensionless mobility ratio,

$$M_{12} = \frac{(k/u)_1}{(k/u)_2} \tag{A7}$$

Dimensionless dispersion ratio,

$$\omega_{12} = \frac{(\varphi C_t)_1}{(\varphi C_t)_2} \tag{A8}$$

Dimensionless distance,

$$x_D = \frac{x}{L}, y_D = \frac{y}{L}, r_D = \frac{r}{L} \tag{A9}$$

Dimensionless location for the origin,

$$x_{wD} = \frac{x_w}{L}, y_{wD} = \frac{y_w}{L}, r_{wD} = \frac{r_w}{L} \tag{A10}$$

References

1. Yang, C.; Wang, T.; Li, Y.; Yang, H.; Li, J.; Qu, D.; Xu, B.; Yang, Y.; Daemen, J. Feasibility analysis of using abandoned salt caverns for large-scale underground energy storage in China. *Appl. Energy* **2015**, *137*, 467–481. [CrossRef]
2. Matos, C.R.; Carneiro, J.F.; Silva, P.P. Overview of large-scale underground energy storage technologies for integration of renewable energies and criteria for reservoir identification. *J. Energy Storage* **2019**, *21*, 241–258. [CrossRef]
3. Jensen, J.H.; Poulsen, J.M.; Andersen, N.U. From coal to clean energy. *Nitrogen Syngas* **2011**, *310*, 34–38.
4. Weijermars, R.; Drijkoningen, G.; Heimovaara, T.J.; Rudolph, E.S.J.; Weltje, G.J.; Wolf, K.H.A.A. Unconventional gas research initiative for clean energy transition in Europe. *J. Nat. Gas Sci. Eng.* **2011**, *3*, 402–412. [CrossRef]
5. Qyyum, M.A.; Naquash, A.; Haider, J.; Al-Sobhi, S.A.; Lee, M. State-of-the-art assessment of natural gas liquids recovery processes: Techno-economic evaluation, policy implications, open issues, and the way forward. *Energy* **2022**, *238*, 121684. [CrossRef]
6. Zhang, J.; Tan, Y.; Zhang, T.; Yu, K.; Wang, X.; Zhao, Q. Natural gas market and underground gas storage development in China. *J. Energy Storage* **2000**, *29*, 101338. [CrossRef]
7. Chen, J.; Yu, J.; Ai, B.; Song, M.; Hou, W. Determinants of global natural gas consumption and import–export flows. *Energy Econ* **2019**, *83*, 588–602. [CrossRef]
8. Crow, D.J.; Giarola, S.; Hawkes, A.D. A dynamic model of global natural gas supply. *Appl. Energy* **2018**, *218*, 452–469. [CrossRef]
9. Demirel, N.Ç.; Demirel, T.; Deveci, M.; Vardar, G. Location selection for underground natural gas storage using Choquet integral. *J. Nat. Gas Sci. Eng.* **2017**, *45*, 368–379. [CrossRef]
10. Wang, J.; Zhang, J.; Xie, J.; Ding, F. Initial gas full-component simulation experiment of Ban-876 underground gas storage. *J. Nat. Gas Sci. Eng.* **2014**, *18*, 131–136. [CrossRef]
11. Wang, J.; Feng, X.; Wanyan, Q.; Zhao, K.; Wang, Z.; Pei, G.; Xie, J.; Tian, B. Hysteresis effect of three-phase fluids in the high-intensity injection–production process of sandstone underground gas storages. *Energy* **2022**, *242*, 123058. [CrossRef]
12. Lv, J.; Li, Z.; Fu, J.; Tang, J.; Wei, X.; Qie, X. Compositional variation laws of produced gas from underground gas storage tanks rebuilt from sour gas reservoirs: A case study of Shaan 224 UGS in the Ordos Basin. *Nat. Gas Ind.* **2017**, *37*, 96–101.
13. Tang, Y.; Long, K.; Wang, J.; Xu, H.; Wang, Y.; He, Y.; Shi, L.; Zhu, H. Change of phase state during multi-cycle injection and production process of condensate gas reservoir based underground gas storage. *Pet. Explor. Dev.* **2021**, *48*, 395–406. [CrossRef]
14. Lewandowska-Śmierzchalska, J.; Tarkowski, R.; Uliasz-Misiak, B. Screening and ranking framework for underground hydrogen storage site selection in Poland. *Int. J. Hydrogen Energy* **2018**, *43*, 4401–4414. [CrossRef]

15. Cedigaz. Underground Gas Storage in the World—2017 Status. 2017. Available online: http://www.cedigaz.org/resources/free-downloads.aspx/ (accessed on 1 March 2018).
16. Lee, J. Well testing. *Soc. Pet. Eng.* **1982**.
17. Meehan, D.N.; Horne, R.N.; Ramey, H.J. Interference testing of finite conductivity hydraulically fractured wells. In Proceedings of the SPE Annual Technical Conference and Exhibition, San Antonio, TX, USA, 8 October 1989.
18. Malekzadeh, D.; Tiab, D. Interference testing of horizontal wells. In Proceedings of the SPE Annual Technical Conference and Exhibition, Dallas, TX, USA, 6–9 October 1991.
19. Awotunde, A.A.; Al-Hashim, H.S.; Al-Khamis, M.N.; Al-Yousef, H.Y. Interference Testing Using Finite-Conductivity Horizontal Wells of Unequal Lengths. In Proceedings of the SPE Eastern Regional/AAPG Eastern Section Joint Meeting, Pittsburgh, PA, USA, 11–15 October 2008.
20. Al-Khamis, M.; Ozkan, E. Analysis of interference tests with horizontal wells. *SPE Res. Eval. Eng.* **2005**, *8*, 337–347. [CrossRef]
21. Spivey, J.P.; Lee, W.J. *Applied Well Test Interpretation*; Society of Petroleum Engineers: Richardson, TX, USA, 2013; Volume 13, pp. 187–229.
22. Warren, J.E.; Root, P.J. The behavior of naturally fractured reservoirs. *Soc. Pet. Eng. J.* **1963**, *3*, 245–255. [CrossRef]
23. Onur, M.; Serra, K.V.; Reynolds, A.C. Analysis of pressure-buildup data from a well in a multiwell system. *SPE Form. Eval.* **1991**, *6*, 101–110. [CrossRef]
24. Fokker, P.A.; Verga, F. A semianalytic model for the productivity testing of multiple wells. *SPE Reserv. Eval. Eng.* **2008**, *11*, 466–477. [CrossRef]
25. Lin, J.E.; Yang, H.Z. Analysis of well-test data from a well in a multiwell reservoir with water injection. In Proceedings of the SPE Annual Technical Conference and Exhibition, Anaheim, CA, USA, 11–14 November 2007.
26. Izadi, M.; Yildiz, T. Transient flow in discretely fractured porous media. *SPE J.* **2009**, *14*, 362–373. [CrossRef]
27. Wei, C.; Liu, Y.; Deng, Y.; Cheng, S.; Hassanzadeh, H. Analytical well-test model for hydraulically fractured wells with multiwell interference in double porosity gas reservoirs. *J. Nat. Gas Sci. Eng.* **2022**, *103*, 104624. [CrossRef]
28. Chu, H.; Liao, X.; Chen, Z.; John Lee, W.J. Rate-transient analysis of a constant-bottomhole-pressure multihorizontal well pad with a semianalytical single-phase method. *SPE J.* **2020**, *25*, 3280–3299. [CrossRef]
29. Chu, H.; Chen, Z.; Liao, X.; Zhao, X.; Lee, W.J. Pressure transient analysis of a multi-horizontal-well pad by a semi-analytical model: Methodology and case study. *J. Pet. Sci. Eng.* **2022**, *208*, 109538. [CrossRef]
30. Du, G.; He, L.; Zhang, G.; Wang, Y.; Zhang, S.; Zhou, N.; Yang, D.; Zhou, X.; He, H.; Lin, J. Well Testing Analysis Method for a Well in Hutubi Multi-well Underground Gas Storage Reservoir. In Proceedings of the International Field Exploration and Development Conference 2017, Chengdu, China, 12 July 2018; Springer: Singapore; pp. 1870–1882.
31. Uniper. 2022. Available online: https://www.uniper.energy/energy-storage-uniper/gas-storage-technology (accessed on 11 December 2021).
32. Ozkan, E.; Raghavan, R. New solutions for well-test-analysis problems: Part 1-analytical considerations (includes associated papers 28666 and 29213). *SPE Form. Eval.* **1991**, *6*, 359–368. [CrossRef]
33. Stehfest, H. Algorithm 368: Numerical inversion of Laplace transforms [D5]. *Commun. ACM* **1970**, *13*, 47–49. [CrossRef]
34. Zheng, Y.; Sun, J.; Qiu, X.; Lai, X.; Liu, J.; Guo, Z.; Wei, H.; Min, Z. Connotation and evaluation technique of geological integrity of UGSs in oil/gas fields. *Nat. Gas Ind. B* **2020**, *7*, 594–603. [CrossRef]

Numerical Simulation on Borehole Instability Based on Disturbance State Concept

Daobing Wang [1,2,*], Zhan Qu [1], Zongxiao Ren [1], Qinglin Shan [3], Bo Yu [2], Yanjun Zhang [1] and Wei Zhang [2]

[1] The Key Laboratory of Well Stability and Fluid & Rock Mechanics in Oil and Gas of Shaanxi Province, Xi'an Petroleum University, Xi'an 710065, China
[2] School of Mechanical Engineering, Beijing Institute of Petrochemical Technology, Beijing 102617, China
[3] School of Energy and Mining Engineering, Shandong University of Science and Technology, Qingdao 266510, China
* Correspondence: upcwdb@bipt.edu.cn; Tel.: +86-10-81292036

Abstract: This paper carries out a study on the numerical simulation of borehole instability based on the disturbance state concept. By introducing the disturbance damage factor into the classical Mohr–Coulomb yield criterion, we establish a finite element hydro-mechanical coupling model of borehole instability and program the relevant field variable by considering elastic–plastic deformation in borehole instability, the distribution of the damage disturbance area, the variation of porosity and permeability with the disturbance damage factor, etc. Numerical simulation shows that the borehole stability is related to the action time of drilling fluid on the wellbore, stress anisotropy, the internal friction angle of rock, and borehole pressure. A higher horizontal stress difference helps suppress shear instability, and a higher rock internal friction angle enhances shear failure around the borehole along the maximum horizontal principal stress. When considering the effect of the internal friction angle of rock, the rock permeability, disturbance damage factor, and equivalent plastic strain show fluctuation characteristics. Under the high internal friction angle of rock, a strong equivalent plastic strain area and disturbance damage area occur in the direction of the maximum horizontal principal stress. Their cloud picture shows the mantis shape, where the bifurcation corresponds to the whiskers of the shear failure area in borehole instability. This study provides a theoretical basis for solving the problem of borehole instability during drilling engineering.

Keywords: borehole stability; disturbance state concept; elastic–plastic deformation; finite element method; numerical simulation

1. Introduction

In the oil and gas industry, 50–80% of exploration and development costs are spent on drilling. The downhole accidents such as wellbore shrinkage, sticking, formation collapse, and leakage caused by borehole instability lead to an increase in the drilling cycle, the damage of downhole equipment, and an increase in the cost, which has restricted oil and gas exploration and development [1–3]. The core of high-quality, safe, efficient, and low-cost drilling is to evaluate the downhole surrounding rock environment, study the mechanism of borehole instability, and control the borehole instability, and it is of great theoretical and practical significance [4–6].

The disturbance state concept (DSC) was proposed by Desai in 1974 in the United States [7], and a relatively complete theoretical system was formed. The DSC has been applied in metal, welding material, soil, rock, oil sand, concrete, and electronic packaging materials. Research on the DSC started from the material mechanical response, hierarchical single surface model, and numerical simulation method. In 1992, Desai et al. established a DSC-based unified constitutive model for study on the static behavior of rock joints and interfaces [8]. In 1995, Katti et al. established a DSC-based clay constitutive model for the prediction of the response of stress–strain and pore pressure of undrained saturated clay under cyclic loading, as well as the response of soil under earthquake [9]. Desai et al. introduced the viscoplastic constitutive relation into the DSC to describe the response of the material in the relatively intact state [10]. In 1996, Desai et al. established a constitutive model based on stress–strain and non-destructive behavior and used DSC to describe the crack density [11]. In 1998, Desai et al. used the DSC-based numerical simulation method to establish the governing equation, disturbance function, and finite element equation of the relatively complete state and fully adjusted state [12]. In 1996, Pal integrated DSC and computer methods to describe the mechanical behavior of the solid and contact face [13]. In 2016, Fan et al. established a general compression model of metal-rich clay based on the general DSC compression model [14]. In 2017, Ouria et al. used the DSC function to describe the coefficient of compressibility of structural soil [15]. In 2018, Ghazavi Baghini et al. applied DSC to simulate the behavior of the pile under the axial load [16].

In China, some research progress has been made on DSC since 2000. Wu et al. applied the DSC to establish the nonlinear constitutive model and elastic–plastic constitutive model of rock [17]. Zheng et al. developed the method of describing the triaxial compression response of rock and the stress anisotropic response of soil based on DSC [18] and proposed a evolution equation of the disturbance factor through a mesoscopic analysis of the DSC established by the hardening model [19]. Zhang et al. established a creepage model of structural soft soil based on DSC [20]. Fu et al. proposed two methods of disturbance factor evolution based on the conventional triaxial test curve and the volumetric strain threshold, and the limit state of deviator strain energy [21]. Yang et al. applied the DSC hardening parameters to establish a structural clay boundary surface model [22]. Huang et al. established a creepage disturbance factor model with time as an independent variable [23]. Zou et al. established a stress–strain model of hydrated soil with the DSC method to describe the process of the failure of the cement structure [24]. The application of the DCS method in borehole stability is still not reported [25–27].

In previous mechanical theory, it was supposed that the cracks and damage inside the borehole rock have no strength [27–40]. In the DCS, it is proposed that the cracks and damaged parts are caused by the continuous merging and integration of defects in the internal complex microstructure, and they still have a certain strength and reflect softening and weakening caused by the propagation of crack and failure and hardening and strengthening caused by continuous compression. The DCS reveals the mechanism of the mechanical response of the borehole wall. Moreover, the DCS suggests that various forces cause the disturbance of the material microstructure, and the self-adjustment of the material internal microstructure includes relative motion that leads to damage, softening, or compression hardening of the material and macroscopically obvious disturbance. A description of disturbance through macroscopic observation provides the method of a cross-scale analysis of the micro-response of internal complex microstructure and the macroscopic behavior of yield failure in borehole rock. In the DCS, the material is considered as a random mixture under the relatively intact stage and the fully adjusted state, which correspond to the undamaged part and the damaged part in previous models. Material deformation and failure caused by disturbance is a process of transition from a relatively intact state to a fully adjusted state through self-adjustment and self-organization.

To overcome the defects and limits of conventional methods such as fracture mechanics, damage mechanics, and configuration mechanics, we carried out a numerical simulation of borehole instability based on the DSC by considering microscopic to macroscopic effects and the multi-regional response of borehole rock. We revealed the mechanism and evolution of borehole instability and developed a system for DSC-based study on borehole stability.

The paper is organized as follows. Section 2 introduces the mechanical theories and methods of borehole stability, including the mechanical equilibrium equation, seepage equation, the theory of borehole instability in a disturbed state, and model verification. Section 3 introduces the finite element model for borehole instability, mesh division, boundary conditions, and secondary development of subroutine. Section 4 discusses the numerical simulation results and analyzes the effects of action time of drilling fluid on the wellbore;

stress anisotropy; internal friction angle; and borehole pressure on the equivalent plastic strain, permeability, borehole wall stress, and disturbance damage factor. The main conclusions of this study are summarized in the last section.

2. Theory and Method

2.1. Mechanical Equilibrium Equation

According to rock mechanics, the mechanical equilibrium equation of rock borehole stability is expressed as [41–43]:

$$\sigma_{ij,j} + X_i = 0 \tag{1}$$

where σ_{ij} is the stress tensor component; X_i is the body force vector component.

Assuming small rock deformation, the geometric equation of borehole stability is expressed as:

$$\varepsilon_{ij} = \frac{1}{2}(u_{i,j} + u_{j,i}) \tag{2}$$

where ε_{ij} is the strain tensor component; u_i is the displacement vector component.

According to the effective stress of porous media, we have [44]:

$$\sigma'_{ij} = \sigma_{ij} - \alpha p \delta_{ij} = C_{ijkl} : \varepsilon_{kl} \tag{3}$$

where C_{ijkl} is the stiffness tensor component; p is the pore pressure; α is the Biot constant; σ'_{ij} is the effective stress tensor component; δ_{ij} is the Kronecker symbol, which is 0 when $i = j$ and 0 when $i \neq j$.

2.2. Seepage Equations

According to the theory of seepage mechanics, the fluid seepage equation in the borehole surrounding rock is expressed as [45]:

$$\nabla \cdot \left(\frac{k}{\mu} \nabla p \right) = \frac{1}{M} \frac{\partial p}{\partial t} - \alpha \frac{\partial \varepsilon_v}{\partial t} \tag{4}$$

where k is the permeability tensor; ε_V is the volumetric strain component; M is the Biot modulus; and t is the time of the drilling fluid action on the borehole wall.

2.3. Theory of DSC Borehole Instability

The basic principle of the DSC is to consider the material under stress disturbance as a random mixture in an undisturbed state and a completely disturbed state, and its mechanical response is determined by a weighted average of the mechanical response of the part in an undisturbed state and the part in a completely disturbed state. The undisturbed state refers to that the material is at the idealized state or the undisturbed and little disturbance state which is defined subjectively. For example, a stable material with a hardening response is in an undisturbed state. The completely disturbed state refers to the limit of the material under stress disturbance. According to the DSC principle, the basic elements include the undisturbed state, the completely disturbed state, and the disturbance function.

The establishment of the model requires a definition of the basic elements, where the undisturbed state corresponds to the non-crack (non-damaged) part in the damage mechanics constitutive model, and the completely disturbed state corresponds to the damaged part; the disturbance function corresponds to the damage function. The crack (damage) part has no strength; the completely disturbed state part has specific stress–strain and strength; and the disturbance function characterizes softening (damage), strengthening, and hardening.

Damage to the rock affects the effective shear strength parameters c^* and ϕ^*, which are the function of the disturbance state. Under the action of disturbance and pore pressure, the Mohr–Coulomb criterion of rock failure is expressed by effective stress, pore pressure, and effective shear strength as follows:

$$\frac{\tau_n}{1-D} = c^* + \frac{\sigma_n + Dp_w}{1-D}\tan\phi^* \quad (5)$$

where D is the disturbance damage factor; τ_n is the shear stress; and p_w is the pore pressure.

Assuming that the rock's uniaxial compressive strength is σ_c and the uniaxial compressive strength of the damaged rock is $\sigma_c^* = (1-D)\sigma_c$, the relationship between the shear strength and uniaxial compressive strength is expressed as

$$\sigma_c^* = (1-D)\sigma_c = (1-D)\frac{2c^*\cos\phi^*}{1-\sin\phi^*} \quad (6)$$

By solving the above two equations, the effective shear strength parameters c^* and ϕ^* are expressed as a function of stress σ_n and τ_n on the failure surface, the compressive strength σ_c of non-damaged rock, and disturbance damage factor D.

When the equivalent plastic strain of a rock element exceeds the limit plastic strain $\bar{\varepsilon}_{pmax}$, plastic deformation and failure occur. The relationship between the disturbance damage factor and the equivalent plastic strain satisfies the first-order exponential decay function, and the equivalent plastic strain is normalized as:

$$D = A_0 e^{-\bar{\varepsilon}_{pn}/a} + B_0 \quad (7)$$

where $\bar{\varepsilon}_{pn}$ is the normalized equivalent plastic strain and the material parameter a is a constant, which is equal to 0.2 in the simulation. $A_0 = \frac{1}{e^{-1/a}-1}$ and $B_0 = -\frac{1}{e^{-1/a}-1}$. The parameter a reflects the rate of the disturbance damage factor evolution with the plastic strain.

In the hydro-mechanical coupling system, the solid phase is expressed as $S = U_n + D_a$, where U_n is the undamaged phase, D_a is the damaged solid phase, and L is the liquid phase. The D_a component cannot support the shear load, and the U_n component can support the shear stress and hydrostatic pressure. Therefore, the load capacity of the rock is reduced, i.e., damage has occurred. Assuming that the volume of the porous medium is V, the damaged volume is expressed as:

$$V_D = V(1-n)D \quad (8)$$

where n is the rock porosity and D is the disturbance damage factor.

According to the cubic law of seepage [44], the rock permeability coefficient is evolved as follows:

$$k = (1-D)k^M + Dk^D\left(1+\varepsilon_v^{pF}\right)^3 \quad (9)$$

where k^M and k^D are the permeability coefficients of non-damaged and fractured rock, respectively, and ε_v^{pF} is the plastic volumetric strain of the damage phase.

Assuming that no damage occurs during the elastic deformation of the rock, and plastic deformation and damage occur simultaneously, ε_v^{pF} is expressed as:

$$\varepsilon_v^{pF} = D\varepsilon_v^p \quad (10)$$

where ε_v^p is the plastic volume strain.

2.4. Model Validation

According to rock mechanics, there is an analytical solution for the stress field around the borehole in the homogeneous formation. The analytical solution and finite element solution of the stress field component S_{xx} are calculated by setting the bottom hole pressure as 30 MPa, 40 MPa, and 50 MPa [41,45] (Table 1 and Figure 1a), and the solutions have a good agreement, which verifies the reliability of the finite element solution under the hydro-mechanical coupling conditions.

Table 1. Input parameters.

Parameters	Value
Porosity/decimal	0.05
Poisson's ratio/decimal	0.25
Elastic modulus/GPa	34.5
Rock density/kg/m^3	2500
Rock permeability/mD	0.001
Tensile strength/MPa	6.04
Uniaxial compressive strength/MPa	100
Internal friction angle of rock/°	33.7
Element damage evolution factor/decimal	2
Fluid density/kg/cm^3	1020
Fluid compression coefficient/1/Pa	2×10^{-10}
Fluid viscosity/mPa·s	1.8
Initial formation pressure/MPa	28
Maximum horizontal principal stress/MPa	40
Minimum horizontal principal stress/MPa	30
Borehole radius/m	0.1
Injection time/s	60

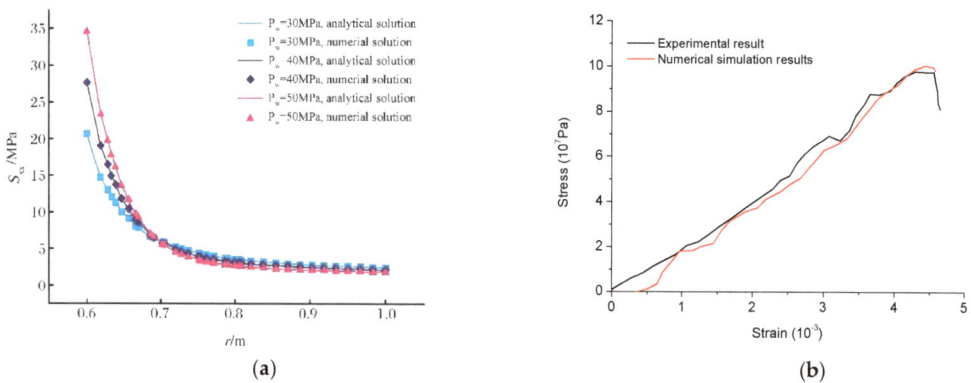

Figure 1. Validation examples: (a) numerical solution and analytical solution of S_{xx} stress component; (b) comparison of numerical simulation and experimental results.

To validate our DCS theory, we compare the numerical results with experimental results, as shown in Figure 1b. The cohesive force is 30.7 MPa and the friction angle is 27°. The bulk and shear modulus of rock sample are 22 GPa and 16 GPa, respectively. The initial fracture toughness is equal to 12 MPa·mm$^{0.5}$. We observe that the numerical results have a good agreement with the experimental results, which indicate that our DSC model are reliable.

3. Finite Element Model

A 2D 20 m × 20 m finite element model is established, and a borehole with a radius of 0.1m is drilled in the middle of the model (Figure 2). The model is meshed with the structured grid of the plane strain quadrilateral elements (CPE4P) coupled with the degree of freedom of the pore pressure. To simulate the stress concentration around the borehole, the meshes near the borehole are refined locally. The mesh size of the directional quadrilateral elements away from the borehole increases gradually. The finite element model of borehole stability includes a total of 9024 nodes and 8928 CPE4P quadrilateral elements. The basic input parameters are listed in Table 2.

Figure 2. Schematic of the finite element model of borehole stability and mesh division.

Table 2. Input parameters in finite element simulation of rock borehole instability (Base case).

Parameters	Value
Rock elastic modulus/Pa	3×10^8
Poisson's ratio/decimal	0.25
Rock permeability/m²	3×10^{-12}
Porosity/decimal	0.16
Maximum horizontal principal stress/Pa	2.75×10^6
Minimum horizontal principal stress/Pa	1.75×10^6
Vertical stress/Pa	3.5×10^6
Rock cohesion/Pa	3×10^5
Internal friction angle of rock/°	18
Dilation angle of rock/°	0
Initial pore pressure/Pa	1.5×10^6
Material parameter a/decimal	0.2

This finite element simulation of borehole stability is completed in two steps. The first is to establish the stress balance equation, which provides the initial stress field for the DSC-based numerical simulation of borehole instability. The second is to carry out a finite element simulation of borehole instability, and it is operated in a Soils hydro-mechanical coupling solver in ABAQUS. The solver numerically discretizes the time derivative term through an implicit algorithm. The time step is adaptive. The initial time step is 0.1 s. The minimum and maximum are 1×10^{-9} s and 86,400 s. The elastic–plastic deformation of the borehole wall rock is simulated by the Mohr–Coulomb plastic yield criterion. The rock internal friction angle and the dilation angle of rock are listed in Table 1.

As shown in Figure 1, the boundary conditions of this finite element model are set as follows: the normal displacement constraint of the outer boundary is 0, that is, the roller boundary condition is satisfied, and pore pressure is applied to the outer boundary and the inner boundary of the borehole. It is noted that the PORMECH keyword in the ABAQUS input file converts pore pressure into surface force and applies it to the borehole wall to simulate the force of the mud column pressure on the borehole wall.

Based on Section 2.3 of this paper, "Theory of Borehole Instability in Disturbed State", the secondary development is carried out on the commercial finite element software ABAQUS platform, and the USDFLD subroutine is used to realize the porosity, permeability coefficient, disturbance damage factor, and equivalent plastic stress (PEEQ). In this program, the relationships of the permeability coefficient and equivalent plastic stress with the disturbance damage factor are coded by using Equations (8) and (9). The evolution of other parameters is used to obtain the instability process of rock borehole.

4. Results and Analysis

The effects of the action time of drilling fluid on the wellbore, stress anisotropy, internal friction angle of rock, and borehole pressure on borehole stability are simulated with the parameters listed in Table 1. In the cloud picture of equivalent plastic deformation and disturbance damage factor, the horizontal direction is set as the x axis, and the vertical direction is set as the y axis.

4.1. Effect of Action Time of Drilling Fluid on the Wellbore

During drilling, the borehole is filled with the drilling fluid, and the action time of the drilling fluid on the wellbore affects the borehole instability. Here, the effects of the action time of the drilling fluid on the wellbore (i.e., t = 0.675 s, t = 5.126 s, t = 667.2 s, and t = 2210 s) are simulated.

The cloud picture of equivalent plastic strain under different action times of drilling fluid on the wellbore is shown in Figure 3. At the initial stage of the action time of the drilling fluid, the equivalent plastic strain is concentrated around the borehole in the maximum principal stress direction. As the drilling operation continues, the equivalent plastic strain region gradually expands to the periphery of the borehole, showing a typical symmetrical bifurcated feature, which is due to rock shear damage.

The cloud picture of the disturbance damage factor of borehole instability under different action times of drilling fluid on the wellbore is shown in Figure 4. Initially, the rock damage region is concentrated around the borehole in the maximum principal stress direction. Then, the rock damage region develops as the equivalent plastic strain region. The disturbance damage factor gradually expands to the periphery of the borehole and shows symmetrical bifurcation characteristics, indicating the dominant mechanical mechanism of borehole instability as a shear failure.

The rock permeability, disturbance damage factor, and equivalent plastic strain with different distances from the borehole are shown in Figure 5. The node extraction path is shown in Figure 5d. As the distance from the borehole increases, the rock permeability, disturbance damage factor, and equivalent plastic strain value gradually decrease. As the action time of drilling fluid on the wellbore increases, the rock permeability, disturbance damage factor, and plastic strain area increase slightly. At a distance of 0.3 m from the borehole, the permeability, disturbance damage factor, and plastic strain change abruptly, indicating serious damage.

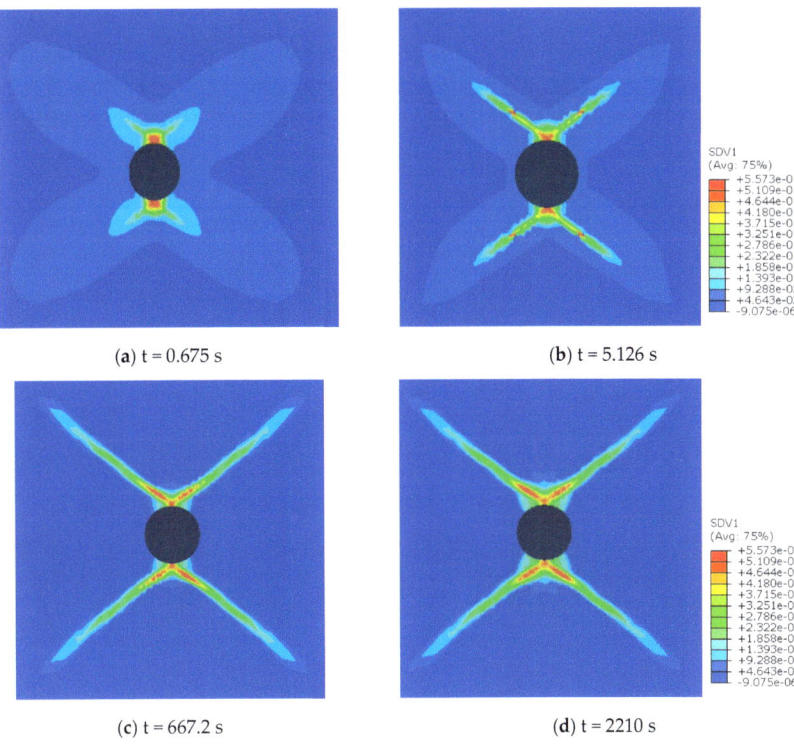

Figure 3. Evolution of equivalent plastic strain region (SVD1 represents equivalent plastic strain).

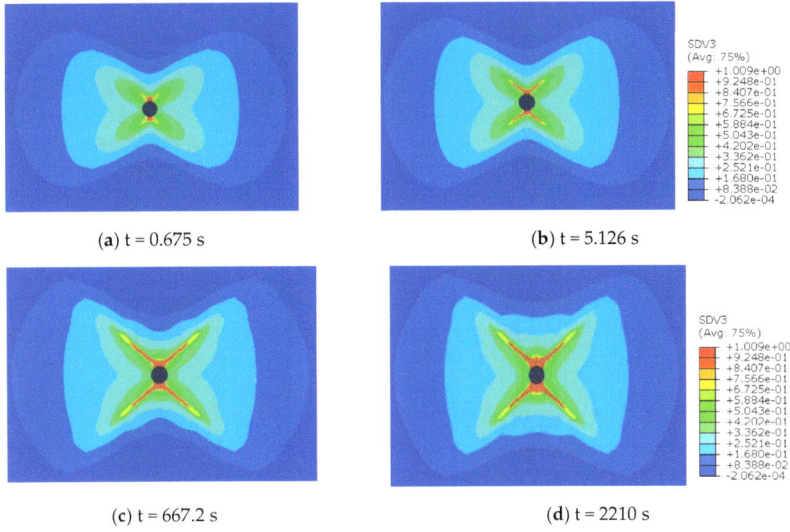

Figure 4. Evolution of damage region (SVD3 represents disturbance damage factor).

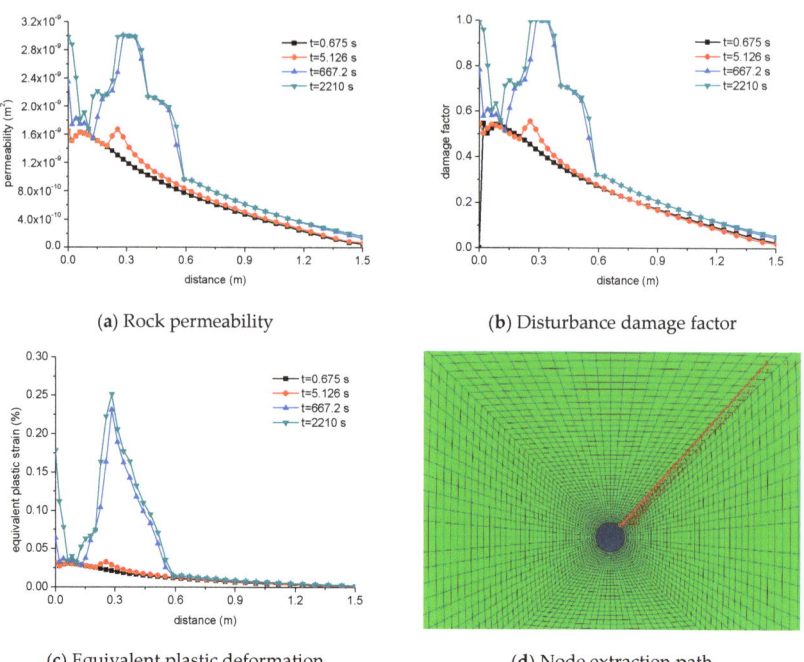

Figure 5. Rock permeability, disturbance damage factor, and equivalent plastic strain with different distances from the borehole.

4.2. Effect of Stress Anisotropy

The rock yield failure is related to its stress state, and the in situ stress affects borehole stability during drilling. The effects of the horizontal stress difference of 0 MPa, 5 MPa, 7.5 MPa, and 10 MPa are simulated.

The cloud picture of equivalent plastic strain around the borehole under various horizontal stress differences is shown in Figure 6. Under the isotropic stress (the stress difference of 0 MPa), the equivalent plastic strain area occurs around the borehole and shows the symmetrical distribution on the x axis and y axis. As the stress anisotropy is enhanced, the equivalent plastic strain region grows in a narrow region in the x direction and grows longer in the y direction. The shape of the equivalent plastic strain cloud picture in Figure 6b,c is similar to the cockroach, and the bifurcation is similar to the whiskers. Under the low stress anisotropy, there are multiple bifurcations on the equivalent plastic strain region, indicating several shear failures. Under the stress anisotropy of 10 MPa, only one bifurcation occurs in the y direction, shear failure is significantly reduced, and the plastic strain zone occurs along the 45° direction. As the stress anisotropy increases, the shear failure area is reduced.

The cloud picture of damage factor distribution around the borehole under different stress anisotropy is shown in Figure 7. As the stress anisotropy is enhanced, the rock damaged area in the x axis is narrowed and elongated in the y axis. As the stress anisotropy reduces, the bifurcation increases. Under the strong stress anisotropy of the stress difference of 10 MPa, only one bifurcation occurs in the y axis, and the shear damage zone is generated along the 45° direction.

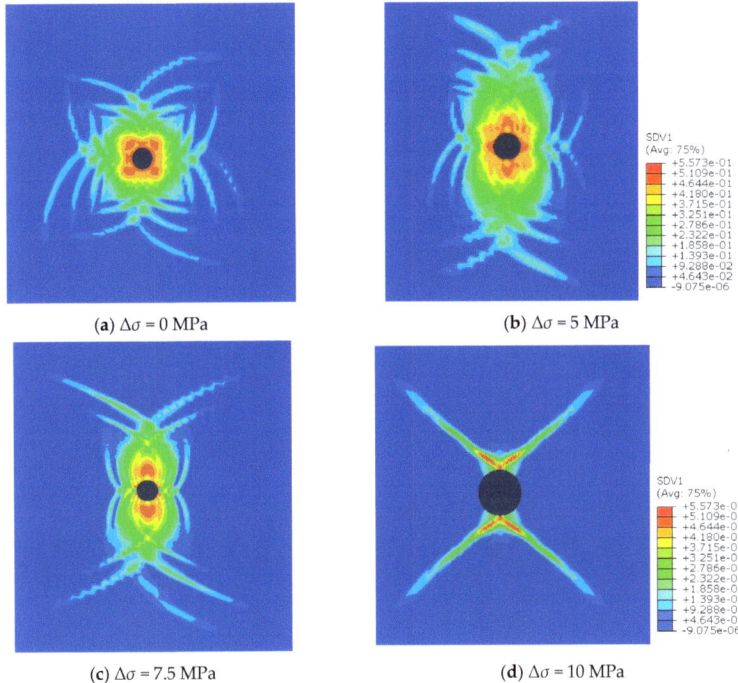

Figure 6. Evolution of equivalent plastic strain with the stress difference.

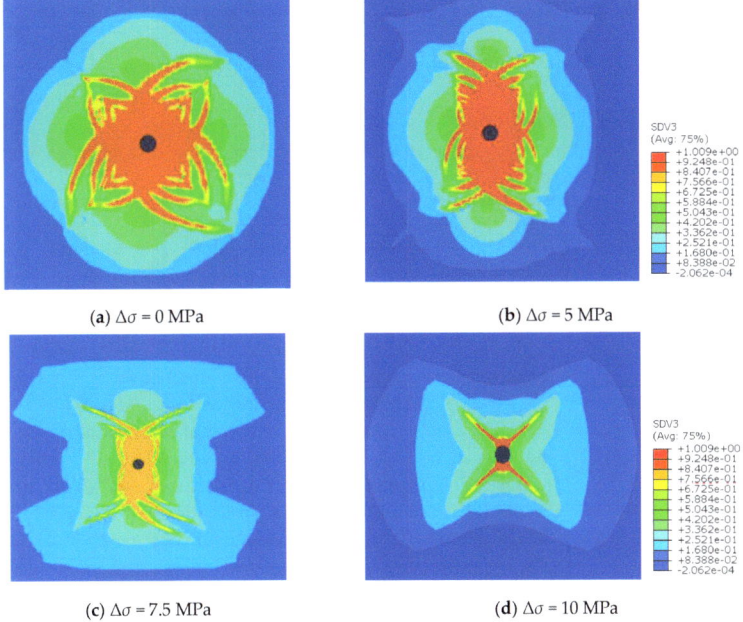

Figure 7. Evolution of disturbance damage factor with different stress differences.

The rock permeability, disturbance damage factor, and equivalent plastic strain with the distance from the borehole during the borehole instability along the direction of the nodal path are shown in Figure 8. As the distance from the borehole increases, the rock permeability, disturbance damage factor, and plastic strain generally show a decreasing trend. Under the low stress anisotropy, the permeability, disturbance damage factor, and plastic strain show fluctuation characteristics, corresponding to multiple bifurcations in Figure 7. As the shear damage increases, the damage region increases. Under the higher stress anisotropy, the rock permeability, disturbance damage factor, and equivalent plastic strain fluctuate at a relatively low level, which is consistent with the condition of one bifurcation in the y direction in Figure 7d.

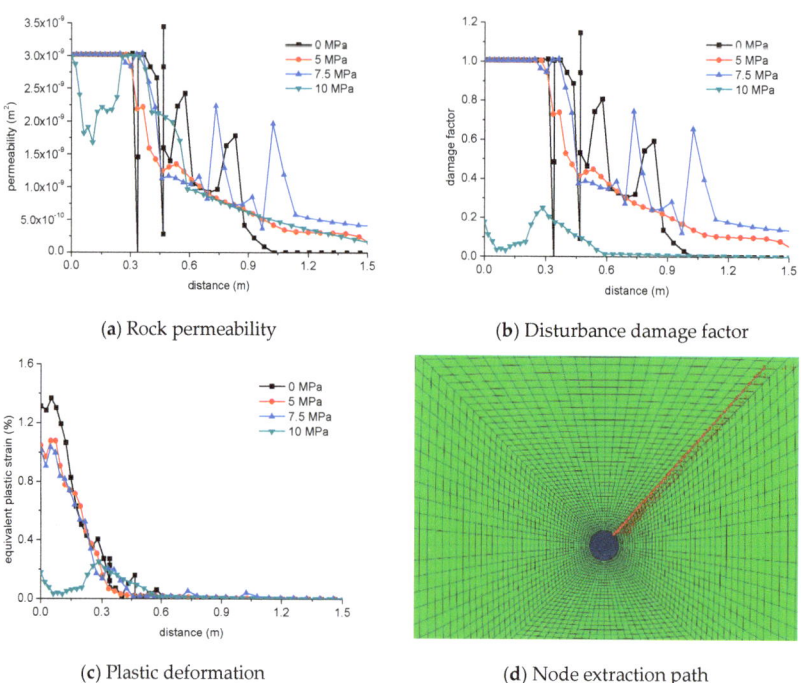

(a) Rock permeability (b) Disturbance damage factor

(c) Plastic deformation (d) Node extraction path

Figure 8. Variation of rock permeability, damage factor, and equivalent plastic strain with different distance.

4.3. Effect of the Internal Friction Angle of Rock

The internal friction angle of rock is a key parameter in the Mohr–Coulomb yield criterion for borehole stability. The effect of the internal friction angle of rock of 13°, 18°, 23°, and 28° on borehole instability is simulated.

The cloud picture of the equivalent plastic strain around the borehole under different internal friction angles of rock is shown in Figure 9. As the internal friction angle increases, the equivalent plastic strain area increases, and the bifurcation is enhanced. Under the internal friction angle of 28°, a strong plastic strain area occurs in the y direction, a 'mantis' shape occurs (Figure 7b,c), and the bifurcation corresponds to the whisker, which is the shear failure area. Under the low internal friction angle of rock, the equivalent plastic strain area shows a chaotic feature, with a weak elongated plastic strain area along the diagonal direction.

The cloud picture of the damage factor around the borehole under different internal friction angles of rock is shown in Figure 10. As the internal friction angle, the damage area increases, and the bifurcation characteristics are enhanced. Under the friction angle of

28°, an obvious bifurcation occurs in the y direction, and the damage degree approaches 1, indicating shear collapse failure around the borehole. Under the low internal frictional angles, a narrow and long damaged area occurs in the sub-diagonal direction.

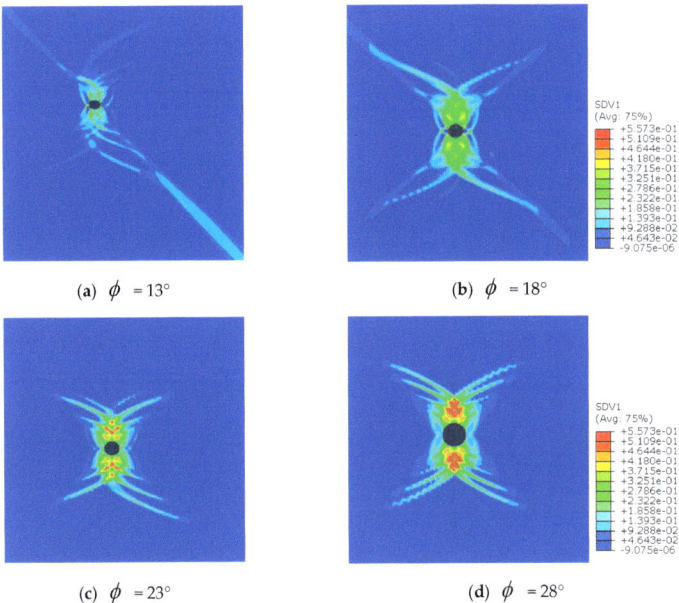

Figure 9. Evolution of equivalent plastic strain region.

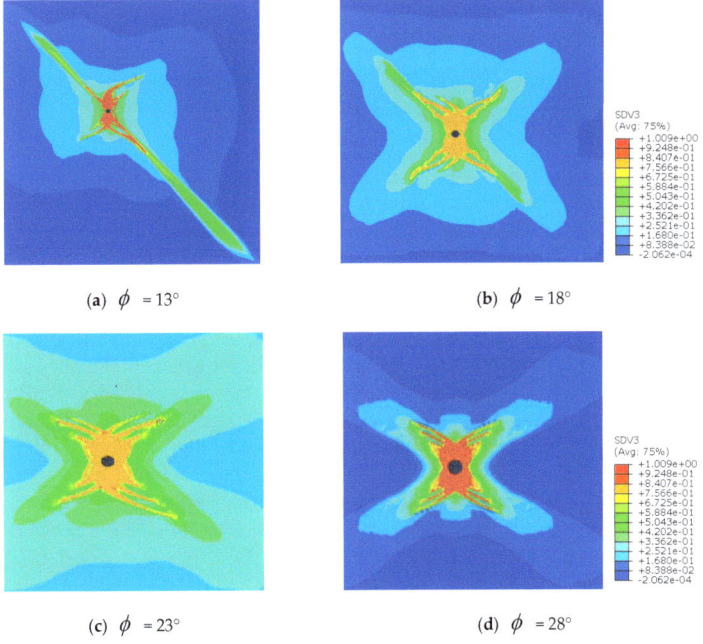

Figure 10. Evolution of disturbance damage factor with the internal friction angle of rock.

The rock permeability, disturbance damage factor, and equivalent plastic strain along the nodal path are shown in Figure 11. As the distance from the borehole increases, the rock permeability, disturbance damage factor, and equivalent plastic strain show a decreasing and fluctuation trend, indicating the heterogeneous damage features.

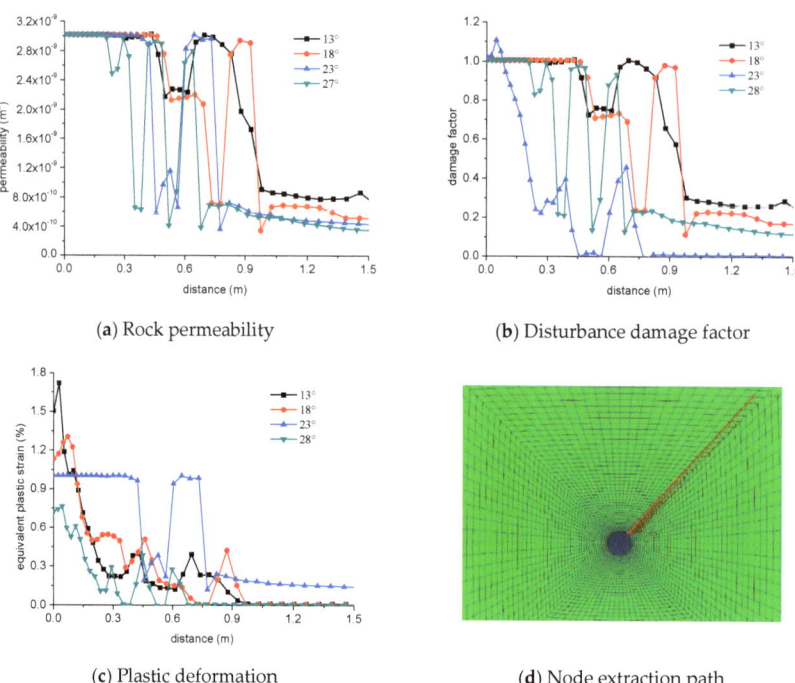

(a) Rock permeability (b) Disturbance damage factor

(c) Plastic deformation (d) Node extraction path

Figure 11. Variation of rock permeability, damage factor, and equivalent plastic strain with different distance.

4.4. Effect of Borehole Pressure

During drilling, the drilling fluid within the borehole generates hydrostatic pressure on the borehole wall and causes compression stress on the borehole wall. The effect of the borehole pressure of 3.5 MPa, 4 MPa, 4.5 MPa, and 5 MPa on borehole instability is simulated.

The cloud picture of equivalent plastic strain around the borehole under drilling fluid static pressure is shown in Figure 12. As the hydrostatic pressure increases, the equivalent plastic strain area is enlarged, and the bifurcation characteristics are enhanced. When the hydrostatic pressure is 5 MPa, several bifurcated plastic strain regions occur in the y axis. Under the low borehole pressure, the bifurcation occurs only in the y direction, and shear failure occurs along the y axis.

The cloud picture of the disturbance damage factor around the borehole under drilling fluid column pressure is shown in Figure 13. As the drilling fluid static pressure increases, the damaged area is enlarged, and the bifurcation characteristics are enhanced. When the drilling fluid hydrostatic pressure is 5 MPa, multiple bifurcated damage zones occur in the y direction, indicating that increasing the drilling fluid density promotes shear damage near the borehole and borehole instability.

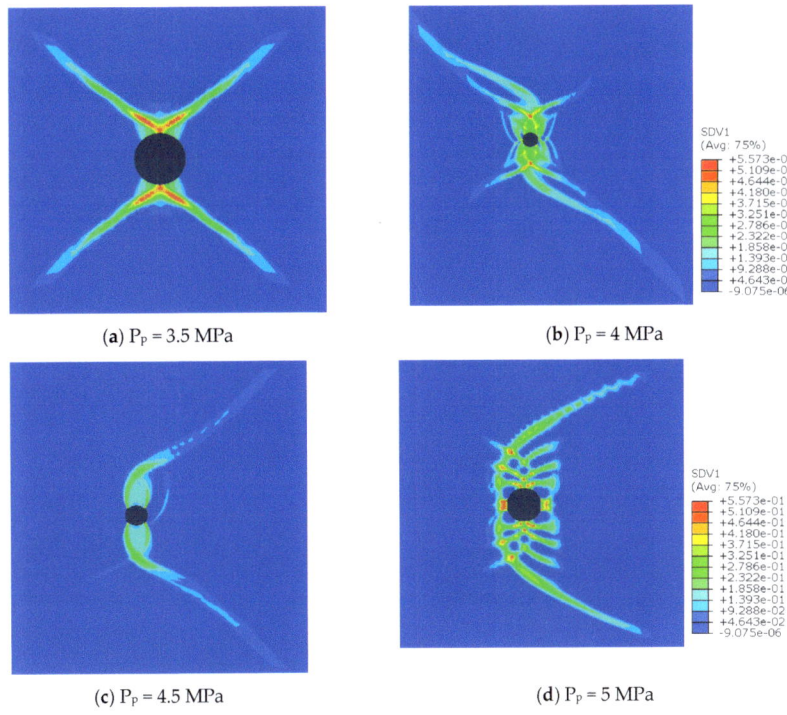

Figure 12. Evolution of equivalent plastic strain.

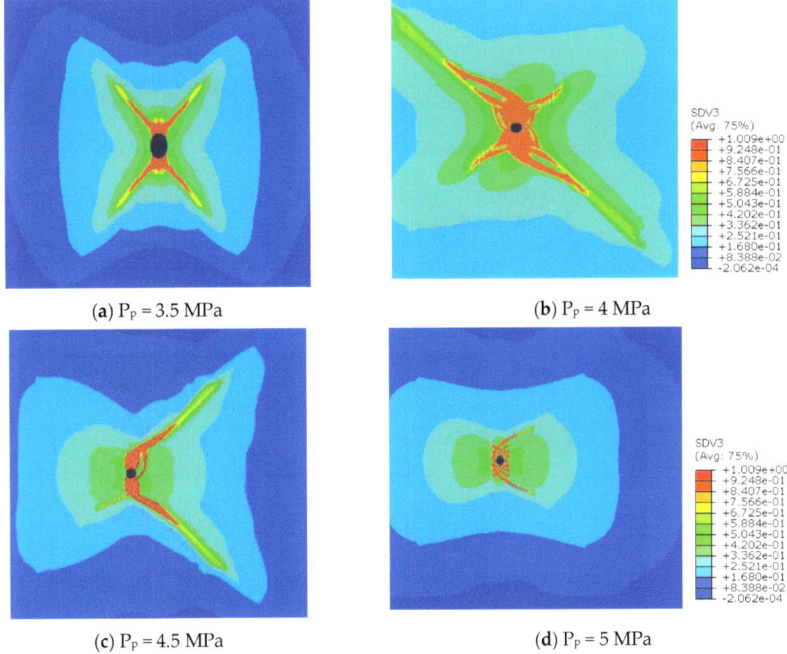

Figure 13. Evolution of damage factor with borehole pressure.

The rock permeability, disturbance damage factor, and plastic strain area with distance from the borehole along the direction of the node path are shown in Figure 14. As the distance from the borehole increases, the permeability, disturbance damage factor, and plastic strain gradually decrease. Under the low hydrostatic pressure of drilling fluid, the rock permeability, disturbance damage factor, and equivalent plastic strain fluctuate slightly, and the damage position is determined. Under the high borehole pressure, the rock permeability, disturbance damage factor, and plastic strain area fluctuate significantly. The degree of damage varies in different locations, corresponding to multiple bifurcation positions on the cloud picture.

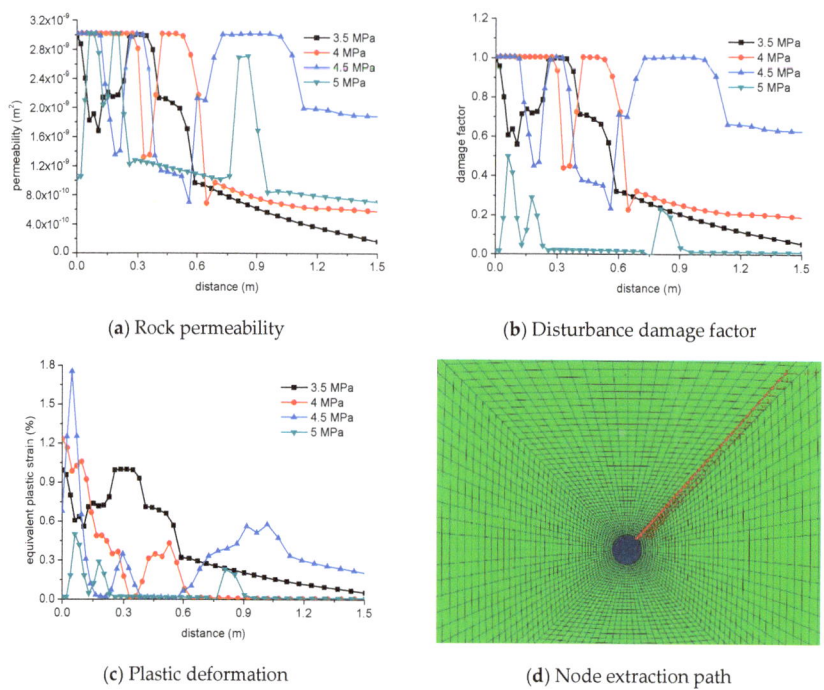

Figure 14. Variation of rock permeability, damage factor, and equivalent plastic strain with different distance.

5. Conclusions

Based on the DSC, we carried out a finite element hydro-mechanical coupling model of borehole instability by introducing the disturbance factors into the Mohr–Coulomb yield criterion and writing the subroutine for the field variables. The model considers elastic–plastic deformation, the damage distribution area, and the variation of rock porosity and permeability with the disturbance area in borehole instability. The following conclusions can be drawn:

(1) The finite element numerical simulation results show that borehole stability is related to the action time of drilling fluid on the wellbore, stress anisotropy, internal friction angle of rock, and borehole pressure. Excessive drilling fluid density and long action time between the drilling fluid and the borehole should be avoided. Under the small stress anisotropy, shear failure occurs often around the borehole. A high horizontal stress difference restricts shear instability around the borehole. The high internal friction angle of rock enhances shear failure around the borehole in the direction of the maximum horizontal principal stress.

(2) The equivalent plastic strain zone has a good agreement with the borehole instability disturbance damage zone, and they show the same characteristics. A high internal friction angle of rock, low stress anisotropy, and long action time of the drilling fluid on the wellbore enlarge the plastic zone and disturbance damage zone around the borehole.

(3) The model of borehole stability considers the variation of rock permeability, rock porosity, and equivalent plastic strain with the disturbance damage factor. Under the large borehole pressure and the low stress anisotropy, the rock permeability, the disturbance damage factor, and the equivalent plastic strain show fluctuation characteristics, which is due to the different damage magnitudes. When considering the internal friction angle of rock, the rock permeability, disturbance damage factor, and equivalent plastic strain area show fluctuation characteristics.

(4) Under the large internal friction angle of rock, a strong equivalent plastic strain zone and a disturbance damage zone occur in the direction of the maximum horizontal principal stress, and they correspond to the mantis shape. The bifurcation corresponds to the whisker, which is the shear failure area. Under the low internal friction angle of rock, the equivalent plastic strain and disturbance damage region show chaotic features, and an elongated equivalent plastic strain region occurs along the diagonal.

Author Contributions: Conceptualization, Z.Q. and B.Y.; methodology, D.W. and Q.S.; software; validation, W.Z.; data curation, Y.Z. and Z. R; writing—original draft preparation, D.W.; writing—review and editing, D.W. and Z.R.; supervision, Z.Q. All authors have read and agreed to the published version of the manuscript.

Funding: The authors would like to give their sincere gratitude to the Beijing Natural Science Foundation Project (No. 3222030); the National Natural Science Foundation Project (No. 51936001, No. 51974255 and No. 51804258); the Scientific Research Project of Beijing Educational Committee (KZ202110017026); and The Key Laboratory of Well Stability and Fluid & Rock Mechanics in Oil and Gas of Shaanxi Province (No. WSFRM20210201002) for their financial support.

Institutional Review Board Statement: Not applicable.

Informed Consent Statement: Not applicable.

Data Availability Statement: Datasets related to this article can be found by contacting the corresponding author.

Conflicts of Interest: The authors declare no conflict of interest.

References

1. Yu, M.; Chen, G.; Chenevert, M.E.; Sharma, M.M. Chemical and Thermal Effects on borehole stability of Shale Formations. In Proceedings of the SPE Annual Technical Conference and Exhibition, New Orleans, LA, USA, 30 September–3 October 2001.
2. She, H.; Hu, Z.; Qu, Z.; Zhang, Y.; Guo, H. Determination of the Hydration Damage Instability Period in a Shale Borehole Wall and Its Application to a Fuling Shale Gas Reservoir in China. *Geofluids* **2019**, *2019*, e3016563. [CrossRef]
3. Han, Q.; Qu, Z.; Ye, Z. Research on the Mechanical Behaviour of Shale Based on Multiscale Analysis. *R. Soc. Open Sci.* **2018**, *5*, 181039. [CrossRef] [PubMed]
4. Freij-Ayoub, R.; Tan, C.; Clennell, B.; Tohidi, B.; Yang, J. A borehole stability Model for Hydrate Bearing Sediments. *J. Pet. Sci. Eng.* **2007**, *57*, 209–220. [CrossRef]
5. Zhang, J.; Bai, M.; Roegiers, J.-C. Dual-Porosity Poroelastic Analyses of borehole stability. *Int. J. Rock Mech. Min. Sci.* **2003**, *40*, 473–483. [CrossRef]
6. Salehi, S.; Hareland, G.; Nygaard, R. Numerical Simulations of borehole stability in Under-Balanced-Drilling Wells. *J. Pet. Sci. Eng.* **2010**, *72*, 229–235. [CrossRef]
7. Desai, C.C. *A Consistent Finite Element Technique for Work-Softening Behavior*; University of Texas: Austin, TX, USA, 1974.
8. Desai, C.S.; Ma, Y. Modelling of Joints and Interfaces Using the Disturbed-State Concept. *Int. J. Numer. Anal. Methods Geomech.* **1992**, *16*, 623–653. [CrossRef]
9. Katti, D.R.; Desai, C.S. Modeling and Testing of Cohesive Soil Using Disturbed-State Concept. *J. Eng. Mech.* **1995**, *121*, 648–658. [CrossRef]
10. Desai, C.S.; Samtani, N.C.; Vulliet, L. Constitutive Modeling and Analysis of Creeping Slopes. *J. Geotech. Eng.* **1995**, *121*, 43–56. [CrossRef]

11. Desai, C.S.; Toth, J. Disturbed State Constitutive Modeling Based on Stress-Strain and Nondestructive Behavior. *Int. J. Solids Struct.* **1996**, *33*, 1619–1650. [CrossRef]
12. Desai, C.; Park, I.; Shao, C. Fundamental yet Simplified Model for Liquefaction Instability. *Int. J. Numer. Anal. Methods Geomech.* **1998**, *22*, 721–748. [CrossRef]
13. Pal, S.; Wathugala, G.W. Disturbed State Model for Sand-Geosynthetic Interfaces and Application to Pull-out Tests. *Int. J. Numer. Anal. Methods Geomech.* **1999**, *23*, 1873–1892. [CrossRef]
14. Fan, R.-D.; Liu, M.; Du, Y.-J.; Horpibulsuk, S. Estimating the Compression Behaviour of Metal-Rich Clays via a Disturbed State Concept (DSC) Model. *Appl. Clay Sci.* **2016**, *132–133*, 50–58. [CrossRef]
15. Ouria, A. Disturbed State Concept–Based Constitutive Model for Structured Soils. *Int. J. Geomech.* **2017**, *17*, 04017008. [CrossRef]
16. Ghazavi Baghini, E.; Toufigh, M.M.; Toufigh, V. Analysis of Pile Foundations Using Natural Element Method with Disturbed State Concept. *Comput. Geotech.* **2018**, *96*, 178–188. [CrossRef]
17. Wu, G.; Zhang, L. Analysis on post-failure behaviors of rock in uniaxial compression using disturbed state concept theory. *Chin. J. Rock Mech. Eng.* **2004**, *10*, 1628–1634. [CrossRef]
18. Zheng, J.; Ge, X.; Sun, H. Application of disturbed state concept to issues in geotechnical engineering. *Chin. J. Rock Mech. Eng.* **2006**, *25*, 3456–3462.
19. Zheng, J.; Ge, X.; Sun, H. Meso analysis for rationality of disturbed state concept theory on utilization of hardening model for softening response depiction. *Rock Soil Mech.* **2007**, *28*, 127–132.
20. Zhang, X.; Wang, C. Study of creep constitutive model of structural soft soil based on the disturbed state concept. *China Civ. Eng. J.* **2011**, *44*, 81–87.
21. Fu, P.; Chu, X.; Yu, C.; Xu, Y.; Qu, W. Simulation of Strain Localization of Granular Materials Based on Disturbed State Concept. *J. South China Univ. Technol. Sci. Ed.* **2014**, *42*, 59–69+76.
22. Yang, J.; Yin, Z.; Huang, H.; Jin, Y.; Zhang, D. Bounding surface plasticity model for structured clays using disturbed state concept-based hardening variables. *Chin. J. Geotech. Eng.* **2017**, *39*, 554–561.
23. Huang, M.; Jiang, Y.; Wang, S.; Deng, T. Identification of the creep model and its paramters of soft rock on the basis of disturbed state concept. *Chin. J. Solid Mech.* **2017**, *38*, 570–578.
24. Zou, Y.; Wei, C.; Chen, H.; Zhou, J.; Wan, Y. Elastic-plastic model for gas-hydrate-bearing soils using disturbed state concept. *Rock Soil Mech.* **2019**, *40*, 2653–2662.
25. Cao, W.; Deng, J.; Liu, W.; Yu, B.; Tan, Q.; Yang, L.; Li, Y.; Gao, J. Pore Pressure and Stress Distribution Analysis around an Inclined borehole in a Transversely Isotropic Formation Based on the Fully Coupled Chemo-Thermo-Poroelastic Theory. *J. Nat. Gas Sci. Eng.* **2017**, *40*, 24–37. [CrossRef]
26. Liang, C.; Chen, M.; Jin, Y.; Lu, Y. borehole stability Model for Shale Gas Reservoir Considering the Coupling of Multi-Weakness Planes and Porous Flow. *J. Nat. Gas Sci. Eng.* **2014**, *21*, 364–378. [CrossRef]
27. Kang, Y.; Yu, M.; Miska, S.Z.; Takach, N. Borehole Stability: A Critical Review and Introduction to DEM. In Proceedings of the SPE Annual Technical Conference and Exhibition, New Orleans, LA, USA, 4–7 October 2009.
28. Chen, G.; Chenevert, M.E.; Sharma, M.M.; Yu, M. A Study of borehole stability in Shales Including Poroelastic, Chemical, and Thermal Effects. *J. Pet. Sci. Eng.* **2003**, *38*, 167–176. [CrossRef]
29. Zeynali, M.E. Mechanical and Physico-Chemical Aspects of borehole stability during Drilling Operations. *J. Pet. Sci. Eng.* **2012**, *82–83*, 120–124. [CrossRef]
30. Gao, L.; Shi, X.; Liu, J.; Chen, X. Simulation-based three-dimensional model of wellbore stability in fractured formation using discrete element method based on formation microscanner image: A case study of Tarim Basin, China. *J. Nat. Sci. Eng.* **2022**, *97*, 104341. [CrossRef]
31. Ma, T.; Zhang, Y.; Qiu, Y.; Liu, Y.; Li, Z. Effect of parameter correlation on risk analysis of wellbore instability in deep igneous formations. *J. Pet. Sci. Eng.* **2022**, *208*, 109521. [CrossRef]
32. Cao, W.; Liu, W.; Liu, H.; Lin, H. Effect of formation strength anisotropy on wellbore shear failure in bedding shale. *J. Pet. Sci. Eng.* **2022**, *208*, 109183.
33. Liu, D.; Deng, H.; Zhang, Y. Research on the Wellbore Instability Mechanism of Air Drilling Technology in Conglomerate Formation. *Fresen. Environ. Bull.* **2020**, *29*, 600–606.
34. Liu, H.; Cui, S.; Meng, Y.; Li, Z.; Yu, X.; Sun, H.; Zhou, Y.; Luo, Y. Rock mechanics and wellbore stability of deep shale during drilling and completion processes. *J. Pet. Sci. Eng.* **2021**, *205*, 108882. [CrossRef]
35. Aslannezhad, M.; Kalantariasl, A.; Keshavarz, A. Borehole stability in shale formations: Effects of Thermal-Mechanical-Chemical parameters on well design. *J. Nat. Gas Sci. Eng.* **2021**, *88*, 103852. [CrossRef]
36. AlBahrani, H.; Morita, N. Risk-Controlled Wellbore Stability Criterion Based on a Machine-Learning-Assisted Finite-Element Model. *SPE Drill. Completion* **2022**, *37*, 38–66. [CrossRef]
37. Li, J.; Qiu, Z.; Zhong, H.; Zhao, X.; Liu, Z.; Huang, W. Effects of water-based drilling fluid on properties of mud cake and wellbore stability. *J. Pet. Sci. Eng.* **2022**, *208*, 109704. [CrossRef]
38. Liu, W.; Lin, H.; Luo, C.; Wang, G.; Deng, J. Numerical Investigation of Wellbore Stability in Deepwater Shallow Sediments. *Geofluids* **2021**, *2021*, 5582605. [CrossRef]
39. Cui, S.; Liu, H.; Meng, Y.; Zhang, Y.; Tao, Y.; Zhang, X. Study on fracture occurrence characteristics and wellbore stability of limestone formation. *J. Pet. Sci. Eng.* **2021**, *204*, 108783. [CrossRef]

40. Ding, Y.; Liu, X.; Luo, P. Investigation on influence of drilling unloading on wellbore stability in clay shale formation. *Pet. Sci.* **2021**, *17*, 781–796. [CrossRef]
41. Wang, D.; Zhou, F.; Ding, W.; Ge, H.; Jia, X.; Shi, Y.; Wang, X.; Yan, X. A Numerical Simulation Study of Fracture Reorientation with a Degradable Fiber-Diverting Agent. *J. Nat. Gas Sci. Eng.* **2015**, *25*, 215–225. [CrossRef]
42. Wang, D.; Zlotnik, S.; Díez, P. A Numerical Study on Hydraulic Fracturing Problems via the Proper Generalized Decomposition Method. *CMES Comput. Model. Eng. Sci.* **2020**, *122*, 703–720. [CrossRef]
43. Wang, D.; Ge, H.; Wang, X.; Wang, Y.; Sun, D.; Yu, B. Complex Fracture Closure Pressure Analysis During Shut-in: A Numerical Study. *Energy Explor. Exploit.* **2022**, *40*, 014459872210773. [CrossRef]
44. Wang, D.; Dong, Y.; Sun, D.; Yu, B. A Three-Dimensional Numerical Study of Hydraulic Fracturing with Degradable Diverting Materials via CZM-Based FEM. *Eng. Fract. Mech.* **2020**, *237*, 107251. [CrossRef]
45. Jaeger, J.C.; Cook, N.G.W.; Zimmerman, R. *Fundamentals of Rock Mechanics*; John Wiley & Sons: Hoboken, NJ, USA, 2009.

Article

Experimental Study on the Hydraulic Fracture Propagation in Inter-Salt Shale Oil Reservoirs

Yunqi Shen [1,2,3], Zhiwen Hu [4,*], Xin Chang [5] and Yintong Guo [5]

1. State Key Laboratory of Shale Oil and Gas Enrichment Mechanisms and Effective Development, Beijing 100083, China
2. National Energy Shale Oil Research and Development Center, Beijing 100083, China
3. Petroleum Exploration and Production Research Institute, SINOPEC, Beijing 100083, China
4. State Key Laboratory for Coal Mine Disaster Dynamics and Control, Chongqing University, Chongqing 400044, China
5. State Key Laboratory of Geomechanics and Geotechnical Engineering, Institute of Rock and Soil Mechanics, Chinese Academy of Sciences, Wuhan 430071, China
* Correspondence: 202020131066@cqu.edu.cn

Citation: Shen, Y.; Hu, Z.; Chang, X.; Guo, Y. Experimental Study on the Hydraulic Fracture Propagation in Inter-Salt Shale Oil Reservoirs. Energies 2022, 15, 5909. https://doi.org/10.3390/en15165909

Academic Editor: Eric James Mackay

Received: 18 April 2022
Accepted: 27 July 2022
Published: 15 August 2022

Publisher's Note: MDPI stays neutral with regard to jurisdictional claims in published maps and institutional affiliations.

Copyright: © 2022 by the authors. Licensee MDPI, Basel, Switzerland. This article is an open access article distributed under the terms and conditions of the Creative Commons Attribution (CC BY) license (https://creativecommons.org/licenses/by/4.0/).

Abstract: In response to the difficulty of fracture modification in inter-salt shale reservoirs and the unknown pattern of hydraulic fracture expansion, corresponding physical model experiments were conducted to systematically study the effects of fracturing fluid viscosity, ground stress and pumping displacement on hydraulic fracture expansion, and the latest supercritical CO_2 fracturing fluid was introduced. The test results show the following. (1) The hydraulic fractures turn and expand when they encounter the weak surface of the laminae. The fracture pressure gradually increases with the increase in fracturing fluid viscosity, while the fracture pressure of supercritical CO_2 is the largest and the fracture width is significantly lower than the other two fracturing fluids due to the high permeability and poor sand-carrying property. (2) Compared with the other two conventional fracturing fluids, under the condition of supercritical CO_2 fracturing fluid, the increase in ground stress leads to the increase in inter-salt. (3) Compared with the other two conventional fracturing fluids, under the conditions of supercritical CO_2 fracturing fluid, the fracture toughness of shale increases, the fracture pressure increases, and the fracture network complexity decreases as well. (4) With the increase in pumping displacement, the fracture network complexity increases, while the increase in the displacement of supercritical CO_2 due to high permeability leads to the rapid penetration of inter-salt shale hydraulic fractures to the surface of the specimen to form a pressure relief zone; it is difficult to create more fractures with the continued injection of the fracturing fluid, and the fracture network complexity decreases instead.

Keywords: inter-salt shale oil; fracture propagation; fracturing fluid viscosity; in-situ stress; pumping displacement; supercritical CO_2 fracturing

1. Introduction

With the increasing demand for energy, the scale of unconventional oil and gas resource extraction has been gradually expanded. Shale gas is a typical unconventional natural gas, produced in very low-permeability, organic-rich, shale-based reservoir rock systems [1–3]. The conditions of inter-salt shale are unique, as the upper and lower compartments are salt rocks with complex lithology and low reservoir permeability, making it difficult to fracture and transform the reservoir [4–11]. Therefore, it is extremely important to explore the law of fracture expansion through the layer of inter-salt shale reservoirs, and to master the degree of reservoir transformation fracture network development and its influencing factors for inter-salt shale oil reservoir transformation [12–14].

Domestic and international scholars have carried out relevant studies on inter-salt shale oil reservoirs and the fracture expansion pattern of hydraulic fracturing. Chizhi

Xian [11] revealed the lithological characteristics of the inter-salt reservoirs in the depression through the lithological description of dolomite in the Qianjiang inter-salt shale oil reservoirs, combined with the geochemical characteristics of inter-salt muddy dolomite. Shizhao Dai [15,16] found that the inter-salt formation is a sediment located in the relative desalination conditions of salt lake waters in the salt-bearing rhyolite formation in the south–central part of the depression, and the salt-bearing rhyolite formation is a mixture of carbonate, sulfate and mudstone, frequently interacting with the inter-salt formation. The inter-salt layer is doubly influenced by chemical deposition and mechanical deposition, the chemical rocks and clastic rocks are alternately interbedded at millimeter or even micron level, and the rock types are more complex. Chen Bo [17] compared the deposition and salt rhythm characteristics of the Qianjiang Formation in the Qianjiang Depression and the Shashi Formation in the Jiangling Depression, and obtained that the salt formation of the Qianjiang Formation is a gray salt rock with a small amount of associated minerals such as aragonite, and the rock types of the inter-salt layer include mud dolomite, dolomitic mudstone, aragonite-bearing dolomite, mudstone and oil shale. Zhengming Yang et al. [18] conducted a study and evaluation related to salt dissolution in salt rock compartments in inter-salt shale and found that the pore volume, porosity and permeability of salt rock compartments increased after salt dissolution. Several different model compounds were also selected to simulate the chemical changes in crude oil components and to propose the reaction mechanism. N.R. Warpinski [19,20] et al. studied the influence law of hydraulic fracture through the interlayer. S. C. Blair [21] et al. suggested that when a hydraulic fracture expands in a vertical discontinuity, the fluid will first penetrate along the interface, and after penetrating a certain distance on the interface, the hydraulic fracture will break through the interface and continue to expand in the original direction. T. L. Blanton [22,23] investigated the effect of natural fractures on hydraulic fracture extension by indoor triaxial hydraulic fracturing tests in shales containing natural fractures, and concluded that hydraulic fractures would continue to extend through natural fractures only under high stress differences and large approach angles, while, in most cases, hydraulic fractures would stop or turn at natural fractures. A. A. Daneshy [24] conducted a theoretical and experimental study on the expansion law of hydraulic fractures in layered formations, pointing out that strong interfaces do not prevent fracture expansion, while the law of weak interfaces preventing fracture expansion does not change with the change in the nature of the strata on both sides of the interface. Relatively few domestic scholars have studied this area. Chen Zhixi [25] and others applied the theory and method of rock mechanics to establish a numerical model of hydraulic fracture vertical extension in laminated media, and the research results showed that the ground stress profile is the main factor affecting the range and direction of fracture vertical extension, the fracture toughness of the rock formation has a significant fracture-stopping effect on fracture vertical extension, and the flow pressure drop in the fracturing fluid along the direction of the fracture height has a large influence on the fracture height. Under certain stratigraphic conditions, whether the fracture extends to the compartment and the size of the extension mainly depend on the operating pressure. By introducing an additional enhancement term in the interpolation function of displacement to achieve fracture extension independent of the grid boundary, Su-Ling Wang [26] et al. obtained the extension law of fractures through the sand/mudstone interface. Moreover, the fracture extension at the sand/mudstone interface of a low-permeability reservoir was tracked in real time using a white light scattering experiment, and the fracture extension process was in good agreement with the numerical simulation process, indicating that the extended finite element method is an effective means to quantitatively analyze the fracture extension, while the influencing factors of the fracture extension through the interface were also analyzed.

From the above studies, it can be seen that most of the current research on inter-salt shale oil reservoirs is focused on the microscopic mineral composition and lithological characteristics, while most of the studies on the hydraulic fracture expansion law are focused on sandstone and mudstone interfaces, and there are fewer studies on the expansion law of hydraulic fractures through layers in inter-salt shale. To this end, it is necessary to carry out real triaxial hydraulic physical simulation experiments with actual cores downhole, and to systematically study the hydraulic fracture expansion law and influence mechanism of inter-salt shale hydraulic fractures with fracturing fluid viscosity, ground stress and fracturing fluid discharge as variables, so as to optimize the hydraulic fracture extraction design of inter-salt shale oil reservoirs and improve the reservoir recovery rate.

2. Materials and Methods

2.1. Equipment

Based on the conventional three-axis rock mechanics test machine, a set of cylindrical fracturing physical simulation test systems was implemented, as shown in Figure 1.

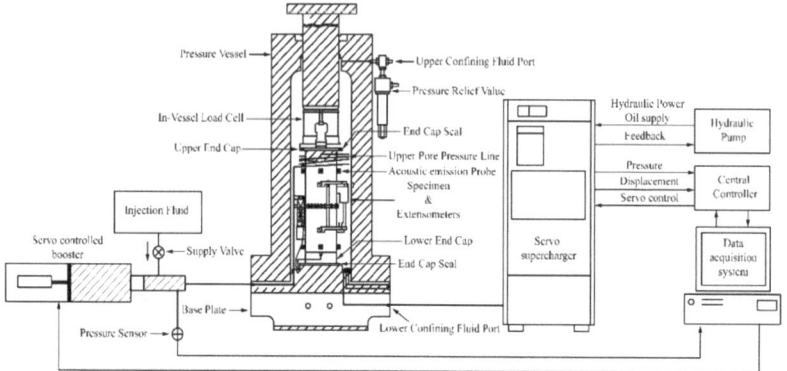

Figure 1. Schematic diagram of a full diameter core fracturing physical simulation experimental device.

The test system is mainly composed of a rock stress servo loading system, a fracturing fluid injection system and a sound emission detection system. Among them, the rock stress servo loading system is mainly used to provide the stress state in deep shale, simulating the real stress environment of the formation; the fracturing fluid injection system is mainly injecting the fracturing in the sample to simulate a wellbore in a constant voltage/constant flow mode liquid, and realize the real-time acquisition of pump pressure and displacement; the acoustic emission detection system is mainly used to detect micro-crack signals in the rock hydraulic fracturing process and obtain hydraulic cracks and spatial features in real time.

2.1.1. Rock Stress Servo Loading System

The rock stress servo loading system is mainly composed of a full digital electro-hydraulic servo rock triaxial test system, as shown in Figure 2. The system mainly includes the following: core high pressure triaxial chamber, servo hydraulic source, servo supercharger, door character rigid frame, radial/axial strain, warming system, and digital acquisition system. The maximum output axial force is 2000 KN, and the maximum working confining pressure is 140 MPa, which can better meet the requirements of the stress of the shale oil reservoir; at the same time, the sample is equipped with a sample heating device, the constant temperature system, and the upper limit temperature is 100 °C. Maximum allowable specimen size is 100 mm × height of 200 mm in diameter and equipped with axial/radial strain regulations, resolution 0.0001 mm.

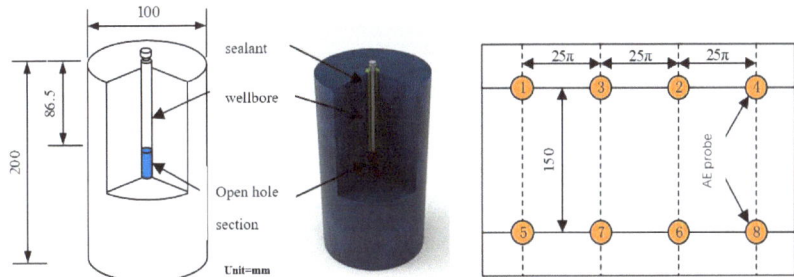

Figure 2. Schematic diagram of full sample and acoustic emission probe.

2.1.2. Fracturing Fluid Pump Injection System

The fracturing liquid pump is mainly used to provide a high-pressure fracturing fluid of constant flow. In order to provide more accurate flow control than most reciprocating pumps, this test system uses the US Teledyne (Thousand Oaks, CA, USA) Isco-260HP high-precision high-pressure plunger pump as the power source of fracturing fluid. Isco-260HP plunger pump maximum capacity 266 mL, output displacement 0.001–107 mL/min, measurement accuracy ± 0.5% (maximum leakage amount is 0.50 μL/min); the maximum output pressure is 65.5 MPa, and the measurement accuracy is 0.1%.

2.1.3. Acoustic Emission Detection System

The acoustic transmitting detection system is mainly composed of a host, sensor, preamplifier, acquisition card, and AEWIN signal acquisition and analysis software with five major modules.

The selection of the acoustic emission sensor has an important influence on the detection result. The acoustic emission probe selected in this test is the SR40M low-frequency narrow-bandwidth probe produced by Beijing Sonhua Xingye Technology Co., LTD (Beijing, China). The performance parameters are as follows: frequency range 15 KHz–75 KHz, resonant frequency 400 KHz, and sensitivity peak >75 dB.

Since the acoustic transmission sensor has a higher output impedance, the output signal current is weak, and it is not suitable for long-distance transmission. Therefore, it is necessary to connect the front amplifier to enlarge the voltage signal of the sensor to the capture card. The acoustic emission preamplifier used in this trial is a 2/4/6 amplifier produced by the American Physical Acoustic Corporation (PAC) (Lansing, MI, USA), and the gain is 20,40,60 dB three-stop, which is powered by a PAC sound emission card.

The core components of the acoustic emission detection system consist of 8 PCI-2-type sound acquisition cards. PCI-2 is a high-performance/low-cost position-transmitting capture card developed by PAC, which is 40 MHz, with an 18-bit A/D converter, which can perform real-time analysis and higher signals. Treatment accuracy, minimum noise threshold value 17 dB. In addition, the PCI-2 system has a unique waveform stream data storage function that continuously deposits the hard disk at a rate of 10 m per second. A layer of acoustic coupling agent is applied to the sound emission probe during the test, ensuring that the probe is closely coupled to the surface of the sample.

2.2. Preparation of Materials

The samples used in this experiment are the core and shale outcrop in the 4th dip of well Bangye Oil 2, which contains a large amount of salt rock interlayer bands. First, as shown in Figure 2, the core is processed into a cylindrical sample of $\varphi 100$ mm × 200 mm specification and the outer diameter of the sample is processed evenly by a lathe. The sample center drills a blind hole having a diameter of 12 mm as the simulated wellbore, and is buried in an outer diameter of 6.0 mm, with a stainless steel pipe of a wall thickness of 1.5 mm as an analog sleeve, and the annular gap is subjected to sealing treatment.

2.3. Experimental Method

In this experiment, the influence of fracturing fluid viscosity, in-situ stress, and fracturing fluid displacement on the hydraulic fracture propagation law of an inter-salt shale oil reservoir is analyzed with the control variable method. This experiment has carried out 10 sets of cylindrical sample water pressure tests. The specific parameters of each sample are shown in Table 1, and sample numbers and test conditions are shown in Table 2.

Table 1. Sample-specific parameters.

No.	Sample No.	Length/mm	Diameter/mm	Quality/g
1	2418	49.88	98.94	500.94
2	2985	49.54	108.64	536.084
3	2988	49.24	101.42	490.268
4	3665	49.12	108.02	519.248
7	1933	49.06	100.3	490.36
8	1956	49.24	99.86	491.832
9	25-SC	98.88	201.124	3955.308
10	26-SC	98.88	199.28	3949.955
11	27-SC	99.6	198.93	4031.648
12	28-SC	99.08	196.28	3845.258

Table 2. Full diameter core three-axis fracturing test experiment.

No.	Sample No.	Fracturing Medium	Axial Stress/MPa	Perimum Pressure/MPa	Solid Discharge (mL/min)	Remark
1	2418	Slippery	29	25	4.8	
2	2985	Slippery	29	25	7.2	
3	2988	Melter liquid	29	25	4.8	BY-2
4	3665	Melter liquid	29	25	7.2	
7	1933	Supercritical CO_2	29	25	4.8	
8	1956	Supercritical CO_2	29	25	7.2	
9	25-SC		28	25	6.0	
10	26-SC	Supercritical CO_2	31	28	6.0	Shale outcrop
11	27-SC		18	15	6.0	
12	28-SC		28	25	1.2	

The specific steps of this experiment are as follows:

(1) Description before hydraulic fracturing: We observe the sample layer and natural crack development with the naked eye, and identify the weak distribution of the original structure by chalk or marker.

(2) Gas tight inspection: Because the shale layer is more developed, it is very prone to shear damage during the preparation of the sample, causing wellbore sealing damage, so, before the experiment, we pump a certain amount of fluid into the wellbore, and the pressure is maintained at around 0.5 MPa, for 3–5 min, and we observe whether there is a fracturing liquid in the sample surface.

(3) Acoustic emission probe installation: Eight acoustic emission probes are installed on the surface of the sample, and the probes are tightly bonded with the sample with a coupling agent, while gently tapping each probe to ensure that each probe can work normally.

(4) Stress loading: We place the sample in a three-axis high-pressure chamber, sequentially load the peripheral pressure and axial pressure to the specified value, and maintain for 10–20 min to ensure that the internal force is uniform.

(5) Fracturing fluid injection: We start the Isco plunger pump, inject the fracturing fluid inside the sample according to the set displacement, and synchronously operate the sound emission detection; when the pump pressure has dropped significantly, we

close the plunger pump and sound emission detection system, and remove the axial pressure and peripheral sequential.

(6) Description after hydraulic fracturing: At the end of the experiment, a digital camera is first used to photograph the sample surface and record the surface cracks. Then, we select a partial sample to conduct a CT sweep surface, quantitatively characterize the spatial distribution of hydraulic cracks, and finally take the test sample after the test, and the hydraulic crack is displayed; then, we summarize the hydraulic crack expansion.

3. Results

3.1. Slippery Water Fracturing Fluid

Two sets of indoor slick water fracturing physical simulation experiments were carried out with core samples from Well Bengyeyou 2. The experimental results are shown in Table 3.

Table 3. Summary of physical simulation test results of slick water fracturing.

No.	Sample No.	Fracturing Medium/MPa	Axial Stress/MPa	Solid Discharge (mL/min)	Burst Pressure/MPa	Pump Pressure Curve Characteristics	Crack Propagation Pattern
1	2418	29	25	4.8	27.57	It rises rapidly before the peak, and then quickly falls to the confining pressure after the peak.	Form 1 horizontal bedding seam.
2	2985	29	25	7.2	30.97	Rapid rise before the peak, obvious fluctuation after the peak, and gradually reduced to confining pressure.	Open multiple horizontal bedding seams.

The number of sample 1 is 2418, which is taken from Well Bengyeyou 2 and is a shale core with well-developed bedding. The photos and CT scan results of the sample before the experiment are shown in Figure 3. From the CT scanning results of the core before the experiment, it can be seen that the inter-salt shale is extremely developed due to the bedding, and some beddings are opened due to the stress release of the downhole core. Before the experiment, a layer of black epoxy resin glue was evenly spread on the surface of the core to prevent damage to the sample caused by external disturbance.

Figure 3. Core photos and CT scan results of sample 1 before the experiment.

Under the conditions of confining pressure of 25 MPa, axial force of 29 MPa and displacement of 4.8 mL/min, the hydraulic fracturing physical simulation experiment was carried out by using slick water fracturing fluid with viscosity of 3 MPa·s, and the fracture pressure of the sample was 27.57 MPa. Figure 4 shows the fracture morphology photos and CT scan results of sample 1 after the test. During the fracturing process, the hydraulic fractures started from the bottom of the well and expanded, and then extended to the open bedding surface near the bottom of the well and then turned to the edge of the well. As the bedding plane expands, a horizontal bedding fracture is formed near the bottom of the well.

Figure 4. Core photos and CT scan results of sample 1 after the experiment.

The acoustic emission and pump pressure curves of sample 1 are shown in Figure 5. The pump pressure curve can be divided into three distinct stages: 0–760 s is the stable injection stage; 760–890 s is the fracture initiation and development stage; 890–980 s is the post-peak fracture expansion stage. In the initial stage, with the injection of fracturing fluid, some micro-cracks at the bottom of the well were damaged and a small amount of acoustic emission events occurred, and the energy was lower than 100 mv·ms. When the injection time lasted 760 s, the acoustic emission energy curve began to rise rapidly, indicating that the initiation point appeared and the hydraulic fracture began to develop. When the bottom hole pressure continued to rise to the fracture pressure of 27.57 MPa, the macroscopic damage of the sample occurred, and then the pump pressure curve rapidly decreased to the size of the confining pressure. Since the fracturing is slightly smaller than the axial stress, it shows that the hydraulic fractures are linked to the bedding fractures that were originally opened during the expansion process. As fracturing fluid continues to be injected, more bedding surfaces expand and open. Due to the limitation of confining pressure, the energy accumulated in the early stage is rapidly released to the surroundings in the form of elastic waves, and the peak of acoustic emission energy begins to appear intensively.

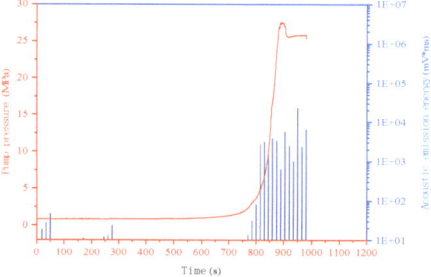

Figure 5. Pump pressure and acoustic emission characteristic curve of sample 1.

Sample 2 is numbered 2985, which is taken from Well Bengyeyou 2 and is a shale core with well-developed bedding. Figure 6 shows the photo of the sample before the experiment. As in the case of sample 1, some beddings of the downhole core were opened due to stress release. Before the experiment, a layer of black epoxy resin glue was evenly applied to the surface of the lower half of the core to prevent external disturbance from damaging the sample and thus affecting the experimental results.

Under the condition that the confining pressure is 25 MPa, the axial force is 29 MPa, and the displacement is 7.2 mL/min, the hydraulic fracturing physical simulation experiment is carried out with slick water fracturing fluid with a viscosity of 3 MPa·s, and the fracture pressure of the sample is obtained, which is 30.97 MPa. The characterization of crack morphology after the test of sample 2 is mainly described based on the acoustic emission localization results. Figure 7 shows the fracture morphology photo and AE positioning effect of sample 2 after the test. The acoustic emission events are mainly concentrated at the bottom of the well, and several acoustic emission event points are scattered vertically, indicating that, during the hydraulic fracturing process of sample 2, the hydraulic fractures start from the bottom of the well and expand longitudinally. During the expansion process, the opened bedding plane turns to expand along the direction of the bedding plane and continues to expand vertically, opening a number of horizontal bedding fractures.

Figure 6. Core photos and CT scan results of sample 2 before the experiment.

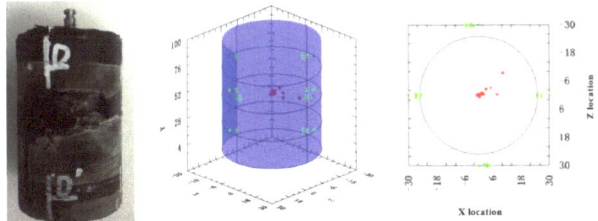

Figure 7. Core photos and AE positioning renderings of sample 2 after the experiment.

The acoustic emission and pump pressure curves of sample 2 are shown in Figure 8. Here, 0–150 s is the stable fluid injection stage, and the fracturing fluid is continuously injected into the bottom of the well, resulting in pressure hold-up at the bottom of the well. At 150 s, the pump pressure curve began to rise rapidly, and at 200 s, the bottom hole pressure reached the peak rupture pressure, the rupture pressure was 30.97 MPa, and the sample suffered macroscopic damage. Moreover, 200–500 s is the crack propagation stage. After the peak, the pump pressure curve did not drop rapidly to the confining pressure, but showed a sawtooth fluctuation. At the same time, the acoustic emission energy continued to maintain a high level, which was due to the macroscopic damage of the sample. After this, the continuous injection of fracturing fluid resulted in the continuous opening of a large number of bedding fractures and micro-fractures. The bedding cracks and micro-cracks opened in 500–640 s extended to the surface of the sample, and the pump pressure curve quickly dropped to the confining pressure.

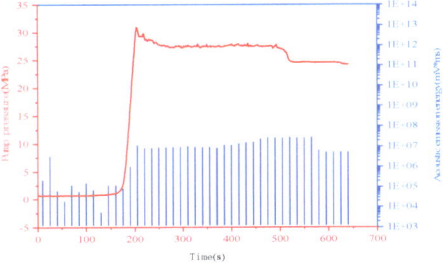

Figure 8. Pump pressure and acoustic emission characteristic curve of sample 2.

3.2. Guar Gum Fracturing Fluid

An indoor guar gum fracturing physical simulation experiment was carried out using cores from Well Bengye 2 and Diving 4. The experimental results are shown in Table 4.

Table 4. Summary of results of guar gum fracturing physical simulation experiments.

No.	Sample No.	Fracturing Medium/MPa	Axial stress/MPa	Solid Discharge(mL/min)	Burst Pressure/MPa	Pump Pressure Curve Characteristics	Crack Propagation Pattern
3	2988	29	25	4.8	27.89	The pre-peak rises approximately linearly, and the post-peak rapidly drops to the confining pressure.	Open multiple horizontal bedding seams.
4	2985	29	25	7.2	32.35	The pre-peak rises approximately linearly, and the post-peak gradually decreases to the confining pressure.	Open multiple horizontal bedding seams.

Sample 3 is No. 2988. This sample is taken from Well Bengyeyou 2. It is a bedded shale core. The photos and structure of the sample before the experiment are shown in Figure 9. Some thin salt rocks can be seen on the surface of the sample in the form of sandwich strips.

Figure 9. Core photo of sample 3 before experiment.

Under the condition that the confining pressure is 25 MPa, the axial force is 29 MPa, and the displacement is 4.8 mL/min, the hydraulic fracturing physical simulation experiment is carried out with guar gum fracturing fluid with a viscosity of 120 MPa·s, and the rupture pressure of the sample is obtained, which is 27.89 MPa. The crack morphology characterization of sample 3 after the test is mainly described based on the acoustic emission localization results. Figure 10 shows the fracture morphology photos and AE positioning effect of sample 3 after the test. The acoustic emission events are mainly concentrated at the bottom of the well, and some acoustic emission event points appear along the bedding plane of the sample, indicating that sample 3 is in the process of hydraulic fracturing; hydraulic fractures start from the bottom of the well and expand longitudinally. During the expansion process, the bedding plane formed by the thin interlayer strip of salt rock turns to expand in the direction of the bedding plane, while the hydraulic fracture continues longitudinally, expanding and communicating with more bedding planes, and opening up multiple horizontal bedding seams.

The acoustic emission and pump pressure curves of sample 3 are shown in Figure 11. The pump pressure curve can be divided into three distinct stages: 0~540 s is the stable injection stage; 540~570 s is the fracture initiation and development stage; 570~720 s is the post-peak fracture expansion stage. In the initial stage, the fracturing fluid is continuously injected at the bottom of the well to form pressure suppression, the pump pressure curve changes gently, and the micro-fractures at the bottom of the well are damaged and a large number of acoustic emission events occur. When the injection time continued to 540 s, the pump pressure curve began to rise rapidly, indicating that the crack initiation point appeared, and the hydraulic fractures began to develop and expand; when the bottom hole pressure continued to rise to the fracture pressure of 27.89 MPa, the macroscopic damage of the sample occurred, and then after the peak, the pump pressure dropped rapidly to the level of the confining pressure. The fracturing of sample 3 is also slightly smaller than the axial stress, which also shows that the hydraulic fractures communicate with the bedding fractures formed by the thin interlayer strips of the salt rock during the expansion process; the pump pressure curve remains at the level of the confining pressure, and the energy accumulated in the early stage is rapidly released to the surroundings in the form of elastic waves, and high acoustic emission energy still appears.

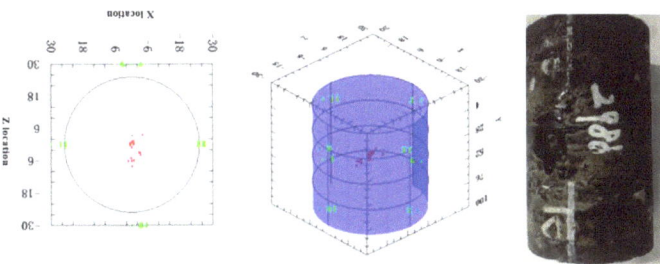

Figure 10. Core photos and acoustic emission location results of sample 3 after the experiment.

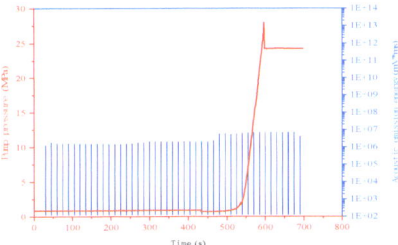

Figure 11. Pump pressure and acoustic emission characteristic curve of sample 3.

Sample 4 is No. 3665. This sample is taken from Well Bengyeyou 2. It is a bedded shale core. The photo of this sample before the experiment is shown in Figure 12. On the same surface as in sample 3, some thin salt rocks can be seen with the naked eye, in the form of sandwich strips.

Figure 12. Core photos of sample 4 before the experiment.

Under the condition that the confining pressure is 25 MPa, the axial force is 29 MPa, and the displacement is 7.2 mL/min, the hydraulic fracturing physical simulation experiment is carried out with guar gum fracturing fluid with a viscosity of 120 mPa·s, and the fracture pressure of the sample is obtained, which is 32.35 MPa. The fracture morphology and acoustic emission location after the experiment are shown in Figure 13. The acoustic emission events are mainly concentrated at the bottom of the well, and some acoustic emission event points appear along the bedding plane, indicating that the fractures start from the bottom of the well and expand along the longitudinal direction. At the same time, the direction and expansion occurred along the bedding plane, and the communication opened up multiple bedding seams. The CT scan results of sample 4 after pressing are shown in Figure 14. The same verification shows that multiple horizontal bedding seams are opened inside the sample after pressing.

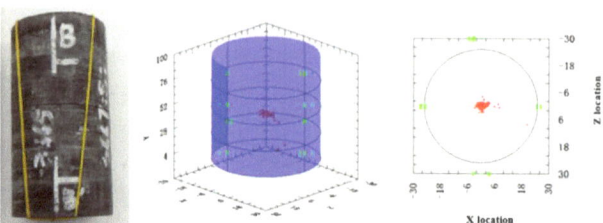

Figure 13. Core photos and acoustic emission location results of sample 4 after the experiment.

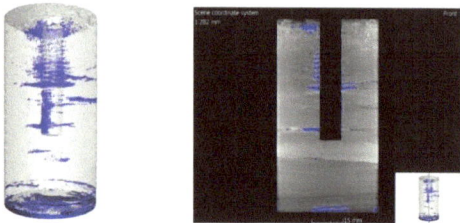

Figure 14. CT scan results of sample 4 after the experiment.

Figure 15 shows the acoustic emission and pump pressure curves of sample 4. The pump pressure curve of 0~55 s is flat. The fracturing fluid is injected at the bottom of the well to form pressure suppression, and the micro-fractures at the bottom of the well are damaged, resulting in a large number of acoustic emission events. When the injection time lasted for 55 s, the pump pressure curve rose rapidly to the rupture pressure of 32.35 MPa, and the sample was macroscopically damaged, and then the pump pressure quickly dropped to the confining pressure, and the fracturing fluid continued to be injected. Acoustic emission energy continues at high levels.

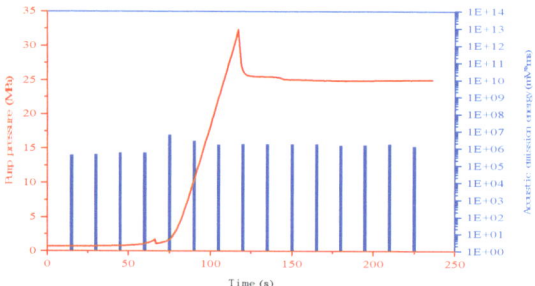

Figure 15. Pump pressure and acoustic emission characteristic curve of sample 4.

3.3. Supercritical Carbon Dioxide Fracturing Fluid

Indoor supercritical carbon dioxide fracturing physical simulation experiments were carried out with cores and shale outcrops from Well Bengye 2, respectively. The experimental results are shown in Table 5.

Table 5. Summary of results of physical simulation experiments of supercritical carbon dioxide fracturing.

No.	Sample No.	Fracturing Medium /MPa	Axial Stress /MPa	Solid Discharge (mL/min)	Burst Pressure /MPa	Pump Pressure Curve Characteristics	Crack Propagation Pattern
5	1933	29	25	4.8	37.55	The pre-peak pump pressure curve rises slowly, and the post-peak curve rapidly drops to the confining pressure.	Forming a curved longitudinal crack.
6	1956	29	25	7.2	50.77	The pre-peak pump pressure curve rises slowly before the peak, and the post-peak curve rapidly drops to the confining pressure.	Forming a curved longitudinal crack.
7	25-SC	28	25	6.0	37.4	The pre-peak rises slowly, and the post-peak curve rapidly drops to the confining pressure.	Generate 1 longitudinal main fracture and open 2 bedding fractures.
8	26-SC	31	25	6.0	52.6	The pre-peak rises slowly, and the post-peak curve rapidly drops to the confining pressure.	Generate 1 longitudinal main fracture and open 2 bedding fractures.
9	27-SC	18	15	6.0	31.6	The pre-peak rises slowly, and the post-peak curve rapidly drops to the confining pressure.	Generate 1 longitudinal main fracture and open 2 bedding fractures.
10	28-SC	28	25	1.2	34.5	The pre-peak rises slowly, and the post-peak curve rapidly drops to the confining pressure.	Generate 1 longitudinal main fracture and open 2 bedding fractures.

Sample 5 is No. 1933. This sample is taken from Well Bengyeyou 2 and is argillaceous dolomite. The photo of this sample before the experiment is shown in Figure 16. The surface of the sample can be seen with glauberite and other mineral-filled belts, which are not observed. There is an obvious bedding seam, and the integrity of the sample is good.

Figure 16. Core photos of sample 5 before the experiment.

The experimental confining pressure of the sample is set to 25 MPa, the axial force is 29 MPa, the displacement is 4.8 mL/min, supercritical CO_2 is used as the fracturing medium, and the sample rupture pressure is 37.55 MPa.

The crack morphology and acoustic emission location after the experiment are shown in Figure 17. Since there is no obvious bedding surface in the sample, the horizontal bedding crack is not opened after the sample is pressed, a curved longitudinal through-crack is formed, and the sample is split. The crack propagation path is mainly affected by the stress conditions; the glauberite transition zone is developed in the lower part of the sample, and the crack propagation direction is bent due to the lithological heterogeneity. In addition, the crack surface of the sample was visually observed, and the crack surface was rough, showing the characteristics of obvious tensile failure.

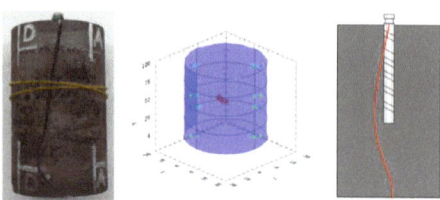

Figure 17. Core photos and acoustic emission location results of sample 5 after the experiment.

The acoustic emission and pump pressure curves of the samples are shown in Figure 18. The pump pressure curve of sample 5 can be divided into the following three stages, as follows. ① 0~200 s—in the initial stage of fracturing, due to the existence of wellbore cavities and primary fractures, the initial carbon dioxide preferentially enters such large spaces; at this stage, the number of acoustic emission events is small, and it can be considered that no new pores and cracks are generated. ② 200~1550 s—in the wellbore pressure holding stage, CO_2 changes from the wellbore filling stage to the wellbore pressure holding stage; due to the strong compressibility of carbon dioxide, the growth trend of the pressure–time curve is relatively gentle. In order to shorten the pressure holding time, in this stage, we set the CO_2 displacement to 20 mL/min. In the middle of this stage, there were basically no AE events, and AE events did not occur until the pump pressure exceeded 7.38 MPa in 1280 s. This was mainly because CO_2 entered the supercritical state under this pressure, its diffusion capacity was significantly improved, and the primary fractures were affected by CO_2. At the same time, new micro-cracks are generated, and the pump pressure also increases rapidly as CO_2 enters the supercritical state. ③ 1550~1980 s—during the fracturing failure stage, the CO_2 injection displacement was set to 4.8 mL/min. During this stage, the pump pressure rapidly increased to the fracture pressure. At the same time, accompanied by a large number of acoustic emission events, the acoustic emission energy count rate also increased significantly. Moreover, stacking appeared, indicating that the cracks extended and expanded rapidly at this stage. In addition, due to the ultra-low viscosity and zero interfacial tension of supercritical carbon dioxide, the post-peak pump pressure curve quickly dropped to the confining pressure.

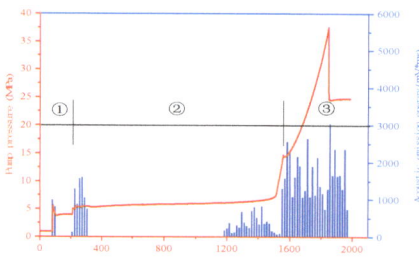

Figure 18. Pump pressure and acoustic emission characteristic curve of sample 5.

Sample 6 is No. 1956, which is taken from Well Bengyeyou 2 and is argillaceous dolomite. The photo of this sample before the experiment is shown in Figure 19. The core bedding is not developed, and the integrity of the sample is good.

The experimental confining pressure of the sample was set to 25 MPa, the axial force was 29 MPa, the displacement was 7.2 mL/min, supercritical CO_2 was used as the fracturing medium, and the sample rupture pressure was 50.77 MPa. Figure 20 shows the fracture morphology and acoustic emission location after the sample test. The acoustic emission events are mainly concentrated at the bottom of the well, indicating that the sample cracked from the bottom of the well and expanded. From the fracture morphology diagram after the sample test, it can be seen that the sample expands longitudinally after bottom-hole fracture initiation. Because the sample bedding is not developed and the glauberite transition zone is developed in the lower part, the lithological heterogeneity causes the fracture to expand longitudinally to the lower part and bend, and finally only a curved longitudinal through-fracture is formed.

Figure 19. Core photo of sample 6 before experiment.

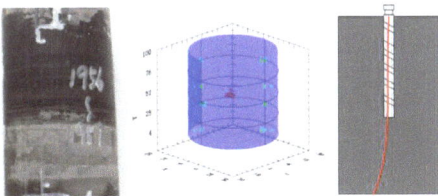

Figure 20. Core photos and acoustic emission location results of sample 6 after the experiment.

The pump pressure and acoustic emission curves are shown in Figure 21. The pump pressure curve of sample 6 can be divided into the following three stages. ① 0~120 s is the initial stage of fracturing. With the increase in the CO_2 injection rate, its inertia is significantly enhanced, and CO_2 gas rapidly fills into the non-pore space of the sample, accompanied by the generation of partial acoustic emission events. ② 120–1120 s is the wellbore pressure holding stage. At this stage, fluid pressure accumulation is still relatively gentle, but the number of AE events increases significantly with the increase in the wellbore cumulative fluid injection amount, indicating that fracture initiation and propagation of the sample's primary fractures begin under the action of CO_2, but no coherent macro-fractures are formed at this stage. ③ 1120~1420 s is the fracturing failure stage. With the continuous increase in the liquid injection volume, the pressure of supercritical CO_2 also increases rapidly. When the wellbore pressure reaches 50.77 MPa, the sample is unstable and fails. Due to the ultra-low viscosity and zero interfacial tension of supercritical CO_2, the pump pressure curve rapidly drops to the confining pressure.

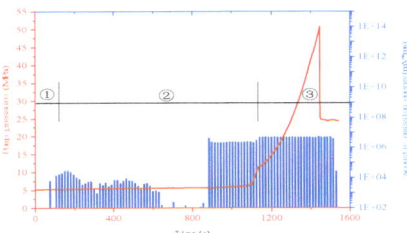

Figure 21. Pump pressure curve of sample 6.

Sample 7 is No. 25-SC; this sample is a shale outcrop, the coring direction is perpendicular to the bedding plane, the confining pressure is set to 25 MPa, the axial force is 28 MPa, the displacement is 6.0 mL/min, supercritical CO_2 s used as the fracturing medium, and the sample rupture pressure is 37.4 MPa.

The photo of the sample after the sample test is shown in Figure 22. The compression fracture is mainly composed of one longitudinal main fracture and two horizontal bedding fractures. The longitudinal main fractures are symmetrically distributed in the middle and lower parts of the sample, undergoing stratification and termination.

Figure 22. Core photos and acoustic emission location results of sample 7 after the experiment.

The sample pump pressure curve is shown in Figure 23. According to the pump pressure–time curve, the fracture pressure was 37.4 MPa, and after reaching the fracture pressure, it dropped rapidly to around 25 MPa, roughly equivalent to the confining pressure, indicating that the pressure crack was fully opened and penetrated the sample surface, and then dropped rapidly to approximately 20 MPa. This is due to the subsequent injection of supercritical CO_2 through the fracture to the surface of the sample along the two levels, coupled with ultra-low viscosity and zero interfacial tension of supercritical CO_2; the sand carrying capacity of supercritical CO_2 is weak and the support effect is poor, and the pressure in the fracture gradually decreases.

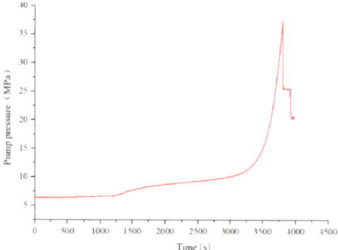

Figure 23. Pump pressure curve of sample 7.

Sample 8 is No. 26-SC; this sample is a shale outcrop, the coring direction is perpendicular to the bedding plane, the confining pressure is set to 25 MPa, the axial force is 31 MPa, the displacement is 6.0 mL/min, supercritical CO_2 is used as the fracturing medium, and the sample rupture pressure is 52.6 MPa.

The photo of the sample after the test is shown in Figure 24. One longitudinally symmetrical main crack and two bedding seams were observed on the surface of the sample. One of the bedding seams was completely opened, and the other was less than half open. The main crack has a large opening and ends at the bedding crack in the lower part of the sample during the longitudinal expansion.

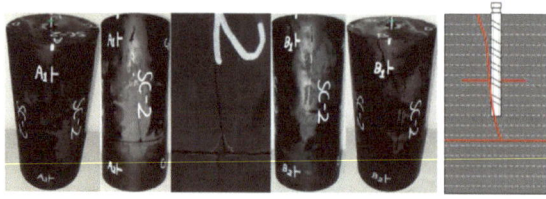

Figure 24. Core photos and acoustic emission location results of sample 8 after the experiment.

The sample pump pressure curve is shown in Figure 25. From the pump pressure–time curve, it can be seen that the rupture pressure of the sample is 52.6 MPa, and after reaching the rupture pressure, it quickly drops to around 25 MPa, which is roughly equivalent to the confining pressure, indicating that the pressure cracks are fully opened and have penetrated to the surface of the sample.

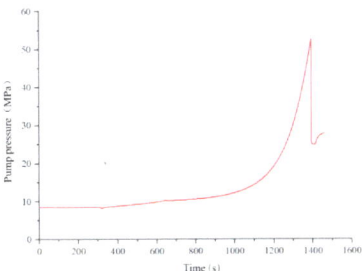

Figure 25. Pump pressure curve of sample 8.

Sample 9 is No. 27-SC; this sample is a shale outcrop, the coring direction is perpendicular to the bedding plane, the confining pressure is set to 15 MPa, the axial force is 18 MPa, the displacement is 6.0 mL/min, supercritical CO_2 is used as the fracturing medium, and the sample rupture pressure is 31.6 MPa.

The photo of the sample after the test is shown in Figure 26. A longitudinal crack is formed on the surface of the sample. The crack extends downward to the bedding plane and turns and expands along the bedding plane, and the crack opening is small.

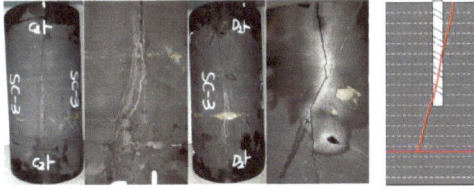

Figure 26. Core photos and acoustic emission location results of sample 9 after the experiment.

The sample pump pressure curve is shown in Figure 27. It can be seen from the pump pressure curve that the rupture pressure of the sample is 31.6 MPa. After reaching the rupture pressure, the pump pressure curve rapidly drops to the confining pressure of approximately 15 MPa, and the sample is damaged, forming a longitudinal bedding joint and a horizontal bedding joint. All penetrated to the surface of the sample, and then dropped rapidly to around 7 MPa. This is because of the subsequent injection of supercritical CO_2 along the two through-fractures to the surface of the sample, coupled with the ultra-low-viscosity and zero interfacial tension of supercritical CO_2; the supercritical CO_2 has a weak sand-carrying ability and poor supporting effect, and the pressure in the fracture gradually decreases.

Sample 10 was designated 28-SC. The sample is a shale outcrop, the coring direction is perpendicular to the bedding plane, the confining pressure is set to 25 MPa, the axial force is 28 MPa, the displacement is 1.2 mL/min, supercritical CO_2 is used as the fracturing medium, and the sample rupture pressure is 34.5 MPa.

The photo of the sample after the test is shown in Figure 28. The pressure crack is composed of one unilateral main seam and three bedding seams. The main seam is in the form of a single wing, which only appears on one side of the sample, and no cracks are found on the symmetrical side. Among the bedding seams, only one is fully open, and the other two are open in a small area. The opening of the main crack is small, and the crack is relatively fine.

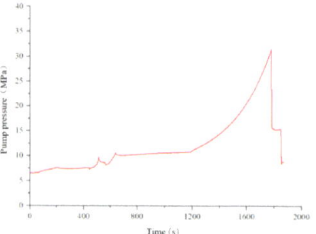

Figure 27. Pump pressure curve of sample 9.

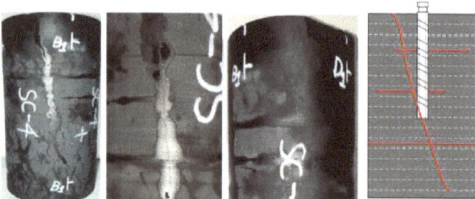

Figure 28. Core photos and acoustic emission location results of sample 10 after the experiment.

The sample pump pressure curve is shown in Figure 29. It can be seen from the pump pressure curve that the rupture pressure of the sample is 34.5 MPa. After reaching the rupture pressure, the pump pressure curve decreases slowly, and finally drops to around 25 MPa, which is the same as the confining pressure. This is because, after the longitudinal cracks and horizontally opened bedding fractures of the sample extend to the surface of the sample, there are still other horizontal bedding fractures opened in a small range, so the pump pressure curve decreases slowly.

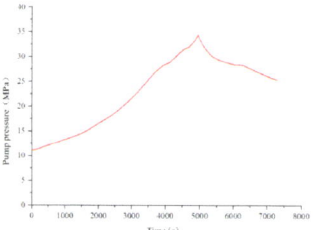

Figure 29. Pump pressure curve of sample 10.

4. Discussion

4.1. Fracturing Fluid Viscosity

Figures 4, 10 and 17 show the post-test crack propagation characterization pictures of sample 1, sample 3, and sample 5. By comparing the post-pressing photos, CT scan results, and AE acoustic emission localization results of the three samples, it can be seen that, under low-displacement conditions, the high-viscosity guar gum fracturing fluid opened many horizontal fractures with larger fracture widths, while the low-viscosity slick water fracturing fluid and the lower-viscosity supercritical carbon dioxide fracturing fluid only opened one horizontal bedding fracture and one curved longitudinal fracture, and the fracture width gradually decreased. This is due to the influence of the development degree of bedding planes in the actual core used in the experiment. Sample 1 is a bedded shale, but only contains one. The strips open the bedding plane, so only one horizontal bedding seam is opened. Sample 5 is argillaceous dolomite. The surface of the sample is filled with glauberite and other minerals, but no obvious bedding fractures are observed. In addition, the supercritical carbon dioxide has a weak sand-carrying ability, and some fractures with small widths may be compressed. Closure occurs, so only one longitudinal expansion seam with a curved lower half is formed after pressing. However, it can be seen from the CT scan and AE acoustic emission location points that both samples contain a large number of expanding horizontal bedding fractures. Even if the geological conditions of the actual core itself are excluded, it can still be shown that the low-viscosity fracturing fluid is easier to open. There are lots of bedding cracks. The larger fracture width of the high-viscosity fracturing fluid is mainly because the sand-carrying ability and fracture-supporting ability increase with the increase in fracturing fluid viscosity, and the fracture width increases after fracturing. The supercritical carbon dioxide fracturing fluid has low viscosity and poor sand-carrying ability, and it has the weakest ability to support fractures, resulting in the smallest fracture width and even closure. The acoustic emission and pump pressure curves of the three samples are shown in Figures 5, 11 and 18. The supercritical carbon dioxide fracturing fluid, slick water fracturing fluid, and guar gum fracturing fluid reach the rupture pressure for 1780 s, 860 s, and 580 s, respectively, and the achieved burst pressures are 37.55 Mpa, 27.57 Mpa, and 27.89 Mpa, respectively. By comparison, it can be seen that under the condition of low displacement, with the increase in fracturing fluid viscosity, the time for macroscopic failure of the sample gradually decreases, and the fracture pressure during failure gradually increases. This is because, as the viscosity of the fracturing fluid increases, the permeability decreases, and the pressure rise time of the fracturing fluid in the bottom hole and fractures is shortened, so the time for reaching the bottom-hole fracture initiation pressure and promoting fracture expansion to reach the failure pressure is shortened; however, due to the viscosity, the increase leads to a faster rate of pressure rise, which is also accompanied by an increase in rupture pressure when the specimen fails. Although supercritical carbon dioxide fracturing fluid has the characteristics of low viscosity and high diffusivity, it is easy to activate the natural bedding surface, but due to its poor sand-carrying ability and strong permeability, the pressure holding time is the longest, and it also reaches the time of failure. The rupture pressure is also the highest. The failure of the sample causes the pump pressure curve to drop rapidly to the confining pressure after the fracture. Due to the ultra-low viscosity of supercritical carbon dioxide, zero interfacial tension, poor sand-carrying ability, and strong permeability, the drop is the fastest and the largest. At the same time, the acoustic emission events of the slick water and supercritical carbon dioxide fracturing fluid are mainly concentrated in the stage before and after the fracturing pressure, while the acoustic emission events of the guar fracturing fluid are relatively dense throughout the pump pressure curve, indicating that the fracturing fluid under low-viscosity conditions causes hydraulic fractures to be formed and expanded after the bottom is held back to a certain extent, while the high-viscosity fracturing fluid can communicate and generate a large number of micro-cracks at the beginning of fracturing fluid injection.

Figures 7, 14 and 20 show the post-test crack propagation characterization pictures of sample 2, sample 4, and sample 6. By comparing the post-pressing photos, CT scan results, and AE acoustic emission localization results of the three samples, it can be seen that under the condition of high displacement, the high-viscosity guar gum opened many horizontal fractures after the hydraulic fracturing, and the low-viscosity guar gum opened many horizontal fractures. After the slick water fracturing hydraulic pressure, multiple horizontal fractures were opened, while the lower-viscosity supercritical carbon dioxide fracturing fluid only had one curved vertical fracture, which is the same as the case of sample 5, both of which are due to the same geological downhole core due to constraints. Excluding actual core conditions, it can be shown that both low-viscosity fracturing fluids and high-viscosity fracturing fluids can open a large number of bedding fractures. At the same time, the acoustic emission and pump pressure curves of the three samples shown in Figures 8, 15 and 21 are compared. At 1380 s, 220 s, and 117 s, the achieved burst pressures were 50.77 Mpa, 32.35 Mpa, and 30.97 Mpa, respectively. According to the analysis, we can derive the same law as the viscosity change of the fracturing fluid under low-displacement conditions, while, under high-displacement conditions, the slick water and supercritical carbon dioxide fracturing fluid have some intensive acoustic emission events in the initial stage of fracturing fluid injection, due to increased displacement.

4.2. Ground Stress

As the influence of ground stress on the law of hydraulic fracture propagation of conventional slick water fracturing fluid and guar gum fracturing fluid is relatively clear, supercritical carbon dioxide, with unique properties, was selected as the fracturing fluid in this experiment to study the influence of ground stress on the law of hydraulic fracture propagation. Sample 7, sample 8, and sample 9 are shale outings of Well Bangyeyou 2, and the three groups of in-situ stress combinations are axial pressure = 28 Mpa and confining pressure = 25 Mpa; axial pressure = 31 Mpa, confining pressure = 25 Mpa; axial pressure = 18 Mpa, confining pressure = 15 Mpa, respectively. The post-test fracture characteristics of the three groups of samples are shown in Figures 22, 24 and 26. After the compression of sample 7, the fractures extend longitudinally along the bottom of the hole to connect the upper and lower salt rock thin interlayers, and then turn along the salt rock bedding plane. They expand to form two horizontal bedding seams running through. After the fracturing of sample 8, the fractures started from the bottom of the well and extended longitudinally to connect the upper and lower salt rock thin interlayers, and then turned downward and expanded along the salt rock bedding plane, and turned upward and expanded along the salt rock bedding plane while continuing to rise through the interlayer. They expanded and finally formed a longitudinal fracture penetrating upwards, a horizontal bedding fracture penetrating and a horizontal bedding fracture in the extended part. After the fracturing of sample 9, the fractures from the bottom of the well extended longitudinally to the upper and lower salt rock thin interlayers, crossed the salt rock bedding plane upwards, turned downward and expanded along the salt rock bedding plane, and finally formed a vertical line penetrating upward, with cracks and one horizontally penetrating bedding joint. From the comparison between Figures 22 and 26, it can be seen that in the case of high in-situ stress combination, the fracture will turn longitudinally to the bedding plane, and it will expand along the bedding plane and cannot pass through the bedding plane. In the case of low in-situ stress, the fractures grow longitudinally along the direction of the bedding plane, and at the same time, they attempt to pass through the bedding plane to communicate with the surface of the sample; thus, the in-situ stress decreases, and the development of the fracture network is more complex. This is mainly because with the increase in confining pressure, the brittle ductility characteristics of rock will change, the vertical stress difference coefficient will decrease, and the fracture toughness will increase significantly, resulting in an increase in the difficulty of crack initiation and the ability to prevent crack propagation; the vertical bedding plane is more difficult to penetrate, so it still leads to a decrease in fracture complexity. From the comparison of Figures 22 and 24, it can be seen that under the same confining pressure, with the increase in axial pressure, the crack can open the horizontal bedding fracture and continue to expand upward through the bedding surface and penetrate the surface of the sample. The main reason is that the axial pressure increases, the vertical stress difference coefficient increases, the fracture's ability to penetrate the bedding plane in the longitudinal direction increases, and the fracture complexity increases. Figures 23, 25 and 27 are the pump pressure curves of sample 7, sample 8, and sample 9. The rupture pressures of the three groups of samples are 37.4 Mpa, 52.6 Mpa, and 31.6 Mpa, respectively. It can be seen from the comparison that with the increase in in-situ stress, the rupture pressure gradually increases, which is mainly due to the increase in fracture toughness, which leads to the improvement in the difficulty of crack initiation and the ability to prevent crack expansion, as well as an increase in the rupture pressure.

4.3. Pump Displacement

The comparison between Figures 4, 6, 13, 16, 17 and 20, shows that, excluding the influence of the bedding plane distribution of the sample core itself, with the increase in pumping capacity, the more bedding planes propagate through and open fractures, the more bedding cracks are formed, and the complexity of the fracture network also increases. Figures 22 and 28 are the characterization diagrams of fracture propagation after the fracturing of sample 7 and sample 10. When the fracturing fluid is supercritical carbon dioxide, the fracture longitudinally expands to connect the upper and lower salt rock thin interlayers under the condition of high pumping displacement. The diversion occurs and expands along the salt rock bedding plane. In order to pass through the bedding plane, two horizontal bedding fractures are formed. Under the condition of low pump injection rate, the fractures extend longitudinally to connect the upper and lower bedding planes, turn to expand along the bedding planes, and, at the same time, pass upward through the bedding planes and penetrate to the surface of the sample, forming three main horizontal bedding fractures and a longitudinal crack. By comparison, it can be seen that under the condition of supercritical carbon dioxide fracturing fluid, with the increase in pumping displacement, the complexity of the fracture network decreases instead. This is due to the low viscosity and high diffusion characteristics of supercritical carbon dioxide, which can easily activate natural bedding fractures. With the increase in pumping displacement, the longitudinal fractures expand to communicate with the bedding fractures and then turn to rapidly expand to the surface of the sample. The high permeability causes most of the fracturing fluid to flow out along the through-fractures, making it difficult to continue to open new fractures and form a complex fracture network. By comparing Figure 5 with Figures 8, 13, 15, 18 and 29, it can be seen that as the displacement increases, the rupture pressure gradually increases, and at the same time, some acoustic emission events appear in the early stage of the pump pressure curve; this is due to the fact that increasing the displacement leads to a faster rate of pressure hold-up at the bottom of the well at the initial stage of fracturing fluid injection, and a large number of micro-fractures are opened.

5. Conclusions

(1) Low-viscosity slick water fracturing fluid can more easily open bedding fractures and induce complex fracture networks in the reservoir. Guar gum fracturing has high viscosity and good sand-carrying performance, so high-viscosity guar fracturing fluid can still open bedding fractures around the wellbore. Supercritical carbon dioxide has low viscosity and high diffusion characteristics, which can easily activate natural bedding fractures and induce complex fracture networks in the formation, but the fracture width is generally small. In addition, due to the weak sand-carrying capacity of supercritical carbon dioxide, the fracture support is insufficient. At the same time, with the increase in viscosity, the occurrence time of fracturing pressure decreases gradually, and the fracturing pressure gradually increases. However, the fracturing pressure of supercritical carbon dioxide is much higher than that of other types of fracturing fluid due to its strong permeability and poor sand-carrying properties.

(2) Under the condition of pumping supercritical carbon dioxide fracturing fluid, as the in-situ stress increases, the fracture pressure of the inter-salt shale gradually increases, the brittle–ductile characteristics of the shale will change, and the fracture toughness will be significantly enhanced. The difficulty of crack initiation and the ability to prevent crack propagation increase, and the complexity of cracks decreases.

(3) With the increase in pumping displacement, the fracture pressure of inter-salt shale gradually increases, and a large number of micro-fractures will be opened in the early stage of fracturing fluid injection. At the same time, the greater the number of cracks passing through layers and opening bedding planes, the more bedding fractures are formed, and the complexity of the fracture network also increases. However, due to the low viscosity, high diffusion characteristics, and high permeability of supercritical carbon dioxide fracturing fluid, as the displacement increases, the fractures rapidly expand to the surface of the sample to form through-fractures, and most of the fracturing fluid flows out from the through-fractures to form pressure relief zones; it is difficult to continue to open new cracks to form a complex network of cracks.

Therefore, the complexity of the seam network decreases with the increase in the displacement of supercritical carbon dioxide.

(4) When hydraulically fracturing the inter-salt shale reservoirs in the field, a fracturing fluid with moderate viscosity should be selected as much as possible, while the pumping displacement should be increased to open more laminar fractures and form a complex fracture network. However, an excessively high pumping displacement will lead to higher fracture pressure and affect the safety of on-site construction, so a follow-up study will be carried out to investigate the influence of pumping methods and other effects on the expansion pattern of hydraulic fractures and the degree of fracture network development in inter-salt shale.

Author Contributions: Data curation, Y.S.; formal analysis, Z.H.; methodology, Z.H. and X.C.; writing—original draft, Z.H.; writing—review and editing, Y.G. and X.C. All authors have read and agreed to the published version of the manuscript.

Funding: The State Energy Center for Shale Oil Research and Development (grant no. G5800-18-ZS-KFNY001).

Institutional Review Board Statement: Not applicable.

Informed Consent Statement: Not applicable.

Data Availability Statement: Not applicable.

Acknowledgments: The authors would like to acknowledge the funding support of the State Energy Center for Shale Oil Research and Development (grant no. G5800-18-ZS-KFNY001).

Conflicts of Interest: The authors declare no conflict of interest.

References

1. Vikram, V.; Mohd, R.; Bankim, M.; Pradhan, S.P.; Singh, T.N. Temperature effect on the mechanical behavior of shale: Implication for shale gas production. *Geosyst. Geoenviron.* **2022**, *1*, 100078.
2. Mohyuddin, S.G.; Radwan, A.E.; Mohamed, M. A review of Pakistani shales for shale gas exploration and comparison to North American shale plays. *Energy Rep.* **2022**, *8*, 6423–6442.
3. Owusu, E.B.; Tsegab, G.H. The potential of shale gas resources in Peninsular Malaysia. *IOP Conf. Ser. Earth Environ. Sci.* **2022**, *1003*, 012024. [CrossRef]
4. Liu, H.; Xu, S.; Zhu, B.; Zhou, L.; Huang, Y.; Li, B. Research and practice of volumetric fracturing technology for inter-salt shale oil. *Spec. Oil Gas Reserv.* **2022**, *29*, 149–156.
5. Shu, H. Characteristics and Comparative Analysis of Oil Reservoirs in Jianghan Inter-salt Shale. *J. Jianghan Pet. Staff. Univ.* **2021**, *34*, 31–33.
6. Li, Y. Characteristics of inter-salt shale oil reservoirs in the Jianghan Basin and countermeasures for their development. *J. Jianghan Pet. Staff. Univ.* **2019**, *32*, 24–26.
7. Li, B. Analysis of key factors affecting fracturing construction and rejection characteristics of inter-salt shale reservoirs. *J. Jianghan Pet. Staff. Univ.* **2018**, *31*, 18–21.
8. Zhang, Z. Evaluation of the Effectiveness of Artificial Fractures in Inter-Salt Shale Oil Reservoirs. China University of Petroleum Beijing. 2018. Available online: https://cdmd.cnki.com.cn/Article/CDMD-11414-1019927289.htm (accessed on 17 April 2022).
9. Liu, J. Study on the Method of Evaluating the Compressibility of Inter-Salt Shale Oil Reservoirs. Chongqing University, Chongqing, China. 2018. Available online: http://cdmd.cnki.com.cn/Article/CDMD-10611-1018853435.htm (accessed on 17 April 2022).
10. Zhang, X.; Bi, Z.; Chang, X.; Wang, L.; Yang, H. Experimental Investigation on Hydraulic Fracture Morphology of Inter-Salt Shale Formation. *Frontiers* **2021**, *9*, 893. [CrossRef]
11. Wang, S.; Nie, H.; Ma, S.; Ding, Y.; Li, H.; Liang, W. Evaluation of inter-salt shale oil resources and sweet spot prediction in the Paleocene Qianjiang Formation, Qianjiang Depression, Jianghan Basin. *Pet. Exp. Geol.* **2022**, *44*, 94–101.
12. Wen, H.; Lu, S.; Xue, H.; Wang, W.; Li, J.; Hu, Y.; Zhang, P.; Li, J. Main controlling factors of physical development of shale oil reservoirs in The Xingouzui Formation of Jianghan Basin. *Oil Gas Geol.* **2016**, *37*, 56–61.
13. Zheng, Y. Key Technology for Enrichment Mechanism and Dessert Prediction of Onshore Shale Oil in Salt Lake Basin. Research Institute of Exploration and Development, Jianghan Oilfield Branch. *China Pet. Chem. Corp.* **2020**, *12*, 1.
14. Zhi, X.; Shu, X.; Sang, L.; Zhang, Y. Analysis of formation patterns of inter-salt argillaline dolomite in Qianjiang Formation of Qianjiang Sag. *J. Yangtze Univ. Nat. Sci. Ed.* **2012**, *9*, 19–23+4–5.
15. Shi, Z.; Jiang, J. Prediction of shallow oil and gas reservoirs in Jianghan Basin. *Oil Gas Geol.* **1988**, 204–208.
16. Shi, Z.; Pan, G. Hidden reservoirs and their exploration in Qianjiang Sag in Jianghan Salt Lake Basin. *Daqing Pet. Geol. Dev.* **1984**, 119–129. [CrossRef]

17. Chen, B.; Xiao, Q.; Cao, J.; Zhao, H. Comparison of exploration potential of non-sandstone oil and gas reservoirs between Qianjiang Formation and Shashi Formation in Jianghan Basin. *Pet. Explor. Dev.* **2017**, 190–196.
18. Yang, Z.; Li, R.; Li, H.; Luo, Y.; Chen, T.; Gao, T.; Zhang, Y. Experimental evaluation of the salt dissolution in inter-salt shale oil reservoirs. *Pet. Explor. Dev. Online* **2020**, *47*, 803–809. [CrossRef]
19. Warpinski, N.R.; Clark, J.A.; Schmidt, R.A.; Huddle, C.W. Laboratory investigation on the-effect of in-situ stresses on hydraulic fracture containment. *Soc. Pet. Eng.* **1982**, *22*, 333–340. [CrossRef]
20. Anderson, G.D. Effects of friction on hydraulic fracture growth near unbonded interfaces in rocks. *Soc. Pet. Eng.* **1981**, *21*, 21–29. [CrossRef]
21. Blair, S.C.; Thorpe, R.K.; Heuze, F.E.; Shaffer, R.J. Laboratory observations of the effect of geological discontinuities on hydrofracture propagation. In Proceedings of the 30th US Symposium on Rock Mechanics, Morgantown, WV, USA, 19–22 June 1989; pp. 433–450.
22. Blanton, T.L. An experimental study of interaction between hydraulically induced and pre-existing fractures. In Proceedings of the SPE Unconventional Gas Recovery Symposium, Pittsburgh, PA, USA, 16–18 May 1982; pp. 1–13.
23. Blanton, T.L. Propagation of hydraulically and dynamically induced fractures in naturally fractured reservoirs. In Proceedings of the SPE Unconventional Gas Technology Symposium, Louisville, KY, USA, 18–21 May 1986; pp. 1–15.
24. Daneshy, A.A. Hydraulic fracture propagation in layered formations. *Soc. Pet. Eng. J.* **1978**, *18*, 33–41. [CrossRef]
25. Chen, Z.; Chen, M.; Huang, R.; Shen, Z. Vertical expansion of hydraulic cracks in layered media. *J. Univ. Pet. Nat. Sci. Ed.* **1997**, 25–28+34+116.
26. Suling, W.; Yiming, Z.; Minmin, J.; Yang, L. Study on the propagation mechanism of cracks in non-uniform rock formations. *Mech. Pract.* **2012**, *34*, 38–41+45.

Article

A Core Damage Constitutive Model for the Time-Dependent Creep and Relaxation Behavior of Coal

Tingting Cai [1], Lei Shi [1], Yulong Jiang [2,3,*] and Zengchao Feng [3]

[1] College of Safety and Emergency Management Engineering, Taiyuan University of Technology, Taiyuan 030024, China; ctttyut@163.com (T.C.); shilei1290@link.tyut.edu.cn (L.S.)
[2] College of Mining Engineering, Taiyuan University of Technology, Taiyuan 030024, China
[3] Key Laboratory of In-Situ Property Improving Mining of Ministry of Education, Taiyuan University of Technology, Taiyuan 030024, China; zc_fengg@163.com
* Correspondence: 13485368423@163.com

Abstract: The creep and stress relaxation behaviors of coal are common in coal mining. The unified constitutive model is suitable to describe and predict both the creep and relaxation evolution characteristics of rocks. The generalized Kelvin model is the core element for traditional and improved component models to reflect both the nonlinear creep and relaxation. In this paper, an improved core damage model, which could both reflect the creep and stress relaxation in relation to the damage evolution, was established based on a comparison of the traditional and improved component models, and the responding constitutive equations (creep and stress relaxation equation) at constant stress/strain were deduced. Then, the core damage model was validated to the uniaxial compressive multistage creep and stress relaxation test results of coal, showing that the model curves had great accordance with the experimental data. Moreover, the model comparisons on accuracy, parameter meaning, and popularization among the core damage model, hardening-damage model, and the fractional derivative model were further discussed. The results showed that the parameters in the core damage model had clear and brief physical significances. The core damage model was also popularized to depict the time-dependent behaviors of other rocks, showing great accuracy.

Keywords: rock mechanics; creep; stress relaxation; damage; constitutive model; coal

Citation: Cai, T.; Shi, L.; Jiang, Y.; Feng, Z. A Core Damage Constitutive Model for the Time-Dependent Creep and Relaxation Behavior of Coal. *Energies* **2022**, *15*, 4174. https://doi.org/10.3390/en15114174

Academic Editors: Wenchao Liu, Hai Sun, Daobing Wang and Manoj Khandelwal

Received: 27 April 2022
Accepted: 2 June 2022
Published: 6 June 2022

Publisher's Note: MDPI stays neutral with regard to jurisdictional claims in published maps and institutional affiliations.

Copyright: © 2022 by the authors. Licensee MDPI, Basel, Switzerland. This article is an open access article distributed under the terms and conditions of the Creative Commons Attribution (CC BY) license (https://creativecommons.org/licenses/by/4.0/).

1. Introduction

Studies of the long-term time-dependent creep and stress relaxation behaviors of rocks are common in geotechnical engineering. Laboratory testing data are always needed for the prediction of time-dependent creep and stress relaxation behavior, and many studies on the time-dependent behaviors of rocks at the laboratory-scale have been reported [1–3]. For example, Mishra and Verma [4] undertook many single and multistage creep tests on laminated shale in both uniaxial compression and triaxial compression to study the creep characteristics in roof falls. Cong and Hu [5] performed creep tests on the Jurassic sandstones in the Majiagou landslide in triaxial compression under different low confining pressures, and proposed a modified Burges model to reflect the time-dependent creep behavior of sandstone. With regard to stress relaxation studies, Paraskevopoulou et al. [6] conducted many triaxial compressive stress relaxation tests on limestones and investigated the characteristics of the three relaxation stages to predict and evaluate the stress relaxation behaviors of the limestone samples. Tian et al. [7] performed triaxial compressive stress relaxation tests under different confining stresses on argillaceous sandstone samples and adopted an empirical formula in the expression of a power-law to study the impact of confining pressure on relaxation.

With respect to the creep behavior and stress relaxation behavior description and prediction, the strain/stress evolution is always analyzed, thus correspondingly, many models have been proposed or developed [8,9]. The component models with specific physical meanings of parameters are the most commonly and widely used. Many component models can flexibly reflect the time-evolving creep behaviors and stress relaxation behaviors of rocks, and plenty of modified component models have been proposed, supplementing nonlinear viscous elements or viscous–elastic elements or other elements on traditional models [10–13]. Unlike empirical models and component models, damage models have been put forward on the basis of crack propagation and damage evolution. For example, Wang et al. [14] established a creep–damage model to reflect the complete creep–damage curves of rock salt, and the evolution of the deformation and damage in the three creep phases were deduced. Yang et al. [15] analyzed the damage process during creep and introduced a damage factor into the creep model and then constructed a nonlinear creep damage model. The damage models could directly reflect the damage evolution and reveal the rheology mechanism in rock. Nevertheless, the models are confusingly used in the creep behavior and stress relaxation behavior descriptions of rocks; specifically, there is a common phenomenon where one constitutive model is developed for one rock's creep behavior, but another model is built for this rock's stress relaxation behavior. For example, Li et al. [16] proposed an improved nonlinear Burgers model for the creep behavior of silty mudstone, while Yu et al. [17] took a modified generalized Maxwell model for the stress relaxation behavior of silty mudstone. Both testing samples in the two studies were silty mudstones and both the creep tests and the stress relaxation tests were conducted in the triaxial compression state, but the two constitutive models they developed were completely different. Creep behaviors and the stress relaxation behaviors are both critically essential rheology behaviors of rocks. It is of great necessity to build a unified constitutive model that can both reflect the creep and stress relaxation characteristics to describe the time-dependent behavior of rocks (creep and stress relaxation).

Coal is a special kind of rock with great heterogeneity. It usually contains many pores and fractures with different sizes and many mineral crystal grains distributed in the coal matrix. The stress–strain response and time-dependent stress/strain evolution may be quite different from other hard rocks [18]. In our previous work, the time-dependent creep and relaxation behaviors of raw lean coal were preliminarily discussed. Similarly, for mechanism reflection and better model accuracy, we once developed an improved harden-damage model for lean coal's creep behavior [19], but built another fractional derivation generalized Kelvin model for the stress relaxation behaviors of lean coal [20]. In essence, it is much more appropriate to use a unified constitutive model, which can both reflect and predict the time-dependent creep and stress relaxation behavior of rocks, instead of the respective models to describe and predict the creep or stress relaxation behaviors of coal, separately. However, many models have been improved in creep or stress relaxation separately for better accuracy, while few efforts have been attempted to make these models unified, concise, and of specific physical meaning.

In this paper, a core damage constitutive model was established on the basis of model development, which can both reflect the nonlinear creep and stress relaxation behaviors of rocks, and then the data from the creep tests and the stress relaxation tests of lean coal in the uniaxial compressive state were used for model validation. Additionally, the model comparison on the accuracy and parameter meanings between the core damage model, traditional classic models, and modified models were discussed, and the core damage model was further popularized for the creep behavior and the stress relaxation behavior depiction of other rocks. The results of this work can help to understand the long-term time-dependent behaviors of rocks and provide references for their long-term stability in geotechnical engineering.

2. Model Development

Traditional component models are composed of different elastic, viscous, and plastic elements, which are combined in series or in parallel. The traditional component models can well flexibly reflect the time-dependent elastic, viscous, and plastic behaviors (creep and stress relaxation) of rocks. Many modifications in the improved models have been made based on them. The traditional and some improved models are listed and compared in Table 1.

Table 1. The traditional and modified model comparisons.

Category	Models	Sketch	Characteristics
Two-element model	St. Venant model		no creep, no relaxation
	Ideal viscous–plastic model		linear creep, no relaxation
	Kelvin model		nonlinear creep, no relaxation
	Maxwell model		linear creep, relaxation
Three-element model	Bingham model		linear creep, relaxation
	Poyting–Thomson model		nonlinear creep, relaxation
	Generalized Kelvin model		nonlinear creep, relaxation
Four-element model	Burgers model		nonlinear and steady creep, relaxation
Five-element model	Nishihara model		nonlinear and steady creep, relaxation
Modified model	Fahimifar et al. model [10]		nonlinear and steady creep, relaxation
	Zhou et al. model [11]		nonlinear and steady creep, relaxation

As can be seen concerning the two-element models, the St. Venant model failed to depict the creep behavior of rocks, nor the stress relaxation behaviors. As for the ideal viscous–plastic model and the Kelvin model, these could reflect the time-dependent creep but failed to describe the stress relaxation behavior, while the Maxwell model could only describe the linear creep behavior and stress relaxation behavior of rocks, being unable to depict the nonlinear time-dependent behaviors. Therefore, we came to the conclusion that the two-element models could not both reflect the nonlinear time-dependent creep behaviors and the stress relaxation behaviors of rocks. When the models expanded to three elements, the combinations of the elastic and viscous elements in series or in parallel could both reflect the nonlinear creep and stress relaxation behaviors well such as the Poyting–Thomson model and the generalized Kelvin model (GK model). Furthermore, when the model expanded to more elements such as the Burgers model and the Nishihara model, both of them could reflect the nonlinear creep and stress relaxation behavior of rocks from the perspective of model elements, which is because the two models are improvements on the generalized Kelvin model by supplementing the viscous element and the ideal elastic–viscous element, respectively. In essence, the Burgers model and the Nishihara model are the further extension versions of the generalized Kelvin model, so the two models can well describe the nonlinear creep behaviors and the stress relaxation behaviors of rocks, let alone those improved models modified with specific elements supplemented by the Burgers model or the Nishihara model. These modified and improved models are still, in essence, further versions of the generalized Kelvin model (GK model).

Therefore, we can conclude that the core component elements for the traditional and modified models that best reflect the nonlinear time-dependent behaviors of rocks (creep and stress relaxation) are a combination of elastic and viscous elements in series or in parallel. The generalized Kelvin model has such typical core component elements. A core unified constitutive model was established to depict the nonlinear time-dependent behaviors of rocks (creep and stress relaxation) based on the generalized Kelvin model.

3. A Unified Core Damage Constitutive Model

3.1. Model Establishment

As stated, the generalized Kelvin model is a typical core component for the traditional and modified models to both reflect the nonlinear time-dependent creep behaviors and stress relaxation behaviors of rocks. The model establishment and component are analyzed below. Generally, at the initial time of stress/strain loading, the rock specimens generally initially respond, behaving as instantaneous strain/stress, which means that there should be an independent elastic element in the model to reflect the initial responding of samples. In creep, the strain of the specimens grows gradually, and the stress on the specimens in stress relaxation decreases over time, showing remarkably nonlinear time-evolving characteristics, which indicates that there should be a combination of elements in series or in parallel in the model to reflect the nonlinear time-dependent strain/stress evolution characteristics [21]. The generalized Kelvin model contains the independent elastic element and the combination of elastic and viscous elements, and it meets the above component analysis and can well reflect the nonlinear time-dependent strain/stress evolution, so a core unified constitutive model was established to depict the nonlinear time-dependent behaviors of coal (creep and stress relaxation) based on the generalized Kelvin model.

In our former multi-stage creep tests, there was a creep start stress threshold in the stress [21]. At low stress, the specimens only behaved as instantaneous strain, and the specimens did not show time-dependent creep strain until the stress on the specimens reached or exceeded the critical creep start stress threshold. Therefore, a plastic element, specifically, a stress-triggering mode, was introduced into the model to make the plastic element work or not at different stresses.

In the long-term time-dependent behavior of rocks, especially in heterogenous coal specimens, internal structural adjustments always occur, and the internal damage affects the rock viscosity remarkably. In the creep behaviors and the stress relaxation behaviors of rocks, it is the internal damage evolution that makes the viscosity change and the time-dependent strain/stress evolution show [14,22]. Therefore, a damage factor was introduced to the time-dependent viscous element to reflect the internal time-evolving damage evolution. The time-evolving damage factor is always adopted in an expression of exponential law [23,24], as seen below.

$$D = 1 - e^{-\beta t} \tag{1}$$

where D is damage factor; t is time; and β is a material constant.

After introducing the damage factor into the generalized Kelvin model, the viscosity of the element evolves and is damaged with time, and the viscous coefficient can be written as

$$\eta_1(D,t) = \eta_1(1-D) \tag{2}$$

where $\eta_1(D,t)$ is the viscosity evolving with damage factor and time after damage factor is introduced, and η_1 is the viscosity before the damage factor is introduced.

Moreover, to reflect the nonlinear accelerating creep phase in coal, a nonlinear viscous–plastic element is supplemented to the core unified generalized Kelvin model [25]. The constitutive equation of the viscous–plastic element is:

$$\varepsilon = \frac{\sigma - \sigma_{s2}}{\eta_2} \cdot t^n \tag{3}$$

where ε is the strain; σ is the stress; σ_{s2} is the critical stress for the viscous element to work; η_2 is the viscosity; and n is a material constant.

Consequently, a core unified damage model that can both reflect the nonlinear damage evolution in the creep and stress relaxation of coal is obtained (Figure 1). In this model, in Section I, there is a core component element combination containing damage evolution to both reflect the nonlinear creep behavior and the stress relaxation behavior of coal, and in

Section II, there is a viscous–plastic element to reflect the nonlinear accelerating creep. This core unified damage model can both reflect the creep and stress relaxation behaviors of coal with damage evolution.

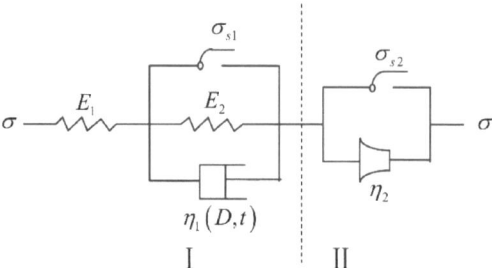

Figure 1. The unified core damage model.

3.2. Creep Equation

If the stress on the specimens does not reach the value of the critical creep start stress, that is, $\sigma < \sigma_{s1}$, the elastic element and the viscous element in parallel in Section I do not work, let along the plastic–viscous element in Section II, thus, there is only an elastic element in the model. The constitutive relation at such stress is:

$$\varepsilon = \frac{\sigma}{E_1} \quad (4)$$

where E_1 is the elastic coefficient.

If the stress on the specimens is greater than the critical creep start stress σ_{s1} but less than the critical stress for the viscous element σ_{s2}, that is, $\sigma_{s1} \leq \sigma \leq \sigma_{s2}$, the elastic, viscous, and plastic elements in Section I work in the model, and the unified core damage model turns into a model that evolves with the time and damage factor. The constitutive relation in the model in this circumstance can be written as,

$$\sigma - \sigma_{s1} + \frac{E_2}{E_1}\sigma + \frac{\eta_1(D,t)}{E_1}\dot{\sigma} = E_2\varepsilon + \eta_1(D,t)\dot{\varepsilon} \quad (5)$$

where E_2 is the elasticity; $\dot{\sigma}$ is the stress rate; and $\dot{\varepsilon}$ is the strain rate.

When the stress is held constant, at the time of $t = 0$, the initial strain $\varepsilon(0) = \frac{\sigma}{E_1}$, and then the initial strain and Equation (2) are substituted into the above equation, so we obtain

$$\varepsilon = \frac{\sigma}{E_1} + \frac{\sigma - \sigma_{s1}}{E_2}\left(1 - e^{\frac{E_2}{\beta\eta_1}(1-e^{\beta t})}\right) \quad (6)$$

Similarly, when $\sigma \geq \sigma_{s2}$, the creep of the unified damage model can be obtained as,

$$\varepsilon = \frac{\sigma}{E_1} + \frac{\sigma - \sigma_{s1}}{E_2}\left(1 - e^{\frac{E_2}{\beta\eta_1}(1-e^{\beta t})}\right) + \frac{\sigma - \sigma_{s2}}{\eta_2} \cdot t^n \quad (7)$$

Consequently, we obtain the creep equation at different stresses as below.

$$\varepsilon = \begin{cases} \frac{\sigma}{E_1}, & \sigma < \sigma_{s1} \\ \frac{\sigma}{E_1} + \frac{\sigma-\sigma_{s1}}{E_2}\left(1 - e^{\frac{E_2}{\beta\eta_1}(1-e^{\beta t})}\right), & \sigma_{s1} \leq \sigma \leq \sigma_{s2} \\ \frac{\sigma}{E_1} + \frac{\sigma-\sigma_{s1}}{E_2}\left(1 - e^{\frac{E_2}{\beta\eta_1}(1-e^{\beta t})}\right) + \frac{\sigma-\sigma_{s2}}{\eta_2}\cdot t^n & \sigma \geq \sigma_{s2} \end{cases} \quad (8)$$

Equation (8) can well describe the instantaneous strain, steady creep, and accelerating creep phase.

3.3. Stress Relaxation Equation

As stated before, the two plastic elements in the unified core damage model are both of the stress-triggering mode. Similarly, when the stress on the specimens does not reach the value of the critical creep start stress, that is, $\sigma < \sigma_{s1}$, there is only an elastic element in the model. Under these circumstances, when the strain is held constant, there is no stress relaxation in the core damage model. When $\sigma \geq \sigma_{s2}$, the stress is quite great and specimens are prone to fail at such great strain levels. Such conditions are quite rare for model use. When the stress on the specimens is greater than the critical creep start stress σ_{s1}, but less than the critical stress for viscous element σ_{s2}, that is, $\sigma_{s1} \leq \sigma \leq \sigma_{s2}$, this stress is suitable for the most stress relaxation behaviors of rocks to occur and we can obtain the stress relaxation equation under such circumstances. At this time, the constitutive equation of the core damage model is still Equation (5) and this model is appropriate to depict the stress relaxation behaviors at most circumstances.

At the constant strain, Equation (5) is

$$\sigma - \sigma_{s1} + \frac{E_2}{E_1}\sigma + \frac{\eta_1(D,t)}{E_1}\dot{\sigma} = E_2\varepsilon \tag{9}$$

Substitute Equation (2) into Equation (9), and the integral is taken. Additionally, in the initial stage, the responding stress is $\sigma(0) = E_1\varepsilon$ at the time of $t = 0$, so we obtain

$$\sigma = \frac{E_1}{E_1 + E_2}(E_1\varepsilon - \sigma_{s1})e^{\frac{E_1+E_2}{\beta\eta_1}(1-e^{\beta t})} + \frac{E_1\sigma_{s1} + E_1 E_2 \varepsilon}{E_1 + E_2} \tag{10}$$

The above stress relaxation equation is appropriate for most stress relaxation behaviors.

4. Model Validation

In this section, the multistage creep and stress relaxation test results of lean coal in uniaxial compression were used for model validation [19,20]). The samples were raw lean coal from the Heshun Mine in the Qinshui Coalfield in Yangquan, Shanxi, North China. Cylindrical core samples with a size of ϕ50 mm × 100 mm were prepared as suggested by the International Society for Rock Mechanics (ISRM) [26]. Basic mechanical parameter tests of the raw lean coal were performed and the results are shown in Table 2. In creep, the axial stress at the first stress level was set at 30~40% of the uniaxial compressive strength of the coal samples, and the maximum axial stress was set at 70~80% of the uniaxial compressive strength. The middle stress levels were designed with equal or close intervals. Each stress was applied to the samples quickly, and then kept constant for no less than 2~3 days. When the axial displacement of the sample was steady, the next stress level was applied to the sample until failure was reached.

Table 2. The basic mechanical parameter values of the raw lean coal.

Parameter	Uniaxial Compressive Strength/MPa	Elastic Modulus/GPa	Cohesion/MPa	The Angle of Internal Friction/°	Poisson's Ratio
Average value	12.26	3.27	1.76	22.41	0.23

The core damage model was validated to the multi-stage creep data and the multi-stage stress relaxation data to check the model's accuracy. In the model validation to the creep data, σ_{s1} is the critical creep start stress threshold for the time-dependent strain evolution to occur and σ_{s2} is the critical stress threshold for the accelerating creep phase; both stress parameters were acquired as the testing data showed. Distinctly, in the unified core damage model and as seen in Equation (8), E_1 is the independent elasticity, and it reflects the initial response of the elastic modulus in the specimens. E_1 was calculated by the stress and the initial instantaneous strain.

Regarding the other parameters in the core damage model, in creep, E_1, σ, σ_{s1}, and σ_{s2} are substituted into Equation (8), and then the other parameters of E_2, β, and η can be determined by the least-squares fitting in Origin software. The fitting results of the core damage model curves to the experimental creep data of the five different lean coal specimens at different stresses are shown in Figure 2. The corresponding parameters after calculation and identification are listed in Table 3.

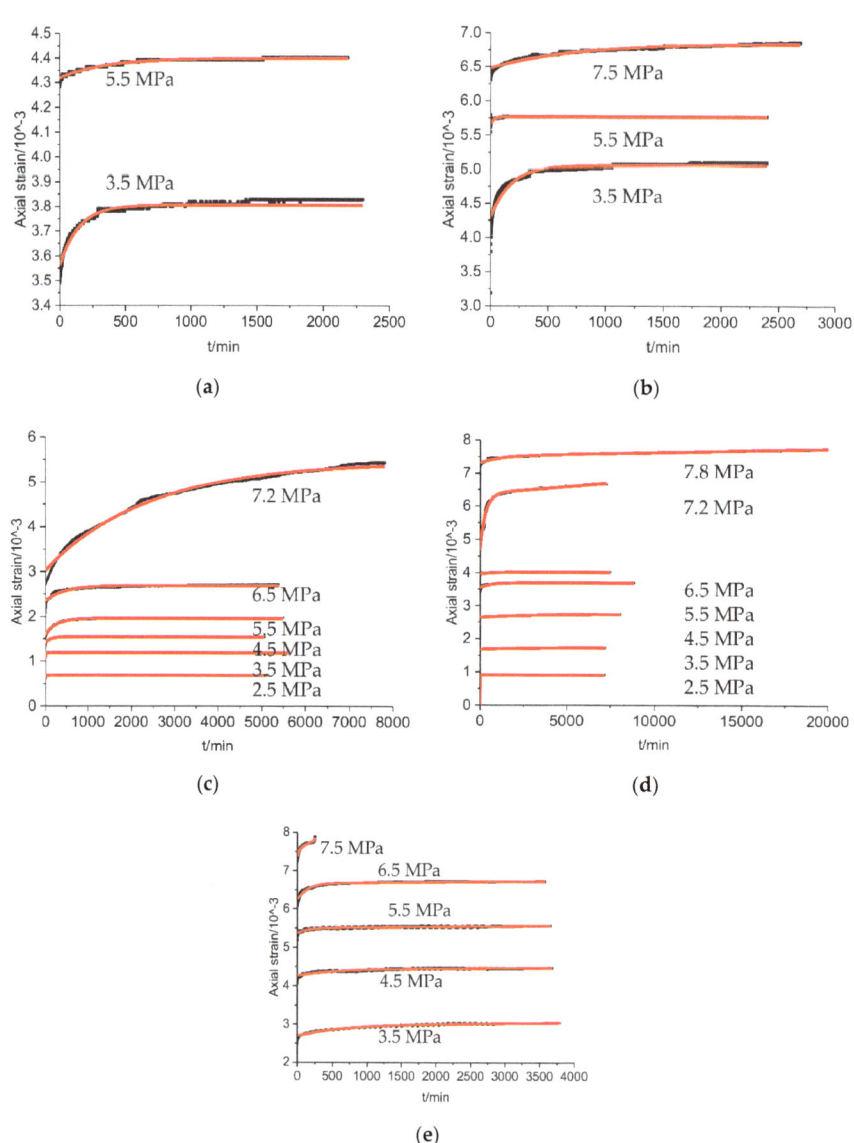

Figure 2. The model validation curves compared to the creep data of lean coal. Black lines refer to the experimental data and the red lines denote the model curves. (**a**) c-1, (**b**) c-2, (**c**) c-3, (**d**) c-4, (**e**) c-5.

Table 3. The parameters identified in the creep tests in the core damage model.

Samples	Stress/MPa	E_1/GPa	E_2/GPa	η_1/GPa·h	β	η_2/GPa·h	n	r^2
c-1	3.5	0.982	4.098	1.559	0.316	—	—	0.973
	5.5	1.272	4.390	1.258	0.328	—	—	0.966
c-2	3.5	0.813	3.175	1.099	0.351	—	—	0.949
	5.5	1.972	3.519	1.312	0.333	—	—	0.959
	7.5	1.162	3.989	1.330	0.329	—	—	0.963
c-3	2.5	3.625	—	—	—	—	—	0.979
	3.5	3.292	—	—	—	—	—	0.955
	4.5	3.480	3.546	0.985	0.341	—	—	0.967
	5.5	3.333	2.604	0.811	0.389	—	—	0.977
	6.5	3.947	2.831	0.809	0.318	—	—	0.933
	7.2	2.377	2.506	0.728	0.352	—	—	0.991
c-4	2.5	2.744	—	—	—	—	—	0.966
	3.5	2.009	—	—	—	—	—	0.968
	4.5	2.689	3.480	1.334	0.394	—	—	0.931
	5.5	3.550	3.333	0.892	0.317	—	—	0.972
	6.5	2.638	4.290	0.911	0.329	—	—	0.959
	7.2	2.546	3.988	0.936	0.351	—	—	0.991
	7.8	2.066	2.714	0.821	0.376	—	—	0.986
c-5	3.5	1.204	0.120	4.322	0.288	—	—	0.944
	4.5	1.394	0.149	3.956	0.254	—	—	0.935
	5.5	1.424	0.118	3.371	0.316	—	—	0.966
	6.5	1.493	0.130	4.962	0.341	—	—	0.959
	7.5	1.322	0.113	4.332	0.299	0.708	0.415	0.962

Similarly, in stress relaxation, Equation (10) is appropriate to depict the stress relaxation behaviors at different strains. At the initial start of strain loading, the instantaneous responding stress is $\sigma(0) = E_1\varepsilon$ at the time of $t = 0$, and from Equation (10), the initial stress is $\sigma = \frac{E_1}{E_1+E_2}(E_1\varepsilon - \sigma_{s1}) + \frac{E_1\sigma_{s1}+E_1 E_2\varepsilon}{E_1+E_2}$. Additionally, the final stable stress approaches $\frac{E_1\sigma_{s1}+E_1 E_2\varepsilon}{E_1+E_2}$, so E_1 can be obtained and the relation between E_2, σ, and ε are obtained and substituted into Equation (10), so the other parameters can be obtained through the least-squares fitting. The model curves of the unified core damage model were compared to the stress relaxation data of the six lean coal specimens at different strains, and the comparisons are shown in Figure 3. The corresponding parameters after calculation and identification are listed in Table 4.

In Figures 2 and 3, the black curves are the experimental strain/stress data in creep/stress relaxation and the red ones are the core damage model curves. As seen in the two figures, the model curves were in high accordance with the time-dependent creep data at different stresses and the stress relaxation data at different stains. Remarkably, the correlation coefficients of the fitting results to the creep testing data and the stress relaxation testing data were quite high, as much as 0.993, which indicate that this unified core damage model can very well depict the nonlinear creep data at different stresses and the stress relaxation data at different strains of lean coal in relation to damage evolution.

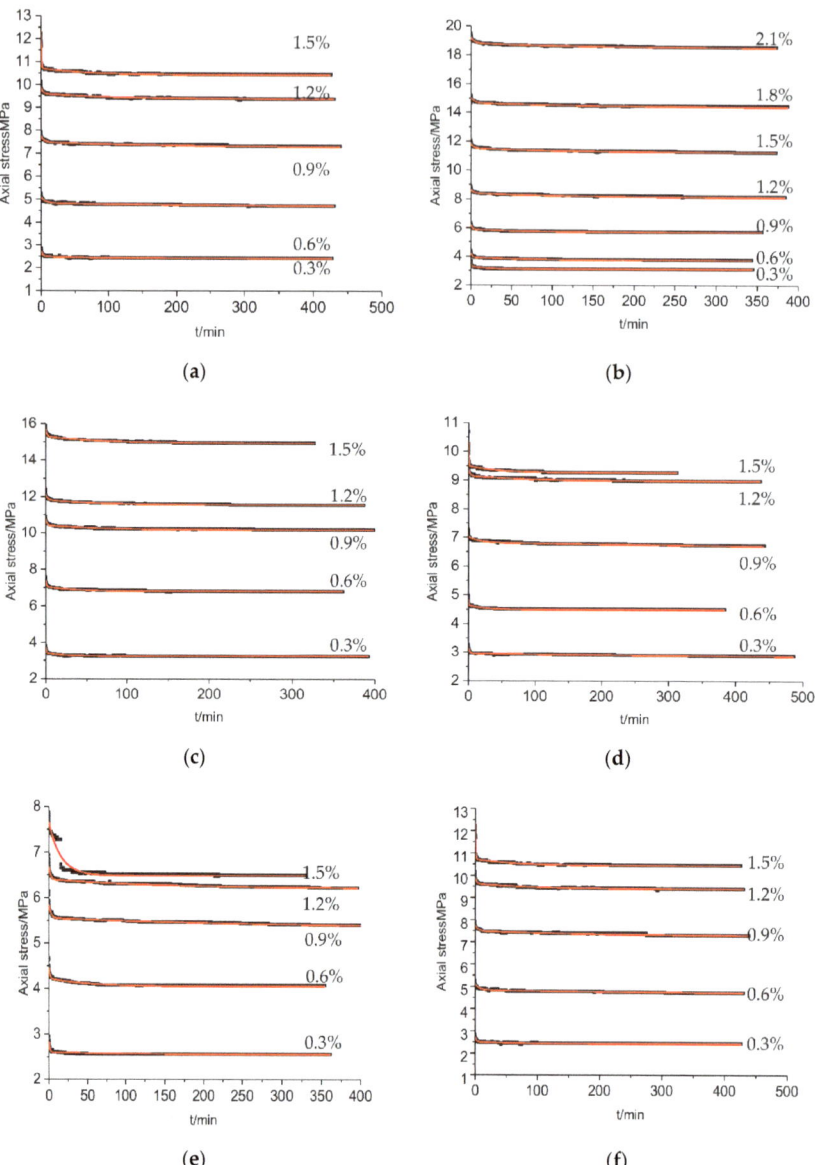

Figure 3. The model validation curves compared to the stress relaxation data of lean coal. Black lines refer to the experimental data and the red lines denote the model curves. (**a**) r-1, (**b**) r-2, (**c**) r-3, (**d**) r-4, (**e**) r-5, (**f**) r-6.

Table 4. The parameters identified in the stress relaxation in the core damage model.

Samples	Strain	E_1/GPa	E_2/GPa	η_1/GPa·h	β	r^2
r-1	0.3%	1.115	0.290	4.113	0.219	0.981
	0.6%	1.209	0.366	4.156	0.189	0.983
	0.9%	1.433	0.299	4.264	0.193	0.961
	1.2%	1.572	0.345	4.369	0.205	0.993
	1.5%	1.596	0.294	4.223	0.211	0.972
r-2	0.3%	1.117	0.326	4.109	0.203	0.983
	0.6%	1.133	0.305	4.084	0.211	0.977
	0.9%	1.299	0.324	4.197	0.206	0.979
	1.2%	1.594	0.337	4.129	0.211	0.963
	1.5%	1.603	0.339	4.294	0.208	0.969
	1.8%	1.712	0.356	4.256	0.213	0.977
	2.1%	1.638	0.389	4.218	0.216	0.969
r-3	0.3%	1.013	0.298	3.882	0.207	0.963
	0.6%	1.089	0.315	4.241	0.217	0.967
	0.9%	1.397	0.337	4.139	0.203	0.982
	1.2%	1.684	0.348	4.155	0.197	0.970
	1.5%	1.789	0.308	4.321	0.218	0.971
r-4	0.3%	0.629	0.231	2.909	0.216	0.991
	0.6%	0.711	0.249	2.803	0.208	0.985
	0.9%	0.829	0.267	2.931	0.209	0.949
	1.2%	0.717	0.255	3.007	0.199	0.957
	1.5%	0.646	0.287	3.026	0.219	0.966
r-5	0.3%	0.591	0.287	2.905	0.198	0.972
	0.6%	0.684	0.273	2.913	0.205	0.985
	0.9%	0.710	0.292	2.902	0.206	0.982
	1.2%	0.569	0.284	3.154	0.213	0.979
	1.5%	0.564	0.296	3.133	0.221	0.968
r-6	0.3%	0.933	0.233	2.899	0.199	0.988
	0.6%	0.947	0.249	2.941	0.203	0.969
	0.9%	1.021	0.255	2.884	0.215	0.977
	1.2%	0.959	0.297	3.016	0.207	0.967
	1.5%	0.941	0.306	3.231	0.216	0.964

5. Discussions

5.1. Model Comparison

The core damage model was put forward based on the generalized Kelvin model. In the generalized Kelvin model, the creep equation at the constant stress and the stress relaxation equation at the constant strains are listed as:

$$\varepsilon = \frac{\sigma}{K_1} + \frac{\sigma}{K_2}\left(1 - e^{-\frac{K_2}{\eta}t}\right) \tag{11}$$

$$\sigma = \left(K_1\varepsilon - \frac{K_1 K_2 \varepsilon}{K_1 + K_2}\right)e^{-\frac{K_1+K_2}{\eta}t} + \frac{K_1 K_2 \varepsilon}{K_1 + K_2} \tag{12}$$

where K_1 and K_2 are the elastic coefficients in the generalized Kelvin model; η is the viscosity coefficient; and the other parameters are of the same physical meaning as stated previously.

The model validation accuracy of the two models to the experimental creep data and stress relaxation data were compared. The axial strain of sample #c-4 in creep at 7.2 MPa was taken as an example as well as the stress relaxation data of specimen #r-4 at the strain of 1.2%. The unified core damage model and the traditional generalized Kelvin model were validated to the sample testing data. Consequently, the comparison results are shown in Figure 4. Remarkably, the unified core damage model accorded with the experimental data much better than the traditional generalized Kelvin model. This can mainly be attributed to the introduced damage factor. It is the introduced damage factor that makes the unified core damage model depict the nonlinear time-dependent behavior more flexibly, showing much better accuracy.

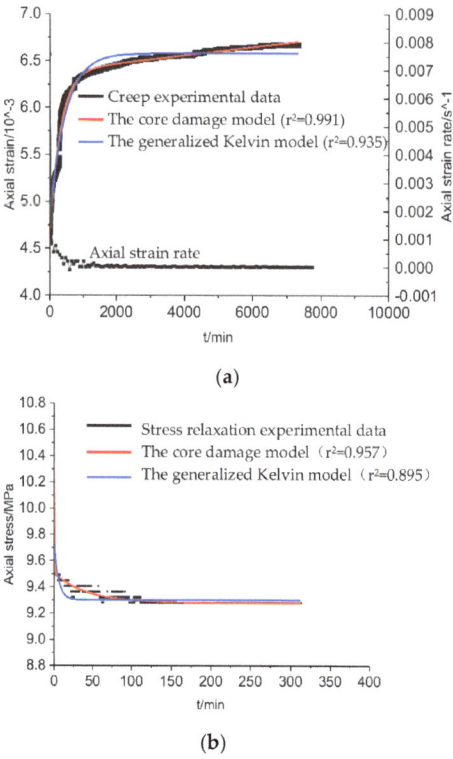

Figure 4. The generalized Kelvin model and the core damage model curves compared to the experimental data of lean coal. (**a**) Axial strain and rate of sample c-4 at the stress of 7.2 MPa, (**b**) Stress relaxation of sample r-4 at the strain of 1.2%.

Regarding the improved models, in creep, the core damage model was similar to the hardening-damage creep constitutive model proposed in Cai's work [19]. The hardening-damage model was proposed based on the hardening-damage mechanism of lean coal, in which the creep equation is:

$$\varepsilon = \begin{cases} \dfrac{\sigma^{1-m}}{E_{10}}, & \sigma < \sigma_s \\ \dfrac{\sigma^{1-m}}{E_{10}} + \dfrac{\sigma-\sigma_s}{E_2}\left[1 - e^{\frac{E_2}{\alpha\eta}(1-e^{\alpha t})}\right], & \sigma \geq \sigma_s \end{cases} \quad (13)$$

where E_{10} and E_2 are the elastic coefficients in the hardening-damage model; α is the damage factor; m is the material constant in relation to rocks; and the other parameters have the same physical meaning as stated before.

Comparing Equation (8) and Equation (13), when the stress on specimens was greater than the critical creep start stress σ_{s1} but less than the critical stress for viscous element σ_{s2}, that is, $\sigma_{s1} \leq \sigma \leq \sigma_{s2}$, the two equations were similar to exponential expressions. Both models could well reveal the damage evolution characteristics in the nonlinear creep of lean coal. In essence, is there is no obvious difference in the model accuracy, then the fitting results of the two models to the experimental data and the identified parameter of the two models would be the same. From the perspective of model component, the hardening effect was only reflected in the instantaneous strain phases; once the specimen showed time-dependent nonlinear strains, the core damage model and the hardening-damage model had the same depiction and accuracy in the experimental creep data. Furthermore, the hardening-damage model failed to depict the nonlinear accelerating creep phase, while the core damage model supplemented a viscous–plastic element, which could well overcome this shortcoming.

Regarding the stress relaxation, we developed an improved model with a fractional derivative Abel dashpot [18], in which the stress relaxation equation is:

$$\sigma = K_1 \varepsilon \cdot E_{\gamma,1}\left(-\frac{K_1 + K_2}{\eta}t^\gamma\right) + \frac{K_1 K_2 \varepsilon}{\eta} \cdot t^\gamma \cdot E_{\gamma,\gamma+1}\left(-\frac{K_1 + K_2}{\eta}t^\gamma\right) \qquad (14)$$

where γ is the fractional derivative order of the dashpot; $E_{a,b}(\cdot)$ is the Mittag–Leffler function; and the other parameters had the same physical meanings as stated before.

Similarly, the unified core damage model and the fractional model were validated to the experimental stress relaxation data for the comparison of the model accuracy. The stress relaxation data of specimen r-4 is taken as an example and the model validation accuracy comparison results were shown in Figure 5. As can be seen, the fractional model showed much greater accuracy in the time-dependent stress relaxation description of raw lean coal. Such a difference in accuracy is mainly attributed to the fractional derivation dashpot, in which the parameters η and γ can adjust to the nonlinear stress relaxation data as closely as possible, while the introduced damage factor in the unified core damage model is analogous to the traditional viscous element. Therefore, the fractional model showed much better accuracy while the unified core damage model showed limited improvement. However, though the core damage model was less accurate in the nonlinear stress relaxation curve description, the parameters in the unified core damage model had clear physical significances. There were elastic, plastic, and viscous elements in the models and the viscosity $\eta(D,t)$ evolving with damage and time well revealed the time-dependent damage evolution inside the samples, while in the Abel dashpot in the fractional derivative model, the viscous coefficient was a nonlinear function of fractional derivation, and the physical meaning of parameter γ was unclear and ranged with the experimental data. Moreover, the equation of the fractional derivative model is complicated with many parameters and is difficult to popularize in practical engineering.

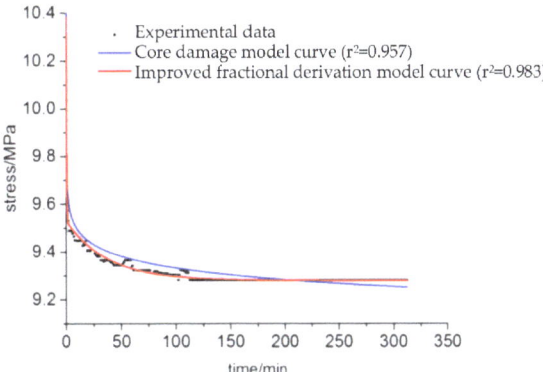

Figure 5. The comparisons of the core damage model and the fractional derivative model to the relaxation data of lean coal sample r-4 at the strain of 1.2%.

5.2. Model Popularization

To further validate and popularize this core damage model, the creep testing data of a marble at a stress of 69 MPa in the uniaxial state [27] and the stress relaxation test data of a Cobourg limestone (Cbrg_16R) under uniaxial compression [6] were taken as examples, and the model curve and the experimental data are shown in Figure 6a,b. As seen, the unified core damage model curves could well describe the axial strain of marble at a stress of 69 MPa and the stress relaxation of the Cobourg limestone (Cbrg_16R) and the correlation coefficients were as high as 0.996. Generally, this core damage model is not only suitable to describe the creep and relaxation behavior of raw lean coal, but can also be used to describe the time-dependent damage evolution behavior of the other rocks.

In model popularization, we should also note the model limitations. In the establishment of our model, the special initial elastic phase of coal was taken into consideration when the stress did not reach the creep start initial stress, as can be seen in the first situation in Equation (8). Consequently, there are three situations or two situations (the first and second situations without accelerating phase) to describe the creep behavior of coal that may take place. For other rocks, there may be two other situations (the second and third situation) and the first elastic phase can be ignored. When using this model to describe the rheological behaviors of other rocks, we can start from the second situation.

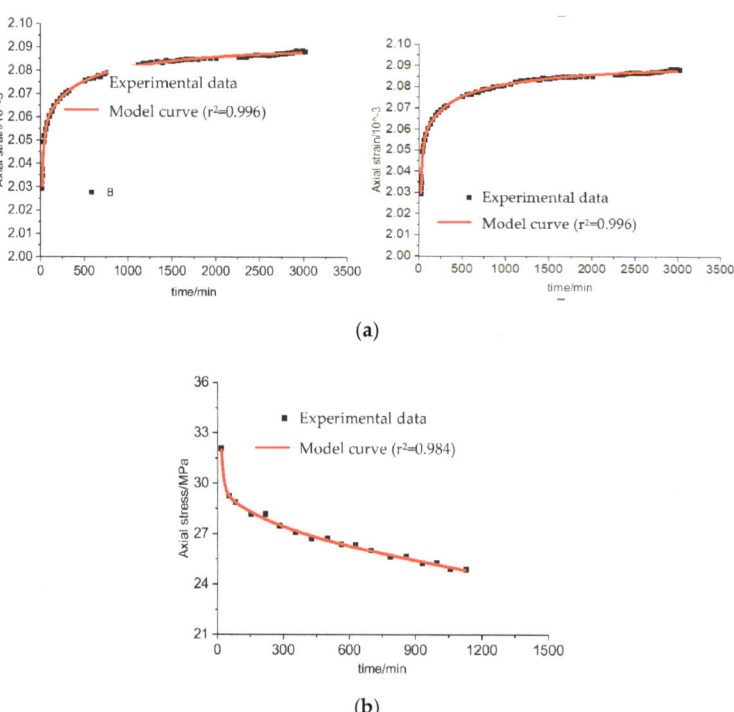

Figure 6. Cont.

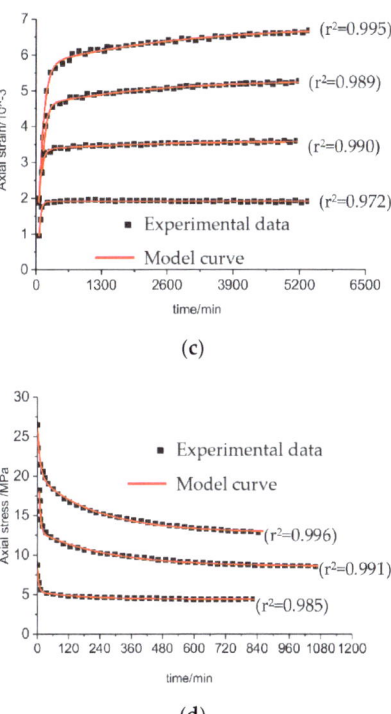

Figure 6. The core damage model curve compared to the creep and stress relaxation behavior of the other rocks. (**a**) Axial strain of a marble at the stress level of 69 MPa under uniaxial compression [27]. (**b**) Axial stress of Cobourg limestone (Cbrg_16R) in the stress relaxation tests [6]. (**c**) Axial strain of a silty mudstone at the stresses of 3, 9, 15, and 17.7 MPa in triaxial compression [16]. (**d**) Axial stress of the silty mudstone at strains of 0.2%, 0.6%, and 1.27% in the triaxial compressive stress relaxation tests [17].

The unified core damage model is established in the 2D state, so to popularize the model, the equations of creep and stress relaxation were induced in the 3D state, as follows:

$$\varepsilon' = \begin{cases} \frac{\sigma_1+2\sigma_3}{9K} + \frac{\sigma_1-\sigma_3}{3G_1} & \sigma < \sigma_{s1} \\ \frac{\sigma_1+2\sigma_3}{9K} + \frac{\sigma_1-\sigma_3}{3G_1} + \frac{\sigma_1-\sigma_3-\sigma_{s1}}{3G_2}\left(1 - e^{-\frac{G_2}{\beta\eta_1}(1-e^{\beta t})}\right) & \sigma_{s1} < \sigma < \sigma_{s2} \\ \frac{\sigma_1+2\sigma_3}{9K} + \frac{\sigma_1-\sigma_3}{3G_1} + \frac{\sigma_1-\sigma_3-\sigma_{s1}}{3G_2}\left(1 - e^{-\frac{G_2}{\beta\eta_1}(1-e^{\beta t})}\right) + \frac{\sigma_1-\sigma_3-\sigma_{s2}}{3\eta_2} \cdot t^n & \sigma > \sigma_{s2} \end{cases} \quad (15)$$

$$S_{i,j}(t) = \frac{2e_{ij}}{\sqrt{P_1^2 - P_2^2}} \cdot \left[(-q_1 + q_2 c)e^{-ct} + (q_1 - q_2 d)e^{-dt}\right] \quad (16)$$

The above equations were used to validate the creep and stress relaxation data of silty mudstone in triaxial compression. The model depiction curves compared to the experimental creep and stress relaxation data are shown in Figure 6c,d. As seen, when the model was expanded to the 3D state, the unified core damage model could still well describe the nonlinear time-dependent creep and stress relaxation behaviors of the other rocks.

5.3. Model Further Improvement

In further studies, this core damage model could be modified and improved for higher accuracy. It can be inferred that these modifications and improvements can be made in the following two parts:

(1) Modifications and improvements may be made in Section I in the model, and some viscous elements, elastic elements, or nonlinear elastic-viscous elements can be added. However, as stated previously, this model is core and concise to describe the time-dependent behaviors of rocks (creep and stress relaxation). These modifications may improve the model accuracy, but they may also make the physical meanings of the elements unclear. Furthermore, the model expressions will be much more complicated with more parameters to be identified, so the modified models would have limited guidance in engineering practice.

(2) The other modifications may be made in Section II in the model. The viscous–plastic element to reflect the accelerating creep phase could be modified into any nonlinear viscous combination. Those modified models may have a more accurate description in the nonlinear accelerating creep phase, but for the transient creep and steady creep phases, the expressions of the models after modification were the same as that in the core damage model proposed in this paper.

The core damage model can approximately depict the time-dependent behaviors of rocks (creep and stress relaxation) and the model is suitable for popularizing in practical engineering. Further model modification on this core damage model may have limited improvements.

6. Conclusions

(1) The generalized Kelvin model is the core element for traditional models and improved models to reflect both the nonlinear creep and stress relaxation behaviors of rocks. A unified core damage model was established based on the generalized Kelvin model, and the creep and stress relaxation equations were deduced.

(2) The unified core damage model was validated to the multistage creep data and stress relaxation data of the lean coal in uniaxial, and the validation to the experimental data showed that the unified core damage model could well depict the time-dependent creep behavior and stress relaxation behavior of lean coal.

(3) The core damage model showed much better accuracy than the traditional generalized Kelvin model in accordance with the experimental creep and stress relaxation data. The core damage model was the further modified version of the hardening-damage model. The parameters in the core damage model had much clearer and brief physical significances than the fractional derivative model in depicting the stress relaxation behavior.

(4) The core damage model was popularized to the time-dependent behaviors of other rocks. When using this model to describe the rheological behaviors of other quite heterogenous rocks, we can start from the second situation. The model validation results show that the core damage model can well depict the creep behavior and stress relaxation behavior of the other rocks.

Author Contributions: Conceptualization, Z.F. and T.C.; Investigation, T.C. and Y.J.; Methodology, L.S. and T.C.; Writing and revising, T.C. All authors have read and agreed to the published version of the manuscript.

Funding: This research was supported by the National Natural Science Foundation of China (51904197, 52104097) and the Applied Basic Research Foundation of Shanxi Province (201901D211031, 20210302124352, 20210302123147).

Institutional Review Board Statement: Not applicable.

Informed Consent Statement: Not applicable.

Data Availability Statement: Not applicable.

Acknowledgments: The authors would also like to express their sincere gratitude to the editors and anonymous reviewers for their valuable comments, which have greatly improved this paper.

Conflicts of Interest: The authors declare no conflict of interest.

References

1. Phienwej, N.; Thakur, P.K.; Cording, E.J. Time-Dependent Response of Tunnels Considering Creep Effect. *Int. J. Geomech.* **2007**, *7*, 296–306. [CrossRef]
2. Barla, G.; Debernardi, D.; Sterpi, D. Time-Dependent Modeling of Tunnels in Squeezing Conditions. *Int. J. Geomech.* **2012**, *12*, 697–710. [CrossRef]
3. Ering, P.; Babu, G.L.S. Slope Stability and Deformation Analysis of Bangalore MSW Landfills Using Constitutive Model. *Int. J. Geomech.* **2016**, *16*, 04015092. [CrossRef]
4. Mishra, B.; Verma, P. Uniaxial and triaxial single and multistage creep tests on coal-measure shale rocks. *Int. J. Coal Geol.* **2015**, *137*, 55–65. [CrossRef]
5. Cong, L.; Hu, X. Triaxial rheological property of sandstone under low confining pressure. *Eng. Geol* **2017**, *231*, 45–55. [CrossRef]
6. Paraskevopoulou, C.; Perras, M.; Diederichs, M.; Amann, F.; Löw, S.; Lam, T.; Jensen, M. The three stages of stress relaxation-Observations for the time-dependent behaviour of brittle rocks based on laboratory testing. *Eng. Geol.* **2017**, *216*, 56–75. [CrossRef]
7. Tian, H.M.; Chen, W.Z.; Yang, D.S.; Dai, F. Relaxation behavior of argillaceous sandstone under high confining Pressure. *Int. J. Rock Mech. Min.* **2016**, *88*, 151–156. [CrossRef]
8. Wang, D.K.; Wei, J.P.; Yin, G.Z.; Wang, Y.G.; Wen, Z.H. Triaxial creep behavior of coal containing gas in laboratory. *Procedia Eng.* **2011**, *26*, 1001–1010. [CrossRef]
9. Zhang, Y.; Xu, W.Y.; Shao, J.F.; Zhao, H.B.; Wang, W. Experimental investigation of creep behavior of clastic rock in Xiangjiaba Hydropower Project. *Water Sci. Eng.* **2015**, *1*, 55–62. [CrossRef]
10. Fahimifar, A.; Karami, M.; Fahimifar, A. Modifications to an elasto-visco-plastic constitutive model for prediction of creep deformation of rock samples. *Soils Found* **2015**, *55*, 1364–1371. [CrossRef]
11. Zhou, H.W.; Wang, C.P.; Han, B.B.; Duan, Z.Q. A creep constitutive model for salt rock based on fractional derivatives. *Int. J. Rock Mech Min.* **2011**, *48*, 116–121. [CrossRef]
12. Kang, J.H.; Zhou, F.B.; Liu, C.; Liu, Y.K. A fractional non-linear creep model for coal considering damage effect and experimental validation. *Int. J. Nonlin. Mech* **2015**, *76*, 20–28. [CrossRef]
13. Pramthawee, P.; Jongpradist, P.; Sukkarak, R. Integration of creep into a modified hardening soil model for time-dependent analysis of a high rockfill dam. *Comput. Geotech.* **2017**, *91*, 104–116. [CrossRef]
14. Wang, G.J.; Zhang, L.; Zhang, Y.; Ding, G.S. Experimental investigations of the creep-damage-rupture behavior of rock salt. *Int. J. Rock Mech. Min.* **2014**, *66*, 181–187. [CrossRef]
15. Yang, S.Q.; Xu, P.; Ranjith, P.G. Damage model of coal under creep and triaxial compression. *Int. J. Rock Mech. Min.* **2015**, *80*, 337–345. [CrossRef]
16. Li, Y.L.; Yu, H.C.; Liu, H.D. Study of creep constitutive model of silty mudstone under triaxial compression. *Rock Soil Mech.* **2013**, *33*, 2035–2040.
17. Yu, H.C.; Li, Y.L.; Liu, H.D. Study on stress relaxation model of silty mudstone under triaxial compression. *J. China Coal Soc.* **2011**, *36*, 1258–1263.
18. Xu, T.; Tang, C.A.; Zhao, J.; Li, L.C.; Heap, M.J. Modelling the time-dependent rheological behavior of heterogeneous brittle rocks. *Geophys. J. Int.* **2012**, *189*, 1781–1796. [CrossRef]
19. Cai, T.T.; Feng, Z.C.; Jiang, Y.L. An improved hardening-damage creep model of lean coal: A theoretical and experimental study. *Arab. J. Geosci.* **2018**, *11*, 645. [CrossRef]
20. Cai, T.T.; Feng, Z.C.; Jiang, Y.L.; Zhang, X.Q. Anisotropy characteristics of stress relaxation in coal: An improved fractional derivative constitutive model. *Rock Mech. Rock Eng.* **2019**, *52*, 335–349. [CrossRef]
21. Cai, M.F.; He, M.C.; Liu, D.Y. *Rock Mechanics and Engineering*; Science Press: Beijing, China, 2009.
22. Chen, L.; Wang, C.; Liu, J.; Liu, Y.; Liu, J.; Su, R.; Wang, J. A damage-mechanism-based creep model considering temperature effect in granite. *Mech. Res. Commun.* **2014**, *56*, 76–82. [CrossRef]
23. Lin, Q.; Liu, Y.; Tham, L.G.; Tang, C.; Lee, P.K.K.; Wang, J. Time-dependent strength degradation of granite. *Int. J. Rock Mech. Min.* **2009**, *46*, 1103–1114. [CrossRef]
24. Yang, S.Q.; Tang, J.Z.; Elsworth, D. Creep Rupture and Permeability Evolution in High Temperature Heat-Treated Sandstone Containing Pre-Existing Twin Flaws. *Energies* **2021**, *14*, 6362. [CrossRef]
25. Zhao, B.Y.; Liu, D.Y.; Zheng, Y.R.; Liu, H. Uniaxial compressive creep test of red sandstone and its constitutive model. *J. Min. Saf. Eng.* **2013**, *30*, 744–747.
26. Ulusay, R. *The ISRM Suggested Methods for Rock Characterization, Testing and Monitoring: 2007–2014*; Springer: Berlin/Heidelberg, Germany, 2014; Volume 1, pp. 47–48.
27. Chen, W.L.; Kulatilake, P.H.S.W. Creep Behavior Modeling of a Marble Under Uniaxial Compression. *Geotech. Geol. Eng.* **2015**, *33*, 1183–1191. [CrossRef]

MDPI AG
Grosspeteranlage 5
4052 Basel
Switzerland
Tel.: +41 61 683 77 34

Energies Editorial Office
E-mail: energies@mdpi.com
www.mdpi.com/journal/energies

Disclaimer/Publisher's Note: The title and front matter of this reprint are at the discretion of the Guest Editors. The publisher is not responsible for their content or any associated concerns. The statements, opinions and data contained in all individual articles are solely those of the individual Editors and contributors and not of MDPI. MDPI disclaims responsibility for any injury to people or property resulting from any ideas, methods, instructions or products referred to in the content.